Graduate Texts in Physics

For further volumes:
www.springer.com/series/8431

Graduate Texts in Physics

Graduate Texts in Physics publishes core learning/teaching material for graduate- and advanced-level undergraduate courses on topics of current and emerging fields within physics, both pure and applied. These textbooks serve students at the MS- or PhD-level and their instructors as comprehensive sources of principles, definitions, derivations, experiments and applications (as relevant) for their mastery and teaching, respectively. International in scope and relevance, the textbooks correspond to course syllabi sufficiently to serve as required reading. Their didactic style, comprehensiveness and coverage of fundamental material also make them suitable as introductions or references for scientists entering, or requiring timely knowledge of, a research field.

Series Editors

Professor Richard Needs

Cavendish Laboratory
JJ Thomson Avenue
Cambridge CB3 0HE, UK
rn11@cam.ac.uk

Professor William T. Rhodes

Department of Computer and Electrical Engineering and Computer Science
Imaging Science and Technology Center
Florida Atlantic University
777 Glades Road SE, Room 456
Boca Raton, FL 33431, USA
wrhodes@fau.edu

Professor Susan Scott

Department of Quantum Science
Australian National University
Science Road
Acton 0200, Australia
susan.scott@anu.edu.au

Professor H. Eugene Stanley

Center for Polymer Studies Department of Physics
Boston University
590 Commonwealth Avenue, Room 204B
Boston, MA 02215, USA
hes@bu.edu

Professor Martin Stutzmann

Walter Schottky Institut
TU München
85748 Garching, Germany
stutz@wsi.tu-muenchen.de

Anwar Kamal

Nuclear Physics

 Springer

Anwar Kamal *(deceased)*
Murphy, TX, USA

ISSN 1868-4513 ISSN 1868-4521 (electronic)
Graduate Texts in Physics
ISBN 978-3-662-51218-0 ISBN 978-3-642-38655-8 (eBook)
DOI 10.1007/978-3-642-38655-8
Springer Heidelberg New York Dordrecht London

Dedicated to my parents

Preface

This is an introductory textbook of nuclear physics for upper undergraduate students. The book is based on lectures given at British, American and Indian Universities over several years.

The idea of writing a text book on this subject was born some forty years ago. It is attempted to survey the major developments in nuclear physics during the past 100 years. In Rutherford's time and early 1950's, only a few Elementary particles were known and the existence of the neutrino was taken for granted. The development of the subject is so fascinating that we were inclined to present the historical facts in chronological order.

The prerequisites for the use of this book are the elements of quantum mechanics comprising Schrodinger's equation and applications, Born's approximation, the golden rule, differential equations and Vector Calculus. Basic concepts are explained with line diagrams wherever required. An attempt is made to strike a balance between theory and experiment. Theoretical predictions are compared with latest observations to show agreement or discrepancies with the theory.

The subject matter is developed in each chapter with the necessary mathematical details. Feynman diagrams are used extensively to explain the fundamental interactions. The subjects of various chapters are so much intimately connected that the logical sequential presentation of various topics became a vexing problem. For example, from the point of view of introducing quarks, the logical sequence would be strong, electromagnetic, weak and electroweak interactions, but from the point of view of introducing Feynman's diagrams, the desirable sequence would be electromagnetic, weak, electroweak and strong interactions, which is why one finds some variance in sequences for particle physics in various textbooks. The only remedy is to make cross references to the chapters which were previously studied and to those in which the relevant material is anticipated.

The size of the book did not allow to also include applied nuclear physics and cosmic rays. At the end of each chapter, a set of questions is given. A large number of worked examples is additionally presented. A comparable number of unworked problems with answers helps the student to test the understanding. The examples and problems are not necessarily of plug-in type but are given to explain the underlying physics. Useful appendices are provided at the end of the book.

Murphy, TX, USA   Anwar Kamal

Note: These two volumes are the last books by my father Dr. Ahmad Kamal, the work he had conceived as his dream project and indeed his scientific masterpiece. Unfortunately, he passed away before he could see his manuscript in print. While we have tried our best to bring the publishing process to as satisfactory conclusion as possible, we regret any errors you may discover, in particular, that some of the references could not be as completely specifically cited as would otherwise be the case. We trust that these errors however do not compromise the quality or standard of the content of the text.

Suraiya Kamal
Daughter of Dr. Ahmad Kamal

Acknowledgements

I am grateful to God for helping us to complete the dream project of Dr. Ahmad Kamal after his demise, which seemed very difficult and even impossible at times.

I would like to thank Springer-Verlag, in particular Dr. Claus Asheron, Mr. Donatas Akmanavičius, Ms. Adelheid Duhm and Ms. Elke Sauer for their constant encouragement, patience, cooperation, and for bringing the book to its current form.

This project would not be complete without the constant support and encouragement of Mrs. Maryam Kamal, wife of Dr. Ahmad Kamal. Her determination kept us all moving to get these books done. My sincere thanks is due to the family, friends and well-wishers of the author who helped and prayed for the completion of these books.

Suraiya Kamal
Daughter of Dr. Ahmad Kamal

Contents

Chapter 1
Passage of Charged Particles Through Matter

1.1 Various Types of Processes

When charged particles pass through matter, the following processes may take place:

(1) Inelastic collisions with the bound electrons of the atoms of the medium, in which case the particle energy is spent in the excitation or ionization of atoms and molecules. The energy losses of this kind of collisions are called ionization losses (collision losses) to distinguish them from radiation losses that are concerned with the generation of bremsstrahlung.
(2) Inelastic collisions with nuclei, leading to the production of bremsstrahlung quanta, to the excitation of nuclear levels, or to the nuclear reactions.
(3) Elastic collisions with nuclei, in which part of the kinetic energy of the incident particle is transferred to the recoil nuclei. However, the total kinetic energy of the colliding particles remains unchanged. A particular type of elastic scattering is the Rutherford scattering which results from the interaction of a charged particle with the Coulomb field of the target nucleus in single encounters. When thick materials are used, cumulative single scatterings give rise to the phenomenon of multiple scattering.
(4) Elastic collisions with bound electrons.
(5) Cerenkov effect, i.e. emission of light by charged particles passing through matter with a velocity exceeding the velocity of light waves in the given medium.

1.2 Kinematics

1.2.1 Laboratory (Lab) System (LS) and Centre of Mass System (CM)

In order to describe the motion of particles in the collision problem one must choose a definite frame of reference (co-ordinate system). Two frames of reference are im-

A. Kamal, *Nuclear Physics*, Graduate Texts in Physics,
DOI 10.1007/978-3-642-38655-8_1, © Springer-Verlag Berlin Heidelberg 2014

Fig. 1.1 The position vectors m_1 and m_2 and their centre of mass is shown

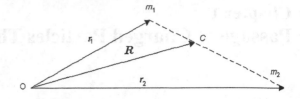

portant, one is the lab system (LS) and the other one is centre of mass system (CMS). In the lab system, the observer who is at rest in the lab views the collision process. In the CM system the centre of mass is at rest initially and always. Observations are usually made in the lab system but theoretical calculations are made in the CM system. It is of great interest to find out how various quantities like velocity, angle of scattering, etc. are related in these two systems. It is easier to perform calculations in the CM system rather than in the lab system. For, the great merit of CM system is that the total linear momentum of particles is always zero so that in the two-body process particles move directly towards each other before the collision and they recede in the opposite direction after the collision.

The collision process in the CM system may be visualized as the one in which a particle of reduced mass $\mu = m_1 m_2/(m_1 + m_2)$ moving with initial velocity u_1 collides with a fixed scattering centre. Here, u_1 is the initial velocity of m_1 moving towards the target particle of mass m_2 at rest.

1.2.2 Total Linear Momentum in the CM System Is Zero

In Fig. 1.1, the position of the centre of mass of two particles m_1 and m_2 is shown by C. The position of masses m_1 and m_2 are indicated by the position vectors r_1 and r_2 and that of the centre of mass by R. By definition

$$R = \frac{m_1 r_1 + m_2 r_2}{M} \quad \text{or}$$

$$MR = m_1 r_1 + m_2 r_2$$

Differentiating with respect to time

$$M\dot{R} = m_1 \dot{r} + m_2 \dot{r} \quad \text{or}$$

$$M v_c = m_1 u_1 + m_2 u_2$$

where u_1 and u_2 are the initial velocities of particles 1 and 2, respectively and v_c is the CM velocity. Since m_2 is initially at rest, $u_2 = 0$, and the centre of mass which is located at $M = m_1 + m_2$, must move in the lab system towards m_2 with velocity

$$v_c = \frac{m_1 u_1}{m_1 + m_2} \tag{1.1}$$

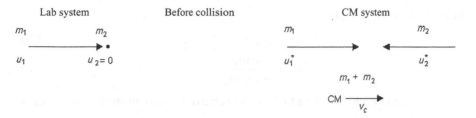

Fig. 1.2 Collision in the LS and CMS are shown

In the lab system, let m_1 move from left to right with initial velocity u_1, m_2 being initially at rest as in Fig. 1.2. As m_2 is initially at rest, its initial velocity in the CMS must be just equal to v_c in magnitude but oppositely directed. Denoting the velocities in the CMS by asterisk (*) we get

$$u_2^* = v_c = \frac{m_1 u_1}{m_1 + m_2} \tag{1.2}$$

$$u_2^* = -v_c \tag{1.3}$$

The initial velocity of m_1 in the CMS is reduced by an amount equal to v_c

$$u_1^* = u_1 - v_c$$

$$u_1^* = u_1 - \frac{m_1 u_1}{m_1 + m_2} = \frac{m_2 u_1}{m_1 + m_2}$$

where we have used (1.2). Total initial linear momentum of m_1 and m_2 in the CMS is

$$P^* = P_1^* + P_2^* = m_1 u_1^* + m_1 u_2^* = \frac{m_1 m_2 u_1^*}{m_1 + m_2} - \frac{m_2 m_1 u_1^*}{m_1 + m_2} = 0 \tag{1.4}$$

where we have used (1.2), (1.3) and (1.4). Thus total linear momentum of particles in the CMS is zero before the collision and by conservation of momentum, this must be so after the collision.

1.2.3 Relation Between Velocities in the LS and CMS

Lab system	CM system	
$m_1 : u_1,$	$u_1^* = \dfrac{m_2 u_1}{m_1 + m_2}$	(1.5)
$m_2 : u_2 = 0,$	$u_2^* = \dfrac{m_1 u_1}{m_1 + m_2}$	(1.6)

For elastic collisions, both momentum and kinetic energy must be conserved. This implies that the respective velocities of the particles before and after the collisions in the CMS must be equal

$$u_1^* = v_1^*; \qquad u_2^* = v_2^* \tag{1.7}$$

$$v_1^* = \frac{m_2 u_1}{m_1 + m_2} \tag{1.8}$$

Observe that in both the LS and CMS, the relative velocity of the two particles is equal to u_1. We know

$$u(\text{rel}) = u_1^* + u_2^* = \frac{m_2 u_1}{m_1 + m_2} + \frac{m_1 u_1}{m_1 + m_2} = u_1$$

Using (1.2) and (1.8),

$$\frac{v_c}{v_1^*} = \frac{m_1}{m_2} = \gamma. \tag{1.9}$$

It is seen that if $m_1 < m_2$, then $v_c < v_1^*$ and if $m_1 > m_2$, $v_c > v_1^*$.

1.2.4 Relation Between the Angles in LS and CMS

Figure 1.3 shows the scattering and recoil angles in the LS and CMS.

The lab velocity v_1 of m_1 after the collision is obtained by combining vectorially its velocity v_1^* in the CMS and the CM velocity v_c (Fig. 1.4)

$$v_1 = v_1^* + v_c$$

Let m_1 be scattered at an angle θ as seen in the LS, its corresponding angle in the CMS being θ^*. In the velocity triangle (Fig. 1.4) resolving the velocities along the x-axis and y-axis, we get

$$v_1 \sin \theta = v_1^* \sin \theta^* \tag{1.10}$$

$$v_1 \cos \theta = v_1^* \cos \theta^* + v_c \tag{1.11}$$

Dividing (1.10) by (1.11)

$$\tan \theta = \frac{v_1^* \sin \theta^*}{v_1^* \cos \theta^* + v_c} = \frac{\sin \theta^*}{\cos \theta^* + v_c/v_1^*} = \frac{\sin \theta^*}{\cos \theta^* + m_1/m_2} \tag{1.12}$$

where we have used (1.9).

Special cases

(i) $m_1 \ll m_2; \theta \simeq \theta^*$. Here $v_c \to 0$ and the CMS is reduced to the LS.
    Example: α-gold nucleus scattering.
(ii) $m_1 \gg m_2; \theta \simeq 0°$.
    Example: nucleus-electron scattering.
(iii) $m_1 = m_2; \tan \theta = \frac{\sin \theta^*}{\cos \theta^* + 1} = \tan \frac{1}{2}\theta^*$ so that $\theta = \frac{1}{2}\theta^*$.
    Example: proton-proton scattering.

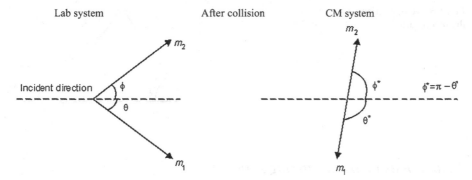

Fig. 1.3 Relation between the angles in LS and CMS

Fig. 1.4 Velocity triangle for
the scattered particle

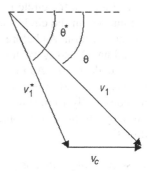

1.2.5 Recoil Angle

Let m_2 recoil with velocity v_2 at an angle ϕ with the incident direction in the LS. Let its velocity be v_2^* at angle ϕ^* in the CMS. From the velocity triangle in Fig. 1.5, we get

$$v_2 \sin \phi = v_2^* \sin \phi^* \tag{1.13}$$

$$v_2 \cos \phi = v_2^* \cos \phi^* + v_c \tag{1.14}$$

Dividing (1.13) by (1.14)

$$\tan \phi = \frac{v_2^* \sin \phi^*}{v_2^* \cos \phi^* + v_c}$$

but by (1.1), (1.6) and (1.7), $v_2^* = v_c$

$$\therefore \quad \tan \phi = \frac{\sin \phi^*}{\cos \phi^* + 1} = \tan \frac{\phi^*}{2} \quad \text{or}$$

$$\phi = \phi^*/2 \quad \text{(regardless of the ratio } m_1/m_2\text{)} \tag{1.15}$$

Fig. 1.5 Velocity triangle for
the recoil particle

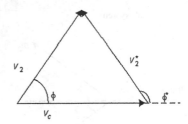

1.2.6 Limits on the Scattering Angle θ

Case (i) $m_2 > m_1$, or $\gamma < 1$; i.e. $v_1^* > v_c$.

In Fig. 1.6, the circle is drawn with O as the centre and radius $OP = v_1^*$. A is
a point within the circle such that $AO = v_c$, and the line AOB represents the incident direction. As before, the lab velocity of m_1 is v_1 which is obtained by com-
pounding v_1^* and v_c vectorially. The lab angle $\theta =$ angle PAO and the CM angle
$\theta^* =$ angle POB. As the point P moves counterclockwise on the circumference,
θ^* increases and so does θ. When P approaches P', $\theta^* = \theta = \pi$. Thus, θ increases
monotonically from 0 to π, and in this case there is no restriction on the scatter-
ing angle in the LS. In other words, m_1 can be scattered in completely backward
direction.

Case (ii) $m_2 = m_1$, or $\gamma = 1$; i.e. $v_1^* = v_c$.

Here A lies on the circumference of the circle (Fig. 1.7). As θ^* increases, θ also
increases. But when P approaches A, PA becomes tangential at A and so $\theta \to \frac{\pi}{2}$.
θ varies from 0 to π. Thus, in this case m_1 can be scattered up to a maximum angle of $\pi/2$ but not beyond. In other words, backward scattering in the LS is not
permissible.

Case (iii) $m_2 < m_1$, or $\gamma > 1$, i.e. $v_1^* < v_c$.

Here A lies outside the circle (Fig. 1.8). There are two positions P and P^l for
which the same scattering angle θ is obtained for two different values of θ^*. As P
moves back on the circumference, θ increases. The maximum angle θ_m is reached
when AP becomes tangent to the circle (Fig. 1.9). In that case

$$\sin\theta_m = \frac{OP}{AO} = \frac{v_1^*}{v_c} = \frac{m_2}{m_1} \quad \text{or}$$

$$\theta_m = \sin^{-1}(m_2/m_1)$$

Thus, there is a limitation on the scattering angle when $m_2 < m_1$. θ first increases
from 0 to a maximum value $\sin^{-1}(1/\gamma)$ which is less than $\pi/2$, as θ^* increases from
0 to $\cos^{-1}(-1/\gamma)$. θ then decreases to 0 as θ^* further increases to π. At a given
angle θ between 0 and $\sin^{-1}(1/\gamma)$, there will be two groups of particles associated
with different velocities corresponding to the two values of θ^*.

Fig. 1.6 Limits on the scattering angle for $m_2 > m_1$, θ max $= \pi$

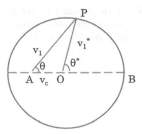

Fig. 1.7 Limits on the scattering angle θ for $m_1 = m_2$, θ max $= \pi/2$

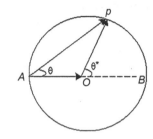

Fig. 1.8 Limits on the scattering angle θ. For $m_2 < m_1$, $\theta(\text{max}) = \sin^{-1}\left(\frac{m_2}{m_1}\right)$

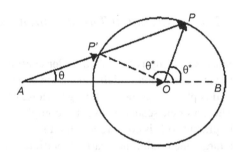

Fig. 1.9 θ_{max} is reached when AP is a tangent to the circle

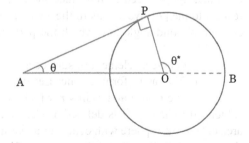

1.2.7 *Limits on the Recoil Angle* φ

Since $v_2^* = v_c$ (always), $\phi = \frac{1}{2}\phi^*$ by (1.15). Since the maximum angle of ϕ^* is π, the maximum angle ϕ_m is $\frac{1}{2}\pi$. In other words, the target particle cannot recoil in the backward hemisphere.

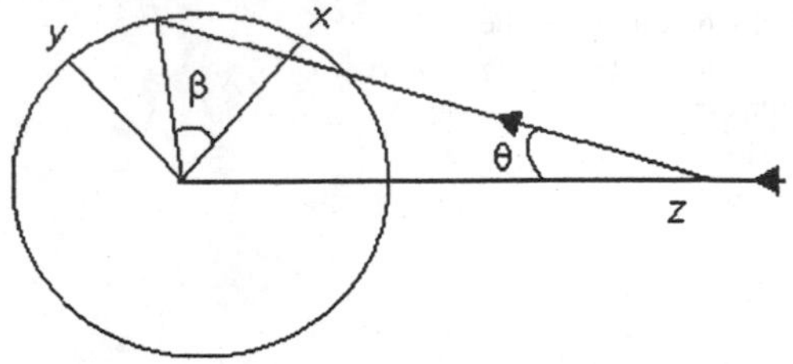

Fig. 1.10 Azimuth angle β is
measured with respect to the
positive axis in the xy plane
$\perp$ to the direction of
incidence

Fig. 1.11 Element of solid
angle $d\Omega = ds/r^2$

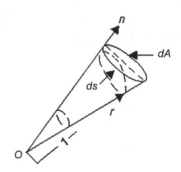

1.2.8 Scattering in Three Dimensions

Since scattering is described under central forces, a particle which is incident on a
target particle and initially moves in a certain plane would be necessarily confined
to this plane after the scattering because of the conservation of angular momentum.
Thus, a single scattering event is completely described in two dimensions. However,
in practice one is concerned with a flux of particles incident say along the z-axis on
a target material. Since various particles proceed in different planes, the scattering
on the whole will be in three dimensions. In order to fix the orientation of the plane
of scattering, we need to introduce the azimuth angle β which is measured with
respect to the positive x-axis in the xy plane, Fig. 1.10. We must also consider the
element of solid angle into which the particles are scattered. This is illustrated in
Fig. 1.11.

Let dA denote an element of surface area and connect all points on the boundary
of dA to O so as to form a cone. Let ds be the area of that portion of a sphere
with O as the centre and radius r which is cut out by this cone. The solid angle
subtended by dA at O is defined as $d\Omega = ds/r^2$ and is numerically equal to the
area cut out by a sphere with centre O and unit radius. From Fig. 1.12, it is seen that
$ds = r^2 \sin\theta d\theta d\beta$ so that $d\Omega = \sin\theta d\theta d\beta = 2\pi \sin\theta d\theta$, where we have integrated
over $d\beta$. When the scattering is independent of the azimuth angle then the area
subtended at O is due to the entire circular strip, $ds = 2\pi r^2 \sin\theta d\theta$ as in Fig. 1.13,
so that the element of solid angle $d\Omega = 2\pi \sin\theta d\theta$. Observe that the maximum
solid angle is 4π since it is given by the entire surface area of a sphere $(4\pi r^2)$
divided by r^2.

Fig. 1.12 Elements of solid
angle for a general case

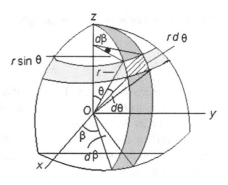

Fig. 1.13 Element of solid
angle for an azimuthal
symmetry

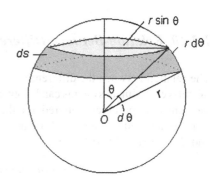

1.2.9 Scattering Cross-Section

In order to describe the angular distribution of particles scattered by target particles
which are initially stationary, the concept of cross-section is introduced. Let a uniform parallel flux of N_0 particles be incident per unit area normal to the direction
per unit time on a group of n scattering centres. Let N particles be scattered per unit
time into a small solid angle $d\Omega$ centred towards a direction which has polar angle
θ and azimuth angle β with respect to the incident direction as polar axis. N will
be proportional to N_0, n and $d\Omega$ provided the flux is small enough to ensure that
the incident particles do not interfere with one another, that there is no appreciable
decrease in the number of scattering centres on account of their being knocked out
due to collisions and that the incident particles are far enough apart so that each
collision is made only by one of them.

The number of incident particles that emerge per unit time in $d\Omega$ can be written
as:

$$N = nN_0\sigma(\theta, \beta)d\Omega \tag{1.16}$$

where the proportionality factor $\sigma(\theta, \beta)$ is called the differential scattering cross-
section. The quantity $\sigma(\theta, \beta)$ is a measure of the probability of scattering in a given
direction (θ, β), per unit solid angle from the given nucleus.

The integral of $\sigma(\theta, \beta)$ over the sphere is called the total scattering cross-section

$$\sigma = \int \sigma(\theta, \beta) d\Omega \tag{1.17}$$

σ has the dimension of area. The unit of σ is a Barn (b). $1\,b = 10^{-24}\,cm^2$. The unit of $\sigma(\theta, \beta)$ is Barn/Steradian, where Steradian (sr) is the unit of solid angle. $\sigma(\theta, \beta)$ is also written as $d\sigma(\theta, \beta)/d\Omega$.

Compared to the scattering in two dimensions the only additional parameter which has been introduced to describe scattering in three dimensions is known as the azimuth angle β.

1.2.10 Relation Between Differential Scattering Cross-Sections

The relation between the differential cross-section in the laboratory and the centre-of-mass co-ordinate systems can be obtained from their definition which implies that the number of particles scattered into the element of solid angle $d\Omega$ about θ, β is the same as are scattered into $d\Omega^*$ about θ^*, β^*. In polar co-ordinates $d\Omega = \sin\theta d\theta d\beta$ and $d\Omega^* = \sin^* d\theta^* d\beta^*$

$$\sigma(\theta, \beta) \sin\theta d\theta d\beta = \sigma(\theta^*, \beta^*) \sin\theta^* d\theta^* d\beta^*$$

but

$$\beta = \beta^*$$

$$\therefore \quad \sigma(\theta) = \sigma(\theta^*) \frac{\sin^* d\theta^*}{\sin\theta d\theta} \tag{1.18}$$

Differentiating (1.12)

$$\sec^2\theta d\theta = \frac{|1 + \gamma\cos\theta^*|d\theta^*}{(\cos\theta^* + \gamma)^2} \tag{1.19}$$

Using (1.12) and the identity, $\sec^2\theta = 1 + \tan^2\theta$ (1.19) is easily reduced to the form:

$$\frac{d\theta^*}{d\theta} = \frac{[1 + 2\gamma\cos\theta^* + \gamma^2]}{|1 + \gamma\cos\theta^*|} \tag{1.20}$$

Also

$$\tan\theta = \frac{\sin\theta}{\cos\theta} = \frac{\sin\theta^*}{\cos\theta^* + \gamma} \tag{1.12}$$

whence

$$\frac{\sin\theta^*}{\sin\theta} = \frac{\cos\theta^* + \gamma}{\cos\theta} \tag{1.21}$$

Also

$$\frac{1}{\cos \theta} = \sec \theta = \sqrt{1 + \tan^2 \theta} \qquad (1.22)$$

Using (1.12) in (1.22) and re-arranging them we get

$$\frac{\cos \theta^* + \gamma}{\cos \theta} = \sqrt{1 + 2\gamma \cos \theta^* + \gamma^2} = \frac{\sin \theta^*}{\sin \theta} \qquad (1.23)$$

Using (1.20) and (1.23) in (1.18)

$$\sigma(\theta) = \frac{(1 + \gamma^2 + 2\gamma \cos \theta^*)^{3/2} \sigma(\theta^*)}{|1 + \gamma \cos \theta^*|} \qquad (1.24)$$

It must be pointed out that the total cross-section is the same for both lab and CM systems, since the occurrence of total number of collisions is independent of the mode of description of the process.

1.2.11 Kinematics of Elastic Collisions

We have to obtain an expression for velocity v_1 as a function of scattering angle θ. From the velocity triangle (Fig. 1.14)

$$v_1^{*2} = v_c^2 - 2v_1 v_c \cos \theta + v_1^2 \qquad (1.25)$$

Substituting for v_c and v_1^* from (1.1) and (1.8), (1.25) becomes

$$v_1^2 - \frac{2m_1 u_1 \cos \theta \, v_1}{m_1 + m_2} + u_1^2 \frac{(m_1 - m_2)}{m_1 + m_2} = 0$$

This is a quadratic equation in v_1 whose solutions are found to be

$$v_1 = \frac{m_1 u_1}{m_1 + m_2} \left[\cos \theta \pm \sqrt{\frac{m_2^2}{m_1^2} - \sin^2 \theta} \right] \qquad (1.26)$$

For the special case, $m_1 = m_2$, (1.26) simplifies to:

$$v_1 = u_1 \cos \theta$$

so that the ratio of kinetic energy T_1 and T_0 of the scattered and incident particle becomes

$$\frac{T_1}{T_0} = \frac{v_1^2}{u_1^2} = \cos^2 \theta$$

with the restriction, $\theta \le 90°$, as pointed out earlier.

Fig. 1.14 Velocity triangle
for the scattered particle

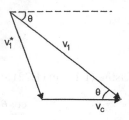

Fig. 1.15 Velocity triangle
for recoil particle

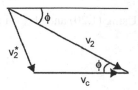

1.2.12 To Derive an Expression for the Recoil Velocity v_2 as a Function of ϕ

From the velocity triangle (Fig. 1.15)

$$v_2^{*2} = v_c^2 + v_2^2 - 2v_c v_2 \cos \phi$$

Since

$$v_2^* = v_c$$

$$v_2 = 2v_c \cos \phi = \frac{2m_1 u_1}{m_1 + m_2} \cos \phi \tag{1.27}$$

where we have used (1.1).

The ratio of kinetic energy of the recoil particle and original kinetic energy of the incident particle is:

$$\frac{T_2}{T_0} = \frac{m_2 v_2^2}{m_1 u_1^2} = \frac{4m_1 m_2}{(m_1 + m_2)^2} \cos^2 \phi \tag{1.28}$$

For the special case $m_1 = m_2$

$$\frac{T_2}{T_0} = \cos^2 \phi \tag{1.29}$$

1.2.13 Available Energy in the Lab System and CM System

Assuming that the target particle is at rest before the collision, total kinetic energy in the lab system is

$$T = T_0, \quad \text{with } T_0 = \frac{1}{2} m_1 u_1^2 \tag{1.30}$$

In the CM system, m_1 has kinetic energy

$$T_1^* = \frac{1}{2}m_1(u_1^*)^2 = \frac{1}{2}m_1\left[\frac{m_2u_1}{m_1+m_2}\right]^2$$

where we have used (1.4).

In the CM system, m_2 has kinetic energy:

$$T_2^* = \frac{1}{2}m_2(u_2^*)^2 = \frac{1}{2}m_2\left[\frac{m_1u_1}{m_1+m_2}\right]^2$$

where we have used (1.2).

Total kinetic energy available in the CM system is:

$$T^* = T_1^* + T_2^* = \frac{1}{2}\frac{m_1m_2}{m_1+m_2}u_1^2 = \frac{1}{2}\mu u_1^2 \qquad (1.31)$$

where μ is the reduced mass.

Formula (1.31) shows that the two-body problem is reduced to a one-body problem by imagining that a particle of mass $\mu = m_1m_2/(m_1+m_2)$ is directed towards a scattering centre, with the velocity u_1. Using (1.30) in (1.31)

$$T^* = \frac{m_2T_0}{m_1+m_2}$$

where $T^* < T_0$.

Thus less energy of motion is available in the CM system. It can easily be shown that the difference in energy in the lab and CM systems is associated with the motion of CM system

$$\Delta T = T_0 - T^* = \frac{1}{2}m_1u_1^2 - \frac{1}{2}\frac{m_1m_2}{m_1+m_2}u_1^2 = \frac{1}{2}\frac{m_1^2u_1^2}{(m_1+m_2)} = \frac{1}{2}(m_1+m_2)v_c^2 \qquad (1.32)$$

where we have used (1.1).

Formula (1.32) shows that the difference of energy goes into the motion of CM of mass (m_1+m_2) with velocity v_c. We conclude that in the CM system energy that is available is always less than that in the lab system, for some energy must go into the motion of CM system.

For the special case $m_1 = m_2$

$$T^* = \frac{1}{4}m_1u_1^2 = \frac{1}{2}T_0$$

This fact has a bearing on production thresholds, i.e. minimum energy that is to be provided in order to produce particles. Consider, for example, the case of pion production in proton-proton collisions. The rest mass of pion is only 140 MeV/c^2. However, this much energy must be available in the CM system. This means that in the lab system, the incident proton must have double this energy viz, 280 MeV in order to produce a pion. Relativistic calculations actually give a value of 290 MeV.

These considerations are also important in the invention of a new class of high energy accelerators in recent years, in which colliding beams of particles are used; i.e. one beam travels in one direction and is intercepted by another beam of similar or dissimilar particles of the same energy moving in the opposite direction. In this case, the CM system is realized in the laboratory itself and lot of energy is made available.

Example 1.1 If a particle of mass m collides elastically with one of mass M at rest, and if the former is scattered at an angle θ and the latter recoils at an angle ϕ with respect to the line of motion of the incident particle, then show that

$$\tan\theta = \frac{\sin 2\phi}{\frac{m}{M} - \cos 2\phi}$$

Hence, show that

$$\frac{m}{M} = \frac{\sin(2\phi + \theta)}{\sin\theta}$$

Solution

$$\tan\theta = \frac{\sin\theta^*}{\cos\theta^* + m/M} \quad \text{but } \theta^* = \pi - \phi^* = \pi - 2\phi$$

$$\therefore \quad \sin\theta^* = \sin(\pi - 2\phi) = \sin 2\phi$$

$$\cos\theta^* = \cos(\pi - 2\phi) = -\cos 2\phi$$

$$\therefore \quad \tan\theta = \frac{\sin\theta}{\cos\theta} = \frac{\sin 2\phi}{m/M - \cos 2\phi}$$

Re-arranging the above we get

$$\frac{m}{M}\sin\theta = \sin\theta\cos 2\phi + \cos\theta\sin 2\phi = \sin(\theta + 2\phi)$$

$$m/M = \frac{\sin(2\phi + \theta)}{\sin\theta}$$

Example 1.2 A particle makes an elastic collision with another particle of identical mass, initially at rest. Prove that after scattering, the lab angle between the outgoing particles is 90°.

Solution

First Method
We use the lab system. Let the particle of mass m, momentum P and kinetic energy T move along the x-axis. After collision the particles have momenta P_1 and P_2 at

Fig. 1.16 Elastic collision in LS for $m_1 = m_2$

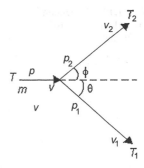

angles θ and ϕ as in Fig. 1.16. Conservation of momentum along the direction of incidence (x-axis) gives

$$P = P_1 \cos\theta + P_2 \cos\phi \tag{1}$$

Conservation of momentum along the perpendicular direction (y-axis) yields

$$P_2 \sin\phi - P_1 \sin\theta = 0 \quad \text{or}$$

$$0 = P_1 \sin\theta - P_2 \sin\phi \tag{2}$$

Squaring and adding (1) and (2) and simplifying we get

$$P^2 = P_1^2 + P_2^2 + 2P_1 P_2 \cos(\theta + \phi) \tag{3}$$

Energy conservation gives:

$$\frac{P^2}{2m} = \frac{P_1^2}{2m} + \frac{P_2^2}{2m} \quad \text{or} \tag{4}$$

$$P^2 = P_1^2 + P_2^2 \tag{5}$$

Using (5) in (3):

$$2P_1 P_2 \cos(\theta + \phi) = 0$$

Since $P_1 \neq 0$; $P_2 \neq 0$; $\cos(\theta + \phi) = 0$ or $\theta + \phi = 90°$.
Second Method (Vector Method)
Momentum of conservation demands that (Fig. 1.17)

$$\boldsymbol{P_1 + P_2 = P}$$

Taking the scalar product

$$\boldsymbol{P \cdot P = (P_1 + P_2) \cdot (P_1 + P_2)}$$

$$\boldsymbol{P \cdot P = P_1 \cdot P_1 + P_2 \cdot P_2 + P_1 \cdot P_2 + P_2 \cdot P_1}$$

$$P^2 = P_1^2 + P_2^2 + 2\boldsymbol{P_1 \cdot P_2} \tag{6}$$

Fig. 1.17 Momentum
triangle for the elastic
collision

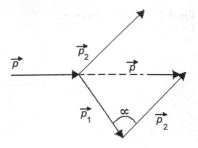

since the scalar product of a vector by itself is the square of the magnitude of the
vectors and the order of scalar product is immaterial.

In view of energy conservation, i.e. with the aid of (5), we find

$$2P_1 P_2 = 0 \quad \text{or} \quad 2P_1 P_2 \cos\alpha = 0$$

where α is the angle between the vectors P_1 and P_2

$$\therefore \quad \alpha = 90°$$

Third Method

Because of the conservation of momentum, P_1, P_2, and P form a closed triangle.
Their magnitudes are indicated in Fig. 1.17. Because of energy conservation we
further have the relation:

$$P_1^2 + P_2^2 = P^2$$

i.e. the triangle must be a right angle triangle. Hence, $\alpha = 90°$.

Fourth Method

We use the following formula for transformation of angles between LS and the
CMS. Set $\frac{m}{M} = 1$ in (1.12)

$$\tan\theta = \frac{\sin\theta^*}{1 + \cos\theta^*} = \tan\frac{\theta^*}{2}$$

$$\therefore \quad \theta = \frac{\theta^*}{2}$$

But $\theta^* = \pi - \phi^*$ and $\phi^* = 2\phi$, always

$$\phi^* = 2\theta = \pi - 2\phi \quad \text{whence } \theta + \phi = \frac{\pi}{2}$$

Example 1.3   Show that if a particle of mass m is scattered by a particle of mass
M initially at rest, then the angle between the final directions of motion in the lab
system is:

$$\frac{\pi}{2} + \frac{1}{2}\theta - \frac{1}{2}\sin^{-1}\left(\frac{m}{M}\sin\theta\right)$$

Hence, show that for particles of equal masses, the angle between final directions of motion is always 90°.

Solution From Example 1.1 we get

$$\frac{m}{M} = \frac{\sin(2\phi + \theta)}{\sin\theta} \quad \text{or}$$

$$\frac{m}{M}\sin\theta = \sin\big[\pi - (2\phi + \theta)\big]$$

$$\sin^{-1}\frac{m}{M}\sin\theta = \big[\pi - (2\phi + \theta)\big]$$

$$\phi + \frac{\theta}{2} = \frac{\pi}{2} - \frac{1}{2}\sin^{-1}\frac{m}{M}\sin\theta$$

The angle between the final directions of motion is

$$\alpha = \phi + \theta = \frac{\pi}{2} + \frac{1}{2}\theta - \frac{1}{2}\sin^{-1}\frac{m}{M}\sin\theta$$

For $m/M = 1$, α reduces to $\frac{1}{2}\pi$.

Example 1.4 At low energies, neutron-proton scattering is isotropic in the C-system. If K is neutron lab energy and σ the total cross section, show that in the lab, the proton energy distribution is

$$d\sigma_p/dK_p = \text{const} = \sigma/K_0$$

Solution In Fig. 1.18, ABC is the momentum triangle. Since the angle between the scattered neutron and recoil proton must be a right angle

$$P_P = P_0\cos\phi$$
$$K_P = P_P^2/2m_P \quad \text{and} \quad K_0 = P_0^2/2m_n$$

but, $m_p \simeq m_n = m$

$$K_P/K_0 = P_P^2/P_0^2 = \cos^2\phi \quad \text{or} \quad K_P = K_0\cos^2\phi$$
$$dK_p = -2K_0\cos\phi\sin\phi d\phi = -K_0\sin2\phi d\phi$$

but, $\phi = \phi^*/2$ and $d\phi = d\phi^*/2$

$$dK_p = -\frac{1}{2}K_0\sin\phi^* d\phi^*$$

Isotropy requires that $d\sigma_p/d\Omega^* = \sigma/4\pi$

$$\frac{d\sigma_p}{dK_p} = -\frac{d\sigma_p}{d\Omega^*}\frac{d\Omega^*}{dK_p} = \frac{\sigma}{4\pi}\frac{2\pi\sin\phi^* d\phi^*}{\frac{1}{2}K_0\sin\phi^* d\phi^*} = \frac{\sigma}{K_0} = \text{const}$$

Fig. 1.18 Momentum
triangle for n–p scattering

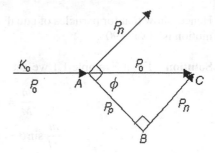

Negative sign is introduced in the last equation because as ϕ^* increases K_p decreases.

Example 1.5 A beam of particles of mass m is elastically scattered by target particles of mass M initially at rest. If the angular distribution is spherically symmetrical
in the centre of mass system, what is it for M in the lab system?

Solution

$$\sigma(\phi^*) = \frac{\sigma}{4\pi} = \text{const}$$

$$\sigma(\phi) = \frac{\sin\phi^* d\phi^*}{\sin\phi d\phi}\sigma(\phi^*)$$

but

$$\phi^* = 2\phi \quad \text{and} \quad d\phi^* = 2d\phi$$

$$\sigma(\phi) = \frac{\sin 2\phi 2d\phi}{\sin\phi d\phi}\frac{\sigma}{4\pi} = \frac{\sigma}{\pi}\cos\phi$$

It may be recalled that ϕ is limited to 90°, i.e. the target particles can recoil only in
the forward hemisphere in the lab system. It is instructive to note that

$$\int\sigma(\phi)d\Omega = \int_0^{\frac{1}{2}\pi}\frac{\sigma}{\pi}\cos\phi 2\pi\sin\phi d\phi$$

$$= 2\sigma\int_0^1\sin\phi d(\sin\phi) = \sigma \quad \text{(as it should)}$$

Example 1.6 Small balls of negligible radii are projected against an infinitely heavy
sphere of radius R. Assuming the balls are elastically scattered and bounce off in
such a way that the angle of reflection (r) is equal to the angle of incidence (i).
Prove that the scattering is isotropic, i.e. $\sigma(\phi)$ is independent of θ and that the total
cross-section is equal to πR^2.

Solution Let b be the impact parameter (perpendicular distance of the line of flight
from the central axis). The angle of incidence and reflection are measured with

Fig. 1.19 Scattering of a
small ball off a heavy sphere

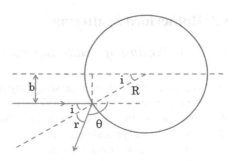

respect to the normal at the point of scattering (Fig. 1.19)

$$\sigma(\theta) = -\frac{b\,db}{\sin\theta\,d\theta} \tag{1}$$

See Eq. (1.56).

From the geometry of Fig. 1.19

$$\theta = \pi - (i + r) = \pi - 2i \tag{2}$$

$$\sin\theta = \sin(\pi - 2i) = \sin 2i = 2\sin i \cos i \tag{3}$$

since

$$r = i, \quad \text{and} \quad d\theta = -2di \tag{4}$$

but

$$b = R\sin i \tag{5}$$

Hence

$$db = R\cos i\,di \tag{6}$$

using (2), (4), (5) and (6) in (1) and cancelling various factors; we get $\sigma(\theta) = R^2/4$.

The right hand side of $\sigma(\theta)$ is independent of θ, the scattering angle. Hence, the scattering is isotropic, i.e. equally in all directions.

The total cross-section is given by

$$\sigma = \int \sigma(\theta)d\Omega = \int_0^\pi \frac{R^2}{4} 2\pi \sin\theta\,d\theta = \pi R^2$$

Observe that the total cross-section σ has the dimension of area, and in the above example it is equal to the projected area of the sphere. It is, therefore, called geometrical cross-section. Formula (5) shows that if $b = 0$ (head-on collision), $i = 0$ and from (2), $\theta = 180°$. Thus, in this case the ball bounces in the opposite direction. Again, when $b = R$, $i = 90°$ and $\theta = 0°$ (glancing collision). Thus, the ball having hit the edge of the sphere, does not suffer any deviation and continues its flight in the incident direction. Of course, if $b > R$, the ball goes undeviated and there is no scattering. The above example shows the concept of $\sigma(\theta)$ and σ.

1.3 Rutherford Scattering

1.3.1 Derivation of Scattering Formula

Here we are concerned with the scattering (deflection) of point charged particles by a massive centre of electric force. The force is assumed to be central, i.e. directed along the line joining the centres of the colliding particles. Rutherford supposed that all the positive charge and hence practically all the mass of the atom is concentrated in a core or nucleus whose volume is very much less than that of the atom. Outside the nucleus is a relatively empty space only occupied by a few electrons. Suppose, an alpha particle is fired against the atom, then it is permitted to penetrate close to the nucleus and owing to the electrical interaction with the nucleus it may suffer a large angle deflection and recede from the nucleus and the effect of widely dispersed electrons can be neglected.

A particle of charge $+ze$ (for alpha particle, $z = 2$) at a distance r from the nucleus of charge $+Ze$ (Z being the atomic number) experiences a repulsive force zZe^2/r^2 (Coulomb's inverse square law) and the corresponding potential energy will be zZe^2/r. When the incident (incoming) particle is at a very large distance, the potential energy will be zero, and the energy is entirely kinetic due to the motion of the particle.

Let the particle of charge $+ze$ and mass m be incident from a very large distance from the nucleus (for example at a point A, Fig. 1.20), with velocity v_0. In the absence of forces between the nucleus (henceforth called target nucleus) at F and the incident particle, the particle would have continued to move along the straight line AOB. Let FQ be perpendicular on AOB. Then $b = FQ$ is called impact parameter. Since the target nucleus is considered infinitely heavy, it does not move during the encounter. The analysis will therefore be made in the lab system. The force is repulsive and central. We shall prove that under the influence of Coulomb's force, the trajectory is a hyperbola with the external focus F at the nucleus.

It is convenient to introduce the polar co-ordinates r, θ. The radial distance r is measured from the focus F and the angle θ with the x-axis, which is arbitrarily chosen. When the particle is near the nucleus, it will be deviated from the rectilinear trajectory under the action of electrical forces and its typical position at some instant would be at some point P with co-ordinates (r, θ) and velocity v. Since the force is repulsive, $v < v_0$. After the complete encounter, the particle is deflected through angle θ_0 and would recede to a remote distance beyond which it would continue along the straight path OD, and at a distant point like D. It would again have the original speed v_0, as the potential energy again approaches zero.

If the original path of the incident particle lies in a plane (here plane of paper) then because angular momentum is conserved, the particle would continue its path in the same plane throughout.

The conservation of energy gives:

$$\frac{1}{2}mv_0^2 = \frac{1}{2}mv^2 + zZe^2/r \tag{1.33}$$

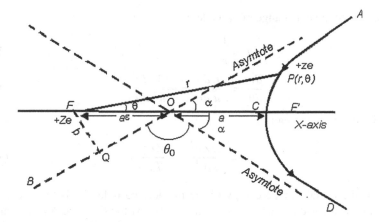

Fig. 1.20 Geometry of Rutherford scattering

In polar co-ordinates, the components of velocity along and perpendicular to r are $\dot{r}$ and $r\dot{\theta}$, respectively, so that

$$v^2 = (\dot{r})^2 + (r\dot{\theta})^2 \tag{1.34}$$

where dot means differentiation with respect to time. Eliminating v between (1.33) and (1.34) and re-arranging we get:

$$1 = \frac{2zZe^2}{mv_0^2 r} + \frac{1}{v_0^2}\left[(\dot{r})^2 + (r\dot{\theta})^2\right] \tag{1.35}$$

Also in the absence of external forces, angular momentum must be conserved. Take the angular momentum about an axis passing through the nucleus and perpendicular to the plane. Initially the momentum mv_0 is in the direction AB and the perpendicular distance FQ is b. Therefore, the initial angular momentum $= (mv_0)b$. At the point P, the component of velocity perpendicular to FP is $r\dot{\theta}$, and the distance $FP = r$. Hence, the angular momentum at P is $m(r\dot{\theta})r$.

Conservation of angular momentum gives,

$$mv_0 b = mr^2\dot{\theta} \quad \text{or} \tag{1.36}$$

$$\dot{\theta} = v_0 b / r^2 \tag{1.37a}$$

Also

$$\dot{r} = \frac{dr}{d\theta}\dot{\theta} = \frac{v_0 b}{r^2}\frac{dr}{d\theta} \tag{1.37b}$$

Using (1.37a) and (1.37b) in (1.35) and dividing by b^2, we find:

$$\frac{1}{b^2} = \frac{2zZe^2}{mv_0^2 b^2 r} + \frac{1}{r^4}\left(\frac{dr}{d\theta}\right)^2 + \frac{1}{r^2} \tag{1.38}$$

It is desirable to have a change of variable, i.e.

$$u = \frac{1}{r}$$ (1.39)

$$\frac{dr}{d\theta} = \frac{dr}{du}\frac{du}{d\theta} = -\frac{1}{u^2}\frac{du}{d\theta}$$ (1.40)

Also, it is convenient to call

$$R_0 = \frac{2zZe^2}{mv_0^2} = \frac{zZe^2}{T_0}$$ (1.41)

Here T_0 is the initial kinetic energy of the incident particle. R_0 is the distance of closest approach for the head-on collision ($b = 0$). At the distance R_0, the particle momentarily comes to rest ($v = 0$) before it makes a sharp U-turn. Using (1.39), (1.40) and (1.41) in (1.38) and re-arranging, we get:

$$\left(\frac{du}{d\theta}\right)^2 = \frac{1}{b^2} - u^2 - \frac{R_0}{b^2}u$$ (1.42)

Above differential equation can be solved easily if it is differentiated once with respect to θ, and bearing in mind that b is a constant for a given encounter,

$$2\frac{d^2u}{d\theta^2}\frac{du}{d\theta} = -2u\frac{du}{d\theta} - \frac{R_0}{b^2}\frac{du}{d\theta}$$

Cancelling the common factor $du/d\theta$ and re-arranging we get:

$$\frac{d^2u}{d\theta^2} + u + \frac{R_0}{2b^2} = 0$$ (1.43)

This has the obvious solution

$$u = A\cos(\theta - \delta) - \frac{R_0}{2b^2}$$ (1.44)

where A and δ are the constants of integration. We may choose $\delta = 0$ to make the trajectory symmetrical about the x-axis. Call

$$g = 2b^2/R_0$$ (1.45)

with $\delta = 0$, we find from (1.44)

$$\frac{1}{u} = r = \frac{g}{gA\cos\theta - 1}$$ (1.46)

This may be compared with the equation for a conic

$$r = \frac{a(\varepsilon^2 - 1)}{\varepsilon\cos\theta - 1}$$ (1.47)

where ε is the eccentricity and 'a' is the semi-major axis. We therefore, identify

$$g = a\left(\varepsilon^2 - 1\right) \tag{1.48}$$

$$\varepsilon = gA \tag{1.49}$$

Using (1.44) and (1.45) in (1.42) and simplifying

$$A^2 = \frac{1}{g^2} + \frac{1}{b^2} \tag{1.50}$$

Eliminating A between (1.49) and (1.50) and using (1.48) we can find ε

$$\varepsilon = \sqrt{1 + \frac{4b^2}{R_0^2}} = \sqrt{1 + \frac{4b^2 T_0^2}{z^2 Z^2 e^4}} \tag{1.51}$$

where we have used (1.41).

It is seen from the above formula that $\varepsilon > 1$ even if the charge is negative and the eccentricity is same in both the cases. The orbit is always a hyperbola and never an ellipse. For a repulsive Coulomb force (positively charged incident particle) the orbit is a hyperbola with the target nucleus at the external focus F, whereas for attractive Coulomb force (negatively charged incident particle) the orbit is a hyperbola with the target nucleus at the inner focus F'.

As $r \to \infty$, the denominator of the right hand side of (1.47) becomes zero, and the limiting angle α is given by:

$$\cos\alpha = \frac{1}{\varepsilon} \quad \text{or}$$

$$\cot\alpha = \frac{1}{\sqrt{\varepsilon^2 - 1}} \tag{1.52}$$

Observe that α is very nearly equal to half of the angle subtended between the asymtotes, since the angle contained between the radius vector r and the x-axis is almost equal to α when $r \to \infty$. The scattering angle θ_0 is equal to angle BOD and is given by $\theta_0 = \pi - 2\alpha$ or, $\alpha = \frac{\pi}{2} - \frac{\theta_0}{2}$. Hence

$$\tan\frac{\theta_0}{2} = \cot\alpha = \frac{1}{\sqrt{\varepsilon^2 - 1}} = \frac{R_0}{2b} \tag{1.53}$$

where we have used (1.52) and (1.51).

Formula (1.53) can be derived by a shorter method by assuming that the trajectory is a hyperbola and by considering the velocity v at the point C which is at distance 'a' from the centre of the nucleus (Fig. 1.20); v being perpendicular to 'a'. Energy conservation gives:

$$\frac{1}{2}mv_0^2 = \frac{1}{2}mv^2 + \frac{zZe^2}{a}$$

which yields

$$\left(\frac{v}{v_0}\right)^2 = 1 - \frac{R_0}{a} \tag{i}$$

Angular momentum conservation gives

$$mv_0 b = mva \quad \text{or}$$

$$\frac{v}{v_0} = \frac{b}{a} \tag{ii}$$

Using (ii) in (i), we find

$$b^2 = a(a - R_0) \tag{iii}$$

From the properties of hyperbola, we know

$$a = b \cot \frac{\alpha}{2} \tag{iv}$$

Eliminating 'a' between (iii) and (iv)

$$\frac{R_0}{b} = \frac{\cot^2 \frac{\alpha}{2} - 1}{\cot \frac{\alpha}{2}} = 2 \cot \alpha$$

$$\cot \alpha = \tan \frac{\theta_0}{2} = \frac{R_0}{2b}$$

Equation (1.53) shows that smaller the impact parameter b, larger is the scattering angle θ_0, and vice versa. Physically a larger value of b implies a weaker force and so a smaller deflection is to be expected.

In particular, $b = \infty$, implies $\theta_0 = 0$ and $b = 0$ implies $\theta_0, = \pi$. Figure 1.21 shows three typical scattering events. They are:

(a) with a large b,
(b) with a moderate value of b, and
(c) for a very small value of b.

Eliminating g between (1.45) and (1.48)

$$a = \frac{2b^2}{R_0(\varepsilon^2 - 1)} = \frac{R_0}{2} \tag{1.54}$$

where we have used (1.51).

For a particular value of b, the closest distance of approach will be FC which is given by putting $\theta = 0$ in (1.47)

$$r(\text{min}) = a(\varepsilon + 1) = \frac{R_0}{2}\left[1 + \sqrt{1 + \frac{4b^2}{R_0^2}}\right] \tag{1.55}$$

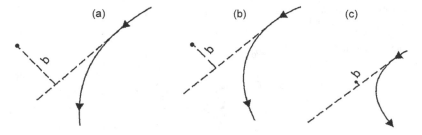

Fig. 1.21 Rutherford scattering for three different parameters b

Considering various scattering events with different b, $r(\min)$ will take on the least value for $b = 0$, i.e. for the head-on collision and in this case $r(\min) = R_0$. Thus the significance of R_0 given by (1.41) is that it represents the least distance of the closest approach. It is also called Collision diameter. This result also follows from very simple considerations. As the positively charged particle approaches the target nucleus, due to the Coulomb's repulsion, it loses kinetic energy and when it is closest to the nucleus in a head-on collision, it would lose all its kinetic energy. Putting $v = 0$ in (1.33) we get

$$r(\min) = \frac{2zZe^2}{mv_0^2} = R_0$$

We can find the numerical value of R_0 in the scattering of 5 MeV alpha particles (typical alpha energy from the radioactive sources) from a gold foil

$$T_0 = 5 \text{ MeV} = 5 \times 1.6 \times 10^{-13} \text{ J} = S \times 10^{-13} \text{ J}$$

$$e = 1.6 \times 10^{-19} \text{ Coul}$$

$$z = 2, \qquad Z = 79$$

For S.I. units, formula (1.41) becomes: $R_0 = \frac{zZe^2}{4\pi\varepsilon_0 T_0}$.

For numerical calculations, $R_0 \text{ (fm)} = \frac{1.44zZ}{T_0 \text{ (MeV)}} = \frac{1.44 \times 2 \times 79}{5} = 45.5 \text{ fm}$.

This value may be compared with the radius of gold nucleus which is 8 fm, a value which is smaller than the minimum distance of closest approach. This then ensures that the alpha particle of 5 MeV stays well outside the gold nucleus in any type of encounter including the head-on collision and that the inverse square law would be valid for all the orbits. For much greater energy, in close encounters, alpha particles may be able to penetrate the target nucleus itself in which case the inverse square law would no longer be valid, and other complications would be introduced into which we shall not enter at the moment.

From (1.53) it is obvious that given the impact parameter b, the scattering angle θ_0 can be determined. But, in practice, it is impossible to know the value of b. However, we can compute the expected angular distribution from the entire range of b's. Consider a uniform beam of particles fired against the target material. The beam intensity I is defined as the number of particles crossing unit area normal to the beam

Fig. 1.22 Particles passing through the ring of radii b and $b + db$ are scattered in the angular interval $\theta_0 + d\theta_0$ and θ_0

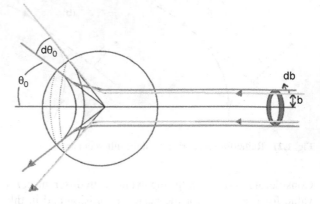

direction per second. Near the centre of force a beam particle bends around and as it escapes from the field of force it once again describes a straight line. Because the force is central, one can expect an azimuthal symmetry in scattering about an axis along the beam direction. Assuming that the scattering is independent of β the element of solid angle becomes $d\Omega = 2\pi \sin\theta_0 d\theta_0$. Consider a uniform beam of particles of same energy directed towards a force centre, with impact parameters b and $b + db$. Such particles are seen to pass through a ring of radii b and $b + db$, the ring being perpendicular to the beam direction and symmetrical about an axis passing through the nucleus, and has an area of $2\pi b db$. Now, particles of given energy and impact parameter have a unique angle of deflection determined by formula (1.53). Therefore, particles passing through this ring must be scattered into the solid angle lying between θ_0 and $\theta_0 + d\theta_0$ (Fig. 1.22). Since the number of particles must be conserved

$$2\pi b db I = -2\pi \sin\theta_0 d\theta_0 I \sigma(\theta_0) \quad \text{or}$$

$$\sigma(\theta_0) = -\frac{b db}{\sin\theta_0 d\theta_0} \tag{1.56}$$

The negative sign is introduced in (1.56) due to the fact that an increase in b implies a decrease in θ_0. Rewriting (1.53)

$$b = \frac{R_0}{2} \cot \frac{\theta_0}{2} \tag{1.57}$$

Hence

$$db = -\frac{R_0}{4} \operatorname{cosec}^2 \frac{\theta_0}{2} d\theta_0 \tag{1.58}$$

Using (1.57) and (1.58) in (1.56) and noting that $\sin\theta_0 = 2\sin\frac{\theta_0}{2} \cdot \cos\frac{\theta_0}{2}$, we get after simplification:

$$\sigma(\theta_0) = \frac{R_0^2}{16 \sin^4 \frac{\theta_0}{2}} = \frac{1}{16} \left[\frac{zZe^2}{T_0} \right]^2 \frac{1}{\sin^4 \frac{\theta_0}{2}}$$

Fig. 1.23 Differential cross-section (in arbitrary units) as a function of scattering angle θ in the LS

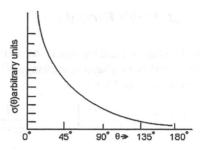

$$\sigma(\theta) = \frac{1}{4}\left[\frac{zZe^2}{mv_0^2}\right]^2 \frac{1}{\sin^4 \frac{\theta_0}{2}} \quad \text{(Rutherford's scattering formula)} \quad (1.59)$$

This is the famous Rutherford's scattering formula. Henceforth, the suffix 0 is dropped off in θ_0. The expected differential cross-section as a function of scattering angle given by (1.59) is shown in Fig. 1.23. Observe that the differential cross-section falls off rapidly with increasing angle, the scattering thus being predominantly in the forward direction.

Formula (1.59) also shows that $\sigma(\theta)$ will be greater for targets and incident particles of higher atomic number and that it will be more important for low energy particles.

For the purpose of numerical calculations (1.59) can be written in the form:

$$\sigma(\theta) = 1.295 \left(\frac{zZ}{T}\right)^2 \frac{1}{\sin^4 \theta/2} \text{ Mb/sr} \quad (1.60)$$

where T is in MeV.

1.3.2 Darwin's Formula

Rutherford's formula which takes into account the recoil of the nucleus is due to Darwin (see Example 1.18)

$$\sigma(\theta) = \left(\frac{zZe^2}{mv^2}\right)^2 \frac{1}{\sin^4 \theta} \frac{[\cos\theta \pm (1 - \gamma^2 \sin^2\theta)^{1/2}]^2}{(1 - \gamma^2 \sin^2\theta)^{1/2}} \quad (1.61)$$

where M is the mass of the target nucleus and m is the mass of the incident particle, and $\gamma = m/M$. If $\gamma < 1$, the positive sign should be used only before the square root. If $\gamma > 1$ the expression should be calculated for positive and negative signs and the results are added to obtain $\gamma(\theta)$. For $\gamma = 1$

$$\sigma(\theta) = \left[\frac{zZe^2}{T}\right]^2 \frac{\cos\theta}{\sin^4 \theta} \quad (1.62)$$

1.3.3 Mott's Formula

If the scattered and the scattering particles are identical (Indistinguishable particles),
the quantum mechanical exchange effects must be taken into account. The scattering
formula due to Mott is

$$\sigma(\theta) = \frac{z^2 Z^2 e^4 \cos\theta}{T^2} \left\{ \frac{1}{\sin^4\theta} + \frac{1}{\cos^4\theta} \begin{bmatrix} +2 \\ -1 \end{bmatrix} \left[\frac{1}{\sin^2\theta\cos^2\theta} - \frac{\cos\gamma^2 Z^2 e^2}{\hbar v} \ln tg^2\theta \right] \right\}$$

(1.63)

where h is Planck's constant. $+2$ is put infront of the square brackets if the particles
have zero spin, and -1, if their spin is $\frac{1}{2}$.

1.3.4 Cross-Section for Scattering in the Angular Interval θ' and θ''

The cross-section $\sigma(\theta', \theta')$ per nucleus for scattering between angle θ' and θ'' is
given by:

$$\sigma(\theta', \theta'') = \int_{\theta'}^{\theta''} \sigma(\theta) d\Omega = 2\pi \int_{\theta'}^{\theta''} \sin\theta\sigma(\theta) d\theta = \frac{2\pi R_0^2}{16} \int_{\theta'}^{\theta''} \frac{\sin\theta d\theta}{\sin^4\theta/2}$$

where (1.59) has been used. As $\sin\theta = 2\sin\frac{\theta}{2}\cos\frac{\theta}{2}$

$$\sigma(\theta', \theta'') = \frac{\pi}{4} R_0^2 \int_{\theta'}^{\theta''} \text{cosec}^2(\theta/2) \cot(\theta/2) d\theta$$

$$= \frac{\pi}{2} R_0^2 \int_{\theta''}^{\theta''} \cot(\theta/2) d\cot(\theta/2)$$ (1.64)

$$\sigma(\theta', \theta'') = \frac{1}{4}\pi R_0^2 \left(\cot^2\theta'/2 - \cot^2\theta''/2 \right)$$

In particular, $\sigma(90°, 180°)$ the cross-section for scattering for angles greater than
$90°$ is given by setting $\theta' = 90°$ and $\theta'' = 180°$ in (1.64)

$$\sigma(90°, 180°) = \frac{\pi R_0^2}{4} = \frac{\pi}{4}\left(\frac{zZe^2}{T_0} \right)^2$$ (1.65)

1.3.5 Probability of Scattering

Consider a box of face area 1 cm^2 and length λ cm, so that its volume becomes
λ cm^3. Let a beam of particles be incident on its face. By definition λ is such a

Fig. 1.24 A box of face area
1 cm^2 and length λ cm
containing n atoms is exposed
to a beam of particles

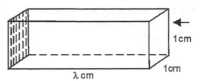

length that on an average the particle suffers the given type of scattering, i.e. λ is the mean-free-path. If there are n number of atoms per cm^3, the number of atoms inside the box of volume λ cm^3 will then the equal to λn. The cross-section arising from all these atoms will then be equal to $\lambda n\sigma\,(\theta',\theta'')$. Imagine all the atoms inside the box to be pushed on the rear surface of the box (Fig. 1.24). The total area corresponding to the cross-section of all the atoms must be such as to completely fill up area of 1 cm^2 since our assumption demands that on an average one scattering of the given type will occur when the incident particle passes through λ cm

$$\therefore \quad \lambda n\sigma\left(\theta'\theta''\right) = 1 \quad \text{or}$$

$$n\sigma\left(\theta',\sigma''\right) = \frac{1}{\lambda}$$

If the foil is only t cm thick, then the probability of scattering between θ' and θ'' will be:

$$P = t/\lambda = nt\sigma\left(\theta,\theta''\right) \tag{1.66}$$

1.3.6 Rutherford Scattering in the LS and CM System

So far, we have considered the scattering of particles from massive target nuclei so that the recoil of the latter can be neglected altogether. However, if a light target be considered then the target nucleus would necessarily recoil due to the collision and the analysis of the collision is rendered fairly complicated when done in the lab system. Figure 1.25 shows for definiteness the elastic scattering of an α-particle ($m_1 = 4$) with a carbon nucleus ($m_2 = 12$) originally stationary seen in the lab system. The α particle moves with velocity u_1, and makes an impact parameter b. Since m_2 is assumed to be stationary, the relative velocity of approach is also u_1.

The centre of mass (indicated by CM in the diagram) moves with constant velocity $v_c = m_1u_1/(m_1 + m_2)$, before, during and after the collision, which is always directed parallel to the incident direction of m_1. In the chosen example, v_c is one-fourth of the initial velocity of the α particle. We have seen that the analytical relationships which connect the scattering angle θ and ϕ with the impact parameter b and with the charges, masses and velocities of m_1 and m_2 are too complicated to be of any general use. Observe that after the collision the initial direction of m_2 is away from that of m_1. This is a simple consequence of Coulomb's repulsion between the two nuclei. It must be pointed out that the trajectories are no longer simple

Fig. 1.25 Scattering of α
particles with a carbon
nucleus in the LS

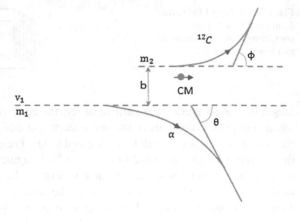

Fig. 1.26 Scattering of α
particles with a carbon
nucleus in the CMS

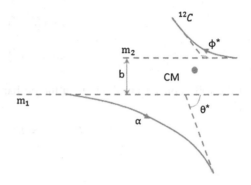

hyperbolas in the lab system. In the CM system, no distinction is made between the
projectile and the target particles, see Fig. 1.26. The relative velocity of the particles
is, $v(\text{rel}) = u_1^* + u_2^* = (u_1 - u_c) + u_c = u_1$, which is identical with that in the lab
system.

There is complete symmetry in the scattering of the particles in the CM system. Both the particles approach each other with equal and opposite momentum
before the collision and recede with equal and opposite momentum after the collision. In the event of elastic scattering, the respective speeds of the particles remain
unaltered before and after the collision. Both are deflected through the same angle
measured with their respective original direction. Their centre of mass remains at
rest throughout the collision. Each of the particles describes a hyperbola. The collision diameter, impact parameter and eccentricity of the orbit are the same for both
the particles, Fig. 1.27. In our example, α particle traverses its hyperbolic path r_1
about the centre of mass, while the carbon nucleus also traverses a similar path,
$r_2 = r_1 m_1 / m_2 = r_1 / 3$, on the other side of the centre of mass. The line joining the
positions of the two particles passes through the centre of mass at all times. The
angular momentum about the centre of mass evaluated in the CM system (Fig. 1.27)

Fig. 1.27 Angular momentum about the centre of mass in the CMS

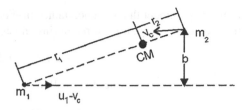

is

$$J = m_1(u_1 - v_c)\frac{r_1 b}{r_1 + r_2} + \frac{m_2 v_c r_2 b}{r_1 + r_2} = \frac{m_1 u_1 r_1 b}{r_1 + r_2} + \frac{v_c b}{r_1 + r_2}(m_2 r_2 - m_1 r_1)$$

$$= \frac{m_1 u_1 r_1 b}{r_1 + r_2}$$

Since $m_1 r_1 = m_2 r_2$

$$\frac{r_1}{r_1 + r_2} = \frac{m_2}{m_1 + m_2}$$

$$J = m_1 u_1 \frac{b m_2}{m_1 + m_2} = \mu u_1 b$$

where μ is the reduced mass. The angular momentum J of this system of two particles is a constant of their motion since no external torques act on the system. The angular momentum taken about the centre of their mass has the same value both in the lab system and CM system since these two systems differ only in regard to the translation velocity of the centre of mass (v_c in the lab system and zero in the CM system).

1.3.7 Validity of Classical Description of Scattering

We must be able to form a wave packet which is narrower than the distance of the closest approach, otherwise there is no way to make sure that the particle experiences a definitely predictable force from which the deflection can be calculated classically. To obtain a rough estimate of the validity of the classical description, we can safely assume that the distance of closest approach is of the same order of magnitude as the impact parameter b. In order to form a wave packet that is smaller than b, it is of course necessary that one uses a range of wavelengths of the order of b or smaller. Thus the first requirement is that the momentum of the incident particles be considerably larger than $p = \hbar/b$. Moreover, in defining the position of this packet will make the momentum of the particle uncertain by a quantity much greater than $\delta p = \hbar/b$. This uncertainty will cause the angle of deflection to be made uncertain by a quantity much greater than $\delta \theta = \delta p/p$. In order that the classical description be applicable, the above uncertainty ought to be a great deal smaller than the deflection itself; otherwise the entire calculation of the deflection by classical method will be meaningless. This requirement is, however, equivalent to the requirement that the uncertainty in the momentum be much smaller than the net momentum, Δp, transferred during the collision, or that

$$\delta p/\Delta p = \hbar/b\Delta p \ll 1 \tag{1.67}$$

Now, for elastic scattering, for small angles

$$\Delta p = 2p\sin(\theta/2) \simeq p\theta \tag{1.68}$$

Also from (1.53)

$$\theta = zZe^2/Tb \tag{1.69}$$

Combining (1.68) and (1.69) and noting that $p/T = 2/v$, the condition is,

$$2zZe^2/\hbar v \gg 1 \quad \text{or} \quad 2zZ/137\beta \gg 1 \tag{1.70}$$

For 5 MeV α and gold nucleus ($Z = 79$) as the target, $\beta = 0.05$, and the left hand side of (1.67) becomes 46, a value which is much greater than unity, so that classical description of scattering is fully valid.

1.3.8 Coulomb Scattering with a Shielded Potential Under Born's Approximation

It is always an abstraction to assume that the Coulomb force continues to be unmodified out to arbitrarily large distance. Thus the Coulomb force resulting from distances of the order of a few atomic radii is screened or shielded by the atomic electrons. The resulting shape of the potential may be approximated by the shielded Coulomb potential of the form

$$V = \frac{zZe^2}{r} \exp(-r/r_0) \tag{1.71}$$

The exponential factor causes the force to become negligible when the factor r/r_0 is much greater than unity. According to the Born approximation, the expression for the differential cross-section is given by

$$\sigma(\theta) = \frac{4\pi^2 m^2}{h^4} [V(p - p_0)]^2 \tag{1.72}$$

where the momentum transfer is

$$|p - p_0| = 2p\sin(\theta/2) \tag{1.73}$$

and

$$V(p - p_0) = zZe^2 \int \exp[i(p - p_0)] \cdot r\hbar \frac{\exp(-r/r_0)dr}{r} = \frac{4\pi zZe^2}{|\frac{p-p_0}{\hbar^2}|^2 + \frac{1}{r_0^2}} \tag{1.74}$$

Fig. 1.28 Rutherford
scattering with a shielded
potential

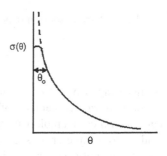

Letting $r_0 \to \infty$, and combining (1.72), (1.73) and (1.74), we get exactly the same expression as (1.59), i.e. Rutherford scattering law. Thus, classical mechanics and quantum mechanics give the same result for the Rutherford scattering.

The general appearance of the cross-section for a shielded Coulomb force as a function of angle is shown in Fig. 1.28. The curve rises steeply with decreasing θ, as is characteristic of the Rutherford cross-section, until

$$\sin \frac{\theta_0}{2} \cong \frac{\hbar}{2pr_0}$$

For angles smaller than θ_0, the rise of $\sigma(\theta)$ is comparatively small. Thus, θ_0, may be regarded as a sort of minimum angle, below which Rutherford scattering ceases, as a result of the shielding effects. With a shielded Coulomb potential, θ_0 will approach zero with increasing b much more rapidly, as soon as b goes beyond the shielding radius. In fact, shortly beyond the shielding radius, the entire scattering effect can be neglected. The minimum angle below which the cross-section ceases to increase is given by setting $b = r_0$ in Eq. (1.69) adapted for small angle approximation, $\theta(\min) = zZe^2/Tr_0$.

1.3.9 Discussion of Rutherford's Formula

The formula fails for indistinguishable particles and also for relativistic particles. The derivation ignores spin interaction. Formula (1.59) predicts pronounced scattering in the forward direction, i.e. small angle scattering is favoured. Spin interaction, however, affects only the large angle scattering. Screening effect of electrons has been ignored which tends to reduce the effective charge of the nucleus. This effect is small for small impact parameters (large θ) and will clearly manifest itself in heavy atoms in which K electrons are very close to the nucleus.

Rutherford's formula is valid only for single scattering. It is, therefore, necessary to use thin foils; otherwise, multiple scattering will result from the superposition of successive single scatterings. Scattering due to orbital electrons may be ignored since maximum angle of scattering will be typically $<10^{-4}$ radians.

If the mass of the incident particle cannot be neglected in comparison with the mass of the target nucleus, then (1.59) is still valid provided the energy T now refers to the centre of mass system, and similarly the angle.

The total cross-section σ is given by

$$\sigma = 2\pi \int_0^\pi \sigma(\theta) \sin \theta d\theta \qquad (1.75)$$

By inserting (1.59) in (1.75), it is seen that $\sigma \to \infty$. This is the direct consequence of the long range character of Coulomb forces. But physically, at ranges (impact parameters) comparable with the atomic radii, the force field is effectively cut-off, leading to a finite cross-section. In other words, divergence in σ which results at $\theta = 0$ does not occur in practice, since infinite impact parameters are not effective owing to the shielding of the nucleus by the orbital electrons.

1.3.10 The Scattering of α Particles and the Nuclear Theory of Atom

The experiments of Geiger and Marsden suggested that if alphas passed through thin foils of platinum then about 1 in 8000 suffered a deflection larger than 90°. From the then existing Thompson's theory of atom (protons and electrons uniformly distributed in the atom), it appeared that the electric fields are weak, much too weak to give rise to such deflections. It was conjectured that a cumulative effect of a large number of small angle scatters might result in a large deflection. But it was not possible to explain deflections of this magnitude. To explain this result, Rutherford postulated that positive charge is concentrated in a very small volume called the nucleus. It was then shown that all the wide angle scatters could be explained due to the existence of the nucleus alone, and that the screening effect due to the orbital electrons could be ignored. Rutherford used the scintillation method to count the scattered alphas, and an excellent agreement was made with the theory. The results were so satisfactory that Rutherford used his formula to verify the atomic numbers of some of the elements that were already known from Mosley's work on X-ray spectra of elements, and an extraordinary agreement was obtained. The dependence of scattering on energy and angle left no doubt as to the validity of Coulomb law of force, the distance of closest approach being 70 to 140 fm (1 fermi $= 10^{-13}$ cm). For alpha scattering against gold foils, the scattering was well represented by the Coulomb law of force when the closest distance of approach was 32 fm. It was, therefore, concluded that the radius of gold nucleus is not greater than this value (actual value being about 8 fm). Thus, any departure from Rutherford's law of scattering is an indication that the incident particle is coming into contact with the target nucleus. Careful experiments can, therefore, give reasonably good estimates of nuclear sizes. For the range of energies considered (alphas from radioactive substances), no departure from Coulomb inverse square law of forces was observed, for very close distance of collision. The alpha scattering was found to be strictly governed by the Rutherford scattering law for targets ranging from copper to uranium. In the mean time, Blacket obtained an independent verification from Wilson chamber photographs by determining the angular distribution at various energies. But experiments with light elements revealed serious departure from the Rutherford scattering law at large angles. Observe that R_0 (minimum distance of approach) is proportional to Z (charge of the target nucleus) while the nuclear radii are roughly proportional to $Z^{\frac{1}{3}}$ for light and medium elements. The latter, therefore, decreases much slower than the former. Consequently, alphas could penetrate the nuclei of light elements. It was also observed that when alphas of larger energies were employed, a more rapid departure from the Rutherford law of scattering was approached, the observed number being greater than the predicted number. These results confirmed the general result that when the particles approach close to the nucleus the simple Coulomb law is no longer obeyed and that no known law concerning the electrical forces could explain the observed scattering. Explanation in terms of magnetic interaction between the alpha particle and the nucleus was ruled out since it was already known that alpha particle has zero spin, and therefore does not have any magnetic moment. Further, the validity of the classical formula of Rutherford could not be questioned since quantum mechanics also gives precisely the same result. The only answer to the anamolous scattering was to postulate the existence of strong short range attractive nuclear forces. For large b the angle θ will be given by pure Coulomb forces. As b decreases θ would increase, but not so rapidly as in the pure Coulomb field. This has the consequence of more particles to be scattered at small angles and fewer at large angles. As b decreases further, the attractive nuclear forces may more than compensate for the repulsive Coulomb forces. The angle of scattering may therefore decrease rather than increase, and in this case there will be a maximum angle of scattering. For very small values of b, the particles may get scattered at random and in some cases may get captured.

Example 1.7 If σ_g is the geometrical cross-section for uncharged particles for hitting a nucleus, show that for positively charged particles, the cross-section will decrease by the factor $(1 - \frac{R_0}{R})$, where $R_0 = zZe^2/4\pi\varepsilon_0 K_0 R = $ radius of the nucleus, $\sigma_g = \pi R^2$ and K_0 is the kinetic energy.

Solution When the charged particle just grazes the nucleus

$$r(\min) = R = \frac{1}{2}R_0\left[1 + \sqrt{1 + \frac{4b^2}{R_0^2}}\right] \tag{1.55}$$

whence we obtain $b^2 = R^2 - RR_0$.

Denoting the cross-section by $\sigma = \pi b^2$, it follows that $\sigma/\sigma_g = \pi b^2/\pi R^2 = 1 - R_0/R$.

Example 1.8 Alphas of 8.3 MeV bombard an aluminum foil. The scattered alphas are observed at an angle of 60°. Calculate the minimum distance of approach in this case.

Solution $T_0 = 8.3$ MeV; $z = 2$, $Z = 13$

$$\frac{R_0}{2b} = \tan\frac{\theta}{2} = \tan\frac{60}{2} = \tan 30° = \frac{1}{\sqrt{3}} \tag{1}$$

$$r(\min) = \frac{R_0}{2}\left[1 + \sqrt{1 + \frac{4b^2}{R_0^2}}\right] \tag{2}$$

Using (1) in (2)

$$r(\min) = 3R_0/2$$

$$R_0 = \frac{1.44zZ}{T_0} = \frac{1.44 \times 2 \times 13}{8.3} = 4.51 \text{ fm}$$

$$r(\min) = 1.5 \times 4.51 = 6.76 \text{ fm}$$

Example 1.9   Find the probability of scattering of alpha particles of energy 5 MeV through an angle greater than 90° in their passage through a foil of gold of thickness 4×10^{-5} cm. Given, Avogadro's number $N_0 = 6 \times 10^{23}$; $A = 196$; $Z = 79$; electronic charge $e = 1.6 \times 10^{19}$ C; $1/4\pi\varepsilon_0 = 9 \times 10^9$ N m^2/C^2, density of gold $= 19.3$ g/cm^3.

Solution

$$N = \text{Number of atoms/cm}^3 = (\text{number of atoms/g}) \left(\text{g/cm}^3\right)$$

$$= \frac{6 \times 10^{23} \times 19.3}{196} = 5.91 \times 10^{22}$$

$$\sigma\left(90°, 180°\right) = \frac{\pi}{4}\left[\frac{1.44zZ}{T_0}\right]^2 = \frac{\pi}{4}\left[\frac{1.44 \times 2 \times 79}{5}\right]^2$$

$$= 1625 \text{ fm}^2 = 16.25 \times 10^{-24} \text{ cm}^2$$

If λ is the mean-free-path, $\frac{1}{\lambda} = N\sigma = 5.91 \times 10^{22} \times 16.25 \times 10^{-24} = 0.961$ cm^{-1}.

Required probability $p = t/\lambda = 4 \times 10^{-5} \times 0.961 = 3.84 \times 10^{-5}$. In other words, one alpha in $1/(3.84 \times 10^{-5})$ or 26000, gets scattered at an angle greater than 90°.

Example 1.10   If the radius of gold nucleus $(Z = 79)$, is 8×10^{-15} m, what is the minimum energy that the a particle should have to just reach it? Give your answer in MeV.

Solution   Set

$$R = R_0 = 8 \text{ fm}$$

$$T = \frac{1.44zZ}{R_0} = \frac{1.44 \times 2 \times 79}{8} = 28.44 \text{ MeV}$$

Example 1.11 The following counting rates (in arbitrary units) were obtained when a particles were scattered through 180° from a thin gold ($Z = 79$) target. Deduce a value for the radius of a gold nucleus from these results.

Energy of α particle (MeV)	8	12	18	22	26	27	30	34
Counting rate		30300	13400	6000	4000	2800	33	4	0.4

Solution Since at a given angle, the counting rate is inversely proportional to the square of energy, we can calculate the expected counting rate for various energies, assuming that at 8 MeV (lowest energy) the counting rate N_8 is in agreement with the expected value. In the table below are displayed the calculated counting rates with the aid of the formula, $N = N_8(8/T)^2$, where T is in MeV.

T in MeV	8	12	18	22	26	27	30	34
N (cal)	30300	13100	5990	4010	2867	2667	2157	1679
N (obs)	30300	13400	6000	4000	2800	33	4	0.4

Comparison between the calculated and observed counting rates indicates that departure from Rutherford scattering begins at 26 MeV. Since scattering angle is $\theta = 180°$, we are concerned with head-on collisions. Hence

$$R_0 = \frac{1.44zZ}{T_0} = \frac{1.44 \times 2 \times 79}{26} = 8.75 \text{ fm}$$

Hence, the radius of the gold nucleus is 8.75 fm.

Example 1.12 (a) If a gold foil is bombarded by 5.4 MeV a particles, determine the distance of closest approach (b) what is the deflection of the alpha particle when the impact parameter is equal to this distance?

Solution

(a) Distance of closest approach

$$R_0 = \frac{1.44zZ}{T_0} = \frac{1.44 \times 2 \times 79}{5.4} = 42.1 \text{ fm}$$

(b)

$$\tan\frac{\theta}{2} = \frac{R_0}{2b} = \frac{R_0}{2R_0} = \frac{1}{2}$$

$$\theta = 53°8'$$

Example 1.13 To what minimum distance will an alpha particle of energy 0.4 MeV approach a stationary Li^7 nucleus in the case of a head-on collision? Take the nuclear recoil into account.

Solution We work out in the CM system. Equating the potential energy at the closest distance of approach R_0 to the initial K.E.

$$\frac{1.44zZ}{R_0} = \frac{1}{2}\mu v^2 = \frac{1}{2}\frac{m_1 m_2 v^2}{(m_1 + m_2)} = \frac{T_0 m_2}{m_1 + m_2} \quad \text{or}$$

$$R_0 = \frac{1.44zZ}{T_0}\left(1 + \frac{m_1}{m_2}\right) = \frac{1.44 \times 2 \times 3}{0.4}\left(1 + \frac{4}{7}\right) = 33.9 \text{ fm}$$

Example 1.14 A narrow beam of alpha particles with kinetic energy $T = 600$ keV falls normally on a golden foil incorporating $n = 1.1 \times 10^{19}$ nuclei/cm^2. Find the fraction of alpha particles scattered through the angles $\theta < \theta_0 = 20°$.

Solution Given $nt = 1.1 \times 10^{19}$ nuclei/cm^2

$$\Delta N/N = 1 - \frac{\pi}{4}R_0^2 \cot^2\frac{\theta}{2} \cdot nt = 1 - \frac{\pi}{4}\left(\frac{1.44zZ}{T_0}\right)^2 \cot^2\frac{\theta}{2} \cdot nt$$

$$= 1 - \frac{\pi}{4}\left(\frac{1.44 \times 2 \times 79}{0.6}\right)^2 \cot^2\frac{20}{2} \times 1.1 \times 10^{19} \times 10^{-26} = 0.6$$

The factor 10^{-26} has been introduced to convert fm^2 into cm^2.

Example 1.15 A narrow beam of protons with kinetic energy $T = 1.4$ MeV falls normally on a brass foil whose mass thickness $\rho t = 1.5$ mg/cm^2. The weight ratio of copper and zinc in the foil is equal to $7 : 3$, respectively. Find the fraction of the protons scattered through the angles exceeding $\theta = 30°$.

Solution

$$\frac{\Delta N}{N} = \frac{\pi}{4}\frac{10^{-26}}{T^2} \times 1.44^2\left(\frac{0.7Z_1^2}{M_1} + \frac{0.3Z_2^2}{M_2}\right)\rho t N_0 \cot^2\frac{\theta}{2}$$

where Z_1 and Z_2 are the atomic numbers of copper and zinc, M_1 and M_2 are their molar masses, N_0 is the Avagadro's number. The factor 10^{-26} is introduced to express fm^2 as cm^2

$$\frac{\Delta N}{N} = \frac{(1.44)^2}{(1.4)^2}\left[\frac{0.7 \times 29^2}{63.55} + \frac{0.3 \times 30^2}{65.38}\right] \times 1.5 \times 10^{-3} \times 6 \times 10^{23}$$

$$\times (3.732)^2 \times 10^{-26}$$

$$= 1.4 \times 10^{-3}$$

Example 1.16   Find the effective cross-section of a uranium nucleus corresponding to the scattering of alpha particles with kinetic energy $T = 1.5$ MeV through the angles exceeding $\theta = 60°$.

Solution

$$\sigma(\theta, \pi) = \frac{\pi}{4}\left(\frac{1.44zZ}{T}\right)^2 \cot^2\frac{\theta}{2} = \frac{\pi}{4}\left(\frac{1.44 \times 2 \times 92}{1.5}\right)^2 \cot^2 30°$$

$$= 73480 \text{ fm}^2 = 735 \text{ b}$$

Example 1.17   The effective cross-section of a gold nucleus corresponding to the scattering of monoergic alpha particles within angular interval from 90° to 180° is equal to $\Delta\sigma = 0.5$ kb. Find (a) the energy of alpha particles (b) the differential cross-section of scattering $\sigma(\theta)$ (kb/sr) corresponding to the angle $\theta = 60°$.

Solution

(a)

$$\sigma(90°, 180°) = \frac{\pi}{4}\left(\frac{1.44zZ}{T}\right)^2$$

$$T = \frac{\sqrt{\pi} \times 1.44zZ}{2 \times \sqrt{\sigma(90°, 180°)}} = \frac{\sqrt{\pi} \times 1.44 \times 2 \times 79}{2 \times \sqrt{5} \times 10^4} = 0.9 \text{ MeV}$$

(b)

$$\frac{d\sigma}{d\Omega} = 1.295\left(\frac{zZ}{T}\right)^2 \frac{1}{\sin^4\frac{\theta}{2}} = 1.295\left(\frac{2 \times 79}{0.9}\right)^2 \frac{1}{\sin^4\frac{6}{2}}$$

$$= 0.638 \times 10^6 \text{ mb/sr} = 0.64 \text{ kb/sr}$$

Example 1.18   Derive Darwin's formula for scattering (modified Rutherford's formula which takes into account the recoil of the nucleus).

Solution

$$\sigma(\theta) = \frac{(1 + \gamma^2 + 2\gamma\cos\theta^*)^{\frac{3}{2}}}{|1 + \gamma\cos\theta^*|}\sigma(\theta^*) \tag{1}$$

Now, Rutherford's formula for CMS is

$$\sigma(\theta^*) = \frac{1}{4}\left(\frac{zZe^2}{\mu v^2}\right)^2 \frac{1}{\sin^4\frac{\theta^*}{2}} \tag{2}$$

Also

$$\sin^4 \frac{\theta^*}{2} = \frac{1}{4} \frac{\sin^4 \theta^*}{(1 + \cos \theta^*)^2} \quad \text{and} \tag{3}$$

$$\mu = \frac{mM}{m + M} = \frac{m}{1 + \gamma} \tag{4}$$

where M and m are the target and incident particle masses respectively and $\gamma = m/M$

$$\tan \theta = \frac{\sin \theta^*}{\gamma + \cos \theta^*} \tag{5}$$

Squaring (5) and expressing it as a quadratic equation and solving it we get

$$\cos \theta^* = \gamma \sin^2 \theta \pm \cos \theta \sqrt{1 - \gamma^2 \sin^2 \theta} \tag{6}$$

combining (1), (2), (3), (4) and (6), and after some algebraic manipulations we get

$$\sigma(\theta) = \left(\frac{zZe^2}{mv^2} \right)^2 \frac{1}{\sin^4 \theta} \frac{[\cos \theta \pm \sqrt{1 - \gamma^2 \sin^2 \theta}]^2}{\sqrt{1 - \gamma^2 \sin^2 \theta}} \tag{7}$$

This is Darwin's formula. For $m \ll M$, $\gamma \to 0$ and (7) reduces to the usual Rutherford formula.

1.4 Multiple Scattering

1.4.1 Mean Scattering Angle

A charged particle in passing through a thick medium is scattered through an angle θ. The observed scattering may be the result of cumulative effect of a number of small deflections produced by different atomic nuclei in the matter traversed, or it may be a single deflection through an angle θ produced by a single nucleus. The first type of scattering is called multiple or plural, according to the number of contributing collisions is large or small. The second type is referred to as single scattering. Which process is mainly operative depends on the nature and velocity of the scattered particle, the matter traversed and the scattering angle. In the simple treatment, it is assumed that Θ is distributed about $\Theta = 0$, according to the Gaussian law, i.e. the probability for scattering in the angular interval Θ and $\Theta + d\Theta$ is

$$P(\Theta)d\Theta = \text{const} \cdot \exp(-K\Theta^2)d\Theta \tag{1.76}$$

Fig. 1.29 Multiple scattering
Θ_i resulting from the
superposition of single
scatterings, $\theta_1, \theta_2, \ldots, \theta_i$

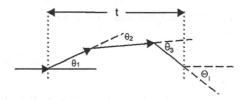

where K is a constant. The single scattering is governed by the Rutherford law of
scattering, which for small angles has the form

$$P(\theta)d\theta = \text{const} \cdot \frac{d\theta}{\theta^3} \tag{1.77}$$

Let θ_i be the deflection in the ith collision. Let there be q collisions in a traversal of
t cm. Since small angles are vectors (Fig. 1.29)

$$\Theta_q = \sum_{i=1}^{q} \theta_i$$

Take the dot product

$$\Theta_q^2 = \sum_{i=1}^{q} \theta_i^2 + \sum_{i \neq j}^{q} \theta_i \theta_j$$

In averaging over many traversals, θ_j is positive as many times as it is negative; and
the second summation drops off. Thus

$$\langle \Theta_q^2 \rangle = \sum_{i=1}^{q} \theta_1^2$$

Since statistically, the individual events do not differ, $\theta_i^2 = \theta^2$ we can therefore write

$$\langle \Theta^2 \rangle = q \langle \theta^2 \rangle \tag{1.78}$$

Now, Rutherford scattering for small angles can be written as

$$\sigma(\theta) = 4\left(\frac{zZe^2}{pv}\right)^2 \frac{1}{\theta^4} \tag{1.79}$$

The probability for the particle to be scattered into solid angle $d\Omega = 2\pi \sin\theta d\theta$, is
given by $\sigma(\theta)2\pi\theta d\theta$ for small θ. Let there be N atoms/cm^3

$$\langle \theta^2 \rangle = \frac{\int_{\theta_{min}}^{\theta_{max}} \theta^2 f(\theta)d\theta}{\int_{\theta_{min}}^{\theta_{max}} f(\theta)d\theta} = \frac{\int \theta^2 f(b)db}{\int f(b)db} = \frac{\int_{b_{min}}^{b_{max}} \frac{4z^2 Z^2 e^4}{p^2 v^2 b^2} 2\pi N t db}{\int 2\pi b db N t}$$

since $\tan \frac{\theta}{2} = \frac{zZe^2}{Mbv^2}$ and for small θ

$$\theta^2 = \frac{4z^2Z^2e^4}{p^2v^2b^2} \tag{1.80}$$

The denominator is nothing but q. We, therefore, find after substituting (1.79) in (1.80)

$$\langle\Theta^2\rangle = q\langle\theta^2\rangle = 8\pi Nt\left(\frac{zZe^2}{pv}\right)^2 \ln\frac{\theta(\max)}{\theta(\min)} \tag{1.81}$$

We have assumed that the particle is sufficiently energetic so that the velocity does not change over the considered traversal. We also ignore the scattering off the electrons, since it is unimportant. Observe that scattering off the nuclei is proportional to Z^2, whilst for electrons, it is proportional to Z. In hydrogen, the scattering off electrons, however, will be important. We shall now consider the limits $\theta(\max)$ and $\theta(\min)$. Nuclear scattering at large distance is reduced by the electrostatic shielding of the nucleus by its electrons. The electrostatic shielding reduces the scattering of distant particles but it does not reduce the energy loss. Thus, a primary particle travelling at a distance such that the fields of electrons and nuclei compensate each other almost completely is still capable of transferring energy to the atom. The physical reason why screening reduces scattering much more than the energy transfer is as follows. A fast particle passing near an atom transfers a certain amount of momentum to each electron and also to the nucleus. The angle of scattering is, however, determined by the transverse component of the total recoil. Due to the opposite sign of charge, the electrons recoil in the opposite direction to the nucleus and if the fast particle passes at a sufficient distance, the transverse components of the recoiling electrons cancel the transverse components of the nuclear recoil and thus no scattering results. The total energy transfer is equal to

$$\sum_{i=1}^{Z} \frac{p_i^2}{2m_e} + \frac{P_{nuc}^2}{2M_{nuc}} \tag{1.82}$$

The energy transfer to any of the electrons or to the nucleus is positive and is not affected by the presence of other particles except for the small effects of the binding forces. In other words, a particle passing near an atom suffers $Z+1$ collisions with the constituents of the atom and loses energy to everyone of them. The $Z+1$ angles of scattering, however, tend to compensate each other and therefore the scattering is reduced strongly by shielding. Now by (1.81), the root mean square multiple scattering angle is given by

$$\sqrt{\langle\Theta^2\rangle} = \sqrt{\frac{8\pi Z^2z^2e^4Nt}{p^2v^2} \ln\frac{b(\max)}{b(\min)}} = K\frac{\sqrt{t}ze}{pv} \tag{1.83}$$

where

$$K = \sqrt{8\pi N Z^2 e^2 \ln \frac{b(\text{max})}{b(\text{min})}} \tag{1.84}$$

is called the scattering constant.

1.4.2 Choice of b(max) and b(min)

We may choose

$$b(\text{max}) = \frac{a_o}{Z^{1/3}} \tag{1.85}$$

where a_0 is the Bohr radius. This is justified by Fermi-Thomas model of the atom, since the right hand side of (1.85) represents the radius of the atom.

The limit on $b(\text{min})$ is dictated by the finite size of the nucleus. Thus, $b(\text{min})$ is greater than $1.3 \times 10^{-13} A^{1/3}$ cm. An alternative criterion would be to avoid counting deflections with $\Theta > 1$ radian. A rough criterion is provided by restricting the individual single scatterings to $\theta < 1$. Now for small scattering angles, formula (1.53) reduces to

$$\theta = \frac{2zZe^2}{mv^2 b} \tag{1.86}$$

Putting $\theta = 1$ radian and $p = mv$, we obtain the rough criterion

$$b(\text{min}) = \frac{2zZe^2}{pv} \tag{1.87}$$

Fortunately, the results are insensitive to the choice of $b(\text{max})$ and $b(\text{min})$

$$\frac{b(\text{max})}{b(\text{min})} \simeq \frac{\text{Atomic dimension}}{\text{Nuclear dimension}} = \frac{10^{-8} \text{ cm}}{10^{-12} \text{ cm}} = 10^4, \quad \text{and so } \ln 10^4 \simeq 10$$

Thus, $\sqrt{\langle \Theta \rangle^2}$ is very insensitive to the logarithmic term in (1.83) which is of the order of 10. The main dependence comes from the factors outside the logarithmic term. Further, in view of (1.87) the scattering constant K is a slow varying function of the particle velocity. Observe that $\sqrt{\langle \Theta \rangle^2}$ is directly proportional to the charge of the scattering nuclei, and the charge of the incident particles, inversely proportional to the energy of the incident particles and directly proportional to the square root of the thickness of the absorber.

1.4.3 Mean Square Projected Angle and the Mean Square Displacement

Consider a charged particle traversing an absorber of thickness t. Assume that the collisions take place at depths $X_1, X_2, \ldots$, resulting in deflections $\theta_1, \theta_2, \ldots$. The azimuth of the deflection will change after each collision, the subsequent azimuths being $\phi_1, \phi_2, \ldots$, the projected angle of deflection is given by

$$\Theta_P = \sum_{i=1}^{q} \theta_i \cos\phi_i$$

We have to average over the parameters of the single collisions. Since the azimuths can be taken as independent, we have

$$\langle \cos\phi_i \cos\phi_j \rangle = \frac{1}{2}\delta_{ij} \tag{1.88}$$

where δ_{ij} is the Kronecker delta. Hence

$$\langle \Theta_P^2 \rangle = \frac{1}{2}q\langle\theta^2\rangle = \frac{1}{2}\langle\Theta^2\rangle \tag{1.89}$$

Observe that the most probable value of Θ or Θ_P is zero. However, $\langle\Theta\rangle$ and $\langle\Theta^2\rangle$ are necessarily positive, whereas $\langle\Theta_P\rangle$ is zero.

Similarly the mean square projected displacement is equal to half the mean square unprojected displacement

$$\langle y^2 \rangle = \frac{1}{2}\langle r^2 \rangle$$

Now, $y = \sum_{i=1}^{q} X_i \theta_i \cos\phi_i$

$$\therefore \quad y^2 = \sum_{i \neq j} x_i x_j \theta_i \theta_j \cos\phi_i \cos\phi_j$$

Since x and ϕ are independent, using relation (1.88)

$$\langle y^2 \rangle = \frac{1}{2}\sum_i \langle x_i^2\rangle\langle\theta^2\rangle$$

Now

$$\langle x_i^2 \rangle = \langle X^2 \rangle = \frac{1}{t}\int_0^t x^2 dx = \frac{t^2}{3}$$

$$\langle y^2 \rangle = \frac{t^2}{6}q; \qquad \langle\theta^2\rangle = \frac{t^2}{6}\langle\Theta^2\rangle \tag{1.90}$$

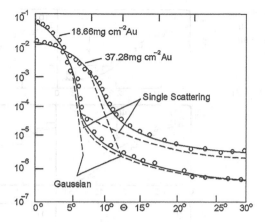

Fig. 1.30 Angular distribution of electrons scattered from AU at 15.7 MeV. *Solid lines* indicate the distribution expected from the Moliere theory for small- and large-angle multiple scattering, with an extrapolation in the transition region: *dashed lines*, the distributions according to the Gaussian and single scattering theories. The ordinate scale gives the logarithm of the fraction of the beam scattered within 9.696×10^{-3} sr (Birkhoff)

Also

$$\langle r^2 \rangle = \frac{t^2}{3} \langle \Theta^2 \rangle \tag{1.91}$$

Expressions (1.90) and (1.91) are of great interest in the cosmic ray shower theory and experiments. Figure 1.30 shows the contribution from multiple scattering (Gaussian) at small angles and single scattering (Rutherford) at large angles. Kamal, Rao and Rao (Fig. 1.31) have compared the experimental distributions of $\overline{D}$ the average of 'Second differences' of 17.2 GeV/c beam tracks in photographic emulsions for cell lengths $t = 4, 6, 8, 10, 20$, and 30 mm and compared with Moliere's theory. The quantity D is related to the projected angular deflection and is obtained from the y-coordinates of the track; $D_i = y_{i+1} - 2y_i + y_{i-1}$, where y_i is the ith coordinate of the track at constant x-intervals called cell length t. In order to avoid very large scattering angles, a $4\overline{D}$ cut-off is usually employed, a procedure in which all deflections larger than four times the mean second difference are eliminated. This is also indicated in the figure. Moliere's probability function (Gaussian function plus the single scattering tail) for the second differences was computed from the work of Scott. A good agreement was found between theory and observations.

In conclusion, we may point out that Rutherford used extremely thin foils for his classical experiments on alpha scattering in order to avoid the contribution of multiple scattering.

From (1.83) it is clear that the determination of root-mean-square angle permits one to estimate the energy of the particle.

Protons and electrons of the same energy will have the same root-mean-square angle of scattering, but their ionization would be different since their velocities would be different. Thus joint measurements of multiple scattering and ionization

Fig. 1.31 Multiple scattering
distribution for various cell
lengths, (**a**) 4 mm; (**b**) 6 mm;
(**c**) 8 mm; (**d**) 10 mm;
(**e**) 20 mm; (**f**) 30 mm
Moliere's Gaussian function
plus single scattering tail [2]

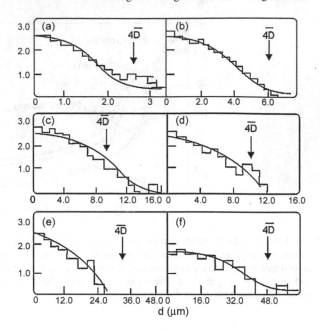

d (μm)

permit us to estimate the mass of the particle and identify it. This method is particularly suitable for particles which are not too energetic and at the same time are not brought to rest within the stack of emulsions.

The existence of multiple scattering can create problems in the curvature measurements of tracks in a cloud chamber. In certain cases the multiple scattering may be so severe that spurious curvatures are observed even in the absence of magnetic fields. In the case of bubble chambers, curvature measurements under magentic fields are rendered difficult when a heavy liquid like xenon is used. It is also implied that curvature measurement in photographic emulsions under pulsed magnetic fields are limited owing to severe multiple scattering by the heavy nuclei of silver and bromine.

The phenomenon of multiple scattering leads to an interesting observation in cosmic ray showers. Owing to multiple scattering in air, the electrons in the shower undergo lateral displacement from the original path through several meters as they traverse down the atmospheric depth (see expressions (1.90) and (1.91)).

1.5 Theory of Ionization

1.5.1 Bohr's Formula

Charged particles in their passage through a medium lose their energy mostly through excitation of atoms and ionization (collision) processes. The collision process is only one of several mechanisms by which charged particles may lose energy.

In the case of electrons it constitutes the most important source of energy loss only for relatively small energies. At energies of the order of 10 to 100 MeV, radiation losses overtake the collision losses, depending on the Z of the absorber. For muons, collision losses remain dominant up to energies of the order of 100 or 1000 MeV. For protons, radiation losses are never significant, but the occurrence of nuclear collisions overshadows the collision losses at energies of the order of 1000 MeV or greater. Energy loss by the emission of Cerenkov radiation is negligible except at very high energies. Thus, in general, collision losses represent the most important source of energy loss only for energies smaller than a certain value that depends on the nature of the particles.

Most of the electrons ejected in ionization processes have energies very small compared with the energy of the primary particle. Nevertheless, they are able to produce several ion pairs before coming to rest. The total specific ionization consists of two parts (i) primary specific ionization which is defined as the average number of ion pairs produced per $g\,cm^{-2}$ (ii) secondary specific ionization which refers to the average number of ion pairs per $g\,cm^{-2}$ by all the secondary electrons and tertiary electrons and radiation. The total ionization implies the sum of (i) and (ii). The total ionization is roughly three times the primary ionization. When the primary particle is absorbed its energy is dissipated in exciting the atoms and in producing secondary electrons partly by collisions and partly by radiation. The secondary electrons radiation will excite more atoms and produce tertiary electrons and photons and so on. It is clear that an electron will continue to lose its energy in elastic collisions as long as its energy is in excess of the lowest excitation potential of the atoms and the photons will be absorbed as long as their energy is greater than the threshold energy for minimum ionization potential. In the event an atom gets into an excited state by an inelastic collision with an electron or by the absorption of a photon it immediately loses the excitation energy by the emission of photons or Auger electrons. The fraction of the initial energy that is used in producing ionization is not appreciably affected by the nature of the primary particle nor by its energy as most of the ionization and excitation processes are caused by electrons of low energy. The classical ionization formula is originally due to N. Bohr. The basic assumptions made in the derivation are (i) electrons are free (ii) the incident particle remains undeviated throughout its motion.

The velocity acquired by the electron in an elastic collision is given by (1.27),

$$v_2 = 2v_c \cos \phi = \frac{2vm_1 \cos \frac{1}{2}\phi^*}{m_1 + m_2}$$

where $v = u_1$. Since $m_2 \ll m_1$, $v_2 \simeq 2v \cos \frac{1}{2}\phi^*$.

The energy imparted to the electron is

$$T = \frac{1}{2}mv_2^2 = 2mv^2 \cos^2 \frac{1}{2}\phi^* \tag{1.92}$$

Fig. 1.32 Atoms lying
within impact parameters b
and $b + db$

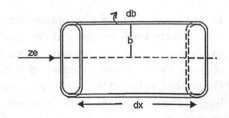

Since

$$\phi^* = \pi - \theta^*, \qquad \cos^2 \frac{1}{2}\phi^* = \frac{1}{1 + \cot^2 \frac{1}{2}\theta^*}$$

But scattering angle in the CMS is related to the impact parameter by $\cot \frac{1}{2}\theta^* = 2b/R_0$. Hence,

$$T = \frac{2z^2 e^4}{mv^2[R_0^2/4 + b^2]} \tag{1.93}$$

where we have used the fact that $R_0 = 2zZe^2/\mu v^2$ and the $Z = -1$ for target electron, and $\mu = m$, since the incident particle is considered much more massive than the electron. Let there be n electron/cm^3 of the medium consider a differential element of length dx along the path of the incident particle. The number of electrons situated between the impact parameters b and $b + bd$ over a length dx is given by $2\pi b \, db \, n \, dx$ (Fig. 1.32). The energy imparted to electrons for this range of b's is given by multiplying this number of electrons by T give by (1.93); but this is equal to the energy lost by the primary particle by traversing the element of length dx. We can, therefore, write:

$$-dE/dx = \int_0^{b(\text{max})} \frac{4\pi n z^2 e^4 b \, db}{mv^2[b^2 + R_0^2/4]} = \frac{4\pi n z^2 e^4}{mv^2} \ln\left[\frac{2b(\text{max})}{R_0}\right] \tag{1.94}$$

The underlying assumption that electrons are free is only approximately correct. Actually they are bound to the atoms and can be considered free only if the collision time is short compared with the period of revolution. On the other hand, if the collision time is long compared with the period of revolution, the electrons do not absorb any energy at all. Let $b(\text{max})$ represent the impact parameter for which the collision time $\tau = 1/\nu$, where ν is the orbital frequency of the electron

$$\text{Impulse} = \int F \, dt = \text{momentum acquired by the electron} \tag{1.95}$$

or

$$\frac{ze^2 \tau}{b^2(\text{max})} = \sqrt{2mT(\text{max})} = \sqrt{4z^2 e^4/v^2 b^2(\text{max})} \tag{1.96}$$

Hence

$$b(\text{max}) = v/2\nu \tag{1.97}$$

$$-dE/dx = \frac{4\pi n z^2 e^4}{mv^2} \ln\left[\frac{mv^3}{2v/ze^2}\right] \quad \text{(Bohr's formula)} \quad (1.98)$$

The negative sign implies that as x increases, E decreases. The quantity $-dE/dx$ is called the linear stopping power and is defined as amount of energy lost per unit length in the absorber. When x is measured in g/cm^2, then this quantity is called the mass stopping power.

Bohr's classical formula is valid provided the particle velocity is larger than the orbital electron velocity. The value of b(max) given in Bohr's classical formula corresponds to such low energy transfers that they are far less than the ionization potential and are therefore incompatible with the acceptable theory of atomic structure. For this reason, the classical theory predicts too great energy loss by high velocity particles.

A quantum mechanical formula which is more exact is due to H.A. Bethe:

$$-dE/dx = \frac{4\pi z^2 e^4 n}{mv^2}\left[\ln \frac{2mv^2}{I} - \ln\left(1 - \beta^2\right) - \beta^2\right] \quad (1.99)$$

where $B = \frac{v}{c}$, I = mean ionization potential of the atoms of the medium; $I = KZ$, $K = 13.5$ volts. The derivation assumes that the particle is a point charge, and that the spin and magnetic moment are disregarded. Observe that the quantity $-dE/dx$ which represents ionization loss per unit length, is a function of velocity of the particle and its charge but is independent of its mass. Bohr's formula (1.98) is not applicable for incident electrons for two reasons: (a) the derivation assumes that the incident particle is undeflected during the collision which is not correct for an electron; (b) for identical particles exchange phenomenon must be considered.

The last two terms in the bracket of (1.99) almost cancel out at low velocities (small β). Since the logarithmic term is quite a slow varying function of velocity, the main variation of $-dE/dx$ comes from the factor $1/v^2$. At very low velocities, the energy loss must go down because of the capture of the electrons by the incident particle. This is not considered in the quantum mechanical formula which can be relied on up to 5 MeV α's or 1.3 MeV protons. As the velocity of the incident particle decreases to very low values, various complicated effects enter the energy loss mechanism. When the incident velocity becomes comparable with the K-shell electron velocity, energy transfer to the K-shell electrons becomes difficult. The electrons effective for energy loss are those with velocities smaller than $v = \sqrt{I/2m}$. At low energies, the charge transfer process becomes more important than ionization process. The atom or the ion formed by capturing an electron may lose the electron. When the particle velocity is significantly greater than Bohr's orbital velocity for the K-shell electron, the electron loss dominates over electron capture. This corresponds to 25 keV proton energy or 400 keV α energy.

At higher velocities, the terms $\ln v^2$ and $\ln(1 - \beta^2)$ in the square brackets of (1.99) become important. The ionization vs velocity curve (Fig. 1.33) passes through a broad minimum as $\beta \to 1$. The origin of the rise of ionization is due to the Lorentz

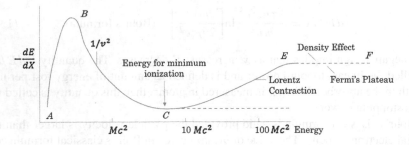

Fig. 1.33 The *curve BCD* gives the $1/v^2$ dependence. At relativistic energies v changes little and CD is asymptotic to $v = c$. At relativistic energies, the log term in $v^2/(1 - \beta^2)$ changes, and increases as $v \to c$, giving the rise at the *curve from C to E*. At very low energies (*region AB*) Eq. (1.99) breaks down because the particle has velocity comparable to that of the orbital electrons in the absorber, and the efficiency of energy exchange is much lower. The particle itself captures electrons and spends part of its time reduced change

contraction of Coulomb field of the incident particle which makes possible the energy transfer to the electrons at greater distance from the particle path. At exceedingly high velocities, however, the 'density effect' limits the energy transfer. This is also called the polarizability effect. In the derivation of (1.99), the atoms have been considered isolated. This is justifiable so long the medium is a gas. In a condensed medium, the atoms may still be considered as isolated in close collisions, but when the impact parameter is larger than the atomic distances, the screening of the electric field due to the simultaneous movement of the electrons of the neighbouring atoms becomes important and this has the consequence of lessening the magnitude of $-dE/dx$. At ultra-relativistic velocities, the curve, therefore, gets saturated to a plateau value called the Fermi plateau (in Fig. 1.33). It may be pointed out that in a cloud chamber the plateau may be higher by 50 percent compared to the trough whilst in photographic emulsions it may be higher only by about 10 percent. This is because the density effect in the former is much less than in the latter.

1.5.2 Range-Energy-Relation

When other types of energy losses are negligible compared with the collision loss, fluctuation in the energy loss are small and in a given material all particles of a given energy have almost the same fixed range. The range is defined as the total distance traversed by the particle along its track till its velocity becomes zero. In principle it should be possible to integrate (1.99) and obtain a relation between the range and the energy of the particle. There are two difficulties with this procedure, first the integration is cumbersome, second at very low velocities the phenomenon of electron capture and the uncertainties in the ionization potential render the calculations exceedingly doubtful. In practice, one uses an empirical relation of the form:

$$E = K z^{2n} M^{l-n} R^n \tag{1.100}$$

where E is the kinetic energy of the particle corresponding to the range R. M is the mass of the particle in terms of proton mass. K and n are empirical constants that depend on the nature of the absorber. The form of (1.100) ensures that the quantity $dE/dR = z^2 f(v)$ and $\neq f(M)$, as desired. Since $-dE/dx = z^2 f(v)$, $dx = -dE/z^2 f(v) = Mf'(v)dv/z^2$, where $f(v)$ and $f'(v)$ are some functions of velocity of the particle. Therefore

$$R = \int_0^R dx = \frac{M}{z^2} \int_0^v f'(v)dv = \frac{M}{z^2} f''(v) \tag{1.101}$$

where $f'(v)$ is still another function of velocity.

Consider two particles of masses M_1 and M_2 having charges z_1 and z_2. Let their initial velocities be the same, and their ranges R_1, and R_2, respectively. It follows from (1.101) that the expected ratio of their ranges would be

$$\frac{R_1}{R_2} = \frac{M_1 z_2^2}{M_2 z_1^2} \tag{1.102}$$

In particular, if the ranges of two tracks of singly charged particles from the point of equal ionization (equal velocity) are known then

$$\frac{R_1}{R_2} = \frac{M_1}{M_2} \tag{1.103}$$

This technique was employed for the mass determination of π meson in the historical experiment of Powell, Occhialini and Lattes, using photographic emulsions. Comparison was made with proton tracks having the same initial ionization.

Example 1.19 The range of a low energy proton is 1500 μm in nuclear emulsions. A second particle whose initial ionization is same as the initial ionization of proton has a range of 228 μm. What is the mass of this particle? (The rate at which a singly ionized particle loses energy E by ionization along its range is given by $dE/dR = K/(\beta c)^2$ MeV μm^{-1} where βc is the velocity of the particle, and K is a constant depending only on emulsions; the mass of proton is 1837 mass of electron.)

Solution Using (1.103)

$$M_2 = \frac{R_2}{R_1} M_1 = \frac{228 \times 1837}{1500} = 279 m_e$$

The particle is identified as π meson (pion).

Example 1.20 α particles and deuterons are accelerated in a cyclotron under identical conditions. The extracted beam of particles is passed through an absorber. Show that the expected range of deuterons is twice that of α particles.

Solution   The condition for a circular orbit in a magnetic field (induction B) is

$$Bzev = mv^2/r$$

Since B and r are same for both d and α

$$v_d = \frac{Bzer}{m_d} \quad \text{and} \quad v_\alpha = \frac{B(2e)r}{m_\alpha}$$

Since $m_\alpha = 2m_d$, it follows that $v_d = v_\alpha$.
    From (1.102)

$$\frac{R_d}{R_\alpha} = \frac{m_d}{m_\alpha}\frac{2^2}{1^2} = 2$$

Example 1.21 α-particles have an initial energy of 8.5 MeV and a range in standard air of 8.3 cm. Find their energy loss per cm in standard air at a point 4 cm distant from a thin source.

Solution   The range-energy-relation is

$$E = Kz^{2n}M^{1-n}R^n \tag{1}$$

$$\frac{dE}{dR} = nKz^{2n}M^{1-n}R^{n-1} = \frac{nE}{R} \tag{2}$$

Let $E_1 = 8.5$ MeV and $R_1 = 8.3$ cm. On moving away 4 cm from the source $R_2 = 8.3 - 4.0 = 4.3$ cm. Let the corresponding energy be E_2

$$dE_2/dR = nE_2/R_2 \tag{3}$$

$$dE_1/dR = nE_1/R_1 \tag{4}$$

Therefore

$$\frac{dE_2/dR}{dE_1/dR} = \frac{E_2 R_1}{E_1 R_2} \tag{5}$$

Also

$$\frac{dE_2/dR}{dE_1/dR} = \frac{1/v_2^2}{1/v_1^2} = \frac{v_1^2}{v_2^2} = \frac{E_1}{E_2} \tag{6}$$

From (5) and (6)

$$\frac{E_1}{E_2} = \sqrt{\frac{R_1}{R_2}} = \sqrt{\frac{8.3}{4.3}} \tag{7}$$

Using (1)

$$\frac{E_1}{E_2} = \left(\frac{R_1}{R_2}\right)^n \tag{8}$$

Comparing (7) and (8), $n = \frac{1}{2}$. From (8) or (7)

$$E_2 = E_1(R_2/R_1)^{\frac{1}{2}} = 8.5\sqrt{\frac{4.3}{8.3}} = 6.12 \text{ MeV}$$

$$\frac{dE_2}{dR} = \frac{nE_2}{R_2} = \frac{0.5 \times 6.12}{4.3} = 0.71 \text{ MeV/cm}$$

1.5.2.1 Range in Air—Geiger's Rule

If we ignore the logarithmic term in the formula for $-dE/dx$, then $dE/dx \propto 1/v^2$ or $R \propto v^4$ for the low energy region. A better approximation is provided by the formula

$$R = \text{const} \cdot v^3 \quad \text{(Geiger's rule)}$$

This formula is valid for 4–10 MeV α particles. At higher energy the exponent changes. A Formula which gives the range of α's in air at 15 °C and atmospheric pressure is

$$R = 0.32 \, (\text{MeV})^{3/2} \text{ cm} \quad \text{(alphas in air)}$$

This formula is correct to about 10 per cent in the low energy region but breaks down for relativistic velocities. Figure 1.34 shows the range energy curves for protons and Fig. 1.35 for alpha particles in air at 15 °C and 760 mm pressure.

1.5.2.2 The Bragg-Kleeman Rule

This rule permits one to convert the range R_1, in medium 1 of known density ρ_1 and atomic weight A_1 to range R_2 in medium 2 of known density ρ_2 and atomic weight A_2

$$\frac{R_2}{R_1} = \frac{\rho_1}{\rho_2}\frac{\sqrt{A_2}}{\sqrt{A_1}} \quad \text{(Bragg-Kleeman rule)} \tag{1.104}$$

This rule is correct to within 15 per cent. As an example, for air $\sqrt{A_1} = 3.81$ and $\rho_1 = 1.226 \times 10^{-4}$ g/cm^3 at 15 °C, 76 cm of Hg. Then $R_2 = 3.2 \times 10^{-4} \times \sqrt{A_2}R(\text{air})/\rho_2$. For aluminum $A_2 = 27$ and $\rho_2 = 2.7$, so that in aluminum the range of α-particles and protons (1–10 MeV) is about 1/1600 of the range in air.

Example 1.22 Compare the stopping power of a 3 MeV proton and a 6 MeV deuteron in the same medium.

Solution

$$v_p = \sqrt{\frac{2E}{m}} = \sqrt{\frac{2 \times 3}{1}} = \sqrt{6}, \quad \text{and} \quad v_d = \sqrt{\frac{2 \times 6}{2}} = \sqrt{6}$$

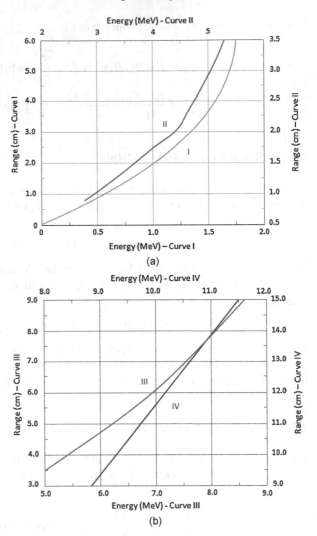

Fig. 1.34 Range-energy relation for protons in air at 15 °C, 760 mm pressure up to 11.8 MeV

Since the velocities are same and also both proton and deuteron are singly charged particles, their stopping powers are the same.

Example 1.23 Show that the specific ionization of a 320 MeV α particle is approximately equal to that of a 20 MeV proton.

Solution

$$-\frac{dE}{dx} \propto \frac{z^2}{v^2} \quad \text{or} \quad \propto \frac{Mz^2}{E}$$

Fig. 1.35 Range-energy relation for alpha-particles in air at 15 °C, 760 mm pressure up to 15 MeV

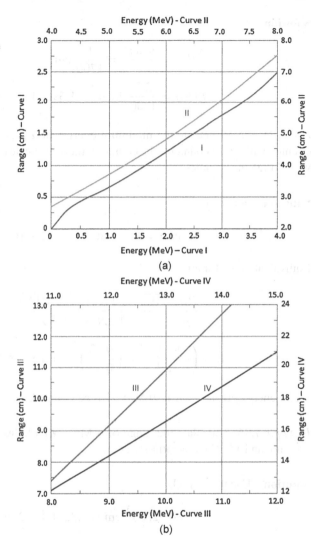

(a)

(b)

$$\text{for } \alpha\text{'s,} \quad -\frac{dE}{dx} \propto \frac{4 \times 2^2}{320} = \frac{1}{20}$$

$$\text{for } p\text{'s,} \quad -\frac{dE}{dx} \propto \frac{1 \times 1^2}{20} = \frac{1}{20}$$

Thus the specific ionization is same.

Example 1.24 If the range is multiplied by density, equivalent thickness in g/cm² is obtained. Calculate the thickness of aluminum that is equivalent in stopping power of 1 cm of air. Given the relative stopping power for aluminum $S = 1700$ and its density $= 2.7$ g/cm³.

Solution

$$R(\text{Al}) = \frac{R(\text{air})}{S} = \frac{1}{1700} \text{ cm}$$

$$R(\text{Al}) = \frac{2.7}{1700} \text{ g/cm}^2 = 1.59 \times 10^{-3} \text{ g/cm}^2$$

Example 1.25 Calculate the minimum energy an α particle can have and still be counted with a GM counter if the counter window is made of stainless steel ($A \approx 56$) with 2 mg/cm^2 thickness.

Solution   For steel

$$R_s \text{ (cm)} = R_s \left(\text{g/cm}^2\right)/\rho_s = 2 \times 10^{-3}/\rho_s$$

Equivalent range for air

$$R_a = \frac{R_s \rho_s \sqrt{A_a}}{\rho_a \sqrt{A_s}} = \frac{2 \times 10^{-3} \times \sqrt{14.5}}{1.226 \times 10^{-3} \times \sqrt{56}} = 0.83 \text{ cm}$$

$$E = \left(\frac{R}{0.32}\right)^{2/3} = \left(\frac{0.83}{0.32}\right)^{2/3} = 1.89 \text{ MeV}$$

α's of energy greater than 1.89 MeV will be counted.

Example 1.26 Calculate the range of 4 MeV α particles in air of 760 mm of Hg pressure and 15 °C temperature.

Solution   Use the formula

$$R = 0.32(E)^{\frac{3}{2}} \text{ cm} = 0.32(4)^{\frac{3}{2}} = 2.56 \text{ cm}$$

Example 1.27 Calculate the range in aluminum of a 5 MeV a particle if the relative stopping power of Aluminum is 1700.

Solution   Relative stopping power $S = R(\text{air})/R(\text{Al})$. But,

$$R(\text{air}) = 0.32(5)^{\frac{3}{2}} = 3.578 \text{ cm}$$

$$R(\text{Al}) = \frac{3.578 \text{ cm}}{1700} = 21 \text{ μm}$$

Example 1.28 The range of 5 MeV a's in air at NTP is 3.8 cm. Estimate the range of 10 MeV a's using Geiger-Nuttal law.

Solution According to Geiger's rule, $R \propto v^3$, or $R \propto E^{\frac{3}{2}}$

$$\frac{R_2}{R_1} = \left(\frac{E_2}{E_1}\right)^{3/2} = \left(\frac{10}{5}\right)^{3/2} = 2\sqrt{2}$$

$$R_2 = 2\sqrt{2}R_1 = (2\sqrt{2})(3.8) = 10.75 \text{ cm}$$

Example 1.29 Mean ranges of a particles in air under standard conditions is defined by the formula R (cm) $= 0.98 \times 10^{-27} v_0^3$, where v_0 (cm/s) is the initial velocity of an alpha particle. Using this formula, find an α-particle with initial kinetic energy 7.0 MeV (a) its mean range (b) the average number of ion pairs formed by the given a-particle over the whole path as well as over its first half, assuming the ion pair formation energy to be equal to 34 eV.

Solution

(a)

$$v_0 = \sqrt{\frac{2T}{m}} = c\sqrt{\frac{2T}{mc^2}} = c\sqrt{\frac{2 \times 7}{3726}} = 0.061c$$

$$R = 0.98 \times 10^{-27} \times \left(3 \times 10^{10} \times 0.061\right)^3 = 6 \text{ cm}$$

(b) (i) Total number of ion pairs $= \frac{7 \times 10^6}{34} = 2.06 \times 10^5$
(ii) For $R = 3$ cm range, $3 = 0.98 \times 10^{-27} v_0^3$, or $v_0 = 1.45 \times 10^9$ cm/s. Corresponding energy at mid path is

$$E = \frac{1}{2}Mv^2 = \frac{1}{2}Mc^2(v/c)^2 = \frac{1}{2} \times 3726 \times (0.048)^2 = 4.39 \text{ MeV}$$

Energy lost in the first half of the path, $\Delta E = 7.0 - 4.39 = 2.61$ MeV.
Number of ion pairs over the first half of the path $= \frac{2.61}{34} \times 10^6 = 7.67 \times 10^4$.

Example 1.30 Assuming that ^{14}C and ^{14}N nuclei are both accelerated to an energy of 40 MeV and are then allowed to pass through a thin foil. If the ^{14}C nuclei lose 2 MeV, how much energy will the ^{14}N nuclei lose?

Solution

$$\frac{-dE}{dx} \propto \frac{z^2}{v^2} \quad \text{or} \quad \propto Z^2\frac{M}{E}$$

As $\frac{M}{E}$ is the same for the nuclei

$$(-dE/dx)_N = \frac{z_N^2}{z_C^2}(-dE/dx)_C = \frac{7^2 \times 2}{6^2} = 2.72 \text{ MeV}$$

Example 1.31   Protons and deuterons have the same kinetic energy when they enter a thin sheet of material. How are their energy losses related?

Solution

$$-\frac{dE}{dx} \propto \frac{z^2}{v^2} \quad \text{or} \quad \propto \frac{M}{E}$$

as both P and d have the same z. Also both have same energy E. Therefore, $(-\frac{dE}{dx})_d = 2(\frac{dE}{dx})_p$.

Example 1.32   If protons and deuterons lose the same amount of energy when they enter a thin sheet of material, how are their energies related?

Solution

$$\left(-\frac{dE}{dx}\right) \propto \frac{M}{E}$$

$$\frac{M_p}{E_p} = \frac{M_d}{E_d}$$

$$E_d = \frac{M_d}{M_p} E_p = \frac{2}{1} E_p = 2E_p$$

1.5.3 *Energy Loss to Electrons and Nuclei*

For fast charged particles the energy loss results more from electron collisions than nuclear collisions. The latter affect stopping mainly for relatively low velocities and large charges of incident particles. For helium ions of energy larger than 0.5 MeV, even in heavy materials like silver and gold, the nuclear collisions do not account for more than 0.5 per cent of the total energy losses. For heavy ions with relatively low velocities, the contribution of nuclear collisions becomes increasingly important with charge. However, in this case too the collisions with electrons is the dominant process for the energy loss. Thus, for example, in the case of quadruply ionized carbon and oxygen ions in metals, nuclear collisions contribute only to the extent of a few per cent of the energy loss.

1.5.4 *Energy Loss of Heavy Fragments*

Heavy ions such as ^{12}C, ^{16}O, ^{40}A, ^{85}Kr are slowed down predominantly by ionization in much the same way as alpha particles. The only difference is that z is replaced by $z_{eff} = f(\beta)z$, where $f(\beta)$ is an increasing function of velocity reaching its limiting value of 1 for $\beta = 2z/137$. At very low incident particle velocities various complicated effects enter the energy loss mechanism. When the velocity approaches that of K-shell electron, energy transfer to K-electrons becomes difficult. The energy at which the energy loss attains maximum value is given by $E(\max) \cong \frac{1}{2}Mc^2(1/137)^2 Z^{2/3}$, where M is the mass of the incident particle, Z being the atomic number of the target. At velocities (v) less than Bohr's orbital velocity (u) for K-electron, the incident particle tends to capture an electron (s) from the atom, resulting in the decrease of the effective charge of the incident particle. This is called 'pick-up' process. It may also lose the captured electron. The pick-up process becomes a highly probable process for velocity $v \approx u$, where $u = zc/137 = 0.22 \times 10^9$ cm/s for protons ($E_p = 25$ keV) and $= 0.44 \times 10^9$ cm/s for alphas ($E_\alpha = 400$ keV). Towards the end of the range, as the velocity decreases, the stopping power increases reaching the maximum value for $\beta = 0.037$ for carbon and 0.059 for argon-40 ions, which correspond to 8 and 65 MeV energy respectively. At lower energy the stopping power decreases as the ions are further slowed down, since the decrease of nuclear charge overcompensates the opposing effect of diminishing velocity. This phenomenon is beautifully demonstrated by the thinning down of very heavy ion tracks just before they are arrested in photographic emulsions. The extreme case is furnished by the fission fragments. Their effective charge is large reaching about $20e$ at the beginning of the range, and nuclear collisions are an important source of energy loss. If a fragment of atomic number z crosses a medium of atomic number Z and nuclear mass m_2, the specific energy loss is

$$\frac{-dE}{dx} \propto \frac{z^2 Z^2}{m_2 v^2} \quad \text{(nuclear)} \qquad (1.105)$$

whereas the loss to electrons is

$$-\frac{dE}{dx} \propto z_{eff}^2 \frac{Z}{mv^2} \quad \text{(electronic)} \qquad (1.106)$$

Equation (1.105) applies to close nuclear collisions where the entire charges of the fragments and the target are effective. In the case of electronic collisions, only the net charge z_{eff} of the fragment is effective, since it carries with it certain number of electrons, and further the target electrons have unit charge. The factor Z in (1.106) arises from the presence of Z electrons/nucleus. The two energy losses may be comparable, but only a few nuclear collisions are responsible for the nuclear component of energy loss whilst in the electronic collisions, the loss is uniformly distributed along the path. The peculiar branches observed in the cloud chamber photographs of fission fragments have their origin in nuclear collisions. The concentration of nuclear energy loss in a limited number of events leads to the enormous spread of ranges of fission fragments of the same energy, a phenomenon called 'straggling'.

It is of interest to point out that heavy ions in passing through crystalline solids lose energy differently depending on the orientation of the trajectory with respect to the axes of a single crystal. For example, 40 keV ^{85}Kr ions are found to penetrate the face centred cubic lattice of aluminum crystals for about 4000 Å in the direction perpendicular to the (101) face but only 1500 Å in the direction perpendicular to the (111) face. This is because the number of atoms encountered in these two cases is not same.

1.5.5 Energy Loss of Electrons

It was pointed out that in the case of heavy ions, ionization is the dominant mode of energy loss. However, for electrons, the energy loss is complicated due to an additional mechanism of loss through radiation, a phenomenon called Bremsstrahlung. At low energies ($E < 2mc^2$) the ionization loss dominates over that due to radiation. The problem of energy loss of electrons by ionization follows similar to that of heavy ions, but the treatment differs in two important respects. They are the identity of particles which participate in the collisions, and secondly their reduced mass.

The formula for non-relativistic electrons is:

$$\frac{-dE}{dx} = \frac{4\pi e^4 n}{mv^2} \left[\ln \frac{mv^2}{2I} - \frac{1}{2} \ln 2 + \frac{1}{2} \right] \qquad (1.107)$$

Except for small differences in the terms within the square brackets, formula (1.107) bears a striking resemblance to (1.99). We, therefore, conclude that the non-relativistic electrons lose their energy by ionization at the same rate as the protons.

The relativistic formula for electrons is

$$-\frac{dE}{dx} = \frac{4\pi e^4 n}{mc^2} \left[\ln \frac{2mc^2}{I} + \frac{3}{2} \ln \gamma + \frac{1}{16} \right] \qquad (1.108)$$

and that for protons

$$-\frac{dE}{dx} = \frac{4\pi e^4 n}{mc^2} \left[\ln \frac{2mc^2}{I} + 2 \ln \gamma - 1 \right] \qquad (1.109)$$

where $\gamma = 1/\sqrt{1 - \beta^2}$ is the Lorentz factor. At equal velocities, formulae (1.108) and (1.109) agree within 10 per cent.

1.6 Delta Rays

1.6.1 Energy Spectrum

In the collision of a charged particle with an atom, one or more electrons are ejected. The more energetic ones of these are called Delta rays and are responsible for the secondary ionization, i.e. the production of further ions due to the collision of delta rays with the atoms of the medium. In what follows we shall be concerned with delta rays of energy larger than the ionization potential of the atoms of the medium. The binding energy of the electron is, therefore, ignored and the collision between the incident particle and the electrons is considered as approximately elastic. From (1.27) the kinetic energy of the ejected electron ($m_2 \ll m_1$)

$$W = 2mv^2 \cos^2 \phi \tag{1.110}$$

where $m = m_2$ is the electron mass, ϕ is the angle of emission of electron, and v is the velocity of the incident particle. The maximum energy, $W(\max) = 2mv^2$ (non-relativistically). Now, for the recoil particle (electron) $\phi = \frac{1}{2}\phi^*$ and $\phi^* = \pi - \theta^*$, and so

$$\cos^2 \phi = \sin^2 \frac{1}{2}\theta^* \tag{1.111}$$

$$W = 2mv^2 \sin^2 \frac{1}{2}\theta^* \tag{1.112}$$

$$dW = mv^2 \sin\theta^* d\theta^* \tag{1.113}$$

But Rutherford's formula for scattering in the CMS is

$$\sigma\left(\theta^*\right) = \frac{d\sigma}{d\Omega^*} = \frac{1}{4} \frac{z^2 e^4}{\mu^2 v^4 \sin^4 \frac{1}{2}\theta^*} \tag{1.114}$$

where we have put $Z = -1$. Since the electron mass is negligible compared to that of the incident particle, $\mu \cong m$. Also, the element of solid angle $d\Omega^* = 2\pi \sin\theta^* d\theta^*$. Therefore,

$$d\sigma = \frac{2\pi \sin\theta^* d\theta^* z^2 e^4}{4m^2 v^4 \sin^4 \frac{1}{2}\theta^*} \tag{1.115}$$

Using (1.112) and (1.113) in (1.115)

$$\frac{d\sigma}{dW} = \frac{2\pi z^2 e^4}{mv^2 W^2} \quad \text{(differential energy spectrum)} \tag{1.116}$$

This gives us the cross-section for finding the delta rays of energy W per unit of energy interval.

1.6.2 Angular Distribution

Using the relations (1.111) and (1.115) and the expression for the element of solid angle in the lab system $d\Omega = 2\pi \sin\phi d\phi$, we obtain the differential cross-section for the delta rays in the LS:

$$\sigma(\phi) = \frac{z^2 e^4}{m^2 v^4 \cos^3 \phi} \tag{1.117}$$

where we have used the relations $\phi = \frac{1}{2}\phi^* = \frac{1}{2}\pi - \frac{1}{2}\theta^*$. It follows that most of the delta rays are emitted at large angles with correspondingly small energy. Note that $\phi(\max) = 90°$ for which $W = 0$. The fact that the delta rays can be emitted only in the forward hemisphere implies that one can find the direction of the primary.

1.6.3 Delta Ray Density

For a 5 MeV proton $W(\max) = 2mv^2 = 4Tm/M = 4 \times 0.51 \times 5/940$ MeV $=$ 10.85 keV. From (1.116), it is evident that the number of delta rays per cm of path is inversely proportional to the primary energy; also it is greater for heavy primaries. The observation of delta ray per cm density is very useful in establishing the charge of heavy nuclei in cosmic radiation. The total number of δ-rays/cm with energy $> W_1$, is given by integrating (1.116) between the limits $2mv^2$ corresponding to the maximum energy of delta rays and some arbitrarily lower value W_1, and multiplying the result by N, the number of electrons per cm^3. This follows from the fact that $n(T, v)$, the number of δ-rays ejected in 1 cm $= 1/\lambda = \Sigma = N\sigma$. We thus have:

$$n(T, v) = \frac{2\pi N e^4 z^2}{mv^2} \left(\frac{1}{W_1} - \frac{1}{2mv^2} \right) \tag{1.118}$$

Below the lower limit W_1, the δ-rays are not recorded. Clearly, $n(T, v)$ is an arbitrary quantity as it depends on the choice of W_1. It follows that for particles of identical velocities but of different charges, $n(T, v)$ varies as z^2 and the distributions of the values of $n(T, v)$ along the tracks of the particles would, apart from statistical fluctuations, be similar in form. It is also seen that at a velocity less than $v_c = \sqrt{(w_1/2m)}$ the primary would not produce δ-rays with energy $> w_1$. Above the critical value, the density would increase at a rate which depends on the variation of the velocity of the particle along the track. The maximum value is attained for $v = \sqrt{(w_1/m)}$ which is simply obtained by maximizing n with respect to v. After this, it varies approximately as $1/v^2$, as the second term in the brackets becomes practically constant. The resulting distribution would thus increase to a maximum and then slowly decrease (Fig. 1.36). The maximum value $n(\max)$ for a given particle of z_2 may be compared with that obtained from similar observations on the tracks of particles of known charge z_1. Thus, the unknown charge z_2 may easily be obtained from the following condition:

$$n_2/n_1 = z_2^2/z_1^2 \tag{1.119}$$

It may be pointed out that this condition is also fulfilled for relativistic particles. It is also possible to determine the mass of the primary particle by measuring the emission angle and the energy of the delta ray caused by the particle whose momentum

Fig. 1.36 Variation of δ-ray density with range for nuclei of charges 2 to 26

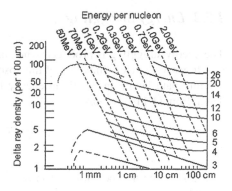

$p = Mv$ is known. We can rewrite (1.110)

$$W = 2mv^2 \cos^2 \phi = \frac{2mp^2}{M^2} \cos^2 \phi \qquad (1.120)$$

From the measurement of W, ϕ and p, the mass M of the primary particle can be deduced. This method is specially suited when the conventional methods do not permit the particles to be identified. For example, in bubble chambers, this method is commonly employed for the estimation of contamination of pions or muons in kaon or antiproton beams.

1.7 Straggling

1.7.1 Theory

Identical charged particles, having the same initial velocity, do not have exactly the same ranges. In other words, for a given energy loss the path length fluctuates. This phenomenon is called *Range straggling*. Also, for a given path length the ionization loss and therefore the energy loss fluctuate. This is called *Energy straggling*. There is an intimate relation between the two. The observed ranges of individual particles from any mono-energetic source will show a substantially normal distribution about the mean range. The standard deviation of this distribution is of the order of 1 per cent for a few MeV alphas in any absorber. The distribution is due to the statistical fluctuations in the individual collisions between the charged particle and atomic electrons, which are finite in number. The nuclear collisions, fewer in number, which may cause substantial loss of energy specially towards the end of the ranges, contribute to the short range tail of the distribution. For small energies, however, this will be a small contribution and the distribution may be taken as approximately symmetrical. The harder collisions account for most of the straggling and because very hard collisions are few in number, the actual distribution is some what asymmetric, with a longer tail in the direction of short ranges and with a mean range slightly shorter than the modal range.

1.7.2 Energy Straggling

The energy straggling is produced when an initially mono-energetic beam of particles traverses a given thickness of the absorber. Let A_x be the number of collisions per unit path length in which an energy between W, and $W + dW$ is transfered. Then, from (1.116) we have

$$A_x = \frac{2\pi N z^2 e^4 W}{m v^2 W_x^2} \qquad (1.121)$$

where N is the number of electrons/cm^3. The energy transfer in a distance Δr is given by

$$\Delta E = \sum_x A_x W_x \Delta r \qquad (1.122)$$

$$\Delta E / \Delta r = \sum_x A_x W_x \qquad (1.123)$$

When the number of collisions is large, we may use integration rather than summation

$$\frac{dE}{dr} = \frac{2\pi N z^2 e^4}{m v^2} \int_{W(\min)}^{W(\max)} \frac{dW}{W_x} \qquad (1.124)$$

The statistical fluctuations in energy loss ΔE arise from fluctuations about the average number of collisions $A_x \Delta r$. We assume that the collisions are randomly distributed and that the S.D. is given by $\sqrt{A_x \Delta r}$. The S.D. of the energy loss is then $W_x \sqrt{A_x \Delta r}$. The variance for all types of collisions is then given by the summation of the individual variances

$$\sigma^2 = \Delta r \sum_x W_x^2 A_x = \frac{2\pi N z^2 e^4 \Delta r}{m v^2} \int_{W(\min)}^{W(\max)} dW$$

$$= \frac{2\pi N Z^2 e^4 \Delta r}{m v^2} \big[W(\max) - W(\min) \big]$$

where we have replaced the summation by integration. Since $W(\min) \ll W(\max) = 2mv^2$

$$\sigma^2 = 4\pi N z^2 e^4 \Delta r \qquad (1.125)$$

If it is assumed that the actual energy loss has a Gaussian distribution around the average value E_0, the use of expression (1.125) for the S.D. in energy loss leads to

$$P(E)dE = \frac{dE}{\sqrt{8\pi^2 z^2 e^4 N t}} \exp - \left[\frac{(E - E_0)^2}{8\pi N z^2 e^4 t} \right] \qquad (1.126)$$

where t is the absorber thickness.

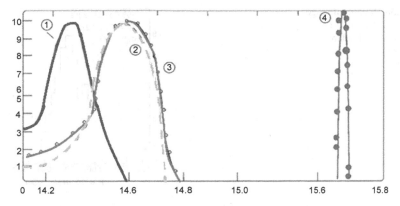

Fig. 1.37 Energy distribution of an 'unobstructed' electron beam and the calculated and experimental distributions of electrons that have passed through 0.86 g cm^{-2} of aluminum. (**1**) Landau theory without density correction; (**2**) Landau theory with Fermi density correction; (**3**) experiment; (**4**) incident beam [1]

In the case of fission fragments large energy losses in individual nuclear collisions give rise to a tail on the side of higher energy losses of the distribution. The straggling effects are much more important for electrons than for heavy particles, because an electron may lose even half its energy in a single elastic collision, where as a heavy particle may lose only a fraction of its energy. Radiation losses add further to the electron straggling. Thus electron straggling reaches values of the order of 0.2 of the total energy loss. Figure 1.37 shows the energy distribution of electrons before they have entered the absorber and after they have traversed 0.86 g/cm^2 thickness of aluminium.

1.7.3 Range Straggling

The fluctuations in range and energy loss are related. Denoting the S.D. of energy and range by σ_E and σ_R respectively, we can use the formula for the propagation of errors and write:

$$\sigma_R^2 = (dR/dE)^2 \sigma_E^2 \tag{1.127}$$

Using (1.125), we get

$$\sigma_R^2 = (dR/dE)^2 4\pi N z^2 e^4 dR \tag{1.128}$$

Writing $dR = (dR/dE)dE$, we get the result:

$$\sigma_R^2 = 4\pi N z^2 e^4 \int_0^{E_0} \left(\frac{dE}{dR}\right)^{-3} dE \tag{1.129}$$

This relation is not applicable to heavy ions and fission fragments that undergo excessive straggling owing to the occurrence of single nuclear collisions. Assuming

Fig. 1.38 Measured ranges
of muons from $\pi-\mu$ decay in
emulsions of standard
composition

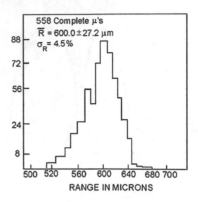

that the ranges of individual particles are distributed about the mean range in a
Gaussian way, the probability that the individual range falls between R and $R + dR$
is

$$P(R)dR = \frac{dR}{\sigma_R\sqrt{2\pi}}\exp-\left[(R-\overline{R})^2/2\sigma_R^2\right] \qquad (1.130)$$

For α particles from Polonium, $E_0 = 5.3$ MeV, $\overline{R} = 3.84$ cm in air, the corresponding $\sigma_R = 0.036$ cm and the ratio $\sigma_R/\overline{R} = 0.9\%$. Figure 1.38 shows the histogram of
ranges of μ mesons produced in the decay of π^+ mesons at rest. Since the π mesons
decay by a two-body process, μ^+ is produced with unique energy (4.27 MeV). The
mean range in photographic emulsions is found to be 600 µm. The S.D. of the range
distribution is found to be, $\sigma_R = 2.7$ µm; this gives $\sigma_R/\overline{R} = 0.045$, or 4.5 percent.

1.7.3.1 The Range Straggling Parameter

This is related to S.D. by

$$\alpha_0 = \sqrt{2}\sigma_R \qquad (1.131)$$

Several common types of particle detectors measure the integrated number of particles. The particles that are still present in the collimated beam having ranges equal
or greater than R is given by

$$n = n_0 - \int_{-\infty}^{R} dn$$

where dn/n_0 is given by the normal distribution

$$\frac{dn}{n_0} = \frac{1}{\alpha\sqrt{\pi}}\exp\left[-(R-\overline{R})^2/\alpha^2\right]dR \qquad (1.132)$$

dn is the actual range between R and $R + dR$, n_0 is the total number of particles
initially present, and α is the half width of the range distribution at $1/e$ of the max-

Fig. 1.39 The extrapolated number-distance range R_n exceeds the mean range $\overline{R}$ by 0.886α, where α is the range-straggling parameter

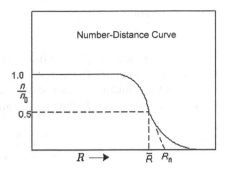

imum. Although the normal distribution is non-integrable, its value can be found from standard tables.

The number-distance curve, n/n_0 against R, is indicated in Fig. 1.39. Its slope $(dn/n_0)/dR$ at the mean range $R = \overline{R}$ is $1/\alpha\sqrt{\pi}$. As the central portion of the number-distance curve is approximately linear, it can be extrapolated to cut the range axis at $R = R_n$. This is called extrapolated range. From Fig. 1.39, we find the relation between R_n and $\overline{R}$

$$\frac{\frac{1}{2}}{R_n - \overline{R}} = \frac{1}{\alpha\sqrt{\pi}} \tag{1.133}$$

whence the mean range, in term of the measured extrapolated range R_n and straggling parameter a, is

$$\overline{R} = R_n - \frac{1}{2}\sqrt{\pi}a = R_n - 0.886\alpha \tag{1.134}$$

1.7.3.2 Deduction of Ranges Parameter

For particles of charge ze and mass M but the same initial velocity v_0 as alpha-particles

$$\sigma_R = \left[4\pi N z^2 e^4 \int_0^{E_0} \left(\frac{dE}{dR}\right)^{-3} dE \right]^{1/2}$$

$$\frac{dE}{dR} = z^2 f(v_0)N, \quad \text{since} \quad \frac{dE}{dR} = \frac{2\pi z^2 e^4 N}{mv^2} \int_{w(min)}^{w(max)} \frac{dw}{w}$$

$$dE = d(Mv^2) = Mvdv \quad \text{and} \quad \sigma_R = \frac{\sqrt{M}}{Nz^2}f(v_0, I)$$

$$R = \frac{Mf'(v_0, I)}{Nz^2} \tag{1.135}$$

$$\frac{\sigma_R}{R} = \frac{1}{\sqrt{M}} f''(v_0, I) \tag{1.136}$$

where f, f', f'' are complicated functions of the initial velocity v_0, mean excitation and ionization potential I. The function f'' and hence σ_R/R are independent of N and of z but is found to decrease slowly with increasing I.

For particles of mass M having the same initial velocity v_0 as the α particles,

$$\frac{(\alpha_0/R)_M}{(\alpha_0/R)_{\alpha\ particle}} = \sqrt{\frac{4}{M}} \tag{1.137}$$

It follows that protons will have about twice the range straggling parameter of α-particles which have the same initial velocity and hence about the same range.

1.8 Cerenkov Radiation

Electromagnetic radiation is emitted when a charged particle passes through a medium in which its velocity $v = \beta c$ exceeds the phase velocity c/μ, where μ is the refractive index of the medium. This observation was discovered by Cerenkov and was explained theoretically by Frank and Tamm. The effect was first observed in the experiments of Cerenkov who was investigating the glow in pure liquids caused by γ rays from radium. Vavilov and Cerenkov showed that the radiation is not due to luminescence (emission from excited atoms and molecules of the medium) but due to the passage of knock-on electrons produced in Compton scattering of γ rays. The radiation is instantaneous and possesses a sharply pronounced spatial symmetry.

When relativistic charged particles are incident on a transparent dielectric, the velocity of the particle is substantially unchanged except for the ionization and radiation losses. On the other hand, the electric field due to the charge of the particle and the magnetic field produced by the moving charge are propagated through the medium with velocity of only c/μ. The resulting electromagnetic radiation is cancelled in all directions if $\beta\mu < 1$; however, if $\beta\mu > 1$, constructive interference can take place in one direction defined by angle θ (Fig. 1.40). When $\beta c > c/\mu$, i.e. the particle velocity exceeds the velocity of light in the medium it is as if the particle runs away from its own slower electromagnetic field, resulting in the emission of all frequencies for which $\beta\mu > 1$. The resulting radiation called Cerenkov radiation is emitted on a conical surface BDA of half angle a_0. Figure 1.40 gives the Huyghens' construction for the electromagnetic waves emitted by the particle along its path. The particle is at A at $t = 0$; and at a later time it moves on to D such that $AD = \beta ct$. The front of the electromagnetic wave lies on the surface of the cone of half angle α. Consequently the corresponding rays of light make an angle θ with the path of the particle. The axis of the cone coincides with the direction of the incident particle and the half angle of the cone is determined by:

$$\sin\alpha = \cos\theta = \frac{(c/\mu)t}{\beta ct} = \frac{1}{\beta\mu} \tag{1.138}$$

Fig. 1.40 Huyghens'
construction for
electromagnetic waves
emitted by a moving charged
particle

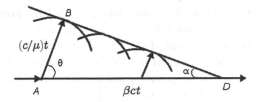

This follows from the condition that the optical difference in the path of the waves emitted by the moving particle at various points of its trajectory is equal to zero. The light is polarized with its electric vector in the plane of the conical surface and radially directed along DB. The conical distribution of the Cerenkov radiation has a natural half width of the order of a few degrees. This is attributed to the occurrence of successive changes in particle velocity when photons are emitted. The phenomenon is analogous to the V-shaped shock wave observed in acoustics when a projectile or an aeroplane travels with supersonic velocity. Apart from (1.138), there are two other conditions that must be fulfilled to achieve coherence. These are (i) pathlength of the particle in the medium must be large compared with the wavelength of the radiation, otherwise diffraction effects become dominant and (ii) velocity of the particle must remain constant during its passage through the medium.

For a medium of a given refractive index μ, there is a threshold velocity $\beta(\min) = 1/\mu$, below which no radiation is emitted. At this critical velocity, the direction of radiation coincides with that of the particle. For glass ($\mu = 1.5$), $\beta(\min) = 0.667$, corresponding to 200 keV electrons or 320 MeV protons. As the refractive index decreases, the threshold velocity increases. For an ultra-relativistic particle, for which $\beta \simeq 1$, there is a maximum angle of emission given by $\theta(\max) = \cos^{-1}(1/\mu)$.

Fermi showed that Cerenkov radiation results from small energy transfers to distant atoms due to the fast moving charged particles which is subsequently emitted as a coherent radiation. Thus the emission of Cerenkov radiation is a particular form of energy loss in extremely soft collisions. The classical theory of Cerenkov effect is originally due to Frank and Tamm and is justified by the quantum theory. Since the radiation in question is believed to be the result of the interaction with the medium as a whole and not due to the interaction of particles with individual atoms, the medium is considered as continuous and is characterized by the macroscopic parameter, the dielectric constant or by the refractive index. It is shown that the rate of energy loss per unit path length is given by:

$$-\frac{dE}{dx} = \frac{4\pi^2 z^2 e^2}{c^2} \int_{\beta\mu > 1} \left[1 - \frac{1}{\beta^2 \mu^2} \right] v dv \text{ ergs/cm} \qquad (1.139)$$

where ze is the charge of the particle and v is the frequency of the emitted radiation. The integration is to be carried over all frequencies for which $\beta\mu > 1$. For glass or Lucite, the energy loss by Cerenkov radiation is of the order of 1 keV/cm, a value which is much less than that incurred in ionization or radiation. Nonetheless the radiation is readily detected as a large number of photons are produced in the visible region. Formula (1.139) shows that the Cerenkov radiation is independent of the rest mass of the moving particle and depends only on the particle's charge and velocity, apart from the refractive index of the medium. The mean number of photons of frequency v and $v + dv$ in the visible region per cm is calculated in Example 1.35, under the assumption that μ is independent of v in the considered range of frequencies. Hence

$$N(v)dv = \frac{4\pi^2 z^2 e^2}{hc^2}\left(1 - \frac{1}{\mu^2\beta^2}\right)dv = \frac{2\pi z^2 dv \sin^2\theta}{137c} \tag{1.140}$$

The radiation has continuous spectrum, with components of all frequencies for which the refractive indices are higher than $1/\beta$. Equation (1.139) shows via the term vdv which is proportional to $d\lambda/\lambda^3$ that the energy per wavelength interval $d\lambda$ is proportional to $1/\lambda^3$. Also, (1.140) shows through the term dv that the number of quanta per cm per wavelength interval is proportional to $1/\lambda^2$. It follows that shorter wavelengths are preferred and the Cerenkov radiation appears visually as bluish white.

The density effect is closely connected with the phenomenon of Cerenkov effect. It was first pointed out by Bohr that the intricate relationship between the density effect and the Cerenkov effect is such that the entire contribution to the most probable energy loss from the minimum out to the beginning of the Fermi plateau in the ionization curve is due to Cerenkov effect.

Example 1.33 Pions and muons each of 160 MeV/c momentum pass through a transparent material. Find the range of the index of refraction of this material over which the muons alone give Cerenkov light. Assume $m_\pi c^2 = 140$ MeV, $m_\mu c^2 = 106$ MeV.

Solution   Momentum, $p = m\beta\gamma c$. Therefore, $\frac{\beta}{\sqrt{1-\beta^2}} = \frac{cp}{mc^2}$

$$\text{Pions:} \qquad \frac{cp}{m_\pi c^2} = \frac{160}{140} = \frac{8}{7} = \frac{\beta}{\sqrt{1-\beta^2}}$$

$$\beta_\pi = 0.7525; \qquad \mu_\pi = \frac{1}{\beta_\pi} = \frac{1}{0.7525} = 1.33$$

$$\text{Muons:} \qquad \frac{cp}{m_\mu c^2} = \frac{160}{106} = \frac{\beta}{\sqrt{1-\beta^2}}$$

$$\beta_\mu = 0.8336; \qquad \mu_\mu = \frac{1}{\beta_\mu} = \frac{1}{0.8336} = 1.2$$

Therefore, the range of the index of refraction of the material over which the muons alone give Cerenkov light is 1.2–1.33.

Example 1.34 A beam of protons moves through a material whose refractive index is 1.6. Cerenkov light is emitted at an angle of 15° to the beam. Find the kinetic energy of the proton in MeV.

Solution

$$\beta = \frac{1}{\mu \cos\theta} = \frac{1}{1.6\cos 15°} = 0.647$$

$$\gamma = \frac{1}{\sqrt{1 - \beta^2}} = \frac{1}{\sqrt{1 - (0.647)^2}} = 1.31$$

$$K.E. = (\gamma - 1)mc^2 = (1.31 - 1) \times 938 = 292 \text{ MeV}$$

Example 1.35 The rate of loss of energy by production of Cerenkov radiation is given by the relation

$$-dW/dl = \frac{z^2 e^2}{c^2} \int \left(1 - \frac{1}{\beta^2 \mu^2}\right) \omega d\omega \text{ erg cm}^{-1}$$

where βc is the velocity of the of charge ze, μ is the refractive index of the medium and $\omega/2\pi$ is the frequency of radiation. Estimate the number of photons emitted in the visible region, per cm of track, by a particle having $\beta = 0.8$ passing through glass ($\mu = 1.5$). The fine structure constant $\alpha = e^2/\hbar c = 1/137$.

Solution   For electron, $z = -1$ and since $\omega = 2\pi v$, the given expression becomes upon integration between the frequencies v_1 and v_2

$$-dW/dl = \frac{4\pi^2 e^2}{c^2}\left(1 - \frac{1}{\beta^2 \mu^2}\right)\frac{(v_2^2 - v_1^2)}{2}$$

where we have assumed that μ is independent of v.

Calling the average photon energy as $h\bar{v} = \frac{1}{2}h(v_1 + v_2)$, the average number of quanta emitted per cm is

$$N = \frac{1}{h\bar{v}}\left(\frac{-dvc}{dl}\right) = \frac{4\pi^2 e^2}{hc^2}\left(1 - \frac{1}{\beta^2 \mu^2}\right)(v_2 - v_1)$$

$$= \frac{2\pi}{137}\left(1 - \frac{1}{\beta^2 \mu^2}\right)\left(\frac{1}{\lambda_2} - \frac{1}{\lambda_1}\right)$$

where $\lambda = c/v$ is the vacuum wavelength and μ is the average refractive index over the wavelength interval from $\lambda_2 = 4000$ Å to $\lambda_1 = 8000$ Å and $\beta\mu = 0.8 \times 1.5 = 1.2$.

$$N = \frac{2\pi}{137}\left(1 - \frac{1}{1.2^2}\right)\left(\frac{1}{4000 \times 10^{-8}} - \frac{1}{8000 \times 10^{-8}}\right) = 175 \text{ photons}$$

1.9 Identification of Charged Particles

In numerous investigations in nuclear physics and particle physics it is necessary to determine the nature and energy of charged particles. In order to identify a particle, its charge and mass must be determined. The charge may be determined by δ-ray density or width measurements in photographic emulsions or pulse height in scintillation counters or proportional counters or solid state detectors.

Assuming that the charge is known, the mass of the particle can be determined from the simultaneous measurements of at least two dynamical quantities such as momentum and velocity, momentum times velocity and $-dE/dX$, $-dE/dX$ and E.

1.9.1 (a) Momentum and Velocity

Momentum can be determined from curvature measurement in a cloud chamber or in a bubble chamber with low Z liquid with known magnetic field or in photographic emulsions with pulsed magnetic field.

Velocity may be estimated by the estimation of ionization through drop density in a cloud chamber, bubble density in a bubble chamber, grain density, blob density or mean gap length in photographic emulsions or by Cerenkov counters or time-of-flight method.

1.9.2 (b) Momentum Times Velocity ($p\beta$) and Velocity

The product $p\beta$ is determined from the mean scattering angle in emulsions for energetic particles. The velocity is measured as in Sect. 1.9.1.

1.9.3 Energy and Velocity

Energy may be determined from range measurements for low energy particles and velocity as in Sect. 1.9.1.

1.9.4 Simultaneous Measurement of dE/dx and E

This method is widely used in the study of nuclear reactions using solid state detectors, since for non-relativistic particles the product EdE/dx is proportional to $z^2 M$. A simultaneous measurement of E and dE/dx and their product permits the separation of the particles according to their masses in a wide range of energy variations.

1.9.5 Energy and Emission Angle

Energy and emission angle measurement of knock-on electron (δ-ray) together with the momentum measurement of the beam particles.

This method is very useful in case the conventional methods are not available. The method is explained in Sect. 1.6.3.

1.10 Bremsstrahlung

In their passage through matter, electrons lose energy in two ways (i) ionization (which was referred to in Sect. 1.5.1) and (ii) radiation or Bremsstrahlung. The electrons undergo radiative collisions mainly with the atomic nuclei of the medium. In the vicinity of the nucleus of charge Ze, the incident particle of charge ze and mass m undergoes acceleration which is proportional to zZ/m. According to electrodynamics, a charged particle undergoing acceleration emits radiation. This is called Bremsstrahlung or braking radiation whose spectrum has the form dE/E where E is the photon energy. The photon energy spectrum extends from low energy to the maximum value equal to the particle energy, with the low energy photons being preferably emitted (see Fig. 1.41 for typical energy spectrum). The radiation intensity is proportional to $z^2 Z^2/m^2$. This then means that under identical conditions, radiation losses are 3×10^6 times as much for electron as for a proton. The total average energy loss per path length dx integrated over all frequencies is given by

$$-(\overline{dE})_{rad} = \frac{4Z(Z+1)}{137} N E r_e^2 \ln \frac{183}{Z^{1/3}} dx \qquad (1.141)$$

where N is the number of nuclei per cm^3, E is the energy of electron and $r_e = e^2/mc^2$ is the classical electron radius.

Since an electron may lose appreciable energy in a single collision, the actual energy loss may vary significantly from the average value given by (1.141). This also implies that the range straggling of electrons would be so great that the definition of mean range would hardly be meaningful. If we define the radiation length X_0 by

$$\frac{1}{X_0} = \frac{4Z(Z+1)}{137} r_e^2 N \ln \frac{183}{Z^{1/3}} \qquad (1.142)$$

we can write from (1.141)

$$\frac{dE}{dx} = -\frac{E}{X_0} \qquad (1.143)$$

Integrating (1.143), we find the average energy of a beam of electrons of initial energy E_0 after traversing a thickness x of medium by the expression

$$\langle E \rangle = E_0 \exp(-x/X_0) \qquad (1.144)$$

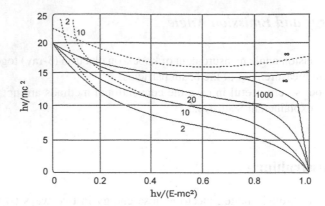

Fig. 1.41 Energy distribution of the radiation emitted by an electron. Ordinate intensity of radiation (quantum energy times number of quanta) per unit frequency interval. Abscissa, energy of emitted quantum as a fraction of the energy of the emitting electron. The *numbers on the curves* indicate the energy of the electron in units of mc^2. *Solid curves* for lead, including effect of screening. *Dotted curves* are without screening, valid for all Z [3]

For $x = X_0$, $\langle E \rangle = E_0/e$, where e is the exponential. This suggests that the radiation length X_0 may be simply defined as that thickness of the medium which reduces the beam energy by a factor of e. Since the thickness x can be measured in cm or g/cm^2 (which is obtained by multiplying the thickness in cm by the density of the medium) X_0, is expressed in corresponding units. At low electron energies ($E \ll mc^2$), the electrons lose their energy predominantly through excitation and ionization, and radiation loss is unimportant. The energy loss by ionization and excitation is proportional to Z and is practically constant at high energies as it increases only logarithmically with energy. On the other hand, radiation losses are proportional to Z^2 and increase linearly with energy. Thus, the radiation loss predominates at high energies. It is apparent that at some energy E_c, called the *critical energy*, $E_{rad} = E_{ion}$. It can be shown that roughly

$$\frac{(dE/dx)_{rad}}{(dE/dx)_{ion}} = \frac{EX}{600} \tag{1.145}$$

so that E_c (in MeV) $= 600/x$. The radiation lengths X_0 and the critical energy E_c, for some of the materials are shown in Table 1.1. Observe that X_0, decreases rapidly with increasing Z.

1.11 Questions

1.1 Why in Rutherford scattering the presence of orbital electrons in the target atom is ignored?

1.2 The total cross-section for Rutherford scattering is infinite. What is the physical reason?

Table 1.1 Radiation lengths and critical energy in different elements

Element	Z	X_0 (g cm^{-2})	E_c (MeV)
Hydrogen	1	58	340
Carbon	6	42.5	103
Air	7.2	36.5	83
Aluminium	11	23.9	47
Iron	26	13.8	24
Lead	82	5.8	6.9

1.3 Why in the famous a-scattering experiment thin foils were used for the target?

1.4 If the incident electron enters the nucleus, would the Coulomb's inverse square law between the charges be still valid? If not, how would it be modified for a nucleus in which the charge is uniformly distributed?

1.5 Why does the ionization fall off for very low particle velocity?

1.6 The inverse square velocity law for ionization would suggest that the rate of energy loss is greater at low speeds, since the time spent by the incident particle in the vicinity of the electron is longer. Is this reasonable? In the same manner would a slow moving heavenly object raise larger tides on approaching close to the earth compared to a fast moving one?

1.7 What is the physical origin of the rise in the $-dE/dx$ curve beyond the minimum?

1.8 At relativistic velocities, the $-dE/dx$ curve saturates to a plateau. What is the origin of the plateau?

1.9 In the cloud chamber studies of ionization, the plateau-to-trough ratio for the $-dE/dx$ curve might be as large as 1.5, but in photographic emulsions it is no more than 1.1. Explain.

1.10 How does the percentage straggling compare for ^{3}H and ^{3}He nuclei of the same initial velocity?

1.11 A cloud chamber photograph shows an alpha track which after certain distance gets thinned down and then disappears. It again re-appears before it stops. What is happening?

1.12 The range of a proton of few MeV is a measure of its initial energy. The energy thus estimated would be close to the actual value within few per cent. However, in the case of electrons of similar energy, the energy thus estimated can hardly be reliable. Explain.

1.13 The tracks of fission fragments often leave peculiar branches before coming to a rest. Explain.

1.14 A water cooled nuclear reactor appears bluish. What could be the origin of this colour?

1.15 A charged particle moves swiftly with uniform velocity in a vacuum. Would it radiate?

1.16 What is the dominant mechanism for energy loss for electrons of energy (a) <1 MeV, (b) 200–500 MeV?

1.12 Problems

1.1 Show that in an elastic collision, the ratio of the kinetic energy K'/K can be expressed through $\alpha = M/m$ and $y = \cos\theta$ as

$$\frac{K'}{K} = (1+\alpha)^{-2}\left[2y^2 + \alpha^2 - 1 + 2y\sqrt{a^2 + y^2 - 1}\right]$$

1.2 A body of mass M rests on a smooth table. Another of mass m moving with a velocity u collides with it. Both are perfectly elastic and smooth and no rotations are set up by the collision. The body M is driven in a direction at an angle ϕ to the previous line of motion of the body m. Show that its velocity is

$$\frac{2mu}{M+m}\cos\phi$$

1.3 A nucleus A of mass $2m$ moving with velocity u collides inelastically with the nucleus B of mass $10m$. After the collision, the nucleus A travels at $90°$ with the incident direction, while B proceeds at an angle $37°$ with the incident direction. (a) Find the speeds of A and B after the collision. (b) What fraction of the initial kinetic energy is gained or lost due to the collision?
[Ans. (a) $v_A = 3u/4$; $v_B = \frac{u}{4}$; (b) 1/8]

1.4 A beam of alphas gets scattered from a hydrogen target. What is the maximum angle of scattering?
[Ans. approximately 15°]

1.5 An alpha particle fired into a cloud chamber undergoes an elastic collision with a nucleus of the gas used to fill the chamber. The collision is recorded photographically as a forked track. Measurements from the photograph show that the collision deviated the alpha-particle at $60°$ and that the struck nucleus recoiled at an angle of $30°$ with the direction of motion of the incident alpha-particle. Assuming that the struck nucleus is initially at rest, calculate:

(a) The mass number of the gas used to fill the chamber.
(b) The ratio of the velocity of projection of the struck nucleus to the velocity of the incident alpha particle.

[Ans. (a) 4; (b) $\sqrt{3}$]

1.6 A particle of mass m makes an elastic collision with a proton, initially at rest. The proton is projected at an angle $22.1°$ whilst the incident particle is scattered through an angle $5.6°$ with the incident direction. Estimate m in atomic mass units. [Ans. 7.8]

1.7 Consider an elastic collision between an incident particle having mass m and a target particle of mass M such that $m > M$. Show that the largest possible scattering angle $\theta(\max)$ in the lab system is given by: $\sin\theta(\max) = M/m$; and that this corresponds to C-system angle $\cos\theta^*(\max) = -M/m$. Also show that the maximum recoil angle $\phi(\max)$ is given by $\sin\phi(\max) = \sqrt{(m-M)/2m}$. Calculate the angle $\theta(\max) + \phi(\max)$ for elastic collisions between the incident deuterons and the target protons.
[Ans. $60°$]

1.8 A billiard ball moving at a speed of 2.5 m/s makes a glancing collision with another identical ball initially at rest. After the collision, one ball is observed to move with a speed 2 m/s at an angle $37°$ with the original direction of motion. Find the speed of the other ball and the angle at which it moves. What is the nature of the collision?
[Ans. 1.5 m/s, $53°$, elastic]

1.9 If a particle of mass m moving with kinetic energy K_0 makes elastic collision with a target particle of mass M initially at rest, such that the scattered particle is deflected at an angle θ in the lab system and has θ^* in the centre of mass system and has a kinetic energy K in the lab system, show that:

$$\frac{K}{K_0} = \frac{1}{(M+m)^2}\left[m\cos\theta + M\cos(\theta^* - \theta)\right]^2$$

1.10 A particle of mass m and initially of velocity u makes an elastic collision with a particle of mass M initially at rest. After the collision m is deflected through lab angle $90°$ with speed $u/\sqrt{3}$. The particle M recoils with speed v at a lab angle ϕ with the incident direction. Find (a) M/m, (b) v/u, (c) ϕ, (d) θ^*, (e) ϕ^*.
[Ans. (a) 2, (b) $1/\sqrt{3}$, (c) $30°$, (d) $120°$, (e) $60°$]

1.11 A deuteron of velocity u strikes another deuteron (twice the mass of proton) initially at rest. As a result of the collision, a proton is produced which moves off at $45°$ with respect to the direction of incidence. The other product of this rearrangement collision is triton (three times the mass of proton). Assuming that this

collision may be approximated to an elastic collision, calculate the speed and direction of triton in the lab and CM system.
[Ans. 0.48 u, 34° in the lab system and $u/2\sqrt{3}$, 111° in the CM system]

1.12 An α-particle from a radioactive source collides with a stationary proton and continues with a deflection of 13.9°. Find the direction in which the proton moves.
[Ans. 30°]

1.13 When α-particles of kinetic energy 30 MeV pass through a gas, they are found to be elastically scattered at angles up to 30° but not beyond. Explain this, and identify the gas. In what way, if any, does the limiting angle vary with energy?
[Ans. Deuterium, does not vary]

1.14 A perfectly smooth sphere of mass m, moving with velocity v collides elastically with a similar but initially stationary sphere of mass m_2 ($m_1 > m_2$) and is deflected through an angle θ_L. Describe how this collision would appear in the centre of mass frame of reference and show that the relation between θ_L and the angle of deflection θ_M, in the centre of mass frame is

$$\tan\theta_L = \frac{\sin\theta_M}{[M_1/M_2 + \cos\theta_M]}$$

Also show that θ_L cannot be greater than about 19.5° if $M_1/M_2 = 3$.

1.15 Show that the maximum velocity that can be imparted to a proton at rest by a non-relativistic alpha particles is 1.6 times the velocity of the incident alpha particle.

1.16 Show that for low energy p–p scattering $\sigma(\theta) = 4\sigma(\theta^*)$ where the differential cross-sections $\sigma(\theta)$ and $\sigma(\theta^*)$ refer to the Lab and CMS, respectively.

1.17 (a) Compute the distance of closest approach in collisions between α-particles of energy 8.9 MeV and nuclei of $^{208}_{82}$Pb.
(b) How is this distance related to the radius of lead nucleus?
(c) What is the deflection of the α-particle when the impact parameter is equal to this distance?
[Ans. (a) 26.5 fm, (b) 7.7 fm, (c) 53°]

1.18 A beam of α-particles of kinetic energy 4.5 MeV passes through a thin foil of $^9_4$Be. The number of alphas scattered between 60° and 90° and between 90° and 120° is measured. What would be the ratio of these numbers?
[Ans. 3]

1.19 If the probability of α-particles of energy 8 MeV to be scattered through an angle greater than θ on passing through a thin foil is 10^{-3} what is it for 4 MeV protons passing through the same foil?
[Ans. 10^{-3}]

1.20   What α-particle energy would be necessary in order to explore the field of force within a radius of 10^{-12} cm of the centre of nucleus of atomic number 80, assuming classical mechanics to be adequate?
[Ans. 30 MeV]

1.21   In an elastic collision with a heavy nucleus, when the impact parameter b is just equal to the collision radius $\frac{1}{2} R_0$, what is the value of the scattering angle θ^* in the CMS?
[Ans. 90°]

1.22   In the elastic scattering of deuterons of 5.9 MeV from $^{208}_{82}$Pb, the differential cross-section is observed to deviate from Rutherford's classical prediction at 52°. Use the simplest classical model to calculate the closest distance of approach d to which this angle of scattering corresponds. You are given that for an angle of scattering θ, d is given by $\frac{1}{2} d_0 (1 + \mathrm{cosec}\, \frac{1}{2}\theta)$, where d_0 is the value of d in a head-on collision.
[Ans. 32.8 fm]

1.23   20000,1 MeV α-particles are incident normally on a 0.004 mm thick copper plate. Using the small angle approximation, calculate the number of α-particles scattered in the angular range 5°–10°. Assume the copper nuclei to act as point charges and neglect nuclear forces. Density of $^{66.6}_{29}$Cu $= 8.9$ g cm^{-3}; Avagadro's number $= 6 \times 10^{23}$ (g molecule), $e = 1.6 \times 10^{-19}$ C; 1 eV $= 1.6 \times 10^{-19}$ J.
[Ans. 7894]

1.24   Given that the angle of scattering is $2\tan^{-1}(a/2b)$, where 'a' is the least possible distance of approach, and b is the impact parameter. Calculate what fraction of a beam of 1.0 MeV deuterons will be scattered through more than 90° by a foil of thickness 10^{-5} cm of a metal of density 5 g cm^{-3} atomic weight 100 and atomic number 50.
[Ans. 1.22×10^{-5}]

1.25   Show that the differential cross-section for the recoil nucleus in the lab system is given by

$$\sigma(\phi) = \left(zZe^2/2T \right)^2 \frac{1}{\cos^3 \phi}$$

1.26   An electron of energy 10 keV approaches a bare nucleus ($Z = 20$) with an impact parameter corresponding to an orbital angular momentum $\hbar$. Sketch the form of the potential energy curve for the electron trajectory and calculate the distance from the nucleus at which this has a minimum (take $\hbar = 10^{-34}$ J s, $e = 1.6 \times 10^{-19}$ C and $m = 10^{-30}$ kg).
[Ans. 0.19 Å]

1.27 A beam of protons of 5 MeV kinetic energy traverses a gold foil. One particle in 5×10^6 is scattered so as to hit a surface 0.5 cm^2 in area at a distance 10 cm from the foil and in a direction making an angle of 60° with the initial direction of the beam. What is the thickness of the foil?
[Ans. 0.0066 μm]

1.28 A narrow beam of protons with velocity $v = 6 \times 10^6$ m/s falls normally on a silver foil of thickness $t = 1.0$ μm. Find the probability of the protons to be scattered into the backward hemisphere ($\theta > 90°$).
[Ans. 0.006]

1.29 A narrow beam of alpha particles with K.E. 0.5 MeV falls normally on a golden foil whose thickness is 1.5 mg/cm^2. The beam intensity is 5×10^5 particles per sec. Find the number of alpha particles scattered by the foil during the time interval of 30 minutes into angular interval 59–61°.
[Ans. 1.6×10^6]

1.30 A narrow beam of alpha particles falls normally on a silver foil behind which a counter is set to register the scattered particles. On substitution of platinum foil of the same mass thickness for the silver foil, the number of alpha particles registered per unit time increases 1.52 times. Find the atomic number of platinum, assuming the atomic number of silver and the atomic masses of both platinum and silver to be known.
[Ans. 78]

1.31 A narrow beam of alpha particles with kinetic energy 1.0 MeV falls normally on a platinum foil which is 1.0 μm thick. The scattered particles are observed at an angle of 60° to the incident beam direction by means of a counter with a circular sensitive area 1.0 cm^2 located at a distance 10 cm from the scattering section of the foil. What fraction of scattered alpha particles enters the counter? Assume the density of platinum as 21.5 g/cm^3.
[Ans. 3.33×10^{-5}]

1.32 Singly charged particles of masses m_1 and m_2 enter a medium with the same velocity. Show that the ratio of their ranges $R_1/R_2 = m_1/m_2$.

1.33 Show that a deuteron of energy E has twice the range of a protons of energy $E/2$.

1.34 If the mean range of 8 MeV proton in a medium in 0.30 mm, calculate the mean range of 16 MeV deuterons and 32 MeV α-particles.

1.35 An alpha particle moving with velocity 2×10^9 cm/sec, loses energy 0.066 MeV/mm by ionization in air and has range 7.86 cm in air. (a) Find the rate of

loss of energy per mm in air for proton and deuteron moving with the same initial velocity as alpha particle. (b) Find the range of proton and deuteron.
[Ans. (a) 0.0165 MeV/mm for both, (b) 7.86 cm, 15.72 cm]

1.36 Estimate by the Bragg-Kleeman rule the mean range of 12 MeV deuterons in cobalt, if their mean range in air at 15 °C, 760 mm Hg is 93 cm. Assume the density of cobalt to be 8.6 g/cm^3.
[Ans. 0.0266 cm]

1.37 Show that the range of α-particles and protons of energy 1 to 10 MeV in aluminium is $1/1600$ of the range in air at 15 °C, 760 mm of Hg.

1.38 Show that the straggling of a beam of ^{4}He is smaller than that of ^{3}He of equal range.

1.39 Compute the energy loss and the approximate number of quanta of visible light ($\lambda = 4000$ to 7000 Å) as Cerenkov radiation by a 20 MeV electron in traversing 1 cm of Lucite. Assume the chemical composition of Lucite to be ($C_5H_8O_2$), and the refractive index $\mu = 1.5$.
[Ans. 660 eV/cm, 270 quanta/cm]

1.40 Show that the order of magnitude of the ratio of the rate of loss of kinetic energy by radiation for a 10 MeV deuteron and a 10 MeV electron passing through lead is 10^{-5}.

1.41 Compute the energy loss and the approximate number of quanta of visible light ($\lambda = 4000$ to 7000 Å) as Cerenkov radiation by a 20 MeV electron in traversing 1 cm of Lucite. Assume the chemical composition of Lucite to be ($C_5H_8O_2$), and the refractive index $\mu = 1.5$.

1.42 Extensive air showers in cosmic rays consist of a 'soft' component of electrons and photons, and a 'hard' component of muons. Suppose at the sea level the central core of a shower consists of a narrow vertical beam of muons of energy 100 GeV which penetrate the interior of the earth. Assuming that the ionization loss in rock is constant at 2 MeV g^{-1} cm^2, and the rock density is 3.0 g cm^{-3}, find the depth of the rock through which the muons can penetrate.
[Ans. 160 m]

References

1. E.L. Goldwasser, F.E. Mills, A.O. Hanson, Phys. Rev. **88**, 1137 (1952)
2. A.A. Kamal, G.K. Rao, Y.V. Rao, Osmania University
3. Proc. R. Soc. A **146**, 96 (1934)

Chapter 2
Passage of Radiation Through Matter

2.1 Kinds of Interaction

Electromagnetic radiation, in its passage through matter, interacts in a variety of ways. The type of interaction depends on (1) photon energy, (2) Z of the material and (3) particle or field with which the photon interacts.

There is a basic difference between the energy loss of photons and that of charged particles. The charged particles lose their energy mainly due to ionization, leaving constant range at a given energy. However, photons lose their energy by a one shot process. This leads to a truly exponential attenuation.

There are a number of processes through which photon can interact with matter. Fano has classified them systematically.

Kinds of interaction	Effects of interaction
(a) Interaction with atomic electrons	(x) Complete absorption
(b) Interaction with nucleus	(y) Elastic scattering (coherent)
(c) Interaction with the electric field	(z) Inelastic scattering (incoherent)
surrounding nuclei or electrons	surrounding nuclei or electrons
(d) Interaction with the nuclear field	

There are 12 ways in which the two columns can be combined. Thus, in principle there are 12 different processes by which photons can be absorbed or scattered. Of these, three major processes are found to be important. They are the *Compton effect* (az), *Photo electric effect* (ax) and *Pair production* (cx). The minor effects are as follows:

Rayleigh scattering (ay)
It occurs with tightly bound electrons. It is coherent and follows the $1/\lambda^4$ law. For large $h\nu$ and small z, Rayleigh scattering is negligible in comparison to the Compton scattering.
Thomson scattering by the nucleus (by)

A. Kamal, *Nuclear Physics*, Graduate Texts in Physics,
DOI 10.1007/978-3-642-38655-8_2, © Springer-Verlag Berlin Heidelberg 2014

Nuclear resonance scattering (bz)

It involves the absorption of incident photon by a nucleus and its re-emission. Unlike the atomic case the resonant absorption and scattering are difficult to achieve. The details are given in Sect. 2.5. The breakthrough came with the discovery of Mossbauer effect.

Del Bruck scattering (cy)

This is a minor effect which involves nuclear potential scattering through virtual electron pair formation in the field of the nucleus.

Photodisintegration of nuclei (bx)

Examples are the reactions of $^9\text{Be}(\gamma, n)^8\text{Be}$ and $^2\text{H}(\gamma, n)p$. The cross-sections are small in comparison to the Compton effect.

Meson production (dx)

$$\gamma + p \rightarrow n + \pi^+$$

The threshold for photomesic production is 150 MeV. The cross-sections, however, are less by two orders of magnitude compared to meson production in NN collisions, and lesser still in comparison with the pair production.

2.2 The Compton Effect

The process of photon scattering by a free electron with reduced frequency is known as Compton scattering, named after Compton [2]. The effect can be explained by the quantum theory.

Here, the incident photon of energy $h\nu_0$ and momentum $h\nu_0/c$ collides elastically with an electron assumed to be stationary. As a result of the collision, the photon is scattered at some angle θ with reduced energy $h\nu$ and momentum $h\nu/c$ and the electron recoils at an angle ϕ with kinetic energy T and momentum p. The incident photon, the scattered photon and the recoil electron are coplanar in order to conserve momentum.

2.2.1 Shift in Wavelength

Energy conservation gives

$$T = h\nu_0 - h\nu \tag{2.1}$$

Taking x-axis as the incident direction, momentum conservation along x- and y-axis gives (Fig. 2.1)

$$\frac{h\nu_0}{c} = \frac{h\nu}{c}\cos\theta + p_e\cos\phi \tag{2.2}$$

$$0 = \frac{h\nu}{c}\sin\theta - p_e\sin\phi \tag{2.3}$$

Fig. 2.1 The Compton effect

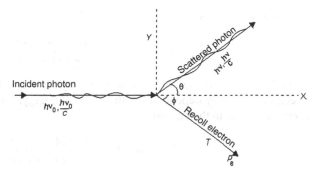

Also, momentum and kinetic energy of an electron are relativistically related

$$c^2 p_e^2 = T^2 + 2Tmc^2 \tag{2.4}$$

where $mc^2 = 511$ keV is the rest mass energy of electron. The angle ϕ can be eliminated between (2.2) and (2.3). Using (2.1) and (2.4) in the resulting equation after some simplification we obtain the change in wavelength

$$\Delta\lambda = \lambda - \lambda_0 = \frac{h}{mc}(1 - \cos\theta) = \frac{2h}{mc}\sin^2\frac{\theta}{2} \tag{2.5}$$

Formula (2.5) shows that the shift in wavelength in Compton scattering in a given direction is independent of the incident energy
   The quantity

$$\lambda_c = \frac{h}{mc} = 2.43 \times 10^{-12} \text{ m} \tag{2.6}$$

is known as the Compton wavelength. Maximum shift in wavelength occurs for $\theta = 180°$, that is, when the photon is scattered completely in the backward direction

$$(\Delta\lambda)_{\text{max}} = \frac{2h}{mc} = 2\lambda_c \tag{2.7}$$

2.2.2 Shift in Frequency

Using the relations

$$\lambda_0 = \frac{c}{\nu_0} \quad \text{and} \quad \lambda = \frac{c}{\nu}$$

In (2.5) the frequency of the scattered photon with $\alpha = h\nu_0/mc^2$ can be calculated. The parameter α measures the photon energy in terms of electron's rest mass energy.

$$\nu = \frac{\nu_0}{1 + \alpha(1 - \cos\theta)} \tag{2.8}$$

and the change in frequency

$$\Delta \nu = \nu_0 - \nu = \frac{\nu_0 \alpha (1 - \cos \theta)}{1 + \alpha (1 - \cos \theta)} \tag{2.9}$$

Formula (2.9) shows that the shift in frequency (or energy) of the photon in a given direction is strongly dependent on the incident energy. From (2.9) we find that the energy imparted to the electron is

$$K = h\nu_0 - h\nu = \frac{\alpha h \nu_0 (1 - \cos \theta)}{1 + \alpha (1 - \cos \theta)} \tag{2.10}$$

Maximum fractional loss of energy is obtained for $\theta = 180°$ in (2.10)

$$\frac{\Delta E_{max}}{E_0} = \frac{2\alpha}{1 + 2\alpha} \tag{2.11}$$

When the incident energy is small ($\alpha \ll 1$), the photon energy is not appreciably changed in the scattering process whereas for high energies ($\alpha \gg 1$), the photon can lose nearly all its energy to the electron. In the high energy limit, the energy of the photon which is scattered completely backward ($\theta = 180°$), approaches a value of $(1/2)mc^2 = 0.25$ MeV. In this case, the electron recoils in the extreme forward direction ($\phi = 0°$), and receives the maximum kinetic energy which is given by

$$K_{max} = \frac{2\alpha h \nu_0}{1 + 2\alpha} \tag{2.12}$$

When photons of fixed energy are used, the electrons have a continuous energy spectrum ranging from zero to the maximum given by (2.12).

2.2.3 Angular Relation

The scattering angle θ and the recoil angle ϕ are related to each other. Eliminating p_e in (2.2) and (2.3) using (2.8), we get,

$$\cot \phi = (1 + \alpha) \tan \frac{\theta}{2} \tag{2.13}$$

A small θ implies large ϕ and vice versa. The photon can be scattered at all angles ($0 < \theta < 180°$). But the electron can recoil only in the forward hemisphere ($0° < \phi < 90°$).

2.2.4 Differential Cross-Section

Applying Dirac's relativistic theory of the electron, Klein and Nishina [9] obtained a formula for the differential cross-section for Compton scattering. The differential

Fig. 2.2 The Compton-
scattering cross-section for
various incident energies. The
polar plot shows the intensity
of the scattered radiation as a
function of the scattering
angle θ [5]

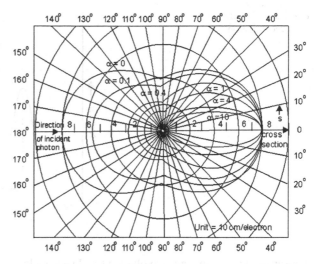

scattering cross-section per electron for unpolarized radiation is given by

$$\frac{d\sigma}{d\Omega} = \frac{r_0^2}{2}\left(\frac{\nu}{\nu_0}\right)^2\left(\frac{\nu_0}{\nu} + \frac{\nu}{\nu_0} - \sin^2\theta\right) \tag{2.14}$$

where ν is given by (2.8), $r_0 = e^2/4\pi\varepsilon_0 mc^2$, is the classical electron radius and $d\Omega = 2\pi\sin\theta d\theta$ is the solid angle in which photons of energy $h\nu_0$ are scattered. Figure 2.2 is a polar plot of (2.14). Observe the tremendous increase m the fraction of backward scattered photons as α increases.

2.2.5 Spectrum of Scattered Radiation

Figure 2.3 shows the spectrum of radiation scattered at 90° from a carbon target, when irradiated by monochromatic X-rays of wavelength $\lambda_0 = 0.707$ A. The spectrum of scattered X-rays shows the unmodified line p and the modified or shifted line s, as observed by Compton.

The presence of the unmodified line in Fig. 2.3 can be explained as follows: in deriving formula (2.5) for the wavelength shift, it was assumed that the incident photon collides with a free electron. However, it is possible that at energy not much greater than the binding energy of the electron, the electron may appear bound in some of the scattering events. In such cases, the target mass m will not be that of the electron but that of the atom as a whole. If we substitute the mass of the atom which is much larger than that of the electron as m in (2.5), we find $\Delta\lambda$ to be negligibly small. This is analogous to the scattering of a gas molecule against a rigid wall in which there is no loss of energy.

At very low frequencies, the scattered photon has the same frequency as that of the incident radiation. This corresponds to scattering from bound electron, known as

Fig. 2.3 Spectrum of
scattered radiation

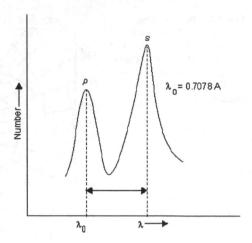

Thompson scattering, and is explained by classical electrodynamics. At frequencies corresponding to energies higher than the binding energy of the electron, Compton scattering starts showing up (scattering from free electron), but scattered radiation of the same wavelength as that of the incident radiation is also present. At higher frequencies ($h\nu_0 \gg B$), Compton scattering dominates and is readily observed. We can thus understand why Compton scattering does not take place with visible light from ordinary materials. It may be pointed out that the Compton shift in wavelength does not depend on the Z of the material.

In deriving various formulae, we have assumed that initially the electron is at rest. However, it will be in motion. Depending on the component of its velocity in the incident direction of the photon, the wavelength shift of the scattered photon will vary. This will, therefore, result in the broadening of the maximum as in Fig. 2.3. This effect will be more important at smaller frequencies. As ν_0 increases further, Compton line tends to become narrower, specially at large angles. Eventually, in any direction of scattering other than that of the incident direction, the unmodified line becomes weaker and the modified line becomes fairly sharp as one would get for scattering off a free electron.

Total Cross-Section   The total collision cross-section is obtained by integrating (2.14) over all permissible values of θ. Tables and graphs of the differential and average cross-sections for Compton collisions have been given by Davisson and Evans [4].

2.2.6 Compton Attenuation Coefficients

If we have a thin absorbing foil with N atoms/cm^3, each with Z electrons/atom then

$$\mu = NZ\sigma_c \tag{2.15}$$

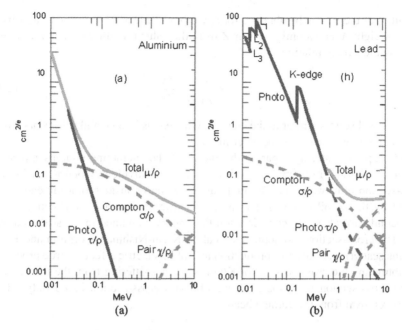

Fig. 2.4 (**a**) Photon mass attenuation coefficients equal to the linear attenuation. (**b**) Coefficients divided by the density (to suppress effects simply due to the number of electrons in the material for the three processes in Al and Pb)

The quantity μ, which has the unit of cm^{-1} is called the linear scattering coefficient. Observe that μ is identical to the macroscopic cross-section Σ. If μ be divided by the density of the absorber, then μ/ρ is called mass-attenuation coefficient (μ_m).

In their passage through a medium, the law of attenuation is

$$n = n_0 e^{-\mu x} \qquad (2.16)$$

where n_0 is the initial number of photons and n is the number of surviving photons after traversing a distance x. If x is expressed in cm, then the linear attenuation coefficient μ is in cm^{-1}. If the distance x is in g/cm^2, then we are concerned with μ_m, the mass attenuation coefficient.

The electronic cross-section will be independent of Z as it is assumed that $h\nu_0$ is much in excess of the binding energy of the electrons. The Compton scattering cross-section per electron as calculated by Klein and Nishina is given as

$$\sigma_c = \pi r_0^2 \left\{ \left[1 - \frac{2(\alpha+1)}{\alpha^2} \right] \ln(2\alpha+1) + \frac{1}{2} + \frac{4}{\alpha} - \frac{1}{2(2\alpha+1)^2} \right\} \qquad (2.17)$$

As before, r_0 is electron radius and $\alpha = h\nu_0/mc^2$.

The total cross-section for an atom is found by multiplying σ_c by Z. The contribution to the total γ-ray absorption coefficient from the Compton effect, as calculated from (2.17) is given in Fig. 2.4. It is seen that σ_c decreases monotonically as $h\nu_0$ increases.

Values of the σ may be obtained for any other elementary material, of density ρ, atomic weight A, and atomic number Z from the value for either lead or aluminium, by using the simple relation

$$\sigma_2 = \sigma_1 \frac{\rho_2}{\rho_1} \frac{A_1}{A_2} \frac{Z_2}{Z_1} \tag{2.18}$$

where the subscripts 1 refer to the element whose σ is known and subscripts 2 refer to the element whose σ is to be determined.

In Compton scattering, a part of the energy of the interacting photons appears as scattered radiation. In many cases it is convenient to divide the total Klein-Nishina cross-section into two parts. The first part takes into consideration the energy absorbed by the recoil electrons and is thus a true absorption cross-section, while the second part takes into consideration the energy contained in the scattered photons. The cross-sections so defined are called Klein-Nishina absorption and Klein-Nishina scattering. Since pair production and photoelectric effect are true processes, by subtracting the cross-section for Klein-Nishina scattering from the total macroscopic cross-section for a material, we obtain a cross-section based only on the energy removal from the gamma beam.

Energy Distribution of Compton Electrons and Photons In certain situations, the energy spectrum of Compton electrons is important. It can be shown that

$$\frac{d\sigma}{dT} = \frac{d\sigma}{d\Omega} \frac{2\pi}{\alpha^2 mc^2} \left[\frac{(1+\alpha)^2 - \alpha^2 \cos^2 \phi}{(1+\alpha)^2 - \alpha(2+\alpha)\cos^2 \phi} \right]^2 \tag{2.19}$$

where $d\sigma/d\Omega$ is given by Eq. (2.14), and $\alpha = h\nu_0/mc^2$.

Figure 2.5 shows the number-energy spectrum of Compton electrons by incident photons of energy $h\nu_0 = 0.51$, 1.2, and 2.76 MeV. The number-energy spectrum of scattered photons can be deduced from Fig. 2.5 because $h\nu = h\nu_0 - T$.

2.3 Photoelectric Effect

At energies below 0.1 MeV, the predominant mode of photon interaction in medium and high-Z elements is known as the photoelectric effect. It can be shown that the photoelectric effect cannot take place with a free electron (see Example 2.10) because energy and momentum cannot be conserved simultaneously. However, total absorption of photon can take place if the electron is bound to an atom or a metal. In that case the momentum can be balanced by the residual atom. The more tightly the electron is bound, the larger is the absorption cross-section. About 80 % of the photoelectric absorption processes occur in the K-shell, provided the incident photon energy $h\nu$ exceeds the K-shell binding energy. In this process a photon interacts with an atom in such a way that its total energy is absorbed and concentrated on

Fig. 2.5 Energy spectrum of electrons recoiling after Compton scattering for various energies of the incident photon. The sharp maximum electron recoil energy is known as the Compton edge [10]

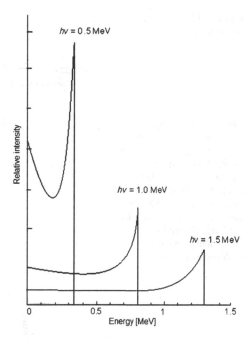

one electron which is thereby expelled. The kinetic energy of the emitted electron equals the photon energy minus the binding energy

$$T = h\nu - B_e \qquad (2.20)$$

where B_e is the binding energy of the ejected electron. The remaining energy appears as X-rays and Auger electrons in the process of filling of the vacancy in the inner shell. The photoelectric interaction is illustrated in Fig. 2.6.

The theoretical treatment of photoelectric effect is complicated by the fact that Dirac's relativistic equation must be applied to a bound electron which does not yield exact solutions.

Whenever the energy of the photon is high enough, the probability of expelling a K-electron is higher than it is for any of the other electrons. At a photon energy equal to the K-electron binding energy, there is a sharp step in the cross-section for photoelectric emission. Similar jumps occur for L- and M-shells, Fig. 2.7.

For example, the binding energy of a K-shell electron in Pb is 88 keV. Incident photons of energy less than 88 keV cannot liberate K-shell photo-electron although they can liberate higher shell electrons that are less tightly bound. When the photon energy increases just above 88 keV, the probability for photo-electron emission suddenly increases. This process is known as absorption edge or K-edge. For energies above the K-absorption edge [6], gives the following cross-section formula for the K-electron emission

$$\sigma_{ph(K)} = \frac{32\pi\sqrt{2}Z^5}{3(137)^4} \left(\frac{mc^2}{h\nu}\right)^{7/2} \qquad (2.21)$$

Fig. 2.6 Photoelectric
interaction

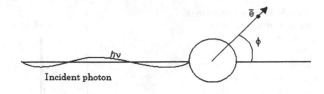

Fig. 2.7 Photoelectric
cross-section in Pb. The
discrete jumps correspond to
the binding energies of
various electron shells; the
K-electron binding energy, in
Pb for example, is 88 keV. To
convert the cross-section to
the linear absorption
coefficient μ in cm^{-1},
multiply by 0.033

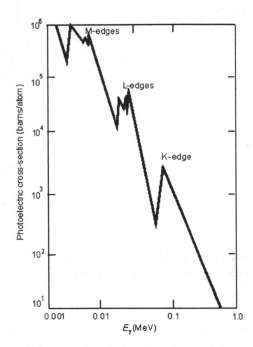

The above formula applies for photon energy much smaller than the rest energy
of an electron ($h\nu \ll mc^2$). Because of Z^5 dependence, photoelectric absorption
becomes significant for heavy elements like tungsten or lead. For fixed values of
$h\nu$, the cross-section is empirically given by

$$\sigma_{ph} \simeq \text{const} \cdot Z^n \qquad (2.22)$$

The exponent n is found to increase from about 4.0 to 4.6 as $h\nu$ increases from 0.1
to 3 MeV, as in Fig. 2.8.

Formula (2.21) shows a dependence of $(h\nu)^{-7/2}$ for low energy photons. However, at high energies, the dependence on $h\nu$ is not so drastic. With the increasing
energy, the exponent of $7/2$ in formula (2.21) decreases until it reaches unity at very
high energies.

The linear attenuation coefficient μ_{ph} for the photoelectric effect is given by

$$\mu_{ph} = \sigma_{ph}N \qquad (2.23)$$

Fig. 2.8 Approximate
variation of the photoelectric
cross-section σ_{ph} (cm^2/atom)
with Z^n for various values
of n (Rasmussen)

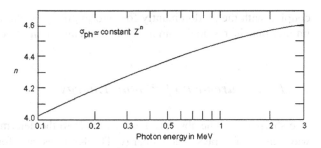

Fig. 2.9 Directional
distribution of photoelectrons
per unit solid angle for
energies as marked. The
curves are not normalized
with respect to each other.
Solid curves are calculated
from Sauter's relativistic
formula; *dashed curve* from
Fischer's non-relativistic
formula (Davisson and
Evans)

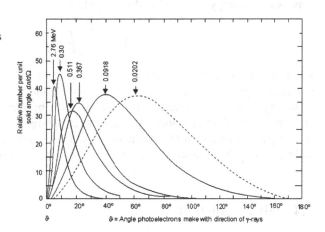

where N is the number of atoms per cm^3 and σ_{ph} is the atomic cross-section in cm^2 per atom. The mass absorption is obtained by dividing μ by density ρ.

Assuming σ_{ph} for one element, the corresponding value for the other element can be found out by the relation

$$\sigma_{ph}(2) = \sigma_{ph}(1)\frac{\rho_2}{\rho_1}\frac{A_1}{A_2}\left(\frac{Z_2}{Z_1}\right)^n \qquad (2.24)$$

where ρ is density, A is atomic weight, Z is atomic number and n is obtained from Fig. 2.8.

Angular Distribution of Photoelectrons   At low photon energies, the photoelectrons tend to be emitted in the direction of the electric vector of the incident radiation, and hence are at right angles to the direction of incidence. As the energy is increased, the angular distribution is pushed forward. Figure 2.9 shows theoretical curves for the number/unit solid angle at various photon energies.

Forward Momentum of Recoil Atom   For $h\nu$ much in excess of binding energy of the electron, the photo-electron will have nearly the same energy as the incident photon. However, because of the finite mass of the electron, its momentum will be much greater than the momentum of the incident photon. This increased momentum coupled with the predominantly forward angular distribution of the photoelectrons implies that the residual atom must recoil on an average in the backward hemisphere.

2.3.1 Measurement of Photon Energy

The sharpness of the absorption edges lends to the determination of the estimation wavelength of γ-rays of low energy. The K-edges vary for different elements. The approximate energy is given by Moseley's law

$$E = 13.6\frac{(Z - \sigma)^2}{n^2} \text{ eV} \tag{2.25}$$

where $n = 1$ for the K-series and $n = 2$ for the L-series etc. The screening constant σ has an approximate value of 3 for the K-shell and 5 for the L-shell. The X-ray wavelengths can be bracketed in small intervals by measuring the absorption in elements with adjacent values of Z and determining between which two K-edges or L-edges, the unknown photon energy lies. This is accomplished by observing the sudden change in absorption with change of K-edge location.

2.4 Pair-Production

At incident photon energies greater than $2mc^2$ (1.02 MeV), the pair-production process becomes increasingly important. In this process the proton is completely absorbed and a positron-electron is produced, the total energy being equal to $h\nu$.

$$h\nu = \left(T_- + mc^2\right) + \left(T_+ + mc^2\right) \tag{2.26}$$

where T_- and T_+ are the kinetic energy of the electron and the positron, respectively, and $mc^2 = 0.511$ MeV is the rest energy of the electron or the positron. The process can occur only in the field of a nucleus and to some degree in the field of an electron. The presence of a particle is necessary for the conservation of momentum. The process is schematically depicted in Fig. 2.10.

Figure 2.11 gives the energy level diagram of the electron as derived from Dirac's relativistic theory. Dirac's equations also give negative energies for the electron, and he further assumed that all these negative energy states are filled and so are not observable. They are not observable because this would imply a possibility of changing their state of motion. This is not possible because of Pauli's principle, the sea of negative states is already filled. The only possibility for the change of state is to cross the barrier of width $2mc^2$ so that the electron enters the domain of empty positive energy states and becomes a free particle with positive energy. The hole produced in the negative state left by the electron is observable as a positron. In the pair-production process, the energy needed to cross the barrier is supplied by the photon as shown in Fig. 2.11.

Fig. 2.10 Schematic
depiction of conservation
of momentum process

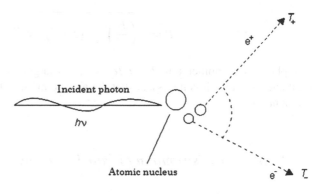

Fig. 2.11 Energy level
diagram of electron from
Dirac's theory

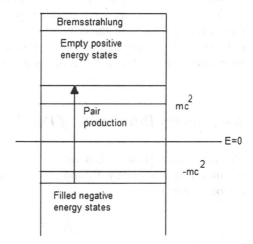

The same figure also indicates the Bremsstrahlung process in which one electron falls down to a lower energy by the emission of a photon.

The pair-production and Bremsstrahlung are intimately connected and are similar, except that the former is concerned with absorption process while the latter is concerned with the emission process. Mathematically, the theories of the two processes are nearly identical. There are similarities in various formulae as well. For example, the cross-sections for both these processes are of the order of $(Z^2/137)(e^2/mc^2)^2$. Compare (2.27) with formula (1.142) for Bremsstrahlung.

Pair-production takes place in the field of a nucleus. No change of state of the nucleus or its atomic electrons is involved, except that the nucleus absorbs some of the momentum of the photon. Pair-production cannot take place in free space because energy and momentum cannot be conserved simultaneously. Suppose, the process takes place in vacuum. The maximum total momentum of the produced pair will occur when the particles move in the same direction with the same velocity. For energy E, the momentum is given by

$$P = \sqrt{\left(\frac{E}{c}\right)^2 - (2mc^2)^2} < \frac{E}{c}$$

The photon has momentum $P = E/c$, which is larger than the pair with the same total energy. Thus, it is necessary that the atomic nucleus be present to balance the momentum.

2.4.1 Angular Distribution of Pair Electrons

At high energies the angular distribution of positron and electron is mainly forward. For $T \gg mc^2$, the mean angle between the electron or positron with the incident direction of photon is of the order of mc^2/T. At incident photon energies of the order of $2mc^2$, the angular distribution is much more complicated and the tendency for projection in the forward direction is less obvious.

2.4.2 Energy Distribution of Pair Electrons

The differential cross-section $d\sigma/dT_+$ cm^2 per nucleus, for the creation of a positron of kinetic energy T_+ and an electron of kinetic energy $h\nu - 2mc^2 - T_+$, can be written as

$$\frac{d\sigma}{dT_+} = \frac{\sigma_0 Z^2 P}{h\nu - 2mc^2} \qquad (2.27)$$

where

$$\sigma_0 = \frac{1}{137}\left(\frac{e^2}{mc^2}\right)^2 = 5.8 \times 10^{-28} \text{ cm}^2/\text{nucleus} \qquad (2.28)$$

and the dimensionless quantity P is a complicated function of $h\nu$ and Z, which varies between 0 for $h\nu < 2mc^2$ and about 20 for $h\nu = \infty$ for all values of Z. Figure 2.12 shows the variation of P with the fraction of the total kinetic energy of both the pair electrons that are carried by the positron $T_+/(h\nu - 2mc^2)$.

2.4.3 Total Pair-Production Cross-Section per Nucleus

The total nuclear pair-production is calculated by integrating the differential cross-section, Eq. (2.27), over all possible energies.

$$\sigma_P = \int d\sigma = \sigma_0 Z^2 \int_0^{h\nu - 2mc^2} \frac{P \, dT_+}{h\nu - 2mc^2} \qquad (2.29)$$

Fig. 2.12 Differential pair-production cross-section, expressed as the dimensionless function P of Eq. (2.27). The curves calculated from the equations of Bethe and Heitler, including screening corrections for photon energies above $10m_0c^2$ (Davisson and Evans)

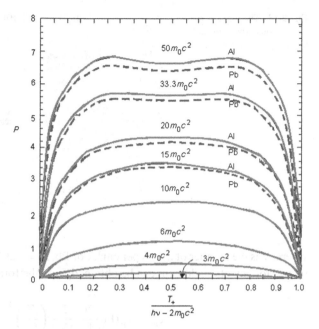

$$\sigma_P = \sigma_0 Z^2 \overline{P} \tag{2.30}$$

where $\overline{P}$ is the mean value of P in Fig. 2.12. Analytical integration of (2.29) is possible only for relativistic cases. When screening of atomic electrons is neglected, that is, the interaction is considered in the field of a bare nucleus, Bethe and Heitler give the formula

$$\sigma_p = \sigma_0 Z^2 \left(\frac{28}{9} \ln \frac{2h\nu}{mc^2} - \frac{218}{27} \right) \tag{2.31}$$

for $mc^2 \ll h\nu \ll 137mc^2 Z^{-1/3}$ (that is about 16 MeV for Pb). It is observed that σ_{pair} increases approximately logarithmically with $h\nu$. At high energies ($h\nu \geq$ 20 MeV) an appreciable contribution to pair-production may come from points outside the K-shell of Pb. For complete screening

$$\sigma_p = \sigma_0 Z^2 \left[\frac{28}{9} \ln \left(183 Z^{-1/3} \right) - \frac{2}{27} \right] \tag{2.32}$$

for $h\nu \gg 137mc^2 Z^{-1/3}$. Formula (2.31) shows that at high energies, say 10 GeV, the pair-production cross-section depends only on the screen and is independent of γ-ray energy.

Pair-Production Linear Attenuation Coefficient The linear attenuation μ_P for pair-production is simply given by

$$\mu_P = \sigma_P N \ \text{cm}^{-1} \tag{2.33}$$

Fig. 2.13 The three γ-ray
interactions processes and
three regions of dominance

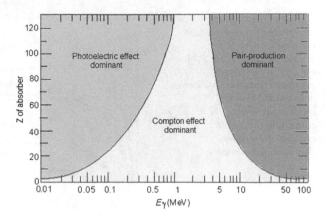

where N is the number of atoms per cm^3. For Pb, $N = 3.3 \times 10^{22}$ atoms/cm^3. Values
of μ_p, for any other element can be obtained from the formula

$$\mu_p = \mu_p(\text{Pb}) \frac{207.2}{11.35} \frac{\rho}{A} \left(\frac{Z}{82} \right)^2 \qquad (2.34)$$

where ρ is the density (11.35 g/cm^3 for Pb), A is the atomic weight (207.2 for
Pb), and Z is the atomic number (82 for Pb). Note that pair-production attenuation is most important in heavy elements and at high photon energies. For example, in Pb the pair-production attenuation exceeds that for Compton scattering at
$h\nu > 4.75$ MeV (Fig. 2.4(b)). Also, pair-production is only one of the three major
processes whose cross-section increases with increasing energy. Due to this fact the
total attenuation coefficient, $\mu = \mu_{ph} + \mu_c + \mu_p$, in heavy elements goes through
a minimum. This gives rise to double-valued energy for the same μ, Fig. 2.4(b). In
light elements such as aluminium, the solution will be single-valued (Fig. 2.4(a)).
This is because the rise in σ_p is more than offset by the decrease in σ_c. In the case
of copper, these two effects cancel over a wide range of photon energies so that one
obtains a fairly flat curve with $\mu = 0.28$ cm^{-1} = const for $h\nu \geq 6$ MeV.

Figure 2.13 shows the relative importance of the three major processes for γ-ray
interaction. The lines indicate the values of Z and $h\nu$ for which the two neighbouring effects are just equal.

Note that pair-production can also take place in the field of an electron, although
to a lesser degree. When the recoil is absorbed by an electron, the threshold required
by the conservation of energy and momentum in the laboratory system is $4mc^2$ (see
Example 2.12).

In this case two electrons and a positron acquire appreciable momentum leaving
three tracks in the forward direction. Such events in which the triplets, also known
as tridents, are formed and are recorded in photographic emulsions, cloud chambers
or bubble chambers.

2.5 Nuclear Resonance Fluorescence

In atomic physics resonance absorption occurs while using the source and the medium of the same material. A familiar example is the sodium absorption spectrum. However, in nuclei, the absorption does not occur in such a straight forward fashion. The centre of the emitted γ-radiation, which is a narrow band, does not coincide with the centre of absorption line because of the recoil energy losses in emission and absorption processes. The incident radiation is thus off resonance and excitation is only possible if the natural width of the level is large compared with the recoil energy loss. For most atomic transitions, this last condition is fulfilled. However, in the case of nuclei, the recoil energy losses effectively prevent the observation of resonance fluorescence when the same isotope is used as the source of radiation and the scattering as the absorbing material. All the attempts to observe this effect had failed for the inability to find wide γ-ray lines, until in (1951) Moon made a successful effort by creating special source conditions.

Resonant scattering of light is a well-known phenomenon which is observed when the energy of the incident photon coincides with the difference between the excitation level and the ground state level of the scattering atom. The incident photon can originate either from an atom of the same kind as the scattering atom or from a continuous spectrum in which case resonant scattering gives rise to absorption lines. Consider the general case of a γ-ray that is scattered by a nucleus of the same kind as that from which it is emitted. Conservation of energy and momentum demand that

$$h\nu + E_R = E_0 \tag{2.35}$$

$$\sqrt{2ME_R} = \frac{h\nu}{c} \tag{2.36}$$

where $h\nu$ and E_R are the energy of the emitted photon and the recoil nucleus, respectively, E_0 is the transition energy and M is the mass of the nucleus. Combining (2.35) and (2.36) we have

$$E_R = \frac{(h\nu)^2}{2Mc^2} = \frac{E_r^2}{2Mc^2} \tag{2.37}$$

where $E_r = h\nu$

$$E_r = E_0 - \frac{E_r^2}{2Mc^2} \tag{2.38}$$

Since the recoil loss occurs both at the emission and absorption processes, total energy displacement amounts to

$$\Delta E = \frac{E_r^2}{Mc^2} \tag{2.39}$$

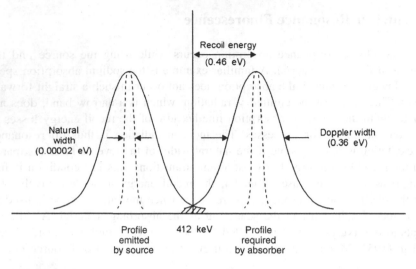

Fig. 2.14 Shift of gamma ray energy in downward and upward direction during emission and absorption processes

If this energy difference is large compared to the width of the level, the system is off the resonance and the cross-section for resonance scattering becomes extremely small. The width of a level Γ can be calculated from the lifetime τ by means of the uncertainty principle.

$$\Gamma\tau \simeq \hbar \tag{2.40}$$

Consider for example the $2.7D$ decay of ^{198}Au, following the β decay to ^{198}Hg from which single γ-ray of energy 412 keV is emitted. If the γ-ray is allowed to fall on the atom of ^{198}Hg in the ground state, there is a possibility for its absorption and excitation to the 412 excited state. Note that the nuclear recoil affects both the emission and the absorption processes. Hence the total energy shift is $2E_R$. The 412 keV excited state of ^{198}Hg has a mean lifetime of 32 ps corresponding to a width of 2×10^{-5} eV.

The recoil energy E_R is $\frac{E_{\gamma}^2}{2Mc^2} = 0.46$ eV. The width of the Doppler broadening due to thermal motion is given by

$$\Delta = 2\sqrt{\ln 2 E_r}\sqrt{\frac{2kT}{Mc^2}} \tag{2.41}$$

where T is Kelvin temperature and k is Boltzmann constant.

Figure 2.14 shows the shift of 4.12 keV γ-ray energy from ^{198}Hg by 0.46 eV downward in the emission process and by 0.46 eV upward in the absorption process. Because of thermal broadening due to Doppler effect (0.36 eV) there is a small overlap (shaded region) between the emission and absorption lines, hence a small probability of resonant excitation.

2.5.1 Restoring Mechanisms

There are various techniques for overcoming the energy difference $2E_R$ between the source and absorber transitions. All of them depend on Doppler broadening. Doppler's formula is

$$\nu = \nu_0 \left(1 + \frac{v}{c} \cos \theta^* \right) \qquad (2.42)$$

Put $\theta^* = 0°$ for forward emission. Then

$$\Delta \nu = \nu - \nu_c = \nu_0 \frac{v}{c} \quad \text{or}$$

$$h \Delta \nu = \Delta E = h \nu_0 \frac{v}{c} = E_0 \frac{v}{c}$$

Compensation requires that

$$E_0 \frac{v}{c} = 2E_R \simeq \frac{E_0^2}{Mc^2}$$

$$v = \frac{E_0}{Mc} \qquad (2.43)$$

2.5.2 Mechanical Motion (Ultra Centrifuge)

Mechanical motion was first proposed by Moon in 1950 as a mechanism for restoring resonance. The energy shift is compensated by the Doppler effect if the emitting nucleus moves toward the scatterer with velocity $v = E_r/Mc$. The experiment was done by attaching the source to the tip of a rotor in a centrifuge spinning at 500–3000 revolutions per second. The apparatus used by Moon (1950) is schematically shown in Fig. 2.15 and the results for the absorption cross-section as a function of source velocity as well as the rotor's speed are indicated in Fig. 2.16. The emission and absorption lines overlap for a source speed of about 670 m/s.

Moon derived an expression for the effective cross-section of resonant scattering taking into account the mechanical velocity of the γ-ray source and the thermal velocities in the source and the scatterer. The curve is fitted for a quadrupole transition between the ground state of spin 1 and an excited level of spin 2 having a width of 2×10^{-5} eV. The intrinsic width corresponds to a half lifetime of 2.2×10^{11} s. The measurement of the σ for resonant scattering is thus a direct determination of the level width Γ and an indirect way of finding the mean lifetime τ which is related to Γ by the uncertainty principle. This method of determining mean lives provides an important technique which is useful for measuring delays longer than 10^{-10} s.

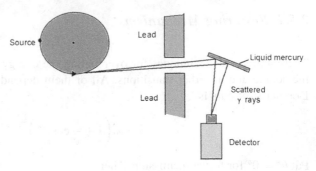

Fig. 2.15 Schematic drawing of the apparatus used by Moon

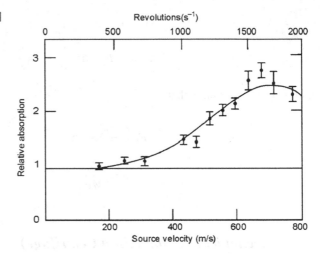

Fig. 2.16 Rotor data from [3]

2.5.3 Thermal Motion

If no mechanical motion is applied, an increased temperature of the source or of the scatterer or of both will result in a broadened energy distribution which increases the probability for two nuclei having the appropriate relative velocities for compensating the recoil.

2.5.4 Preceding β or γ Emission

Kuhn had pointed out that the resonance condition might be restored by a preceding β or γ emission. A nucleus emitting a β particle obtains a recoil velocity in the opposite direction and if the subsequent emission of the γ-ray takes place in such a short time that the nucleus has not lost its velocity, the energy may be compensated by Doppler effect. The resonance condition is that the emitting nucleus approaches the scatterer with velocity $v = E_0/Mc$.

By measuring coincidences between the β-particles and the scattered γ-rays one might expect to find an increase in counting rate when the beta particles are measured at such an angle that the recoiling nucleus has the appropriate velocity component toward the scatterer. Similar arguments apply to $\gamma-\gamma$ coincidence.

This technique has been used by Burgov and others to observe the nuclear resonance fluorescence in ^{24}Mg and to measure the lifetime of 1.38 MeV γ-ray emitted in the decay of ^{24}Na. The 1.38 MeV γ-ray scattered all around from the magnesium bar was detected by the NaI crystal in prompt coincidence with 2.76 MeV γ-ray. At 120° the number of 1.38 MeV γ-rays scattered into NaI crystal in coincidence with 2.76 MeV ray was 2 or 3 times larger than at other angles. For a photon of energy E, the cross-section for resonance fluorescence, that is for the case where the direct γ-transitions to the ground state is the only mode of de-excitation, is given for an isolated level by Bethe and Placzek [1]

$$\sigma_0(E) = \pi \lambda^2 \frac{(2J_1 + 1)}{2(2J_0 + 1)} \frac{\pi^2}{[(E - E_r)^2 + \frac{\Gamma^2}{4}]} \tag{2.44}$$

where J_1 and J_0 are the total angular momenta of the excited state, and the ground state, respectively. E_r is the resonance energy, λ the corresponding wavelength divided by 2π and Γ the natural width of the level. The factor 2 arises due to the two independent polarizations of photons. Taking into account the spins of the ground and the first excited states and the thermal Doppler width, the level width was calculated to be 7×10^{-4} eV corresponding a lifetime of 0.95×10^{-12} s.

2.6 Mossbauer Effect

In 1956 and 1957, R.L. Mossbauer was studying the scattering of γ-rays for his graduation work. He used the metals of ^{191}O and ^{191}Ir as the source and absorber, respectively, both cooled to a low temperature. Mossbauer measured the transmission of ^{191}Ir 129 keV γ-rays through a crystalline natural iridium absorber. His source was ^{191}Os which beta decays with a half life of 16 days to an isomeric state of ^{191}Ir ($T_{1/2} = 5.6$ s). A 42 keV γ-ray is emitted and the iridium nucleus is left in its first excited state (129 keV) which has a lifetime of 1.4×10^{-10} s (Fig. 2.17).

The apparatus used by Mossbauer is shown schematically in Fig. 2.18. Mossbauer measured the transmission of ^{191}Ir keV γ-rays through a natural iridium absorber. He kept both the source and the absorber at 88 K but had the source mounted on a turn table so that the relative velocity of the source and the absorber could be controlled. Figure 2.19 shows the variation of transmission with the turn table speed. The most startling fact which Mossbauer obtained was the turn table effect.

The absorption peak is centred at zero relative speed of the source with respect to the absorber. The half width is about 1 cm (Fig. 2.19). When the abscissa is converted to energy units corresponding to the Doppler shift in the γ-ray energy $\Delta E = v E / c$, the points can be fitted within statistical errors by a Breit-Wigner curve

Fig. 2.17 Decay scheme of
^{191}Os

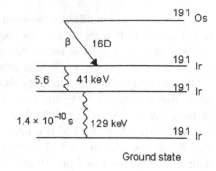

Fig. 2.18 Experimental
arrangement: *A*, cryostat of
absorber; *S*, rotating cryostat
with source; *D*, scintillation
detector; *M*, region in which
the source is seen from
D [11]

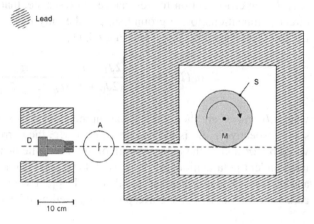

Fig. 2.19 Fluorescent
absorption in ^{191}Ir as a
function of the relative
velocity between source and
absorber. The *upper scale* on
the abscissa shows the
Doppler energy that
corresponds to the velocity on
the *lower scale*.
$T = 88$ °K [11]

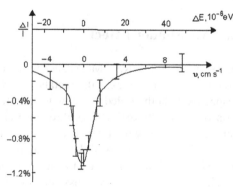

of width $(9.2 \pm 1.2) \times 10^{-6}$ eV (Chap. 7). This is interpreted to be twice the natural
width of the 129 keV level, the factor 2 arising because the observed absorption is
the result of folding an emission spectrum together with the absorption spectrum
each of which has a width Γ.

The other observation of Mossbauer is the temperature effect. Here, he had both
the source and absorber at rest with the absorber at fixed temperature of 88 °K and
the source temperature varied from 88 °K to above room temperature. He measured
the transmission of the 129 keV γ-rays through the absorber as a function of the

Fig. 2.20 Variation of
absorption cross-section with
source temperature

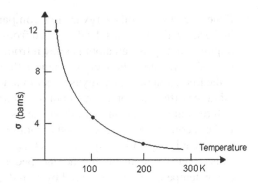

source temperature. The rise in cross-section with decreasing temperature (contrary
to the experiments described earlier under Sect. 2.5) is interpreted as being caused
by the increase in the probability of no recoil emission as the temperature is lowered,
Fig. 2.20.

Mossbauer effect was lost when powder source and absorber were used. In the
early experiments, failure to achieve nuclear resonance was due to the nuclear recoil
which had the inevitable effect of broadening the natural width of γ-ray lines that
is making the indeterminacy in the energy of narrow γ-rays much larger than that
required by the uncertainty principle. In all these methods, though resonance fluorescence is realized, the peak is fairly broad caused by thermal motion and above
all by the fact that they cannot restore the natural widths of the lines from their recoil broadened state. The line width is of great importance in deciding whether a
particular experiment can be successfully undertaken because it is useless to have
Γ larger than ΔE. One should therefore choose isomers of sufficiently long time so
that Γ may be very small ($\Delta E \gg \Gamma$).

2.6.1 Elementary Theory

If the nuclei are rigidly nailed to a crystal, one could immediately understand the
disappearance of a measurable recoil energy. For, if the whole crystal were forced to
recoil together with the emitting or absorbing nucleus the *recoil* energy for a given
momentum transfer would be decreased by a factor equal to the number of atoms
in the crystal. Since a crystal as small as a cubic micron in volume contains about
10^{10} atoms, the recoil loss would be negligible. The fact is however that atoms are
not nailed to a crystal to which they belong. Instead they are arranged in a lattice
which is vibrating. The energy of vibration, which is macroscopically described by
the temperature, is not fixed but the energy put into the system will go into lattice
vibration of the crystal. An essential feature of the picture is that the energy states of
the lattice are quantized, energy can be transferred only in discrete amounts called
phonons. The degree of Mossbauer effect will thus eventually be dependent on the
choice of crystals and the temperature since the modes of vibrations are to be characterized. ^{57}Fe is a very useful isotope as room temperature can be used. To understand
the problem of nuclear recoil in the crystal, three cases must be distinguished:

1. If the free-atom recoil energy is large compared to the binding energy of the atom in the solid, the atom will be dislodged from its lattice site. The minimum energy required to displace an atom is known from radiation damage investigations and is 15–30 eV. Under these circumstances the free-atom analysis is applicable.
2. If the free-atom recoil energy is larger than the characteristic energy of the lattice vibrations (the phonon energy) but less than the displacement energy, the atom will remain in its site and will dissipate its recoil energy by heating the lattice.
3. If the recoil energy is less than the phonon energy, a new effect arises because the lattice is a quantized system which cannot be excited in an arbitrary fashion. This effect is responsible for the unexpected increase in the scattering of γ-rays at low temperature first observed by Mossbauer.

This phenomenon is most readily understood in the case of an Einstein solid, that is one characterized by $3N$ vibrational modes (where N is the number of atoms in the solid) each having the same frequency ω. At a given instant, the solid may be characterized by the quantum numbers of its oscillators. The only possible changes in its state are an increase or decrease in one or more of the quantum numbers. These correspond to the absorption or emission of quanta of energy $\hbar\omega$ which in real solids is characteristically of the order of 10^{-2} eV. The recoil-free fraction f of the decays produce no change in the quantum state of the lattice. In the remaining $1 - f$, an energy $\hbar\omega$ is transferred. The processes with $\Delta n = -1$ and 2 may be neglected.

The emission of a γ-ray is now accompanied by the transfer of integral multiples of this phonon energy $(0, \pm\hbar\omega, \pm 2\hbar\omega + \cdots)$ to the lattice. The depth of the resonance is determined by the fraction (f) of the nuclei in the lattice that emits (or absorbs) energy with no recoil. The calculation of the recoil-free fraction (f) depends on the recoil energy that exceeds the lattice binding energy. At low energies and temperatures, the primary way in which a solid can absorb energy is through lattice vibrations called phonons. These vibrations occur at a spectrum of frequencies, from zero up to a maximum ω_{max}. In the improved theory due to Debye, the energy corresponding to the highest vibrational frequency is expressed in terms of the corresponding temperature called the Debye temperature θ_D through $\hbar\omega_{max} = k\theta_D$, where k is the Boltzmann constant. For typical materials, $\hbar\omega_{max} \sim 0.01$ eV and $\theta_D \sim 1000$ K. The recoilless fraction is

$$f = \exp\left[-k\langle x^2\rangle\right] \tag{2.45}$$

where $\langle x^2\rangle$ is the mean-square vibrational amplitude of the emitting nucleus, $k = (2\pi/\lambda)$, λ being the wavelength of the emitting nucleus. The mean square amplitude is calculated by using the Bose-Einstein distribution and the recoilless fraction is given by

$$
\begin{aligned}
f &= \exp\left\{\frac{-3}{2}\frac{\hbar^2\omega^2/2mc^2}{k\theta_D}\left[1 + 4\left(\frac{T}{\theta_D}\right)^2\int_0^{\theta_D/T}\frac{x\,dx}{e^x - 1}\right]\right\} \\
&\simeq \exp\left\{\frac{-3}{2}\frac{\frac{\hbar^2\omega^2}{2mc^2}}{k\theta_D}\left[1 + \frac{2}{3}\left(\frac{\pi T}{\theta_D}\right)^2\right]\right\} \quad (T \ll \theta_D)
\end{aligned} \tag{2.46}
$$

Fig. 2.21 Fraction of recoilless transitions in iron or rhenium as a function of the temperature [12]

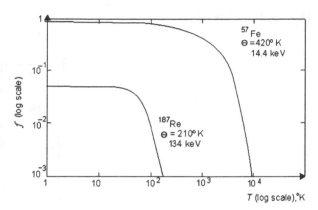

The first term is independent of the temperature and shows that even at absolute zero, the fraction of recoilless decays is large only if the recoil energy of the free nucleus is small compared with $k\theta_D$. The probability of recoilless decay decreases with increasing temperature and it is negligible for temperatures large compared with θ_D (Fig. 2.21). Similar effects occur in the absorption process.

Note that for Fe, $\theta_D \sim 400$ K, $E_r = 14.4$ keV, $E_R = 0.002$ eV and $f = 0.92$, while for Ir, $f \simeq 0.1$. It is therefore not surprising that in Mossbauer's original experiment Ir showed the effect of only 1 % while Fe shows a much larger effect. What is more is that Fe can be used at the room temperature.

2.6.2 Importance of Mossbauer Effect

The importance of the effect lies in the line width of γ-rays. When the lattice is excited in the γ emission process, the effective line width is of the order of the phonon energies. When the lattice is not excited, the widths of the nuclear levels involved in the transitions alone determine the line width of the zero-phonon component. According to uncertainty principle, a nuclear lifetime of 10^{-7} s corresponds to a width of $\sim 10^{-8}$ eV which is six orders of magnitude smaller than that obtained when the lattice is excited. More importantly, this line width is smaller than the characteristic values for the magnetic dipole and electric quadrupole interactions of nuclei with their surrounding electrons. It was widely recognized that these effects could in principle be observed and studied through Mossbauer effect. A measure of accuracy is given by the ratio of the line width to the total energy of the γ-ray. For an energy of 100 keV and the lifetime $\sim 10^{-7}$ s, the fractional line width is 10^{-13} eV. This is equivalent to the statement that the energy of the γ-ray is defined to within one part in 10^{13} which makes it the most accurately defined electromagnetic radiation available for physical experiments. Isotopes with suitable lifetime and energy in the first excited state may be used. ^{57}Fe is the favourite choice for Mossbauer spectroscopy. It has been used in more experiments than all other isotopes to date. Figure 2.22 shows the decay scheme of ^{57}Co into ^{57}Fe. For the ratio $\tau/E_r = 3.1 \times 10^{-13}$ and

Fig. 2.22 Decay scheme
^{57}Co

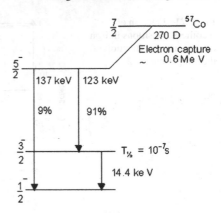

for ^{67}Zn, 5.2×10^{-16}, electromagnetic radiation with comparable stability and line width has not yet been obtained by other means. Even the gas laser which is the best source of narrow-line infrared and visible radiation has not reached the stability and resolution of ^{57}Fe.

2.6.3 Applications

2.6.3.1 Gravitational Red Shift

The gravitational red shift may be thought of most directly as the change in the energy of a photon as it moves from one region of space to another differing gravitational potential. Consider the energy content of a photon moving in a gravitational field. The magnitude of the expected shift can be obtained from a simple argument based only on the energy conservation and the relativistic mass-energy equivalence. Now, the photon carries an inertial and hence also a gravitational mass, given by the expression $h\nu/c^2$, ν being the associated frequency. Consequently, its passage from a point where the gravitational potential is equal to ϕ_1 to a print where the potential is ϕ_2 would entail an expenditure of work given by $h\nu/c^2$ times the potential difference $(\phi_2 - \phi_1)$. This would result in an equivalent decrease in the energy content of the photon and hence its frequency

$$\Delta E = \frac{E}{c^2}(\phi_2 - \phi_1) \tag{2.47}$$

A level difference of H cm near the surface of earth would result in the fractional shift of

$$\frac{\Delta \nu}{\nu} \simeq \frac{gH}{c^2} = 1.09 \times 10^{-18} \, H \tag{2.48}$$

Over a path of 20 m, the expected shift amounts to 2×10^{-15}. In the original experiment of Pound and Rebka [13], ^{57}Fe was used (from a 1-Ci source of ^{57}Co), and the

14.4 eV photons were permitted to travel 22.5 m up the tower of the Jefferson physical laboratory at Harvard to observe the small shift ($\sim 10^{-2}$ of the width of the resonance). Pound and Rebeka concentrated on the portions of the sides of the resonance curve with the largest slope. The systematic errors were reduced by monitoring the temperature of the source and absorber. This is necessary because the temperature difference between source and absorber would cause unequal temperature broadening, thereby shifting the peak. The source and the absorber were periodically interchanged to allow the photons to travel in the opposite direction. After four months of experimentation they obtained the result $\Delta E / E = (4.902 \pm 0.041) \times 10^{-15}$, in close agreement with the expected value of 4.905×10^{-15} for the 45 m round trip. This constitutes one of the most precise tests of the General Theory of Relativity.

2.6.3.2 Atomic Motion

Information about atomic motion and lattice vibration can in fact be obtained from the recoil free γ-ray line which depends upon the ratio of the mean square vibrational amplitude $\langle x^2 \rangle$ of the emitting or scattering atom to the square of the wavelength λ of the scattered radiation. In the equation

$$f = e^{-k2\langle x^2 \rangle} = e^{-4\pi 2\langle x^2 / \lambda^2 \rangle} \tag{2.49}$$

$\langle x^2 \rangle$ is the mean-square amplitude of the vibration in the direction of emission of the γ-ray averaged over an interval equal to the lifetime of the nuclear levels involved in the γ-ray emission process. From (2.49) we note that if $\langle x^2 \rangle$ is not bounded, the recoil-free fraction will vanish. In the light of this conclusion it is clear that the Mossbauer effect can not take place in a liquid where the molecular motion is not restricted. That this is essentially correct has been demonstrated in experiments in which the recoil-free process in ^{119}Sn was studied in metallic tin both below and above the melting point. The liquid phase measurements above the melting point failed to show any resonant effect. Note however that $\langle x^2 \rangle$ must be averaged over a nuclear lifetime. It is conceivable that it may remain sufficiently small in viscous liquids to allow detection of recoilless events. This has been confirmed in the case of a solution of a salt of ^{57}Fe in glycerene. Further, note that (2.49) gives no indication what crystal structure is required for recoil- free emission. It is not surprising that Mossbauer effect is readily observed in glasses. This has been demonstrated, for example, with ^{57}Fe in fused quartz and silicate glass.

Equation (2.46) can be written as

$$f = e^{-[\frac{E_R}{k\theta_D}(\frac{3}{2} + \frac{\pi^2 T^2}{\theta_D^2})]} \tag{2.50}$$

In the limit of low temperatures, f depends only on the ratio of the free atom recoil energy to the Debye temperature

$$F = e^{-(\frac{3E_R}{2k\theta_D})} = e^{-(\frac{3}{4}\frac{E_I^2}{Mc^2 k\theta_D})} \tag{2.51}$$

Characteristic values for f are 0.91 for the 14.4 keV γ-ray of ^{57}Fe and 0.06 for the 129 keV γ-ray of ^{191}Ir.

A simple calculation shows why the Mossbauer effect is limited to low energy γ-rays. For example, for a nucleus of mass number 100 in a lattice with a Debye temperature of 400 K (2.51) becomes

$$f = e^{-(\frac{E(\text{keV})}{64})2}$$ (2.52)

which drops off to low values when $E \gg 64$. To date, Mossbauer effect has not been observed for γ-rays energy greater than 150 keV.

2.6.3.3 Hyperfine Structure

The splitting of nuclear levels of ^{57}Fe both in the ground state and the excited state has been studied. The splitting occurs due to the interaction of the electromagnetic field at the nucleus with the magnetic moment of the nucleus. For a nuclear spin I, a level whose magnetic moment is μ will in an internal magnetic field H be split into $2I + 1$ sublevels separated in energy by $g = \mu H / I$. The ground level is known to have spin $1/2$ and excited level $3/2$. At zero relative velocity between the source and absorber, the six emission components should match perfectly the six absorption components yielding maximum resonance absorption as if there had been no hyperfine splitting. As the source is moved with increasing velocity v, the emission components all receive an additional energy $(v/c)E_0$; they now cease to overlap with the components in the absorber and fluorescence is destroyed. If now the velocity of the source is increased further, a situation will arise in which the energies of source of the emission and absorption will again match resulting in additional fluorescence pips.

For example, if v is such that the energy increment of some of the components is equal to g, the hyperfine energy splitting in the excited states, then the γ-rays resulting from transition labeled I will have correct energy to excite transition 2 in the absorber, 2 will match 3, 4 will match 5 and 5 will match 6. But 3 and 6 are not reasonably absorbed. Similarly, if the source components are given a Doppler energy equal to E_0, then the splitting of the ground state line 2 will match 4, 3 with 5 (Fig. 2.23 and the corresponding Mossbauer spectra is shown in Fig. 2.24). In this manner all the pips are scanned out. We have thus $v = (c\mu H / I E_0)$, where $E_0 = 14.4$ keV, $I = 3/2$. The field in iron is expected to be about 10^5 gauss. From this the value of μ of $(3/2)$ state is calculated to be -0.15 nuclear magneton. Alternatively, if the magnetic moment value be known from magnetic resonance experiments, the field can be known. The unexpectedly small magnetic moments are of interest in solid state physics.

Further, the temperature dependence of the internal magnetic field has been investigated. Results show that the field at the nucleus decreases by 3 % when the temperature is raised from 80 to 300 K and that the plot of the field against temperature follows the classical magnetization curve for iron, heading toward zero at curie temperature.

Fig. 2.23 Hyperfine
structure or ^{57}Fe

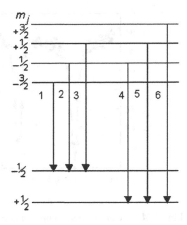

2.6.3.4 Isomer Shift

The nucleus is surrounded and penetrated by electronic charge with which it interacts electrostatically. A change in the s-electron density will result in an altered Coulombic interaction which manifests itself as shift of the nuclear levels. This is called 'Isomer shift' since the effect depends on the difference in the nuclear radii of the ground (gd) and isomeric excited (ex) states. This electrostatic shift of a nuclear level is readily computed from the following model: the nucleus is assumed to be a uniformly charged sphere whose radius is R and the electronic charge density ρ is assumed to be uniform over nuclear dimensions. To simplify the calculation, the difference between the electrostatic interaction of a hypothetical point nucleus and one of actual radius R, both having the same charge is computed. For the point nucleus, the electrostatic potential

$$V_{pt} = \frac{Ze}{r} \tag{2.53}$$

For the finite one, the potential V is

$$V = \frac{ze}{R}\left(\frac{3}{2} - \frac{r^2}{2R^2}\right) \quad \text{for } r \leq R$$

$$= \frac{ze}{r} \quad \text{for } r \geq R \tag{2.54}$$

The energy difference δE is given by the integral

$$\delta E = \int_0^\infty \rho(V - V_{pt})4\pi r^2 dr$$

$$= \frac{4\pi \rho z e}{R} \int_0^R \left(\frac{3}{2} - \frac{r^2}{2R^2} - \frac{R}{r}\right) r^2 dr$$

$$= -\frac{2\pi}{5} z e \rho R^2 = \frac{2\pi}{5} z e^2 |\psi(0)|^2 R^2 \tag{2.55}$$

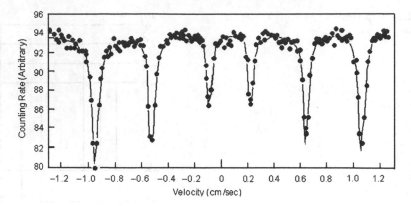

Fig. 2.24 Six individual components are seen for the magnetic dipole transitions in ^{57}Fe with $\Delta m_J = 0$ or ± 1 [14]

where $-e|\psi(0)|^2$ is an alternate expression for the electronic charge density ρ. The expression relates the electrostatic energy of the nucleus to its radius which will in general be different for each nuclear state of excitation or energy level. Observations, however, are made not on the location of individual nuclear levels but on γ-rays resulting from transitions between two such levels. The energy of the γ-ray represents the difference in electrostatic energy of the nucleus in two different states of excitation which in the present model differ only in nuclear radius. The expression for the change in the energy of the γ-ray due to the nuclear electrostatic interaction is therefore the difference of two terms as in (2.55) written for the nucleus in the ground and excited states

$$\delta E_{ex} - \delta E_{gd} = \frac{2\pi}{5} ze^2 |\psi(0)|^2 \left(R_2^{ex} - R_2^{gd}\right) \tag{2.56}$$

If the Mossbauer emitter and absorber are immersed in different materials, with $\psi_e(0)$ for the emitter and $\psi_a(0)$ for the absorber, then a difference between the energy of the line absorbed and emitted can be measured by the Doppler shift. Let E_0 be the photon energy in the absence of isomeric effect. We can then write

$$E_e = E_0 + 2\frac{\pi}{5} ze^2 |\psi_e(0)|^2 \left[R_2^{ex} - R_2^{gd}\right] \tag{2.57}$$

$$E_a = E_0 + \frac{2\pi}{5} ze^2 |\psi_a(0)|^2 \left[R_{ex}^2 - R_{gd}^2\right] \tag{2.58}$$

The isomer shift $(I.S.) = E_a - E_c$

$$I.S. = \frac{2\pi}{5} ze^2 \left[|\psi_a(0)|^2 - |\psi_e(0)|^2\right]\left[R_2^{ex} - R_2^{gd}\right] \tag{2.59}$$

Fig. 2.25 Mossbauer absorption spectra to show the isomeric effect

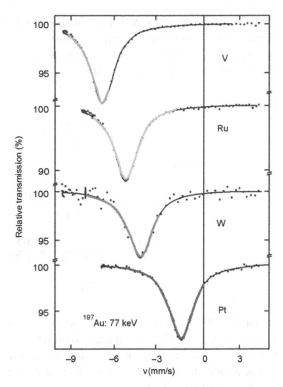

As R_{ex} is expected to be only slightly different from R_{gd}, (2.59) can be written as

$$I.S. = \frac{4\pi}{5}ze^2R^2\left(\frac{\delta R}{R}\right)\left\{|\psi_a(0)|^2 - |\psi_e(0)|^2\right\} \qquad (2.60)$$

The factor in braces is calculated theoretically. Figure 2.25 shows the Mossbauer absorption spectra of the 77 keV gamma rays of ^{197}Au measured at 4.2 K with sources of dilute impurities of ^{197}Au in V, Ru, W, and Pt [7].

The isomeric effect on the Mossbauer spectrum is to shift the centre of the resonance away from zero velocity.

2.6.3.5 Other Applications

Other applications of Mossbauer effect include effects due to accelerated systems, polarization of γ-rays, the lattice properties such as the Debye-Waller factor, specific heats impurities and imperfections, low temperatures, nuclear orientation, superconductivity, recoilless Rayleigh scattering, the quadrupole coupling; the Fe content in hemoglobin etc.

Example 2.1   The 129 keV γ ray transition in ^{191}Ir was used in a Mossbauer experiment in which a line shift equivalent to the full width at half maximum (Γ) was observed for a source speed of 10 mm s^{-1}. What is the value of Γ and the mean lifetime of the excited state in ^{191}Ir?

Solution

$$\frac{\Delta E_r}{E_r} = \frac{v}{c}$$

Put $\Delta E_r = \Gamma$. Then

$$\Gamma = \frac{v}{c} E_r = \frac{1 \text{ cm} \times (129 \text{ keV})}{(3 \times 10^{10})} = 4.3 \times 10^{-6}$$

$$\Gamma = \frac{\hbar}{\Delta E_r} = \frac{\hbar}{\Gamma} = \frac{1.05 \times 10^{-34}}{1.6 \times 10^{-19} \times 4.3 \times 10^{-6}}$$

$$= 1.5 \times 10^{-10} \text{ s}$$

Example 2.2   A linear accelerator produces a beam of excited carbon atoms of kinetic energy 160 MeV. Light emitted on de-excitation is viewed at right angles to the beam and has a wavelength λ'. If λ is the wavelength emitted by a stationary atom, what is the value of $(\lambda' - \lambda)/\lambda$? (Take the rest energies of both protons and neutrons to be 1000 MeV.)

Solution

$$\nu = \nu' \gamma (1 - \beta \cos \theta) = \nu' \gamma \quad (\because \theta = 90°)$$

$$\lambda' = \gamma \lambda$$

$$\frac{(\lambda' - \lambda)}{\lambda} = \frac{\lambda'}{\lambda} - 1 = \gamma - 1 = \frac{T}{Mc^2} = \frac{160}{16 \times 1000} = 0.01$$

Example 2.3   A 60 keV x-ray photon strikes the electron initially at rest and the photon is scattered through an angle of 60°. What is the recoil velocity of the electron?

Solution

$$h\nu = \frac{h\nu_0}{1 + \alpha(1 - \cos \theta)} = \frac{60}{1 + \frac{60}{511}(1 - \cos 60°)}$$

$$= 56.67 \text{ keV}$$

$$T(\text{electron}) = 60.00 - 56.67 = 3.33 \text{ keV}$$

$$\frac{v}{c} = \sqrt{\frac{2T}{mc^2}} = \sqrt{\frac{2 \times 3.33}{511}} = 0.114$$

$$v = 0.114 \times 3 \times 10^8 = 3.42 \times 10^7 \text{ m/s}$$

Example 2.4 The wavelength of the photoelectric threshold for silver is 3250×10^{-10} m. Determine the velocity of electrons ejected from a silver surface by ultraviolet light of wavelength 2500×10^{-10} m.

Solution

$$T = hv - hv_0$$

$$hv = \frac{1241}{250 \text{ nm}} = 4.964 \text{ eV}$$

$$hv_0 = \frac{1241}{325.0 \text{ nm}} = 3.818 \text{ eV}$$

$$T = 4.964 - 3.818 = 1.146 \text{ eV}$$

$$v = c\sqrt{\frac{2T}{mc^2}} = 3 \times 10^8 \sqrt{\frac{2 \times 1.146}{511 \times 10^3}} = 6.35 \times 10^5 \text{ m/s}$$

Example 2.5 Mossbauer found the width, at half of maximum absorption, of the ^{191}Ir 129 keV level, in terms of the velocity spread of the source to be 1.51 cm/s. (a) What is the level width in eV? (b) What is the half-life of the excited state?

Solution

$$h\Delta v = E_0 \frac{v}{c} = \frac{129 \times 10^3 \times 1.51}{3 \times 10^{10}}$$

$$= 6.494 \times 10^{-6} \text{ eV}$$

Level width $= 1/2$(measured width) $= 3.25 \times 10^{-6}$ eV

$$T_{1/2} = 0.693\tau = 0.693 \times \frac{\hbar}{\Delta E} = \frac{0.693 \times 1.05 \times 10^{-27}}{3.25 \times 10^{-6} \times 1.6 \times 10^{-12}}$$

$$= 1.4 \times 10^{-10} \text{ s}$$

Example 2.6 In an experiment, it is necessary to cut down the γ intensity of one MeV γ-rays to 1 %. If water is used as shielding, calculate the length of water column to be used to achieve this, neglect multiple scattering. $\sigma_H = 0.211$ b, $\sigma_0 = 0.169$ b/atom, Avagadro number $= 6.025 \times 10^{23}$.

Solution

$$\mu = (2\sigma_H + 16\sigma_0)\frac{\rho}{A}N_{av}$$

$$= \left(2 \times 0.211 \times 10^{-24} + 16 \times 0.169 \times 10^{-24}\right)\frac{1}{18} \times 6.025 \times 10^{23}$$

$$= 0.1046 \text{ cm}^{-1}$$

$$I = I_0 e^{-\mu x}$$

$$X = \frac{1}{\mu}\ln\frac{I_0}{I} = \frac{1}{0.1046} \times \ln 100$$

$$= 44 \text{ cm}$$

Example 2.7 A collimated beam of 1.6 MeV gamma rays strikes a thin tantalum foil. Electrons of 0.6 MeV energy are observed to emerge from the foil. Are these due to the photoelectric effect, Compton scattering or pair-production? Assume that any electron produced in the initial interaction with the material of the tantalum foil do not undergo a second interaction.

Solution Since the threshold energy for e^+e^- production is 1.02 MeV, the combined kinetic energy of e^+ and e^- will be (1.6–1.02) or 0.58 MeV. The observed electrons of 0.6 MeV cannot be due to this process.

The K-shell ionization potential for silver is under 100 keV. The ejected electrons due to photoelectric effect must have little less than 1.6 MeV. So, photoelectric effect is also ruled out.

In Compton scattering the electron can take up kinetic energy between zero and $T_{max} = \frac{h\nu_0}{1+(1/2\alpha)} = 1.38$ MeV, since $h\nu_0 = 1.6$ MeV and $\alpha = 1.6/0.511$. Thus, the electrons are due to Compton scattering.

Example 2.8 Gamma ray photons when incident upon a piece of uranium, eject photoelectrons from its K-shell. The momentum measured yields a value of Br = 4.8×10^{-3} Weber/m. The binding energy of a K electron in uranium is $B = 115.59$ keV. Determine

(a) The kinetic energy of the photoelectrons.
(b) The energy of the gamma ray photons.

Solution

$$p = 300 \text{ Br MeV}/c$$

$$= 300 \times 4.8 \times 10^{-3} = 1.44 \text{ MeV}/c$$

$$E^2 = (T+m)^2 = p^2 + m^2$$

$$\therefore \quad T = \left(p^2 + m^2\right)^{1/2} - m$$

Putting $p = 1.44$ and $m = 0.511$.

(a) Kinetic energy of the photoelectrons, $T = 1.017$ MeV

(b) The energy of the gamma ray photons,

$$E_r = T + B = 1.017 + 0.115 = 1.132 \text{ MeV}$$

Example 2.9 Ultraviolet light of wavelength λ from a mercury arc falls upon a silver photocathode. Find λ if the threshold wavelength for silver is 325 nm, and the stopping potential is 1.07 V.

Solution

$$h\nu = h\nu_0 + W$$

$$\frac{1241}{\lambda} = \frac{1241}{325} + 1.07 = 4.888 \text{ eV}$$

$$\therefore \lambda = \frac{1241}{4.888} = 253.8 \text{ nm}$$

Example 2.10 Show that photoelectric effect cannot take place with a free electron.

Solution Suppose that photoelectric effect takes place with a free electron.

If $h\nu$ is the photon energy, K is the kinetic energy carried by the electron and p is the electron momentum then

$$K = h\nu \quad \text{(energy conservation)}$$

$$cp = \sqrt{K^2 + 2mc^2 K} = h\nu \quad \text{(momentum conservation)}$$

Eliminating K

$$2mc^2 h\nu = 0$$

Neither mc^2 nor $h\nu$ is zero. Hence we get an absurd result. We conclude that in photoelectric effect with a free electron, both energy and momentum cannot be conserved simultaneously.

Example 2.11 In Compton scattering process, the incident X-ray radiation is scattered at an angle of 60°. If the wavelength of the scattered radiation is 0.200 A, find the wavelength of the incident radiation.

Solution

$$\Delta\lambda = \lambda - \lambda_0 = \frac{h}{mc}(1 - \cos\theta)$$

$$\lambda_0 = \lambda - \frac{h}{mc}(1 - \cos\theta)$$

$$= 0.200 - 0.024(1 - \cos 60°)$$

$$= 1.188 \text{ A}$$

Example 2.12 Show that the minimum photon energy for the production of positron-electron pairs in the field of a free electron is $4mc^2$.

Solution Use the invariance of $E^2 - P^2$ (see [8], Appendix A) and natural units ($c = 1$). There will be $2e^-$ and $1e^+$ in the final state.

$$(E_\gamma + m)^2 - E_\gamma^2 = (3m)^2$$

$$E_\gamma = 4mc^2$$

Example 2.13 Estimate the thickness of lead (density 11.3 g cm^{-3}) required to absorb 99 % of gamma rays of energy 1 MeV. The absorption cross-section for gamma rays of 1 MeV in lead ($A = 207$) is 20 b/atom.

Solution

$$I = I_0 e^{-\mu x}$$

$$\mu x = \ln \frac{I_0}{I} = \ln \frac{100}{1.0} = 4.6$$

$$\mu = \sigma \frac{N_{av}\rho}{A} = \frac{20 \times 10^{-24} \times 6 \times 10^{23} \times 11.3}{207} = 6.55 \text{ cm}^{-1}$$

$$x = \frac{4.6}{6.55} = 0.7 \text{ cm} = 7 \text{ mm}$$

Example 2.14 An X-ray absorption survey of a specimen of silver shows a sharp absorption edge at the expected λ_{ka} value for silver of 0.0485 nm and a smaller edge at 0.0424 nm due to an impurity. If the atomic number of silver is 47, identify the impurity as being ^{44}Rh, ^{45}Rh, ^{46}Pd, ^{48}Cd, ^{49}In or ^{50}Sn.

Solution

$$\lambda_K (\text{Ag}) = 0.0424 \text{ nm}$$

$$E_K (\text{Ag}) = \frac{1241}{0.0424} = 25.588 \times 10^3 \text{ eV}$$

As the screening constant is only approximate, we shall evaluate it

$$25.588 \times 10^3 = 13.6(47 - \sigma)^2$$

$$\sigma = 3.6$$

For the impurity x

$$E_K (x) = 13.6(50 - 3.6)^2$$

$$= 29.28 \text{ keV}$$

The wavelength of 0.0424 nm corresponds to $E = 29.27$ keV. Hence the impurity is ^{50}Sn.

2.7 Questions

2.1 Discuss the conditions necessary to observe resonance fluorescence.

2.2 Compare the resonance fluorescence of an atomic system with that of a nucleus emitting γ-rays in a transition to the ground state. Why is the former easily observable whereas the latter is not?

2.3 Explain the Mossbauer effect with reference to Mossbauer's experiments.

2.4 Explain, giving two examples, how observations of this effect can provide information about the properties of nuclear states and the environment of the emitting or absorbing nucleus in a solid.

2.5 Why is it expected that resonant absorption should decrease at low temperatures and why does it increase in the case of the 129 keV gamma radiation from ^{191}Ir?

2.6 Describe how the Mossbauer effect can be used for nuclear hyperfine measurements and also outline its use to verify the effect of gravity on electromagnetic radiation.

2.7 Given a source of 85 keV γ radiation, what absorbers would you use to bracket this energy and serve to determine it?
[Ans. ^{80}Hg, ^{81}Ti, ^{82}Pb, ^{83}Bi, ^{84}Po]

2.8 When photoelectric effect takes place in a metal, the photoelectrons will have a wide range of energies, from zero to maximum. Why?

2.9 At incident photon energy slightly above the electron binding energy, there will be two lines in the scattered photons. Explain.

2.10 Point out the dependence of μ on $h\nu$ and Z.

2.11 Indicate on a diagram the relative importance of the three major interactions of γ-rays.

2.12 Which process is similar to the pair-production?

2.13 What is the significance of the fine structure constant appearing in the formula for pair-production cross-section?

2.14 Why photoelectric effect cannot take place with a free electron?

2.15 Why cannot pair-production occur in a vaccum?

2.16 When the photoelectric effect takes place in the nuclear field, the nucleus on an average recoils in the backward hemisphere. Why?

2.17 What is the major difference in the energy loss of charged particles and γ-rays?

2.18 Enumerate the similarities between pair-production and Bremsstrahlung.

2.19 What are tridents?

2.20 Under what conditions can one obtain analytical formulae for pair-production cross-section?

2.21 Consider the graphs for total cross-sections (μ), for photons in light and heavy elements at various energies. In heavy element like Pb, there can be the same value of μ for different photon energies; for example, in Pb, $\mu = 0.5$ cm^{-1} for 2.0 and 5.5 MeV. However, in light elements like Aluminium, μ will have single-valued energies. Why?

2.22 If a photon interacts with an atomic electron and is completely absorbed, what is the phenomenon called?

2.23 If a photon interacts with a nucleus and undergoes incoherent scattering, what is the phenomenon called?

2.24 If a high energy photon interacts with the electric field surrounding nuclei or electrons and is completely absorbed, what is the phenomenon called?

2.8 Problems

2.1 Show that the energy E_γ of γ-ray emitted from a nucleus of mass M following a transition from a state of energy E_i to one of energy E_f is approximately given by

$$E_\gamma \simeq \Delta E - \frac{(\Delta E)^2}{2Mc^2}$$

where $\Delta E = E_f - E_i$.

2.2 Calculate the angular velocity of a mechanical rotor that is able to produce resonant absorption of a γ-ray of a given energy emitted from a source placed on its tip at a distance r from the rotation axis.
[Ans. $\omega = E_0/Mcr$]

2.3 Assume that all the ^{137}Cs atoms in a sample move with the same root mean square velocity at a temperature of 15 °C. Calculate the spread in energy of 661 keV internal conversion line of ^{137}Cs due to thermal motion.
[Ans. 666 keV $\pm$ 0.5 eV]

2.4 The Compton electrons ejected from a thin converter in the direction of γ-radiation have a momentum of Br = 0.02 Weber/m. Find the energy of the incident radiation.
[Ans. 5.75 MeV]

2.5 A metal surface is illuminated with light of different wavelengths and the corresponding stopping potentials of the photoelectrons V are found to be as follows:

λ (nm)	600	550	500	450	400	350
V (volts)	0.15	0.33	0.63	0.83	1.26	1.71

Calculate the Planck's constant, the Work function and the threshold wavelength.
[Ans. 6.5×10^{-34} J s, 1.9 V, 650 nm]

2.6 Calculate the maximum wavelength of photon to produce an electron-positron pair.
[Ans. 0.02425 Å]

2.7 Calculate the wavelength of γ-rays emitted in the annihilation of an electron-positron pair at rest.
[Ans. 2.429 p.m.]

2.8 A 4 cm diameter and 1 cm thick NaI is used to detect 660 keV gamma ray emitted by a 100 µc point source of ^{137}Cs placed on its axis at a distance of one metre from its surface. Calculate separately the number of photoelectrons and Compton electrons released in the crystal given that the linear absorption coefficients for the photo and Compton processes are 0.03 and 0.24 per cm, respectively. What is the number of 660 keV gamma rays that pass through the crystal without interacting? (1 curie = 3.7×10^{10} dis. per s.)
[Ans. 10, 78, 282 s^{-1}]

2.9 For Aluminium, and a photon energy of 0.06 MeV, atomic absorption cross-section due to the Compton effect is 8.1×10^{-24} cm^2 and due to the photo-effect is 4.0×10^{-21} cm^2. Calculate how much the intensity of a given beam is reduced

by a 3.7 g cm^{-2} of aluminium and state the ratio of the intensities absorbed due to Compton effect and photo-effect.
[Ans. 0.24]

2.10 Calculate the maximum change in the wavelength of Compton scattered radiation.
[Ans. 0.048 A]

2.11 γ-rays of energy 660 keV from a radioactive source of cesium-137 are elastically scattered from free electrons. Calculate the energy of the scattered photons at (a) 60°, (b) 120° and (c) 180°.
[Ans. (a) 401 keV, (b) 225 keV, (c) 184 keV]

2.12 A 30 keV X-ray photon strikes an electron, initially at rest and the photon is scattered through an angle of 60°. What is the recoil velocity of the electron?
[Ans. 1.73×10^7 m/s]

2.13 The photoelectric thresholds of sodium and zinc occur at wavelengths of 5390 and 2926 A, respectively. Calculate their contact P.D.
[Ans. 2.3, 4.24 V]

2.14 Calculate the wavelength of a γ-ray whose energy is equal to the rest-mass energy of the proton.
[Ans. 1.323 fm]

2.15 A photon incident upon a hydrogen atom ejects an electron from the first excited state. If the kinetic energy of the ejected electron was 12.7 eV, calculate the energy of the photon. What energy would have been imparted to an electron in the ground state?
[Ans. 16.1, 2.5 eV]

2.16 Ultraviolet light of wavelengths, 800 and 700 A, when allowed to fall on hydrogen atoms in their ground state, are found to liberate electrons with kinetic energy 1.8 and 4.0 eV, respectively. Find the value of Planck's constant.
[Ans. 6.57×10^{-34} J s]

2.17 Show that the energy imparted to the electron in Compton scattering is given by

$$T = h\nu_0 \frac{2\alpha \cos^2 \phi}{(1+\alpha)^2 - \alpha^2 \cos^2 \phi}$$

References

1. Bethe, Placzek (1937)

2. Compton (1923)
3. W.G. Davey, P.B. Moon, Proc. Phys. Soc., A **66**, 956 (1953)
4. C.M. Davisson, R.D. Evans, Rev. Mod. Phys. **24**, 79 (1952)
5. R.D. Evans, *The Atomic Nucleus* (McGrawhill, New York, 1955)
6. H. Hall, Rev. Mod. Phys. **8**, 358 (1936)
7. G. Kainill, D. Solomon, Phys. Rev. B **8**, 1912 (1973)
8. A.A. Kamal, *Particle Physics* (2014)
9. Klein, Nishina (1928)
10. Leo (1994)
11. R.L. Mossbauer, Naturwiss. **45**, 538 (1958)
12. R.L. Mossbauer, Ann. Rev Nucl. **12**, 123 (1962)
13. R.V. Pound, G.A. Rebka, Phys. Rev. Lett. **4**, 337 (1960)
14. G.K. Wertheim, *Mossbauer Effect: Principles and Applications* (Academic Press, New York, 1964)

Chapter 3
Radioactivity

3.1 Natural Radioactivity

Definition Radioactivity is defined as the spontaneous disintegration of a nucleus. Natural Radioactivity may manifest itself through one of the following processes: (a) α-decay, (b) β-decay, (c) orbital electron capture, (d) γ-decay. Artificial Radioactivity may include decay via proton, neutron or fission.

3.1.1 The Radioactive Decay Law

The probability for a radioactive nucleus not to decay at time t is given by

$$P(t) = \exp(-\lambda t)$$

This gives the probability for the radioactive nucleus to survive at time t. If N_0 is the number of atoms at $t = 0$ and N the number of atoms at time t, then

$$P(t) = \frac{N}{N_0}$$

$$N = N_0 \exp(-\lambda t) \tag{3.1}$$

Thus, the decay law for a radioactive substance is an exponential one. Figure 3.1 shows the variation of fraction N/N_0 with time. Also, the decay rate is given by

$$\frac{dN}{dt} = -\lambda N_0 \exp(-\lambda t) = -N\lambda \tag{3.2}$$

where we have used (3.1).

A. Kamal, *Nuclear Physics*, Graduate Texts in Physics,
DOI 10.1007/978-3-642-38655-8_3, © Springer-Verlag Berlin Heidelberg 2014

Fig. 3.1 The fraction of
atoms N/N_0 as a function of
time t

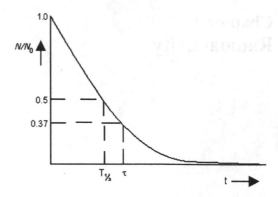

3.1.2 Mean Life and Half-Life

Mean life is the average life time τ_{av}. We can find mean life from the following
equation:

$$\tau_{av} = \frac{\int_0^\infty t \exp(-\lambda t)dt}{\int_0^\infty \exp(-\lambda t)dt} = \frac{1/\lambda^2}{1/\lambda} = \frac{1}{\lambda} = \tau \qquad (3.3)$$

On putting $t = \tau = 1/\lambda$ in (3.1), we get $N/N_0 = 1/e = 0.37$. Thus, in a time equal
to τ, the radioactive atoms are reduced to 37 per cent of the original number.

Another useful parameter is the half-life time $T_{1/2}$. It is that time in which the
number of atoms is reduced to half. Putting $N/N_0 = \frac{1}{2}$ in (3.1), for $t = T_{1/2}$

$$N/N_0 = \frac{1}{2} = \exp(-\lambda T_{1/2}) \quad \text{or}$$

$$T_{1/2} = \frac{\ln 2}{\lambda} = 0.693\tau$$

Clearly, $\tau > T_{1/2}$. The values for τ and $T_{1/2}$ are marked on a typical decay curve as
in Fig. 3.1.

3.1.3 Activity

The activity (A) of a sample is defined as the number of disintegrations per unit
time. Therefore

$$A = \left|\frac{dN}{dt}\right| = N\lambda \qquad (3.4)$$

Specific activity is defined as disintegration rate per unit mass of radioactive substance. A given type of radioactive nucleus may exhibit competitive decay modes,

say e^- capture and β^+ emission. As the probabilities are additive, the total decay constant λ is given by:

$$\lambda = \lambda_1 + \lambda_2 + \cdots \tag{3.5}$$

so that for a mixture of activities

$$\frac{1}{\tau} = \frac{1}{\tau_1} + \frac{1}{\tau_2} + \cdots \tag{3.6}$$

Probability of decay of nucleus during time 0 to t is

$$\frac{N_0 - N}{N_0} = \left[1 - \exp(-\lambda t)\right] \tag{3.7}$$

3.1.4 Units of Radioactivity

(a) *Curie* (Ci)

Curie is measured as 3.7×10^{10} disintegrations per second. Originally, this number was supposed to represent the number of decays observed for 1 gram of radium. 1 mCi $= 10^{-3}$ Ci and 1 μCi $= 10^{-6}$ Ci.

(b) *Becquerel* (Bq)

Becquerel is the SI unit of radioactivity which has replaced Curie. 1 Ci $=$ 3.7×10^{10} Bq.

(c) *Rutherford* (R)

Rutherford is measured as 10^6 disintegrations per second.

(d) *Roentgen*

Roentgen is a unit of Radioactivity applied to X-rays and γ-rays. It is the dose which produces 2.08×10^9 ion pairs of air.

3.1.5 Unit of Exposure and Unit of Dose

The absolute absorbed dose (D) of radiation is the quotient of ΔE by Δm, where ΔE is the energy imparted by ionizing radiation to the matter in a volume element and Δm is the mass of the matter in the volume element Therefore

$$D = \Delta E / \Delta m$$

The special unit of absorbed dose is the rad

$$1 \text{ rad} = 100 \text{ ergs g}^{-1}$$

The SI unit for absorbed dose is Gray which means one joule of absorbed energy per 1 kg of material. 1 Gy $= 100$ rad. In practice a quoted absorbed dose may refer to the whole body average or an average over some particular organ of the body.

The exposure (X) is the quotient of ΔQ by Δm, where ΔQ is the sum of the electrical charges on all the ions of one sign produced in air when all the electrons liberated by photons in volume element of air whose mass is Δm are completely stopped in air

$$X = \frac{\Delta Q}{\Delta m}$$

The special unit of exposure is the roentgen (R)

$$1\,\mathrm{R} = 2.58 \times 10^{-4}\ \mathrm{Coulomb\,kg^{-1}}$$

The roentgen is equivalent to the production of

$$\frac{2.58 \times 10^{-4}}{1.6 \times 10^{-19}} = 1.61 \times 10^{15}\ \mathrm{ion\,pairs\,kg^{-1}}$$

Since an energy of about 34.5 eV is required to produce 1 ion pair in air, the energy deposited due to exposure of 1 R will be

$$1.61 \times 10^{15} \times 34.5 = 5.6 \times 10^{16}\ \mathrm{eV\,kg^{-1}}$$

A more common value for roentgen is

$$1\,\mathrm{R} = 5.6 \times 10^{16} \times 10^{-3} \times 1.6 \times 10^{-12} = 89\ \mathrm{erg\,g^{-1}}$$

The radiation damage to living tissue is not simply proportional to the absolute absorbed dose, but it depends on several other factors, radiation type being one of them. For example, for the same number of Grays, radiation damage is by far greater for α-particles than γ-rays. From the point of view of medicine different types of radiation have been given relative biological effectiveness (RBE) factors. These RBE factors are dimensionless numbers.

The equivalent dose is a measure for the effect of radiation on the human body. It is defined as:

$$D_q = qD$$

where q is a quality factor for the biological effect of different types of radiations on human tissue. The unit of equivalent dose is the Rem (Roentgen equivalent man), 1 Rem $= q \cdot 1$ rad. The quality factors (RBE) are approximately $q = 1$ for X-rays, γ-rays and β-particles, $q = 10$ for neutrons and protons, $q = 20$ for α-particles. Moreover, the RBE factors depend on the particle energy as well.

The *Sievert* is a unit combining the RBE factor with the absolute absorbed dose. The dose equivalent in the Sv equals the dose in Gy multiplied by the appropriate RBE factor. The SI unit is defined as 1 Sievert (Sv) $= 100$ Rem. The dose in Sv indicates the potential harm to living tissue. A dose of 900 rad is certainly fatal.

3.1.5.1 Levels of Radiation and Radiation Hazards

There are three principal natural sources that cause hazards of ionizing radiation. They are:

(a) Cosmic rays
(b) Radioactive nuclei contained in the body
(c) Radioactive elements present in the rocks and soil

The secondary cosmic rays (e^+, e^-, γ-rays, neutrons and muons) produced in the collisions of high energy primary cosmic rays (90 per cent protons) account for a dose of about 0.25 mSv (millisv) per year for the human body at sea level. The actual dose depends on the latitude and increases with altitude. At a height of 4000 m the dose would increase to 2 mSv per year. The most significant radioactive nuclide in the body is ^{40}K. This isotope of potassium has a long mean life of 1.8×10^9 years and constitutes 0.0117 percent of natural potassium. It can undergo all the three types of β-decay. But the most common mode (89 percent) is β^- decay with $E_{max} = 1.32$ MeV, the remaining 11 per cent decays are mainly via electron capture to an excited state of ^{40}Ar which subsequently decays by emitting a 1.46 MeV γ-ray. From these decays the body receives a dose of 0.17 mSv per year. Other radioactive nuclei in the body contribute similarly.

The γ-radiation arising from the decay products of radioactive elements, mainly uranium and thorium deposits in rocks and ground, contribute to a dose between 0.2 and 0.4 mSv per year. However, in the neighbourhood of granite rocks, the dose may be several times greater. A greater hazard is caused by the inhalation of the radioactive gases of isotopes ^{222}Rn and ^{220}Rn. These are the decay products of uranium and thorium and being gases can diffuse into air and finally enter human body. Further, ^{222}Rn and ^{220}Rn give rise to a chain of α-emitters that are solids. The dose received varies widely, depending on the building materials, subsoil, ventilation, etc. The equivalent whole body dose averages about 1.0 mSv per year.

The total natural background thus averages out to 2 mSv per year. Apart from this radiation hazards in hospitals using X-rays, the radioactive fall-out from nuclear weapons, or for that matter in areas near nuclear reactors or accelerators further increase the radiation hazards. In US, maximum permissible dose at present is set at 50 mSv per year.

Gray and Sievert are very large units from the point of view of biological damage. Whole body dose of about 5 Gy may cause death in 50 per cent of cases. Even much weaker doses may result in cancer in many bodies.

Example 3.1 What fraction of the radioactive cobalt nuclei, whose half-life is 71.3 days, decays during a month?

Solution

$$\text{Fraction} = 1 - \exp(-\lambda t) = 1 - \exp(-0.693 t / T_{1/2})$$
$$= 1 - \exp(-0.693 \times 30/71.3) = 1 - e^{-0.29168} = 0.254$$

Example 3.2 How many beta-particles are emitted during one hour by 1.0 μg of ^{24}Na radionuclide whose half-life time is 15 hours?

Solution

$$\left|\frac{dN}{dt}\right| = N\lambda = \frac{6 \times 10^{23}}{24} \times 10^{-6} \times \frac{0.693}{15} = 1.155 \times 10^{15}\beta \text{ particles per hour}$$

Example 3.3 The activity of a certain preparation decreases 2.5 times after 7.0 days. Find its half-life time.

Solution

$$\frac{N}{N_0} = \frac{1}{2.5} = \exp(-\lambda \times 7)$$

$$T = \frac{0.693}{\lambda} = \frac{0.693}{\ln 2.5} \times 7 = 5.3 \text{ D}$$

Example 3.4 Initially the activity of a certain nuclide totalled to be 650 particles per minute. What will be the activity of the preparation after half its half-life time?

Solution

$$650 = \left|\frac{dN_0}{dt}\right| = N_0\lambda$$

$$\left|\frac{dN}{dt}\right| = N\lambda = \left|\frac{dN_0}{dt}\right| \exp(-\lambda t)$$

$$= 650\exp(-0.693t/T_{1/2}) = 650e^{-0.693 \times 0.5}$$

$$= 455 \text{ particles/min}$$

Example 3.5 To investigate the beta-decay of ^{23}Mg radionuclide, a counter was activated at time $t = 0$. It registered N_1 beta-particles by time $t_1 = 2.0$ s, and by time $t_2 = 3t_1$, the number of registered beta-particles was 2.66 times greater. Find the mean life time of the given nuclei.

Solution

$$N = N_0\left[1 - \exp(-\lambda t)\right]$$

$$N_1 = N_0\left[1 - \exp(-\lambda t_1)\right] = N_0\left[1 - \exp(-2\lambda)\right]$$

$$N_2 = 2.66N_1 = N_0\left[1 - \exp(\lambda t_2)\right] = N_0\left[1 - \exp(-6\lambda)\right]$$

$$\frac{N_2}{N_1} = 2.66 = \frac{1 - \exp(-6\lambda)}{1 - \exp(-2\lambda)} = \frac{1 - x^3}{1 - x} = 1 + x + x^2$$

where $x = \exp(-2\lambda) = 0.885$, whence, $\lambda = 0.061$ and $\tau = 16$ s.

Example 3.6 Calculate the time required for 20 % of a sample of thorium to disintegrate ($T_{1/2}$ of thorium $= 1.4 \times 10^{10}$ y).

Solution Decay constant

$$\lambda = \frac{0.693}{T_{1/2}} = \frac{0.693}{1.4 \times 10^{10}} = 0.495 \times 10^{-10} \ y^{-1}$$

Since $N = N_0 \exp(-\lambda t)$

$$\exp(-\lambda t) = \frac{N}{N_0} = \frac{8}{10} \quad \text{or}$$

$$\lambda t = \ln(10/8) = 0.223$$

Thus

$$t = \frac{0.223}{\lambda} = \frac{0.223}{0.495 \times 10^{-10}} = 4.5 \times 10^9 \ y$$

Example 3.7 Calculate the activity of 1 μg of Th $X (T_{(1/2)} = 3.64$ D).

Solution Number of Th X atoms in 1 μg is

$$N = \frac{N_A \times M}{m}$$

$$= \frac{6 \times 10^{23} \times 1.0 \times 10^{-6}}{224} = 2.678 \times 10^{15}$$

$$\lambda = \frac{0.693}{T_{1/2}} = \frac{0.693}{3.64 \times 86400} = 2.2 \times 10^{-6}$$

$$A = \left| \frac{dN}{dt} \right| = N\lambda = (2.678 \times 10^{15})(2.2 \times 10^{-6}) = 5.89 \times 10^9 /s$$

$$= \frac{5.89}{3.7} \times \frac{10^9}{10^{10}} \ Ci = 0.159 \ Ci$$

Example 3.8 A dose of 3.5 mCi of $^{32}_{15}$P is administered intravenously to a patient whose blood volume is 3.5 litres. At the end of one hour it is assumed that the Phosphorous is uniformly distributed. What would be the count rate per millilitre of withdrawn blood if the counter had an efficiency of only 10 %. (a) 1 hour after the injection and (b) 28 days after the injection ($T_{1/2}$ of $^{32}_{15}$P is 14 days).

Solution Dose per millilitre $= \frac{3.5}{3500}$ mCi. Therefore, number of disintegrations $= \frac{3.5}{3500} \times 3.7 \times 10^7$ per second

(a) Assuming no significant decay after 1 hour, disintegrations counted $= \frac{1}{10} \times$ $\frac{3.5}{3500} \times 3.7 \times 10^7 = 3.7 \times 10^3$ per second.
(b) After 28 days, activity $= \frac{1}{4} \times 3.5$ mCi. Therefore, disintegrations counted $= 925$ per second.

Example 3.9 In a laboratory experiment, silver-foil strips are placed near a neutron source. The capture of neutrons by ^{107}Ag produces ^{108}Ag, which is radioactive and decays by β-decay with a half-life of 2.4 min. How long should the foil be irradiated to obtain a maximum activity?

Solution The production of ^{108}Ag is governed by the relation, $N = N_0[1 - \exp(-t/\tau)]$. When $t = 4\tau$, the ratio $\frac{N}{N_0} = 0.98$.

Therefore, after four mean life times, the number of radioactive ^{108}Ag nuclei will be about 98 % of the maximum value. Thus, there is not much gain in irradiating the silver foil longer than four mean life times, i.e.

$$= \frac{4 \times T_{1/2}}{0.693} = \frac{4 \times 2.4}{0.693} = 14 \text{ min}$$

3.1.6 Determination of Half-Life Time

3.1.6.1 Short Lived Source

Determine the decay rate $\frac{dN_0}{dt}$ at $t = 0$ and $\frac{dN_1}{dt}$ at $t = t_1$.

To get an accurate result, t_1 should be large enough so that $\frac{dN_0}{dt}$ and $\frac{dN_1}{dt}$ are significantly different but small enough so that t_1 is not comparable with the half-life time. Now

$$\frac{dN_0}{dt} = -\lambda N_0$$

$$\frac{dN_1}{dt} = -\lambda N_1 = -\lambda N_0 \exp(-\lambda t_1) = \frac{dN_0}{dt} \exp(-\lambda t_1)$$

It follows that

$$T_{1/2} = \frac{0.693}{\lambda} = \frac{0.693 t_1}{\log \frac{dN_0}{dt} - \log \frac{dN_1}{dt}}$$

If several measurements be taken at known time and a graph be plotted as $\log dN/dt$ against t then a straight line is obtained whose slope gives λ.

3.1.6.2 Long Lived Source

Let the substance weigh W grams. If the atomic (molecular) weight is A, then number of atoms

$$N = 6.03 \times 10^{23} \times W/A$$

The actual decay rate dN/dt can be found out after applying solid angle correction for geometrical losses

$$\lambda = \frac{1}{N} \frac{dN}{dt}$$

whence, $T_{1/2}$ can be found out.

3.1.7 Law of Successive Disintegration

Let a radioactive substance A decay into B and B into C, with decay constants λ_A and λ_B, respectively. It is required to investigate the variation of B with time t

$$A \xrightarrow{\lambda_A} B \xrightarrow{\lambda_B} C$$

The rate of decay of parent substance A is given by

$$\frac{dN_A}{dt} = -\lambda_A N_A \tag{3.8}$$

Substance A decays according to the law expressed in the equation $-dN_A = \lambda_A N_A dt$. For every atom of substance A that disintegrates, an atom of substance B is formed. Here the number of atoms of substance B varies for two reasons. It decreases because substance B decays, but it increases because the decay of substance A furnishes new atoms of substance B.

The net change of the daughter is given by

$$\frac{dN_B}{dt} = \lambda_A N_A - \lambda_B N_B \tag{3.9}$$

The first term on the right hand side represents the rate of increase of B (notice the positive sign) and the second term the rate of decrease (notice the negative sign). Here N_A and N_B are the number of atoms of A and B, respectively at time t.

Solution of (3.8) is

$$N_A = N_A^0 \exp(-\lambda_A t) \tag{3.10}$$

Solution of (3.9) is

$$N_B = A \exp(-\lambda_A t) + B \exp(-\lambda_B t) \tag{3.11}$$

where A and B above are the constants, we can find out the constants by the use of initial conditions. At $t = 0$, the initial amount of daughter is zero, i.e. $N_B^0 = 0$. Using this condition in (3.11) gives $B = -A$ and (3.11) becomes

$$N_B = A\left[\exp(-\lambda_A t) - \exp(-\lambda_B t)\right] \qquad (3.12)$$

Also

$$\frac{dN_B}{dt} = -\lambda_A A \exp(-\lambda_A t) + \lambda_B A \exp(-\lambda_B t)$$

but at $t = 0$

$$\frac{dN_B^0}{dt} = \lambda_A N_A^0 = -\lambda_A A + \lambda_B A \quad \text{or}$$

$$A = \frac{\lambda_A N_A^0}{\lambda_B - \lambda_A} \qquad (3.13)$$

Using (3.13) in (3.12) we get

$$N_B = \frac{\lambda_A N_A^0}{\lambda_B - \lambda_A}\left[\exp(-\lambda_A t) - \exp(-\lambda_B t)\right] \qquad (3.14)$$

3.1.7.1 Transient Equilibrium

Case (i) $\lambda_A < \lambda_B$ Here parent amount varies considerably with time and the half-life of parent and daughter are comparable. In such a case the daughter first reaches the maximum and then decreases at a decay rate of the longer lived of the two. The time in which the daughter reaches the maximum may be obtained as follows: Differentiating N_B, with respect to t in (3.14) and setting $\frac{dN_B}{dt} = 0$

$$\frac{\lambda_A N_A^0}{\lambda_B - \lambda_A}\left[\lambda_B \exp(-\lambda_B t) - \lambda_A \exp(-\lambda_A t)\right] = 0 \quad \text{or}$$

$$t_{\max} = \frac{1}{(\lambda_B - \lambda_A)}\ln(\lambda_B/\lambda_A) \qquad (3.15)$$

After this time, the daughter will have the decay rate dependent on the relative values of λ_A and λ_B.

It follows from (3.14) that $\exp(-\lambda_B t)$ will tend to zero faster than $\exp(-\lambda_A t)$, so that (3.14) reduces to

$$N_B = \frac{N_A^0 \lambda_A}{\lambda_B - \lambda_A}\exp(-\lambda_A t) = \frac{N_A \lambda_A}{\lambda_B - \lambda_A} \quad \text{or}$$

$$\frac{N_B}{N_A} = \frac{\lambda_A}{\lambda_B - \lambda_A} \qquad (3.16)$$

Fig. 3.2 Growth and decay
of daughter. Also, atoms
showing the phenomenon of
transient equilibrium

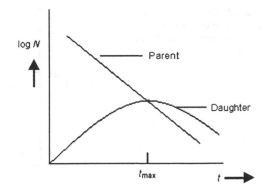

Thus, N_B/N_A is a constant fraction. This means that after considerable time the daughter would decay at the rate of the parent. This is called *transient equilibrium*.

As an example, consider the decay of ^{224}Ra with $T_{1/2} = 3.64$ D which forms ^{220}Rn with $T_{1/2} = 54.5$ s. Starting with pure ^{224}Ra sample, the activity due to ^{220}Rn would increase for several minutes and would then decrease steadily with a 3.64 day ^{224}Ra period (Fig. 3.2). It is easily shown that parent and daughter have the same activity at $t = t_{max}$ (see Example 3.10).

The beginning of the curve up to t_{max} is determined by the shorter life time, i.e. λ_B. But, after this the curve is essentially determined by the longer of the two.

Example 3.10 Show that when

$$\lambda_A < \lambda_B, \quad \text{at } t = t_{max}, \quad A_A = A_B$$

Solution   At $t = t_{max} \frac{dN_B}{dt} = 0$. Hence, from (3.9)

$$\lambda_A N_A = \lambda_B N_B \quad \text{that is, } A_A = A_B$$

3.1.7.2 Secular Equilibrium

Case (ii) $\lambda_A \ll \lambda_B$ Here parent activity does not vary with time. Equation (3.14) reduces to

$$N_B \simeq N_A^0 \frac{\lambda_A}{\lambda_B} \left[1 - \exp(-\lambda_B t)\right] \tag{3.17}$$

After a time t long compared with $T_{1/2}(B)$, a condition is reached for which $\exp(-\lambda_B t)$ tends to zero and consequently the amount of the daughter present is practically constant having the value

$$N_B = N_A^0 \frac{\lambda_A}{\lambda_B} \tag{3.18}$$

Fig. 3.3 Growth of radon
atoms with time when
$\lambda_A \ll \lambda_B$, secular equilibrium
is reached

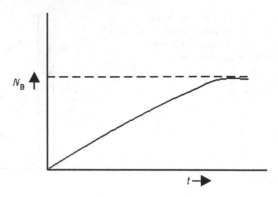

In that case, the daughter is said to be in secular or permanent equilibrium with the
parent. Since the amount of parent is nearly constant

$$N_A^0 = N_A$$
$$\frac{N_B}{N_A} = \frac{\lambda_A}{\lambda_B} = \frac{T_{1/2}(B)}{T_{1/2}(A)} \tag{3.19}$$

Under these conditions the daughter breaks up as fast as it is formed. For example, in the decay of ^{226}Ra ($T_{1/2} = 1620$ y) into Rn (5.5 D) a secular equilibrium is
established, as in Fig. 3.3. When the equilibrium condition is reached and the entire amount of radon is separated from radium then the separated radon will decay
according to the exponential law $\exp(-\lambda_B t)$.

To sum up secular equilibrium can be established when $T_{1/2}(A) \gg T_{1/2}(B)$ and
transient equilibrium can be reached if $T_{1/2}(A) \simeq T_{1/2}(B)$.

Case (iii) $\lambda_A \gg \lambda_B$ In this case, after some time the parent disappears much before the daughter begins to decay at a rate determined by its decay constant. Here
also the rate of growth of the daughter is determined by the shorter $T_{1/2}$ and the
decay by the longer $T_{1/2}$. Formula (3.14) is valid for $\lambda_A > \lambda_B$ as well.

3.1.7.3 Growth of the Grand Daughter Product

Consider the following decay scheme

$$A \xrightarrow{\lambda_A} B \xrightarrow{\lambda_B} C \xrightarrow{\lambda_C}$$

for successive transformations. For the grand daughter C, the net change of atoms
is given similar to (3.9)

$$\frac{dN_C}{dt} = \lambda_B N_B - \lambda_C N_C \tag{3.20}$$

The system of differential equations (3.2), (3.9) and (3.20) can be solved

$$N_A = A \exp(-\lambda_A t) \tag{3.21}$$

$$N_B = B \exp(-\lambda_A t) + C \exp(-\lambda_B t) \tag{3.22}$$

$$N_C = D \exp(-\lambda_A t) + F \exp(-\lambda_B t) + G \exp(-\lambda_C t) \tag{3.23}$$

The constants, A, B, ... are determined from the initial conditions. Let $N_A = N_A^0$, $N_B = 0$, $N_C = 0$ at $t = 0$. This gives us:

$$A = N_A^0 \tag{3.24}$$

$$B + C = 0 \quad \text{or} \quad C = -B \tag{3.25}$$

$$D + F + G = 0 \tag{3.26}$$

Differentiating (3.22) and substituting in (3.9)

$$(\lambda_A A + \lambda_A B - \lambda_B B) \exp(-\lambda_A t) = 0$$

If this equation is to be valid for all values of t, then the first bracket must vanish. Therefore

$$B = \frac{A \lambda_A}{\lambda_B - \lambda_A} \tag{3.27}$$

Differentiating (3.23) and using (3.24) and (3.27) and remembering at $t = 0$, $N_B = N_C = 0$, we get

$$\lambda_A D + \lambda_B F + \lambda_C G = 0 \tag{3.28}$$

Differentiating (3.23) and combining with (3.20) and using the values of N_B and N_C from (3.22) and (3.23)

$$\exp(-\lambda_A t)(-D\lambda_A - B\lambda_B + D\lambda_C) + \exp(-\lambda_B t)(-F\lambda_B + F\lambda_C - C\lambda_B) = 0 \tag{3.29}$$

In order that Eq. (3.29) be valid, the coefficients of the exponentials must vanish separately

$$D = \frac{B \lambda_B}{\lambda_C - \lambda_A} = \frac{N_A^0 \lambda_A \lambda_B}{(\lambda_C - \lambda_A)(\lambda_B - \lambda_A)} \tag{3.30}$$

$$F = \frac{-C \lambda_B}{\lambda_B - \lambda_C} = \frac{N_A^0 \lambda_A \lambda_B}{(\lambda_B - \lambda_C)(\lambda_B - \lambda_A)} \tag{3.31}$$

Using (3.26), (3.30) and (3.31)

$$G = \frac{N_A^0 \lambda_A \lambda_B}{(\lambda_C - \lambda_B)(\lambda_C - \lambda_A)} \tag{3.32}$$

Using (3.30), (3.31) and (3.32), in (3.23)

$$N_C(t) = N_A^0 \lambda_A \lambda_B \left[\frac{\exp(-\lambda_A t)}{(\lambda_C - \lambda_A)(\lambda_B - \lambda_A)} + \frac{\exp(-\lambda_B t)}{(\lambda_C - \lambda_B)(\lambda_A - \lambda_B)} \right.$$
$$\left. + \frac{\exp(-\lambda_C t)}{(\lambda_B - \lambda_C)(\lambda_A - \lambda_C)} \right] \tag{3.33}$$

The method of solution of the general case of n products has been given in a symmetrical form by Bateman. Equation (3.33) has frequent applications as in the chain disintegration in natural radioactivity. It may so happen that one of the members of a radioactive series may have half-life time considerably longer than the succeeding decay products as for uranium ore. Let $\lambda_A \ll \lambda_B$ or λ_C. Then t is long enough $(t \gg 1/\lambda)$, the second and third terms in (3.33) are clearly unimportant and (3.33) reduce to

$$\frac{N_C(t)}{N_A^0} = \frac{\lambda_A}{\lambda_C} \tag{3.34}$$

Thus, the condition for secular equilibrium is reached.

3.1.8 Age of the Earth

Standard methods for the estimation of the age of earth are as follows:

(a) concentration of salt in sea water
(b) the rate of recession of moon from the earth
(c) sedimentation of rocks
(d) the radioactivity method

In natural radioactivity there are four possible series of chain decays:

(a) Radium series represented by $4n + 2$, starting with ^{238}U ($T_{1/2} = 4.56 \times 10^9$ y) and ending up in ^{206}Pb.
(b) Actinium series represented by $4n + 3$, starting with ^{235}U ($T_{1/2} = 0.71 \times 10^9$ y) and ending up in ^{207}Pb.
(c) Radio thorium series, represented by $4n$, starting with ^{232}Th ($T_{1/2} = 1.39 \times 10^{10}$ y) and ending up in ^{208}Pb.
(d) Neptunium series, represented by $4n + 1$, starting with ^{237}Np and ending up with ^{209}Bi does not exist as the members of this series have all half-lives much shorter than the age of earth.

In all the four cases n is a positive integer whose value assigns the mass number of various members of the series. Notice that ordinary lead ^{204}Pb is not a product of radioactive decay.

It must be pointed out that as no half-value period in the existing series is greater than 0.1 per cent of the parent, we can consider that shortly (compared with 10^9 y) after the disintegration of the radioactive atom it becomes directly an isotope of lead without incurring any serious error.

Uranium and Thorium are widely distributed in ordinary rocks in minute quantities and are found occasionally in concentrated deposits in radioactive minerals. Both ultimately decay to lead isotopes.

The age of radioactive minerals can be estimated from the knowledge of uranium and thorium contents, the isotopic composition of lead, and the decay constants using (a) the thorium/lead-208 ratio or (b) uranium/lead-206 ratio, or (c) lead-207/lead-206 ratio.

Nier's method involves (a) an accurate determination of the present-day abundance ratio of the uranium isotopes 235 and 238 using mass spectrographs, (b) determination of relative abundance of lead isotopes in the mineral whose age t is to be estimated and (c) use of the measured decay constants of the two isotopes of uranium.

As the life times of ^{235}U and ^{238}U are different, we expect the ratio $^{206}Pb/^{207}Pb$ to be quite different from uranium isotope ratio at the beginning. Assuming that there are no losses of intervening radioactive products, specially radon which is a gas, we can write the balance equations:

$$^{206}N + {}^{238}N = {}^{238}_0N \tag{3.35}$$

$$^{238}N = {}^{238}_0N \exp(-\lambda_{238}t) \tag{3.36}$$

Combining (3.35) and (3.36)

$$^{206}N = {}^{238}N\left[\exp(\lambda_{238}t) - 1\right] \tag{3.37}$$

Similarly

$$^{207}N = {}^{235}N\left[\exp(\lambda_{235}t) - 1\right] \tag{3.38}$$

where ^{206}N, ^{207}N, ^{238}N, ^{235}N are the number of atoms after the lapse of time t. Equations (3.37) and (3.38) can be used separately for the determination of t. Alternatively, we can make use of the ratio of (3.37) and (3.38)

$$\frac{^{206}N}{^{207}N} = \frac{^{238}N}{^{235}N}\frac{[\exp(\lambda_{238}t) - 1]}{[\exp(\lambda_{235}t) - 1]} \tag{3.39}$$

Using the values, $^{235}N/^{238}N = 1/139$, $\lambda_{235} = 9.72 \times 10^{-10}$ y^{-1}, $\lambda_{238} = 1.52 \times 10^{-10}$ y^{-1}, $^{206}N/^{207}N = 25$ was estimated to be about 2.2×10^9 y. Accepted age of the earth was 3×10^9 y, derived from various sampling and by different methods.

Results would be modified by taking into account the ^{206}Pb atoms which were already present in the sample when the element was formed (primeval abundance). The modified equation would be:

$$^{206}N - {}^{206}_0N = {}^{238}N\left[\exp(\lambda_{238}t) - 1\right] \tag{3.40}$$

There will be a certain primeval abundance ratio among the four lead isotopes, $\text{Pb}^{204,206,207,208}$. Only the amount of ^{204}Pb does not change with time as it alone is not created by radioactive processes. It may be taken as a measure of the original lead in the sample examined. Obviously

$$^{204}N = {}^{204}_0 N$$

We can then rewrite (3.39)

$$\frac{^{206}N}{^{204}N} - \frac{^{206}_0 N}{^{204}_0 N} = \frac{^{238}N}{^{204}N}[\exp(\lambda_{238}t) - 1] \tag{3.41}$$

Corresponding relations can be written for N^{207} and N^{208}. Typical values for the ratio $^{206}N/^{204}N$ are 15–18.

The ratio $^{206}_0 N/^{204}_0 N$ is found by assuming that the lead found in iron meteorites has the same proportion for the four isotopes in the primeval lead, and no alteration in these proportions is possible owing to the extreme smallness of uranium content in these meteorites. Using the ratio $^{206}_0 N/^{204}_0 N \sim 9.4$, t is found to be 4.5×10^9 y. Astronomical observations from recession of galaxies give $t \sim 3.8 \times 10^9$ y (time being measured from the moment all the galaxies were together).

Example 3.11   The present-day abundance ratio of the two isotopes of uranium, ^{238}U and ^{235}U is $137.8 : 1$. Assuming that the abundance ratio could never have been greater than unity, calculate the maximum possible age of the earth. Half-lives of ^{238}U and ^{235}U are 4.5×10^9 y and 7.13×10^8 y respectively.

Solution   If $N(238)$ and $N_0(238)$ refer to present and original numbers of ^{238}U nuclei involved, then

$$N(238) = N_0(238)\exp(-\lambda_8 t)$$

where λ_8 is the decay constant of ^{238}U and t is measured from $t = 0$, i.e. t is the age of the earth. If T_8 refers to the half-life of ^{238}U, then

$$\frac{N(238)}{N_0(238)} = \exp(-0.693t/T_8)$$

Similarly for ^{235}U

$$\frac{N(235)}{N_0(235)} = \exp(-0.693t/T_5)$$

Therefore

$$\frac{N(238)}{N_0(238)}\frac{N_0(235)}{N(235)} = \exp\left[0.693t\left(\frac{1}{T_5} - \frac{1}{T_8}\right)\right]$$

Assuming $N_0(235)/N_0(238) = 1$, the maximum value when $t = t_{max}$, $\log(137.8) = 0.4343 \times 0.693 \times 1.18 \times 10^{-9} \times t_{max}$ or, $t_{max} = 6 \times 10^9$ y.

Example 3.12 In the radioactive series stemming from ^{238}U, intermediate nuclei have negligible mean lives on geological time scale, so that ^{238}U may be assumed to decay directly to ^{206}Pb with a mean life of 6.48×10^9 y. Similarly, ^{235}U decays to ^{207}Pb with a mean life of 1.03×10^9 y. In a certain sample of uranium-bearing rock the proportions of atoms of ^{238}U, ^{235}U, ^{206}Pb and ^{207}Pb were found to be in the ratio of $1000 : 7.19 : 79.7 : 4.85$. The sample had negligible amount of ^{208}Pb. Estimate the age of the rock.

Solution   That the sample contained negligible amount of ^{208}Pb implies that all the lead was formed from the decay of uranium. Suppose that when the sample of rock was formed T years ago, it contained no lead but N_1 atoms of ^{238}U and N_2 atoms of ^{235}U whose mean lives are τ_1 and τ_2 respectively. The rock would now contain $N_1 \exp(-T/\tau_1)$ atoms of ^{238}U. Since each decayed ^{238}U becomes ^{206}Pb the rock now contains $N_1(1 - \exp(-\frac{T}{\tau_1}))$ atoms of ^{206}Pb. Therefore, from the given abundances

$$\frac{N_1(1 - \exp(-T/\tau_1))}{N_1 \exp(-T/\tau_1)} = \frac{79.7}{1000} = 0.0797$$

$$T = \ln(1.0797) \times 6.48 \times 10^9 \text{ y} = 4.97 \times 10^8 \text{ y}$$

Similarly, for ^{235}U and ^{207}Pb, one can obtain $T = 5.31 \times 10^8$ y, the agreement between the two values being reasonably good.

3.1.9 Radiocarbon Dating

^{14}C is an isotope of carbon which is radioactive and has half-life of 5570 years. It decays through the scheme, $^{14}C \rightarrow N^{14} + \beta^- + \nu_e$. The study of ^{14}C content in certain substances provides reliable information about the ages of articles of archaeological interest.

3.1.9.1 Production of ^{14}C-Theory

The incidence of high energy primary cosmic rays, mainly protons, on the top of atmosphere causes occasional spallation of nitrogen or oxygen nuclei. This nuclear break-up produces energetic neutrons amongst other fragments. Fast neutrons get slowed down through successive elastic or inelastic collisions with the nuclei of the components of air. The neutron induced reactions with ^{14}N produce important isotopes ^3H and ^{14}C through

$$n + {}^{14}N \rightarrow {}^{12}C + {}^3H \qquad\qquad (i)$$

$$n + {}^{14}N \rightarrow {}^{12}C + {}^1H \qquad\qquad (ii)$$

The cross-section for reaction (ii) to occur is as high as 1.7 barns at thermal energy, while the corresponding reaction with ^{16}O is only 0.1 per cent of this value.

The reaction $n + {}^{14}N \rightarrow {}^{11}B + {}^{4}He$ is dominant at energies above 1 MeV but even at the most favoured energies, attains cross-sections of only 10 per cent of that of nitrogen for thermal energies. Reaction (ii) is so much more probable that the yield of radiocarbon will be nearly equal to the total number of neutrons generated by the cosmic rays and retained on earth.

If we assume that the cosmic ray intensity has remained constant over the last 20000 or 30000 years with usual daily and annual fluctuations, then the rate of disintegration of ^{14}C will be equal to its rate of formation. Once the radiocarbon atoms are introduced into the air, at a height of some 5 or 6 miles, it seems certain that within a few minutes or hours the carbon atoms would have been burned to carbon dioxide molecules. Radiocarbon dioxide thus formed will be inhaled by plants along with ordinary CO_2. Consequently, all plants will be rendered radioactive and since animals live off the plants, all animals will be rendered radioactive by the cosmic radiation. The time for radiocarbon in air to disperse is not in excess of 500 to 1000 years. Since this time is short compared to the life time of ^{14}C, the distribution of ^{14}C is uniform over all latitudes and longitudes notwithstanding, the fact that the cosmic ray neutron component varies by a factor of 3.5 between equatorial and polar regions.

There are some 8.3 g of carbon in exchange equilibrium with the atmospheric CO_2, for each cm^2 of the earth's surface on the average and since there are some 2.6 ^{14}C atoms formed/cm^2/s, we can expect a specific radioactivity of 2.6/8.3 disintegrations/s/g or 18.8/min/g. This number, which is in agreement with the experimentally observed value of 16.1, gives credence to the theoretical picture discussed above. Living matter yields 15.3 disintegrations/min/g.

If the cosmic ray intensity has remained substantially constant during 20000 to 30000 years, and if the carbon reservoir has not changed appreciably in this time, then we expect a complete balance between the rate of disintegration of radiocarbon atoms and the rate of assimilation of new radiocarbon atoms for all living material. Thus, for example, in a tree or any other living organism radiocarbon assimilated from food would just balance the losses due to its decay in the tissues. However, when the living organism dies, the assimilation processes abruptly come to an end and the disintegration process alone remains.

If I is the specific activity of the dead organism and t is its age in years then

$$I = 15.3e^{-0.693t/5570} \tag{3.42}$$

Here, t is time (years) elapsed since its death.

3.1.9.2 Suitable Materials for Carbon Dating

Suitable materials for carbon dating must be such that the original carbon atoms must be preserved and should not be replaced by other atoms due to chemical changes. Organic materials usually made of macromolecules, such as cellulose and charcoal, are suitable candidates. The recommended materials for carbon dating are charcoal or charred organic materials, such as heavily burned bone, well preserved wood, well preserved shell, antler and similar hairy structures.

3.1.9.3 Technique

The sample is first cleaned and tested with hydrochloric acid or calcium carbonate. This is followed by controlled combustion of the sample, if the sample is organic, to form carbon dioxide, or the addition of hydrochloric acid if the sample is shell, and then purified of radon and other impurities by precipitation method. The resultant CO_2 is reduced to carbon. The metal and its oxide are removed by acid treatment. The carbon which is highly porous is transferred inside the counter. As the maximum energy of β-particles is only 155 keV, the end window of a GM counter would stop most of the particles to be counted. The main difficulty is the small counting rate of β-particles from the material against the background of cosmic radiation and stray radioactivity. The γ-rays (soft component of cosmic rays) background is largely reduced by shielding the instrument with a thick wall of steel and by surrounding the screen-wall counter in anti-coincidence with a tray of Geiger counters, the μ-meson (hard component) background is substantially reduced. It is then possible to measure the radiocarbon radiation to about 1 per cent error in 48 hours in a typical experiment. This corresponds to an error of 80 years. This is the main error. The method is suitable for ages ranging from 2000–20000 y. Other detection techniques employed are proportional counters and scintillation counters.

3.1.9.4 Applications

The method has been found invaluable in archaeology, history and life sciences. Ages can be estimated with only one gram of the substance. Some examples may be cited.

A wood from the tomb of Pharaoh, believed to be 5600 years old, was shown to be in agreement from the specific activity within ±400 y. From the analysis of a piece of charcoal found in a cave in France, it was shown that it was 15500 years old.

Villagers existed in Mexico about 7000 years ago and this may be the date which marks the transition of early American nomads to the life of a farmer and village dwellers. The age of a giant redwood tree felled in 1874 was determined to be 2710 years old, which more or less agreed with the age of 2905 years determined by the number of rings.

Previously botanists had believed that seeds over 200 years old cannot germinate. When a lake in Manchuria had dried up, lotus plants found in the mud were found to be 1000 years old. This set the date for drying up of the lake. These lotus seeds, however, germinated and within four months the first seedling blossomed into a lotus.

Example 3.13 Determine the age of ancient wooden, items if it is known that the specific activity of ^{14}C nuclide in the same sample amounts to 3/5 of that in lately felled trees. The half life of ^{14}C nuclei is 5570 years.

Solution

$$\left|\frac{dN}{dt}\right| = \left|\frac{dN_0}{dt}\right| \exp(-\lambda t)$$

$$\frac{|\frac{dN}{dt}|}{|\frac{dN_0}{dt}|} = \frac{3}{5} = e^{-0.693t/5570}$$

Hence, $t = 4180$ years.

Example 3.14 The carbon isotope ^{14}C is produced in nuclear reactions of cosmic rays in the atmosphere. It is β-unstable. It is found that a gram of carbon extracted from the atmosphere, gives on average 15.3 such radioactive decays per minute. What is the proportion of ^{14}C isotope in the carbon? (Mean life of ^{14}C is 8270 years and mass of carbon is 12.00 u.)

Solution Suppose there are N nuclei of ^{14}C in the sample. Then the mean number of decays per second is

$$\left|\frac{dN}{dt}\right| = N\lambda \quad \text{or} \quad N = \frac{1}{\lambda}\left|\frac{dN}{dt}\right|$$

Hence, $N = \frac{15.3}{60} \times 8270 \times 3.15 \times 10^7 = 6.65 \times 10^{10}$. The mass of carbon is $12.00\,\text{u} = 12.00 \times 1.66 \times 10^{-24}\,\text{g} = 1.994 \times 10^{-23}\,\text{g}$. Therefore, one gram of carbon contains

$$\frac{1}{1.994 \times 10^{-23}} = 5.015 \times 10^{22} \text{ atoms}$$

Hence, the proportion of ^{14}C in one gram of the sample is

$$= \frac{6.65 \times 10^{10}}{5.015 \times 10^{22}} = 1.326 \times 10^{-12}$$

Example 3.15 A total of 16.0 radioactive disintegrations per minute per gram are measured from a sample of living wood. The counter used for measuring the sample is only 5 % efficient. A sample of 8 g of carbon taken from archaeological sample of wood register 9 counts per minute. The background count of 5 per minute is also observed. Calculate the age of the sample ($T_{1/2}$ of ^{14}C $= 5730$ y).

Fig. 3.4 Plot of potential energy between α particle and residual nucleus as a function of distance of separation r

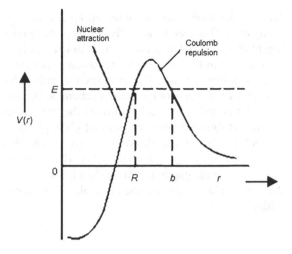

Solution Count rate for archaeological specimen $= 9.0 - 5.0 = 4.0$ counts per minute. Hence, number of disintegrations

$$= 4 \times \frac{100}{5} = 80 \text{ counts per min per 8 g}$$

$$= 10 \text{ counts per min per g}$$

But, $A = A_0 \exp(-\lambda t)$, or $2.303 \log(A_0/A) = \lambda t = \frac{0.693t}{T_{1/2}}$. By the problem, $A_0 = 16.0$, $A = 10$ and $T_{1/2} = 5730$. Therefore, age of the sample is

$$t = \frac{2.303 \log(A_0/A) \times 5730}{0.693} = 3887 \text{ y}$$

3.2 α Decay

It is known that α-energy from radioactive nuclei have much smaller energy than the potential barrier height with respect to the ground level. For example, the decay of $^{238}\text{U} \rightarrow {}^{234}\text{Th} + \alpha$. Figure 3.4 shows the curve of potential energy $V(r)$ between an α-particle and the residual nucleus as a function of distance of separation r.

At large distances from the nucleus α is repelled electrically and will have a potential energy $V(r) = 2Ze^2/r$, where $2e$ is the charge of α and Ze is that of the residual nucleus. When α reaches the nucleus, the repulsion is rapidly overbalanced due to the attractive nuclear forces.

The Coulomb potential is extended inward to a radius R and then arbitrarily cut off for mathematical simplicity. We can calculate the potential barrier height for the decay of ^{238}U

$$V(R) = \frac{1.44 z Z}{R} = \frac{1.44 \times 2 \times 90}{8} = 32.4 \text{ MeV}$$

This is a lot more than the observed kinetic energy of α's, $E = 5$ MeV from the decay of ^{238}U nuclei. Classically α's must have energy at least equal to the barrier height of 32.4 MeV. If α energy E is much less to cross over the repulsive barrier, then according to classical physics α remains trapped inside the nucleus. This was an outstanding paradox. This paradox was solved by Gamow, Gurney and Condon in 1928 by invoking for quantum mechanics. Because of its wave properties, α has a small probability of leaking through the barrier. An optical analogue is the transmission of light between two parallel glass plates facing each other. When the air gap is large, then for angles of incidence greater than the critical angle, no light is transmitted from one glass to the other. However, if the glass plates are brought so close that the air gap is not more than a few wavelengths, then it is observed that the light is able to get into the other plate even for incident angles greater than the critical angle.

3.2.1 Potential Barrier Problem

Consider a stream of particles of energy $E < V$ (region I). The incoming particles can be considered as a wave along positive x-axis (Fig. 3.5) with a plane wave function $A \exp(i k_1 x)$. Here, $k_1^2 = 2mE/\hbar^2$. Part of the beam is reflected at the wall. This wave is represented by the wave function $A \exp(-i k_1 x)$ along the negative x-axis.

In region III we expect only an outgoing wave of the form $D \exp(i k_3 x)$. But $k_3 = k_1$. In region II we must have a linear combination of $\exp(k_2 x)$ and $\exp(-k_2 x)$. Schrodinger's equation is:

$$\frac{d^2 u}{dx^2} + \frac{2m}{\hbar^2}(E - V)u = 0 \tag{3.43}$$

Region I $(x < 0)$; $V = 0$
  As $V = 0$, Eq. (3.43) reduces to

$$\frac{d^2 u}{dx^2} + k_1^2 u = 0 \tag{3.44}$$

The solution is thus

$$u_1 = A \exp(i k_1 x) + B \exp(-i k_1 x) \tag{3.45}$$

Region II $(0 < x < a)$, $V = V_0$
Equation (3.43) becomes

$$\frac{d^2 u_2}{dx^2} - \frac{2m}{\hbar^2}(V_0 - E)u_2 = 0 \quad \text{or} \tag{3.46}$$

$$\frac{d^2 u_2}{dx^2} - k_2^2 u_2 = 0 \tag{3.47}$$

Fig. 3.5 Penetration through a rectangular barrier

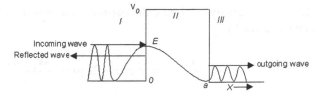

with

$$k_2^2 = \frac{2m}{\hbar^2}(V_0 - E) \tag{3.48}$$

$$u_2 = C\exp(k_2 x) + D\exp(-k_2 x) \tag{3.49}$$

For thin barriers both the exponentials are admissible. Here, for mathematical simplicity we will assume the case of a thick barrier for which we set $C = 0$, so that (3.49) becomes

$$u_2 = D\exp(-k_2 x) \tag{3.50}$$

Region III $(x > a)$
The solution of (3.43) is

$$u_3 = F\exp(ik_1 x) \tag{3.51}$$

A, B, C, D and F are the amplitudes.

Since $k_3 = k_1$. There is no wave proceeding in the opposite direction in region III. For this reason the term $\exp(-ik_3 x)$ is absent.

Continuity conditions for the wave functions require that the amplitude and the first derivatives of the wave functions at the boundary be equal. At $x = 0$, $u_1(0) = u_2(0)$ and $u_1'(0) = u_2'(0)$. These lead to

$$A + B = C \tag{3.52}$$

$$ik_1(A - B) = -k_2 C \tag{3.53}$$

Solving (3.52) and (3.53)

$$2A = C\left[1 - \frac{k_2}{ik_1}\right] \tag{3.54}$$

at

$$x = a, \qquad u_3(a) = u_2(a)$$

$$F\exp(ik_1 a) = C\exp(-k_2 a) \quad \text{or}$$

$$F = C\exp(-ik_1 a)\exp(-k_2 a) \tag{3.55}$$

Relative probability of finding the particle transmitted, called transmission coefficient, is given by

$$T = \frac{|F|^2}{|A|^2} = |C|^2 \exp(-2k_2a) = \frac{4k_1^2 \exp(-2k_a a)}{k_1^2 + k_2^2} \qquad (3.56)$$

where we have used (3.55). Using (3.42) and (3.48) in (3.56) we get

$$T = \frac{4E}{V_0} \exp(-2k_2a) \qquad (3.57)$$

If we do not put the amplitude $C = 0$ and carry out the analysis as above, the following expressions without this approximation would result, the transmission coefficient being given by

$$T = \frac{|F|^2}{|A|^2} = \frac{4k_1^2 k_2^2}{(k_1^2 + k_2^2)^2 \sin h^2 k_2a + 4k_1^2 k_2^2} \qquad (3.58)$$

The probability of reflection, called reflection coefficient, is given by

$$R = \frac{|B|^2}{|A|^2} = \frac{(k_1^2 + k_2)^2 \sin h^2 k_2a}{(k_1^2 + k_2^2)^2 \sin h^2 k_2a + 4k_1^2 k_2^2} \qquad (3.59)$$

Notice that $T + R = 1$, as it should be. There are two interesting situations in which these formulae become easier to understand. Consider the purely formal limit in which $h \to 0$. The quantity $\hbar$ is a physical constant, but we can consider it as a mathematical variable in order to examine the classical limit of the formulae (3.58) and (3.59). As $\hbar \to 0$, k_1 and k_2 approach infinity and $T \to 0$ and $R \to 1$, which is of course the expected behaviour of a classical particle for $E < V_0$.

The other interesting limit occurs when the transmission is small. This happens when the barrier width is large, and $k_2a \gg 1$ or $(V_0 - E) \gg \hbar^2/2ma^2$. This limit also corresponds to having many wavelengths inside the barrier. Then the term $(k_1^2 + k_2^2) \exp(2k_2a)$ dominates the denominator of (3.58) and the transmission coefficient becomes approximately

$$T \simeq \frac{16k_1^2 k_2^2 \exp(-2k_2a)}{(k_1^2 + k_2^2)^2} \quad \text{or}$$

$$T = \frac{16E}{V_0} \left[1 - \frac{E}{V_0} \right] \exp\left[\frac{-2a}{\hbar} \sqrt{2m(V_0 - E)} \right] \qquad (3.60)$$

where we have used (3.42) and (3.48). The expression $16(E/V_0)[1 - (E/V_0)]$ is of the order of unity. Notice that the transmission coefficient is sensitive to the barrier width, energy change and mass of the particle.

Fig. 3.6 Penetration through barrier of arbitrary shape

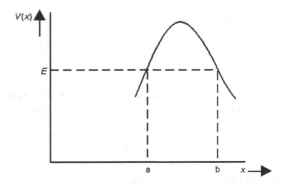

3.2.2 Barrier of an Arbitrary Shape

The order of magnitude of the transparency of an arbitrarily shaped barrier can be calculated by finding the average 'height' and treating it as a rectangular barrier (Fig. 3.6). The transparency given by (3.60) can be approximated to

$$T \simeq \exp\left\{-2\int_a^b \sqrt{\frac{2m}{\hbar^2}[V(x)-E]}dx\right\} \qquad (3.61)$$

where $a = R$ and b define the boundaries of the barrier which is to be crossed. It is assumed that when an α-particle is within the nucleus, it behaves like a free particle moving back and forth hitting the barrier.

Probability of escape/s = (Rate of hitting the barrier) × (Transparency)

$$P = fT = \lambda = \frac{1}{\tau} \qquad (3.62)$$

$$f = v_{(inside\ nucleus)}/R$$

where $R = a$ = nuclear radius

$$f \simeq \frac{10^7\ \text{m/s}}{10^{-14}\ \text{m}} = 10^{21}/\text{s} \qquad (3.63)$$

Combining (3.61), (3.62) and (3.63), we get

$$\lambda = \frac{1}{\tau} \simeq 10^{21}e^{-G} \qquad (3.64)$$

where

$$G = 2\int_a^b \sqrt{\frac{2\mu}{\hbar^2}[V(r)-E]}dr \qquad (3.65)$$

This is called Gamow's factor. Here, we have used the reduced mass μ, rather than m for the mass of the α-particle for greater accuracy.

For the α decay

$$V(r) = \frac{zZe^2}{r} \tag{3.66}$$

with $z = 2$.

The upper limit of the barrier is obtained by setting the kinetic energy E equal to the potential energy when it just leaves the barrier. Hence

$$\frac{1}{2}\mu v^2 = E = \frac{zZe^2}{b} \quad \text{or} \quad b = \frac{zZe^2}{E} \tag{3.67}$$

Using (3.66) and (3.67) in (3.65), we get

$$G = \sqrt{\frac{8\mu zZe^2}{\hbar^2}} \int_R^b \sqrt{\frac{1}{r} - \frac{1}{b}} \, dr \tag{3.68}$$

The integral can be easily evaluated by the substitution, $r = b\cos^2\theta$. Thus, the integral becomes

$$I = b\cos^{-1}\sqrt{R/b} \int (\cos 2\theta - 1)d\theta = \sqrt{b}\left(\cos^{-1}\frac{R}{b} - \sqrt{\frac{R}{b} - \frac{R^2}{b^2}}\right) \tag{3.69}$$

$$G = \sqrt{\frac{8\mu zZe^2 b}{\hbar^2}}\left[\cos^{-1}\sqrt{\frac{R}{b}} - \sqrt{\frac{R}{b} - \frac{R^2}{b^2}}\right]$$

For

$$b \gg R, \quad \cos^{-1}\sqrt{\frac{R}{b}} - \sqrt{\frac{R}{b} - \frac{R^2}{b^2}} \simeq \frac{\pi}{2} - \sqrt{\frac{R}{b}} \cdots - \sqrt{\frac{R}{b}} = \sqrt{\frac{\pi}{2}} - 2\sqrt{\frac{R}{b}}$$

An approximate expression for G is

$$G = \frac{2\pi zZe^2}{\hbar v} - \frac{4}{\hbar}\sqrt{2\mu zZe^2 R} \tag{3.70}$$

The first term is larger than the second one by the factor $(\frac{\pi}{4})\sqrt{b/R}$ and is therefore the dominant term. The first term of (3.70) is sometimes called Gamow exponent and the corresponding approximate value of the barrier transparency for s waves $(l = 0)$ through very high spherical barriers is called the Gamow factor.

$$T \simeq \exp\left(-\frac{2\pi zZe^2}{\hbar v}\right) = \exp\left[-(2\pi Zz/137\beta)\right] \tag{3.71}$$

where we have used the value of the fine structure constant $e^2/\hbar c = 1/137$, and $\beta = v/c$, c being the velocity of light. Expression (3.71) indicates that nuclear barrier will be quite impenetrable if $2zZ/137\beta \gg 1$ and again this inequality represents

Fig. 3.7 Plot of $\log \lambda$ vs
$\log x$ for three naturally
occurring series

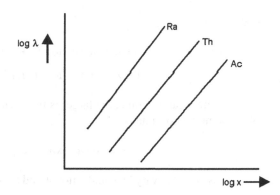

the domain in which classical theory can be expected to be valid. Further, (3.70) shows that emission of α-particles with small velocity, i.e. small energy will be inhibited. Also, the larger the Z (as in the case of fission fragments) the smaller is the barrier transparency.

3.2.3 Determination of Nuclear Radius

If the mean life time and the kinetic energy of an α emitter and Z be known, the nuclear radius R can be found out from (3.70) and (3.64). The values of r_0 are obtained from the formula $R = r_0 A^{1/3}$. The value of r_0 thus deduced is found to be 1.48 fm, which is higher by about 20 per cent obtained by other methods. The corrected radius of the daughter nucleus will be actually slightly smaller than R because of the finite radius of an α, particle. Thus, the effective radius of r_0 will be slightly less. It may be pointed out that larger the value of R, other things being equal, the thinner is the barrier width and hence the greater is the barrier transparency.

3.2.4 Geiger Nuttall Law

From the study of α emitters, Geiger and Nuttall discovered that members of naturally occurring radioactive series (e.g., Ra, Th, Ac) were found to fall along straight lines when $\log \lambda$ is plotted against $\log x$ (here $x =$ range) for each series, see Fig. 3.7. Therefore

$$\log \lambda = k \log x + c \tag{3.72}$$

where k and c are the constants.

Geiger also showed that

$$x = \text{const} \cdot v^3 = \text{const} \cdot E^{\frac{3}{2}} \tag{3.73}$$

so that

$$\log \lambda = \text{const} \cdot \log E + \text{const} \quad \text{or}$$

$$\log \lambda + \text{const} \cdot \log(1/E) = \text{const} \tag{3.74}$$

i.e. smaller the mean life time, the larger is the α energy.

Now Gamow's formula is

$$\lambda = \text{const} \cdot \exp(\text{const}/\sqrt{E})$$

Here, Z is assumed to vary by small amount and absorbed in the constant. Therefore

$$\log \lambda + (\text{const}/\sqrt{E}) = \text{const} \tag{3.75}$$

There is some resemblance between (3.74) and (3.75).

3.2.5 Success of Gamow's Theory

Because of the factor of v in the exponential in Gamow's formula (3.70), the mean life time varies enormously even for small variations in α energies. For example, it is known experimentally that α's from ^{232}Th with half life 1.4×10^{10} y gives α's of 5 MeV energy while Th C' of half life 0.3 μs gives α's of 8 MeV. Thus, a 50 per cent change in energy (E) corresponds to a change of 24 orders of magnitude in half life times. The explanation of the enormous inverse variation of life times with α energy is the brilliant success of Gamow's theory.

The above considerations are valid for transmission of s-waves ($l = 0$) through barrier. Fortunately, a large number of α emitters do enjoy this property. These are the nuclei which have even Z and even A. This will also be true for the daughter nucleus. These nuclei are universally known to have zero angular momentum. From the conservation of angular momentum, it follows that α's would carry s-waves. If $l \neq 0$, the solution of the modified equation becomes complicated. In general $l \neq 0$ has the effect of increasing the barrier height and thereby increasing the mean life time.

3.2.6 Fine Structure of α Spectrum

In a given type of α decay, α particle and the daughter nucleus carry unique energy as it is a two-body decay so that momentum and energy are shared uniquely. Most of the α transitions tend to go predominantly to the ground level of the decay product as the transition energy is the greatest. Transitions to excited level of the decay product, usually but not always, constitute a small fraction of the total transitions

Fig. 3.8 Transitions which give rise to long range and short range α's

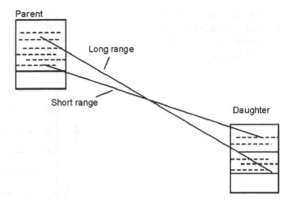

and are confined to low lying excited levels. As the transitions take place to distinct groups of energy levels of the daughter nucleus, the resulting spectrum has a fine structure. In the event the decay of a nucleus by β emission or K-capture leaves the product in an excited state, and the product is an alpha emitter with a very short life time, the additional excitation energy may be carried away in the α decay process rather than by γ emission, resulting in a long range α. On the other hand, if the α emitter is in the ground state or low lying excited state decays to a highly excited level, much less transition energy will be available, resulting in a short range α (Fig. 3.8).

3.2.7 Angular Momentum and Parity in α-Decay

In an α decay for the transitions between the initial nuclear state with spin I_i, to the final nuclear state with spin I_f, the angular momentum of α-particle can range from $I_i + I_f$ to $|I_i - I_f|$. Since α-particle has spin zero, the angular momentum carried by α-particle is purely orbital (l_α). The α-particle wave function is represented by Y_{lm} (spherical harmonics), and $l = l_\alpha$. Thus the parity change associated with α emission is (-1). This leads to the parity selection rule. Conservation of parity requires that if the initial and final states have the same parity, then l_α must be even. If parities are different then l_α must be odd.

As an example, Fig. 3.9 shows α-decay of ^{242}Cm to different excited states of ^{238}Pu characterized by the quantum numbers $0^+, 2^+, 4^+, 6^+, 8^+, \ldots$ for the ground state rotational band. The transitions from 0^+ state of ^{242}Cm to the enumerated states of ^{238}Pu viz. $0^+, 2^+, 4^+$ are allowed by the parity rule. It must be pointed out that the transition rates are quite different.

The intensity depends on the wave functions of the initial and final states and also depends on the angular momentum $l\alpha$. Centrifugal potential barrier $l(l+1)\hbar^2/2mr^2$ has the consequence of raising the thickness of Coulomb's barrier which is to be penetrated. Further, α decay involves different Q values. In this example, for the decay to the ground state, Q is 6.216 MeV. For excited states, the Q value is lowered by an amount equal to the excitation energy. The decay to 2^+ state has less

Fig. 3.9 α decay of ^{242}Cm to different excited states ^{238}Pu. The intensity of each α decay is also indicated

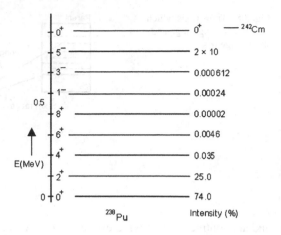

intensity than the decay to the Pu ground state 0^+. This is because the centrifugal potential raises the barrier by about 0.5 MeV and also the excitation energy lowers Q by 0.044 MeV. The decay intensity rapidly decreases as we go up the band to 8^+ state (Fig. 3.9). Once we go up above the ground-state band the intensities drop off to exceedingly low values ($\sim 10^{-7}$ of the total intensity). This results for the mismatch between the initial and final wave functions that are in contrast with the paired vibrationless 0^+ ground state.

Certain types of transitions are absolutely forbidden by parity conservation, for example the 2-states at 0.986 MeV, the 3^+ state at 1.070 MeV and the 4^- state at 1.083 MeV Thus, a $0 \rightarrow 3$ decay must have $la = 3$ which must give a change in parity between the initial and final states. Hence, $0^+ \rightarrow 3^-$ is possible, but $0^+ \rightarrow 3^+$ is forbidden.

Example 3.16 The a particles emitted in the decays of $^{226}_{88}$Ra and $^{226}_{90}$Th have energies 4.9 and 6.5 MeV, respectively. Ignoring the difference in their nuclear radii, find the ratio of their half-life times.

Solution   Using (3.71), $\lambda = 1/\tau = 10^{21} \exp(-2\pi zZ/137\beta)$ with $E_1 = 4.9$ MeV

$$\beta_1 = \sqrt{\frac{2E_1}{Mc^2}} = \sqrt{\frac{2 \times 4.9}{3727}} = 0.05174$$

$$E_2 = 6.5 \text{ MeV}$$

$$\beta_2 = \sqrt{\frac{2 \times 6.5}{3727}} = 0.05906$$

$$\frac{T_{1/2}(1)}{T_{1/2}(2)} = \frac{\tau_1}{\tau_2} = \frac{\exp(\frac{2\pi z}{137}\frac{Z_1}{\beta_1})}{\exp(\frac{2\pi z}{137}\frac{Z_2}{\beta_2})}$$

Putting $z = 2$ for α-particle, $z_1 = 86$ and $z_2 = 88$ for the daughter nuclei, we get

$$\frac{\tau_1}{\tau_2} = \exp\frac{4\pi}{137}\left[\frac{86}{0.05174} - \frac{88}{0.05906}\right] = e^{15.78} = 7.13 \times 10^6$$

Example 3.17 If two α-emitting nuclei, with the same mass number, one with $Z = 82$ and the other with $Z = 84$ had the same decay constant, and if the first emitted α-particles of energy 5.0 MeV, estimate the energy of α-particles emitted by the second.

Solution

$$\lambda_1 = 10^{21}\exp\left(-2\pi z Z_1 e^2/\hbar v_1\right)$$
$$\lambda_2 = 10^{21}\exp\left(-2\pi z Z_2 e^2/\hbar v_2\right)$$

as

$$\lambda_1 = \lambda_2$$
$$\frac{Z_1}{\sqrt{E_1}} = \frac{Z_2}{\sqrt{E_2}}$$
$$E_2 = E_1\left[\frac{Z_2}{Z_1}\right]^2 = 5.3\left[\frac{82}{84}\right]^2 = 5.05 \text{ MeV}$$

Example 3.18 In the α-decay $^8_4\text{Be} \rightarrow \alpha + \alpha$, the kinetic energy released is 0.094 MeV. Estimate the mean life of ^8_4Be ($r_0 = 1.1$ fm, $\tau_0 = 7 \times 10^{-23}$ s and $m_\alpha = 3728.43$ MeV. τ_0 is the mean lifetime without the Gamow factor).

Solution

$$\frac{1}{\tau} = \frac{1}{\tau_0}e^{-G}$$

By (3.70)

$$G = \frac{2\pi z Z}{137\beta}\left[1 - \frac{4}{\pi}\sqrt{\frac{R}{b}}\right]$$

$$b = \frac{zZe^2}{4\pi\varepsilon_0} = \frac{1.44 \times 2 \times 2}{0.094} = 61.27 \text{ fm}$$

$$\beta = \sqrt{\frac{2E}{\mu c^2}} = \sqrt{\frac{2 \times 0.094}{0.5 \times 3727}} = 0.01$$

$$R = 2r_0 A^{1/3} = 2 \times 1.1 \times 4^{1/3} = 3.49 \text{ fm}$$

$$G = \frac{2\pi \times 2 \times 2}{137 \times 0.01}\left(1 - \frac{4}{\pi}\sqrt{\frac{3.49}{61.27}}\right) = 12.74$$

$$\tau = \tau_0 e^G = 7 \times 10^{-23} e^{12.74} = 2.4 \times 10^{-17} \text{ s}$$

Example 3.19   Calculate the energy to be imparted to an α-particle to force it into the nucleus of $^{208}_{82}$Pb ($r_0 = 1.2$ fm).

Solution   The energy required to force an α-particle into a nucleus of charge Ze is equal to the maximum height of the potential barrier, i.e. $\frac{Zze^2}{4\pi\varepsilon_0 R}$, where $R = $ radius of the α-particle + radius of the nucleus, i.e. $R = r_0[A_\alpha^{1/3} + A_{nuc}^{1/3}] = 1.2 \times 10^{-15}[4^{1/3} + 208^{1/3}] = 9.01$ fm. Hence, the energy required is

$$E = \frac{1.44 \times 2 \times 82}{9.01} \text{ MeV} = 26.2 \text{ MeV}$$

Example 3.20   Radium, Polonium and Ra C are all members of the same radioactive series. Given that the range in air at S.T.P. of the α-particles from Radium (half-life 1622 years) is 3.36 cm, whereas from Ra C′ (half-life time 3.3×10^{-9} s) the range is 6.97 cm. Calculate the half-life of Polonium for which the α-particle range at S.T.P. is 3.85 cm assuming the Geiger Nuttall rule.

Solution   Using (3.72)

$$\log \lambda = k \log x + c$$

$$\log \frac{0.693}{T} = k \log x + c$$

$$K \log x + \log T = \log 0.693 - c = C' \tag{i}$$

$$K \log 3.36 + \log(1622 \times 365) = C' \tag{ii}$$

$$K \log 6.97 + \log(3.3 \times 10^{-9}/86400) = C' \tag{iii}$$

Solving (ii) and (iii), $K = 60.57$, $C' = 37.65$. Use the values of K, C', and $x = 3.85$ cm in (i) to find

$$T = 155 \text{ D}$$

3.3 β-Decay

It was known that β spectrum is not discrete but continuous. There was an apparent loss of energy. In many cases γ-rays were not detected. This was explained by Pauli [3] by postulating the existence of neutrino, (ν)—a neutral, massless particle of spin 1/2. A neutrino was assumed to accompany the β decay but could not be detected owing to its extremely feeble interaction with nuclear matter. The decay being

Fig. 3.10 β-ray spectrum

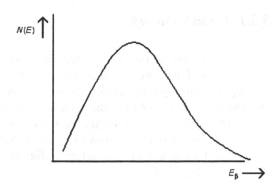

a three-body process accounts for the continuous nature of the spectrum (Fig. 3.10). $N(E)$ is the number of particles at β particle energy E_β.

Energy conservation gives:

$$E_\beta + E_\nu + E_N = Q = \text{const} \tag{3.76}$$

where E_β, E_ν and E_N, are the energies of β particle neutrino and residual nucleus respectively.

Momentum conservation gives

$$\vec{P} + \vec{P_\nu} + \vec{P_N} = 0 \tag{3.77}$$

Now there are variety of ways in which β particle is emitted such that (3.76) and (3.77) are satisfied. The angle between any two particles is not the same. This leads to a variety of configurations. The phenomenon is considered as the decay of one of the nucleons of the nucleus

$$n \to p \to \beta^- + \bar{\nu}_e$$

$$p \to n + \beta^+ + \nu_e$$

The converted nucleon is lodged within the nucleus as the product nucleus recoils. Here we have introduced the suffix e for the neutrino as it accompanies the β-decay to distinguish it from other two types of neutrinos which will be considered in [2], Chaps. 3 and 4. Further, it was found that the anti-neutrino $\bar{\nu}_e$ is different from ν_e. Typical examples of β decay are

$$^3_1\text{H} \to {}^3_2\text{He} + \beta^- + \bar{\nu}_e$$

$$^{27}_{14}\text{Si} \to {}^{27}_{13}\text{Al} + \beta^+ + \nu_e$$

Notice that in the presence of ν, angular momentum is also conserved.

3.3.1 Fermi's Theory

(a) The Coulomb interaction between the β particle and the residual nucleus is neglected. Later, this effect is incorporated in theory. Coulomb interaction is negligible for light nuclei ($Z \leq 10$) and for sufficiently high β particle energy.
(b) The recoil energy (E_R) of the nucleus is neglected.

   The recoil energy is negligible in all cases since the mass of the nucleus is much larger than the mass of electron. Since neutrino is assumed to have zero rest mass, and when it carries negligible energy, maximum total energy, of electron $E_{max} \simeq Q$

$$P_{e(max)} = \sqrt{E_{max}^2 - m_e^2} \tag{3.78}$$

Also, the recoil energy of the nucleus E_R is given by

$$E_R = \frac{P_{N(max)}^2}{2AM} = \frac{P_{e(max)}^2}{2AM} \tag{3.79}$$

as $P_N \simeq P_e$, and M is the mass of nucleon. Combining (3.78) and (3.79) we get

$$E_R \simeq \frac{m_e^2}{2AM}\left[\frac{E^2\max}{m_e^2} - 1\right] \quad \text{or} \tag{3.80}$$

$$\frac{E_R}{E_{max}} = \frac{m_e}{2AM}\left[\frac{E_{max}}{m_e} - \frac{m_e}{E_{max}}\right] \tag{3.81}$$

In a typical case, $A \simeq 20$, $T_{max} = 0.3$ MeV or $E_{max} = 0.8$ MeV, nucleon mass $M = 940$ MeV and electron mass $m_e = 0.51$ MeV. The bracket in (3.81) is of the order of unity and $E_R/E_{max} \sim 1.3 \times 10^{-5}$. Thus E_R is negligible. Similarly, it can be shown that when E_β has minimum value, E_ν has maximum value and once again E_R is negligible. This is also true for intermediate cases.

3.3.1.1 β-Ray Spectrum

Fermi used quantum mechanical time dependent perturbation theory for the β emission analogous to photon emission. Here the particles to be considered are β^- and $\bar{\nu}$ only. Thus

$$E_\beta + E_\nu = Q = E_0 \tag{3.82}$$

As the level of parent nucleus is not sharp, E_0 is not strictly constant. According to perturbation theory, the probability per unit time for the emission of β particle is given by:

$$\lambda = \frac{2\pi}{\hbar}|H_{if}|^2 \frac{dn}{dE_0} \tag{3.83}$$

where H_{if} is the matrix element given by:

$$H_{if} = \int \Psi_f^* H \Psi_i d\tau \tag{3.84}$$

Ψ_i and Ψ_f are initial and final wave functions of the complete system, $d\tau$ is the volume element and H is the operator that describes the weak interaction energy between two parts of the system. dn/dE_0 is the density of final states.

3.3.1.2 Statistical Factor

We postulate that the probability of disintegration leading to a specific accessible volume of phase-space is directly proportional to that volume. Consider the β particle with momentum P_β suddenly appearing at a certain point in phase-space defined by certain Cartesian space coordinates and certain momentum coordinates. Let the particle be restricted to a volume V in actual space with momentum lying between P_β and $P_\beta + \Delta P_\beta$ with an unspecified direction

The volume in phase-space $= (V)\left(4\pi P_\beta^2 d P_\beta\right)$

Number of cells, i.e. number of electron states $= 4\pi P_\beta^2 d P_\beta V / h^3$

Hence, the number of electron states/unit momentum interval

$$\frac{dn}{dP_\beta} = \frac{4\pi P_\beta^2 V}{h^3} \tag{3.85}$$

Number of neutrino states with neutrino momentum between $P\nu$ and $P\nu + dP\nu$

$$= (V)\frac{(4\pi P_\nu^2 d P_\nu)}{h^3} \tag{3.86}$$

Combining (3.85) and (3.86), total number of states is given by:

$$dn = \frac{(4\pi V P^2 d P)}{(h^3)} \frac{(4\pi V P_\nu^2 d P_\nu)}{(h^3)} \tag{3.87}$$

Here, we have dropped the subscript β.

Number of states/unit energy of electron is given by:

$$\frac{dn}{dE_0} = \frac{16\pi^2 V^2}{h^6} P_\nu^2 P^2 d P_\nu \frac{dP}{dE_0}$$

P_ν is determined not only by momentum conservation (being 3 body decay), but also by energy conservation (nuclear recoil energy being negligible). From (3.82)

$$E_v = E_0 - E_\beta \tag{3.88}$$

$$P_v = \frac{E_v}{c} = \frac{E_0 - E_\beta}{c} \tag{3.89}$$

Holding β energy and momentum constant

$$dP_v = \frac{dE_v}{c} = \frac{dE}{c} \tag{3.90}$$

Using (3.89) and (3.90) in (3.87) we obtain

$$\frac{dn}{dE_0} = \frac{16\pi^2 V^2 (E_0 - E)^2 P^2 dP}{h^6 c^3} \tag{3.91}$$

3.3.1.3 Matrix Element

$$H_{if} = \int \Psi_f^* H \Psi_i d\tau \tag{3.92}$$

Fermi put $H = g$. The quantity g is called the coupling constant whose value shows the strength of interaction

$$H_{if} = g \int \Psi_{fN}^* \Psi_\beta^* \Psi_v^* \Psi_{iN} d\tau \tag{3.93}$$

In a typical β decay, the wavelengths of β or v are substantially longer (by an order of magnitude) than nuclear dimensions. We can then approximate their wave functions to those of plane waves

$$\Psi_\beta(r) = V^{-1/2} \exp(i K_\beta \cdot r) = V^{-1/2}[1 + i K_\beta \cdot r + \cdots] \tag{3.94}$$

$$\Psi_v(r) = V^{-1/2} \exp(i K_v \cdot r) = V^{-1/2}[1 + i K_v \cdot r + \cdots] \tag{3.95}$$

where $V^{-1/2}$ is the normalization constant. Replacing the above values at the centre, $(r \rightarrow 0)$

$$\Psi_\beta^*(0) = V^{-1/2}$$

$$\Psi_v^*(0) = V^{-1/2} \tag{3.96}$$

$$|H_{if}|^2 = \frac{g^2 |M_{if}|^2}{V^2}$$

where

$$|M_{if}|_{nucleus} = \int \Psi_{fN}^* \Psi_{iN} d\tau \tag{3.97}$$

and nuclear wave functions are also normalized. M_{if} is called the nuclear matrix element or overlap integral.

Fig. 3.11 β-rays momentum
spectrum

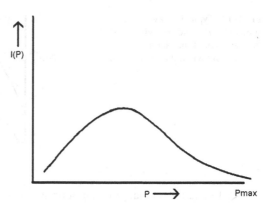

Using (3.91), (3.93) and (3.97) in (3.83), the differential energy spectrum is given
by

$$I(P)dP = \frac{g^2|M_{if}|^2(E_0 - E)^2 P^2 dP}{2\pi^3 \hbar^7 c^3} \tag{3.98}$$

This gives us the number of β particles/sec in the momentum interval P and $P + dP$,
(Fig. 3.11), neglecting coulomb interaction. The spectrum is written in the following
form

$$I(P)dP = K P^2 (E_{max} - E)^2 dP \tag{3.99}$$

where K is constant.

For small momentum, $I(P) \propto P^2$ since the term $(E_{max} - E)^2$ is not very sensitive at small P, i.e. small E; and for large momenta, i.e. E near E_{max}, $I(P) \propto$
$(E_{max} - E)^2$; $I(P)$ vanishes at both ends.

3.3.1.4 Coulomb Interaction

Coulomb interaction between the β particles and the residual nucleus has the consequence of distorting the energy spectrum. If the distortion of the wave function
by the Coulomb field of the nucleus is taken into account, a factor $F(Z, E)$ must be
included in (3.98). Non-relativistic expression for $F(Z, E)$ is

$$F(Z, E) = \frac{2\pi \eta}{1 - \exp(-2\pi \eta)} \tag{3.100}$$

where

$$\eta = \frac{Ze^2}{\hbar v} \quad \text{for electrons}$$

$$= -\frac{Ze^2}{\hbar v} \quad \text{for positrons}$$

Fig. 3.12 Typical β-ray
spectra for $\beta-$ and $\beta+$ with
the inclusion of the factor
$F(z, E)$ and without ($z = 0$)

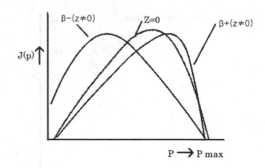

If $\eta \ll 1$ (small Z and large v, but nonrelativistic) $F(Z, E) \to 1$.

Qualitatively, the shift in the curves is due to the Coulomb attraction and repulsion for electrons and positrons respectively (Fig. 3.12).

3.3.1.5 Kurie Plot

It is difficult to determine experimentally the point at which the curve reaches the horizontal axis, because the curve makes a tangent to the axis at $P = P_{\max}$. A greater accurate determination of $E_{\max}$ is given by the Kurie plot. Including $F(Z, P)$ in (3.99), we get

$$I(P) = (E_{\max} - E)^2 p^2 F(Z, P)$$

Here, $F(Z, P)$ includes the constants and the dependence of nuclear charge Z for the daughter nucleus. Rewriting last equation we obtain

$$E_{\max} - E = \sqrt{\frac{I(p)}{P^2 F(z, p)}} \qquad (3.101)$$

Now the plot of the radical against energy should be a straight line whose intercept with the horizontal axis can be reliably determined (Fig. 3.13). Very thin sources should be used, otherwise due to self absorption and back scattering of β particles, Kurie plots would necessarily deviate from a + straight line.

3.3.1.6 The Mass of Neutrino

In Fermi's theory, if we treat neutrino with a finite mass, then the resulting Kurie plot would show deviation at large β energy. Taking into account the finite mass of neutrino m_ν, (3.101) is modified as

$$I(P)dP F(z, P) \propto P^2 (E_0 - E)^2 \sqrt{1 - \left(\frac{m_\nu c^2}{E_0 - E}\right)^2} \, dP \qquad (3.102)$$

Fig. 3.13 Kurie plot

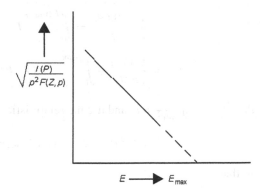

Fig. 3.14 Kurie plot showing kink (*dotted portion of curve*) for $m_\nu \neq 0$

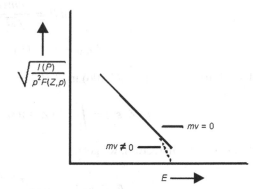

If $m_\nu = 0$, Fermi-Kurie plot is a straight line and intersects the energy E axis at $E_\beta = E_0$ given by Eq. (3.101). If $m_\nu \neq 0$, this plot becomes curved at the maximum end of the energy and intersects the E axis vertically at $E = E_0 - m_\nu c^2$. A determination of the distance between the intersections of extrapolated Fermi-Kurie plot for $m_\nu = 0$ and experimental curve on the energy E axis furnishes a value for the mass of the neutrino.

Figure 3.14 shows a Kurie plot for $m_\nu \neq 0$, which departs from a straight line. One can set an upper limit on neutrino's mass. Observations made with the tritium spectra (super-allowed transition with $T_{max} = 18$ keV) show that $m_\nu < 15$ eV.

3.3.1.7 Comparative Half-Lives

$\lambda_P = I(P)dP$ is the probability per unit time that an electron will be emitted in the momentum interval P and $P + dP$. The probability per unit time that an electron will be emitted with any momentum within limits is just the total disintegration constant and is given by

$$\lambda = \frac{1}{\tau} = \frac{0.693}{T_{1/2}} = \int_0^{P(\max)} I(P)dP$$

$$= \frac{g^2|M_{if}|^2}{2\pi^3\hbar^7c^3}\int_0^{P(\max)} F(Z,P)P^2(E_{\max}-E)^2dP \qquad (3.103)$$

Put $\frac{E_{\max}}{mc^2} = \varepsilon_0$, $\frac{E}{mc^2} = \varepsilon$, and use the relativistic equation

$$c^2P^2 = E^2 - m^2c^4 = (\varepsilon^2-1)m^2c^4$$

so that

$$P = \sqrt{\varepsilon^2-1}\,mc \qquad (3.104)$$

$$dP = \frac{\varepsilon mcd\varepsilon}{\sqrt{\varepsilon^2-1}} \qquad (3.105)$$

$$E_{\max} - E = (\varepsilon_0-\varepsilon)mc^2 \qquad (3.106)$$

Use (3.104), (3.105) and (3.106) in (3.103)

$$F(Z,\varepsilon_0) = \int_1^{\varepsilon_0} F(Z,\varepsilon)(\varepsilon_0-\varepsilon)^2\sqrt{\varepsilon^2-1}d\varepsilon \qquad (3.107)$$

Using (3.107) in (3.103), we get

$$F(Z,\varepsilon_0)T_{1/2} = \frac{2\pi^3\hbar^7}{m^5c^4g^2} \times \frac{0.693}{|M_{if}|^2} \qquad (3.108)$$

The value of the coupling constant is 0.9×10^{-4} MeV fm^3 or 1.4×10^{-49} erg cm^3. The product $fT_{1/2}$ or simply ft is called the comparative half-life. Formula (3.108) shows that $ft \propto |M_{if}|^{-2}$. Assuming that g is known, and that transition is 'allowed', we can use ft value to find M_{if}, the nuclear Matrix element. Since ft is a large value, smaller is M_{if}. The value of M_{if} is uncertain, but is of the order of unity for 'allowed' transitions. Formula (3.108) shows that $ft = $ const if $|M_{if}|^2$ is unchanged. The distribution of ft values for β emitters on the logarithmic scale shows that for values of $ft \sim 10^3$ transitions are allowed since matrix element is large enough and for those for which ft is higher, they are forbidden (Fig. 3.15). Clearly there is no significant grouping. Since $ft \propto 1/|M_{if}|^2$ the matrix elements should be of the same size for allowed decays and successively smaller for forbidden decays of increasing order. It is expected that the experimental ft values fall into groups according to the allowed and first, second, etc. forbidden transitions, so that the ft value of a given β emitter would at once give its order of forbiddeness. But in practice there is no visible clustering of ft values in various groups except for 'super allowed' transitions. The reason for lack of grouping is that the ft value for the given decay depends not only on the degree of forbiddeness, but also on the form of the nuclear wave functions that are not same for different decays.

Fig. 3.15 log ft distribution
for various β emitters

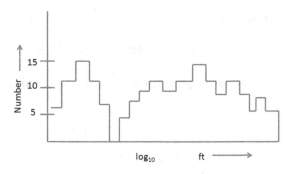

Fig. 3.16 Sargent diagram

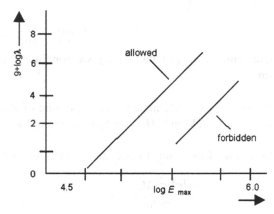

3.3.1.8 Sargent Diagrams

In 1933, Sargent found that the plot of $\log \lambda$ versus $\log E_{\max}$ for naturally occurring β emitters splits up into two straight lines, called Sargent diagram (Fig. 3.16). This was an empirical rule analogous to Geiger-Nuttall Rule in α decay. The upper curve represents allowed transitions (shorter life times) and the lower one represents the forbidden transitions (longer life times). Theory explains the Sargent diagrams. Rewriting (3.108) we get

$$\lambda = \frac{g^2 m^5 c^4 |M_{if}|^2 f(Z, \varepsilon_0)}{2\pi^3 \hbar^7} \tag{3.109}$$

with

$$f(Z, \varepsilon_0) \simeq \int_1^{\varepsilon_0} \varepsilon (\varepsilon_0 - \varepsilon)^2 \sqrt{\varepsilon^2 - 1} \, d\varepsilon$$

There is only one limiting case for which the integral can be evaluated analytically. To this end set $Z = 0$, so that $F(Z, \varepsilon) \to 1$. This result would be valid for low Z and large ε_0. Thus

$$f(0, \varepsilon_0) \simeq \int_1^{\varepsilon_0} \varepsilon (\varepsilon_0 - \varepsilon)^2 \sqrt{\varepsilon^2 - 1} \, d\varepsilon$$

If $\varepsilon_0 \gg 1$

$$f(0, \varepsilon_0) \simeq \int_1^{\varepsilon_0} \varepsilon^2 (\varepsilon_0 - \varepsilon)^2 d\varepsilon$$

where we have approximated $\sqrt{\varepsilon^2 - 1}$ to ε. Ignoring the contribution from the lower limit

$$(\varepsilon_0 \gg 1), \qquad f(0, \varepsilon_0) \simeq \frac{\varepsilon_0^5}{30} \quad \text{(fifth power law)} \qquad (3.110)$$

This relation is valid for $\varepsilon_0 \gg 5$.
For $\varepsilon_0 \ll 0.5$

$$f(0, \varepsilon_0) = \frac{2}{105} \varepsilon_0^7 \qquad (3.111)$$

Combining (3.111) with (3.110), we conclude that $\log \lambda$ vs $\log E_{max}$ is a straight line.

Example 3.21   Determine the half-life time of β emitter ^{6}He whose end point energy is 3.5 MeV and $|M_{if}|^2 = 6$. Take $g = 0.9 \times 10^{-43}$ MeV cm^3.

Solution   Combining (3.109) and (3.110), we obtain

$$\lambda = \frac{m^5 c^4 g^2 |M_{if}|^2}{60\pi^3 \hbar^7} \varepsilon_0^5$$

$$= \frac{g^2 (mc^2)^5 c |M_{if}|^2}{60\pi^3 (\hbar c)^7} \varepsilon_0^5$$

$$\varepsilon_0 = \frac{E_{max} + mc^2}{mc^2} = \frac{3.5 + 0.511}{0.511} = 7.849$$

$$\hbar c = 1.97 \times 10^{-11} \text{ MeV cm}$$

$$\lambda = \frac{(0.9)^2 \times 10^{-86} \times (0.511)^5 \times 3 \times 10^{10} \times 6 \times (7.849)^5}{60\pi^3 (1.97)^7 \times 10^{-77}} = 0.707$$

$$T_{1/2} = \frac{0.693}{\lambda} = \frac{0.693}{0.707} = 0.98 \text{ s}$$

The experimental value is 0.81 s.

3.3.2 Selection Rules

β transitions are classified as

(a) allowed or super-allowed transitions
(b) forbidden transitions

The allowed or super-allowed transitions are characterized by small ft values, and hence relatively short life times, with nuclear spin change, $\Delta I = 0, \pm 1$ and without parity change. Forbidden transitions are characterized by greater values of ft and therefore longer life times, with larger change in nuclear spin $\Delta I = \pm 2, \pm 3$ etc., depending on the degree of forbidness in the transitions with or without change in nuclear parity. The quantum numbers whose changes in a transition are governed by selection rules are total angular momentum and parity. As the life times ultimately depend on the matrix element $|M_{if}|$ which connects the initial and final state, the selection rules are governed by its value.

It has been pointed out that the plane wave functions for electron and neutrino can be expanded by a series of terms (formulae (3.94) and (3.95)). In the series expansion successive terms are of the order of $1/10$ compared to previous terms, R/λ being $\sim 1/10$, $(R/\lambda)^2 \sim 1/100$ and so on. The selection rules for a given order of transition, i.e. for a given term in the expansion give the necessary conditions on the change of the quantum numbers specifying the state of the nucleus so that the term of a given order in the complete matrix element be non-zero. Clearly, different orders in the expansion of the matrix element require different selection rules. This is because different orders correspond to even and odd power of r, which is a polar vector (vector of odd parity).

As far as spin is concerned in the β decay, two leptons of spin $\frac{1}{2}$ each are emitted. The spin orientations of these leptons may be parallel or anti-parallel. The electron and neutrino can be emitted in a singlet state (spins anti-parallel) or in a triplet state (spins parallel).

3.3.2.1 Allowed Transitions

When plane waves are expanded in series it is only the $l = 0$ term that carries the full amplitude of the plane wave at the origin (nucleus). This gives rise to the allowed transitions. Both electron and neutrino have the orbital angular momentum $l = 0$, so that the total orbital angular momentum becomes $L = 0$. If the particles are emitted in the singlet state, then the total angular momentum $J = 0$ and in the triplet state $J = 1$. Of the several possible forms for the matrix element in β-decay, two may be mentioned.

3.3.2.2 Fermi Rule

In the Fermi matrix element, the operator connecting the initial and final nuclear states is a unit operator which is a scalar. Thus,

$$M_F = \int \Psi_{fN}^* \Psi_{iN} d\tau$$

Therefore, no change in spin or parity is involved.

Fermi's selection rule requires $\Delta I = 0$ (i.e. $J_i = J_f$) and the leptons are emitted in the singlet state

$$\left. \begin{array}{c} \Delta I = 0 \\ I_i = 0 \quad \rightarrow \quad I_f = 0 \text{ allowed} \\ \Delta \pi = 0 \end{array} \right\} \quad \text{Fermi rule}$$

An example of pure Fermi transition is

$$^{14}O \rightarrow {}^{14}N^* + \beta^+ + \nu_e$$
$$[0^+] \rightarrow [0^+]$$

where the bracket refers to J^π value.

3.3.2.3 Gamow-Teller (G-T) Rule

The Gamow-Teller nuclear matrix element M_{GT} is a tensor and has the form

$$M_F = \int \Psi_{fN}^* \sigma \Psi_{iN} d\tau$$

where σ is the generalization of Pauli's spin matrices and is a pseudo-vector (axial vector) which does not change sign under space reflection. Hence, the initial and final states have the same parity. In the G-T selection rule, the leptons are emitted in the triplet state. Consequently

$$\left. \text{with} \quad \begin{array}{c} \Delta I = 0, \pm 1 \\ I_i = 0 \quad \rightarrow \quad I_f = 0 \text{ forbidden} \\ \Delta \pi = 0 \end{array} \right\} \quad \text{G-T rule}$$

The transition with $I_i = 0 \rightarrow I_f = 0$ is forbidden because this is not possible for the triplet state. An example of pure G-T transition is

$$^{6}He(0^+) \rightarrow {}^{6}Li(1^+) + \beta^- + \bar{\nu}$$

Following are the examples of mixed G-T and Fermi transitions:

(a) $n(\frac{1}{2}^+) \rightarrow p(\frac{1}{2}^+) + \beta^- + \bar{\nu}$
(b) $^{3}H(\frac{1}{2}^+) \rightarrow {}^{3}He(\frac{1}{2}^+) + \beta^- + \bar{\nu}$

The two types of decay (Fermi and G-T) are observed to proceed with approximately but not exactly at the same rate.

3.3.2.4 Super-allowed Transitions

These transitions occur between a pair of mirror nuclei for which the nuclear wave functions are nearly identical. This leads to a large value for the overlap integral. The $\log ft$ values for such transitions are small, being in the range of 3 to 3.7. These are characterized by

$$\Delta I = 0, \pm 1 \quad \text{and} \quad \Delta\pi = 0$$

The decays $n \to p$ and $^3\text{H} \to {}^3\text{He}$ are examples of super-allowed transitions.

3.3.2.5 Forbidden Transitions

It has been pointed out that for the allowed transitions the first term in the series for the plane waves expansion is considered. In the event the first term is zero, higher order terms corresponding to $l \neq 0$ may be considered. However, the higher orbital angular momentum waves have progressively smaller amplitudes inside the nuclear volume. Hence, the overlap integral is very small when the leptons are emitted with orbital angular momentum other than zero. The effect is to render the life-times of these forbidden transitions much larger on an average compared to those of allowed transitions. Another reason for occurrence of forbidden transitions is the relativistic effect in the nuclear wave functions. These cause a departure from a straight line in the Kurie plot. If the ordinary Kurie plot is not a straight line then the transition is not allowed. The converse, however, is not always true. Special types of forbidden transitions may give rise to straight Kurie plots.

If the transition is caused by the second term in the plane wave expansion leptons are emitted in a wave with $L = 1$ and changing parity. In that case the selection rules become

$$\left.\begin{array}{l} \Delta I = 0, \pm 1 \quad (\text{except } 0 \to 0) \\ \Delta\pi = \text{yes} \end{array}\right\} \quad \text{Fermi rule}$$

$$\left.\begin{array}{l} \Delta I = \pm 2, \pm 1, 0 \\ \Delta\pi = \text{yes} \end{array}\right\} \quad \text{G-T rule}$$

Similar selection rules apply for higher order forbidden transitions. Table 3.1 shows the characteristics of allowed as well as forbidden transitions with examples. The selection rules for parity are reminiscent of those used in optical atomic transitions. The expression e^{irk} is expanded in powers of r/λ. In optical emission $R/\lambda \sim 10^{-3}$ where R is the atomic size. The transition probability depends on the square of the matrix element. $\int \Psi_f^*$ (electric moment) $\Psi_i d\tau$, in case it is a dipole transition, Ψ_f and Ψ_i must be of opposite parity as dipole moment is a polar vector which changes its sign under space inversion. If the first term (dipole) is zero, transition may still proceed via quadrupole radiation but with lesser probability $(R^2/\lambda^2) \sim 10^{-6}$. This is the famous Laporte rule which states that for dipole transitions, odd (or even) terms of initial state combine with even (or odd) terms of the final state.

Table 3.1 Characteristics of allowed and forbidden transitions

Type of transition	ΔI	$\Delta \pi$	$\log ft$	Example
Super allowed	$0, \pm 1$	No	3–3.7	$^3\mathrm{H}(\frac{1}{2}^+) \to {}^3\mathrm{He}(\frac{1}{2}^+)$
Allowed	$0, \pm 1$	No	4–6	$^6\mathrm{He}(0^+) \to {}^6\mathrm{Li}(1^+)$
First forbidden	± 2	Yes	6–9	$^{39}\mathrm{Ar}(\frac{7}{2}^-) \to {}^{39}\mathrm{K}(\frac{3}{2}^+)$
Second forbidden	± 3	No	10–13	$^{22}\mathrm{Na}(3^+) \to {}^{22}\mathrm{Ne}(0^+)$
Third forbidden	± 4	Yes	15–18	$^{40}\mathrm{K}(4^-) \to {}^{40}\mathrm{Ca}(0^+)$
Fourth forbidden	± 4	No	19–23	$^{115}\mathrm{In}(\frac{9}{2}^+) \to {}^{115}\mathrm{Sn}(\frac{1}{2}^+)$

After the discovery of non-conservation of parity in weak interactions, such as β-decay, it was shown that all existing data on β-decay are in excellent agreement with the combination of the vector and axial vector interactions—this is the $V-A$ theory and that Fermi's theory of β decay is basically correct.

3.4 Range-Energy Relation

Although the β rays have a continuous energy spectrum, it is still possible to find a relation between $E_{\max}$, the maximum energy of the spectrum and the corresponding range R. A relation frequently used for a rapid determination of $E(\mathrm{MeV})$ in aluminum is due to Feather

$$R\left(\mathrm{g\,cm}^{-2}\right) = \begin{cases} 0.542E - 0.133 & \text{for } 0.8 < E < 3 \text{ MeV} \\ 0.407E^{1.38} & \text{for } 0.15 < E < 0.8 \text{ MeV} \end{cases} \tag{3.112}$$

Also, the intensity of β particles from the β decay is found to decrease exponentially with absorption thickness d

$$I = I_0 \exp(-\mu d) \tag{3.113}$$

where μ (cm^2/g) is the absorption coefficient.

3.4.1 Double β Decay

It is pointed out in Chap. 6 that for given even A nuclei there exist several isobars which are β stable as the even-odd nuclei on the lower parabola cannot β decay into the adjacent nuclei on the upper parabola since the mass on the latter is heavier. On the other hand, nuclei on the lower parabola are separated by two units of charge and they cannot be transformed into one another by β decay. However, the heavier may decay into the lighter one by a second order process. This process is called *double β decay*. This comes about due to simultaneous β decay of two neutrons or two protons of the same nucleus.

Fig. 3.17 Mass spectrum of
the $A = 76$ isobars

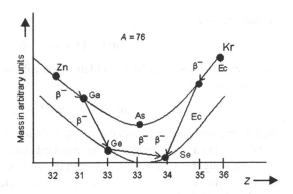

All the potential double beta emitters are even-even nuclei which due to the pairing interaction, have energetically lower ground states than adjacent odd-odd nuclei. Most of the $\beta\beta$ decay involve $0^+ \rightarrow 0^+$ transitions. In certain cases transitions to the first excited state 2^+ and a number of other excited states of daughter nucleus are energetically possible.

Figure 3.17 shows the mass spectrum ($A = 76$) of isobars. It is observed that for ^{76}Ge both the β^- and β^+ or EC decay are energetically forbidden. The only allowed decay mode is the $\beta^-\beta^-$ decay to ^{76}Se. In all, there are about 36 potential $\beta\beta$ emitters.

3.4.1.1 Various Decay Modes

There are two possibilities for the $\beta^-\beta^-$ decay mode

$$(Z, A) \rightarrow (Z + 2, A) + 2e^- + 2\bar{\nu}_e \quad (2\nu\beta\beta) \qquad (3.114)$$

$$(Z, A) \rightarrow (Z + 2, A) + 2e^- \quad \text{Neutrinoless process} \ (0\nu\beta\beta) \qquad (3.115)$$

In (3.114) lepton number is conserved, i.e. $\Delta L = 0$ ([2], Chap. 3). It may be thought of as two consecutive simple β transitions in which the intermediate states are virtual. This process is allowed in the standard model of the electro-weak interaction independent of the nature of ν.

Search of double β decay whose expected half-lives are too long (10^{20} y or more) are confronted with serious difficulties. The muon and neutron components in cosmic radiation and radioactive impurities are reduced by making experiments deep underground.

The first convincing evidence for the double β decay mode $2\nu\beta\beta$ was provided in 1967 by Kirsten in his studies of ^{82}Se, ^{128}Te and ^{130}Te samples using geochemical experiments. But the $2\nu\beta\beta$ decay mode from ^{82}Se was observed directly from counter experiments comparatively recently (1987). The emitted electrons have continuous energy spectrum since it is a four-body decay. Process (3.114) is possible if the parent nucleus is heavier than the product nucleus which differs by two units in the nuclear charge, i.e.

$$m(Z, A) > (Z + 2, A) \tag{3.116}$$

The simple β decay will be forbidden if the following condition on the masses is imposed, i.e.

$$m(Z, A) < m(Z + 1, A) \tag{3.117}$$

The neutrinoless double beta decay ($0\nu\beta\beta$) violates the conservation of lepton number ($\Delta L = 2$) and is forbidden in the standard model. While 2ν mode is confirmed experimentally, neutrinoless double beta decay has not been detected. In the $2\nu\beta\beta$ decay the kinetic energy is distributed over two electrons and two neutrinos. On the other hand $0\nu\beta\beta$ decay is a two body process, the two electrons in the final state share the available energy uniquely. The total (sum) energy spectrum peaks at $Q_{\beta\beta}$ so that this process is easy to distinguish than 2ν mode. In the neutrinoless $\beta\beta$ decay, only two electrons occur in the final state and the phase space as well as the number of final state is larger by 10^6 compared to $2\nu\beta\beta$ mode. Also, the $0\nu\beta\beta$ mode is possible only if (i) $m\nu \neq 0$, (ii) $\nu = \bar{\nu}$ (Majorana particle). The study of double β decay is of great interest in Grand Unification theories ([2], Chap. 8).

3.4.1.2 Theoretical Values for the Half-Lives

The $2\nu\beta\beta$ half-lives are directly related to the nuclear matrix elements and no free particles are involved. Comparison of the predicted decay rates with the experimental values provides a sensitive test for various nuclear structure models. Double beta-decay, $2\nu\beta\beta$ follows from the square of the rate of single-beta decay for the light nuclei. Therefore

$$\lambda_{\beta\beta} \sim \frac{m_e c^2}{\hbar} \left(\frac{G^2 |M_{GT}|^2}{2\pi^3} \right)^2 \left(\frac{\varepsilon_0^5}{30} \right)^2$$

$$T_{1/2}(\beta\beta) = \frac{\ln 2}{\lambda_{\beta\beta}} \sim 3 \times 10^{27} \varepsilon_0^{-10} \tag{3.118}$$

Thus, for ^{76}Ge, $Q_{\beta\beta} = 2.04$ MeV, $\varepsilon_0 = 3.99$, $T_{1/2}(\beta\beta) = 2.9 \times 10^{21}$ y.

The 2ν mode of $\beta\beta$ decay is equivalent to two consecutive G-T transitions. According to the G-T rule the intermediate states can only be 1^+ states since all potential $\beta\beta$ emitters are even-even nuclei with ground state spin 0^+. The half-lives range from 10^{18} years to 10^{25} years.

Other decay modes involve $\beta^+\beta^+$, the Q-value being $4m_e c^2$ which is less than that for double electron capture. That is why electron capture or β^+ emission is generally more favoured than $\beta^+\beta^+$ decay.

Example 3.22 The maximum energy E_{max} of the electrons emitted in the decay of the isotope ^{14}C is 0.156 MeV. If the number of electrons with energy between E

and $E + dE$ is assumed to have the approximate (non relativistic) form:

$$n(E)dE \propto \sqrt{E}(E_{max} - E)^2 dE$$

Find the rate of evolution of heat by a source of ^{14}C emitting 3.7×10^7 electrons per sec.

Solution   The mean energy of electrons:

$$\langle E \rangle = \frac{\int_0^{E_{max}} E n(E)dE}{\int_0^{E_{mm}} n(E)dE}$$

Given $n(E)dE = K\sqrt{E}(E_{max} - E)^2 dE$ where $K = \text{const}$

$$\langle E \rangle = \frac{K \int_0^{E_{max}} E\sqrt{E}(E_{max} - E)^2 dE}{K \int_0^{E_{max}} \sqrt{E}(E_{max} - E)^2 dE} = \frac{E_{max}}{3}$$

Heat evolved/sec = (mean energy) (no. of electrons/s)

$$= \frac{0.156 \times 3.7 \times 10^7}{3} \text{ MeV/s} = 1.92 \times 10^6 \text{ MeV/s}$$

Example 3.23   A radioactive species has a maximum energy of β-rays of 3.5 MeV. Calculate the momentum of the neutrinos accompanying those β particles that have half the possible momentum.

Solution   Total energy of β particle:

$$E = T + mc^2 = 3.5 + 0.511 = 4.011 \text{ MeV}$$

Using the relativistic equation $P_{max} = \sqrt{E^2 - m^2} = \sqrt{(4.011)^2 - (0.511)^2} = 3.978$ MeV/c. Half of this value is 1.989 MeV/c. Corresponding total energy of β particles

$$= \sqrt{(P_{max}/2)^2 + m^2}$$

$$= \sqrt{(1989)^2 + (0511)^2} = 2.053 \text{ MeV}$$

Energy of neutrinos accompanying these β particles is $(4.011 - 2.053) = 1.958$ MeV. The corresponding momentum of neutrinos is also 1.958 MeV/c as neutrino has zero rest mass.

Example 3.24   In the Kurie plot of the decay of the neutron, the end point energy of β particles is 0.79 MeV in the free decay of neutron, calculate the threshold energy for the inverse reaction:

$$\bar{v} + p \rightarrow n + e^+$$

Solution   The threshold energy required in the CMS will be $0.79 + 0.511 = 1.301$ MeV. Use the relativistic invariance of $(E^2 - P^2)$ for the threshold calculations

$$(T + m_p)^2 - T^2 = (m_n + m_e)^2$$

Put $m_p = 938.28$ MeV, $m_n = 939.573$ MeV, $m_e = 0.511$ MeV, $T = 1.8$ MeV.

Example 3.25 $^{108}_{47}$Ag with $J^\pi = 1^+$ is β-unstable and has a mean life of about 3.4 min. It has an excited state of 109 keV with $J^\pi = 6^+$, which is an isomeric state with mean life of 180 years. How can the excited state of a nucleus be more stable than the ground state?

Solution   Transitions from the isomeric state $J^\pi = 6^+$ involve a large change in the value of J. Such transitions are forbidden and hence the excited state has a longer mean life.

Example 3.26 Calculate the ft value for the decay $^{31}_{16}$S $\rightarrow$ $^{31}_{15}$P $+ e^+ + \nu$, for which $T_{1/2} = 2.6$ s, $E_0 = 4.94$ MeV and $F(Z, E) = F(15, 4.94) = 1830$. In the simple shell model, this decay involves a $2s_{1/2}$ proton changing to a $2s_{1/2}$, neutron. Compare this ft value with that of a free neutron (1015 s). Why do the two values differ?

Solution $ft = 1830 \times 2.6 = 4758$ s. In the simple shell model the 1s neutron and proton have identical spatial wave functions if Coulomb distortions are ignored. The spin states are similar to those of a free neutron and free proton. Thus, the predicted ft value is expected to be the same as for the free neutron decay. Since $Z = 15$, the Coulomb distortions cannot be neglected, hence the discrepancy.

Example 3.27 The maximum energy of a β^- spectrum is 1.77 MeV. Find the range of β particles in aluminium.

Solution

$$R = 0.542E - 0.133$$

$$= 0.542 \times 1.77 - 0.133 \text{ g/cm}^2 = 0.826 \text{ g/cm}^2 = \frac{0.826}{2.7} \text{ cm} = 0.306 \text{ cm}$$

Example 3.28 β-particles ($E_{max} = 1.7$ MeV) from ^{32}P are counted by a G.M. counter with a wall thickness of 20 mgl cm^2. Calculate the fraction of particles that are absorbed while passing through the window. Assume $\mu = 10.87$ cm^2/g.

Solution   Fraction of the particles absorbed $f = 1 - \exp(-\mu d)$

$$\mu d = 20 \times 10^{-3} \times 10.87 = 0.2174$$

$$f = 1 - e^{-0.2174} = 0.2$$

Example 3.29 $^{88}_{36}$Kr decays to $^{88}_{37}$Rb with the emission of β-rays with a maximum energy of 2.4 MeV. The track of a particular electron from this nuclear process has a curvature in a field of 10^3 Gauss of 6.1 cm. Determine

(a) the energy of this electron in eV and that of the associated neutrino
(b) the maximum possible kinetic energy of the recoiling nucleus

Solution

(a) Momentum

$$p = 300Hr = (300)(10^3)(6.1)$$
$$= 1.83 \times 10^6 \text{ eV}/c$$
$$= 1.83 \text{ MeV}/c$$

Use the relativistic equation

$$E = \sqrt{p^2 + m^2} = \sqrt{(1.83)^2 + (0.511)^2} = 1.90 \text{ MeV}$$

Kinetic energy of the electron $T = 1.9 - 0.511 = 1.389$ MeV $= 1.389 \times 10^6$ eV. Energy associated with neutrino $= 2.4 - 1.389 = 1.011$ MeV, where we have neglected the energy of the recoiling nucleus.

(b) The maximum Kinetic energy will be carried by the nucleus when it recoils opposite to the β-particle and ν is emitted in the same direction

Momentum conservation gives $\quad P_N = P_{\beta\nu} = \sqrt{T^2 + 2Tm} \quad$ (i)

Energy conservation gives $\quad T_N + T = Q = 2.4 \text{ MeV} \quad$ (ii)

Eliminating T between (i) and (ii) and using $P_N^2 = T_N^2 + 2M_N T_N$, we find $T_N = 50.2$ eV.

Example 3.30 In an experiment to determine the maximum energy of the spectrum of $^{116}_{49}$In, the following results are obtained with aluminium as the absorber after correcting for background:

Cpm	15000	5000	2000	300	70	30	20	19	18
Absorber thickness mg/cm²	0	50	100	200	250	300	400	500	600

Plot these data on a suitable graph and analyze the curve. Calculate the β-energy from the expression $E_{max} = 1.9R + 0.29$ MeV, where R is the range of the β-particles in aluminium in units of g/cm².

Fig. 3.18 Log of counting
rate (CPM) versus absorber
thickness

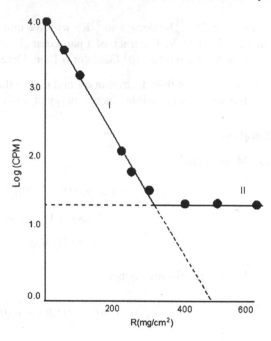

Solution The counting rate (CPM) versus absorber thickness (mg/cm^2) is plotted on log-linear scale, see Fig. 3.18. Up to 300 mg/cm^2 thickness, a straight line (curve I) can be passed. Another straight line (curve II) represents the γ-ray background. The curve II can be extrapolated to zero thickness and its contribution can be subtracted from curve I and a new curve may be drawn for the true absorption of β-particles. However, the γ-ray counting rate is so small that such a procedure is found unnecessary. The curve I, when extrapolated, cuts the range axis at 475 mg/cm^2. Using this value of R in the given formula we obtain

$$E_{\max} = (1.9)(0.475) + 0.29 = 1.19 \text{ MeV}$$

Example 3.31 The nuclide $^{15}_{8}\text{O}$ undergoes β^+ decay to its mirror nuclide $^{15}_{7}\text{N}$. Assuming that the mass difference is entirely due to the Coulomb energy, and that both nuclides have a radius of 3.45 fermis, calculate the maximum kinetic energy of β^+.

The observed half-life $t_{1/2}$ for this decay is 124 s. By inserting this in the expression for the half-life derived from Fermi theory of β decay

$$F(Z,T) \times t_{1/2} = \frac{2\pi^3 (\ln 2)\hbar^7}{m_e^5 G^2 c^4 |M_{if}|^2}$$

Make an estimate of the weak interaction constant G taking the factor $f(Z,T)$ to be given by $Z = 7$ by the empirical form

$$\log_{10} f(7,T) = 0.6 + 4.5 \log_{10} T$$

where T is the maximum kinetic energy of β^+ in MeV. The nuclear wave functions for the mirror pair may be assumed identical.

Solution

$$\frac{0.6e^2}{4\pi\varepsilon_0 R} = \frac{0.6 \times 1.44 \times 15}{3.45} = 3.75$$

$$T = \Delta E_c - m_e - (m_n - m_p) = 3.75 - 0.51 - 1.3$$

$$= 1.94 \text{ MeV}$$

$$\log_{10} f(7, T) = 0.6 + 4.5 \log_{10} T$$

$$\log_{10} f(7, 1.94) = 0.6 + 4.5 \log_{10} 1.94 = 1.895$$

$$\therefore \quad f(1.94) = 78.5$$

$$G^2 = \frac{2\pi^3 (\ln 2)\hbar^7}{m_e^5 c^4 |M_{if}|^2 f(Z, T) \times t_{1/2}}$$

As the transition is super allowed, $M_{if} = 1$

$$G^2 = \frac{2\pi^3 (0.693)(\hbar c)^7}{(m_e c^2)^5 f(Z, T) \times t_{1/2} \times c}$$

$$\hbar c = 197.7 \text{ MeV fm}, \qquad c = 3 \times 10^{23} \text{ fm s}^{-1}$$

$$t_{1/2} = 124 \text{ s}, \qquad f = 78.5$$

$$c = 3 \times 10^{23} \text{ fm/s}$$

$$G = 0.7 \times 10^{-4} \text{ MeV fm}^3$$

Example 3.32 If the β-ray spectrum is represented by

$$n(E)dE \propto \sqrt{E}(E_{max} - E)^2 dE$$

show that the most intense energy occurs at $E = E_{max}/5$.

Solution   Maximizing the expression $\sqrt{E}(E_{max} - E)^2$ and seting the resulting expression to zero, we obtain

$$\frac{1}{2\sqrt{E}}(E_{max} - E)^2 - 2\sqrt{E}(E_{max} - E) = 0$$

whence we get the desired result.

3.5 Electron Capture

Introduction It is a radioactive decay process in which an inner orbital electron of an atom, usually a K-shell electron, is captured by the nucleus indicated as

$$e^- + p \rightarrow n + \nu$$

One of the protons captures an electron and gets converted into a neutron which is lodged within the nucleus. A neutrino accompanies the neutron emission. The available energy is shared between the neutrino and the resulting nucleus where the neutrino is carrying almost the entire energy. The above process may be considered as the inverse process of β^+ decay

$$P \rightarrow n + \beta^+ + \nu$$

An example of electron capture is $e^- + \mathrm{Be}^7 \rightarrow \mathrm{Li}^7 + \nu$.

The criteria for electron capture process are as follows:

(a) If it is energetically possible, the condition is

$$\left[M(A, Z) - M(A, Z - I) \right] c^2 = \Delta c^2 \geq B_k \qquad (3.119)$$

where B_k is the binding energy of K-shell electron.

If $\Delta c^2 < B_k$ but $> B_L$, K-electrons cannot be captured but L-electrons can be.

(b) If the electron wave function is non-zero at the nucleus, the $s_{1/2}$ state for the K-shell and $2s_{1/2}$ state for the L-shell are the only important states for electron capture as only $l = 0$ particles have finite wave functions at the origin (nucleus). Contribution from L_{II} and L_{III} is quite small due to finite extension of the nucleus and relativistic effects.

3.5.1 Decay Constant

The theory for the determination of decay constant proceeds along similar lines as in β-decay. We use the formula from perturbation theory Eq. (3.83)

$$\lambda = \frac{2\pi |H_{if}|^2 dN}{\hbar dE}$$

where the statistical factor refers to the density of ν states only.

If E_0 is the energy released in electron capture

$$E_\nu = E_0 - E_B \qquad (3.120)$$

where E_0 results from the mass difference $+mc^2$, as the electron is consumed.

$$dN = \frac{V4\pi P\nu^2 dp\nu}{\hbar^3} \tag{3.121}$$

But

$$P_\nu = \frac{E_\nu}{c} \quad \text{and} \quad dP_\nu = \frac{dE_\nu}{c} \tag{3.122}$$

Using (3.120) and (3.122) in (3.121), we get

$$\frac{dN}{dE_\nu} = \frac{V4\pi E_\nu^2}{\hbar^3 c^3} = \frac{V(E_0 - E_B)^2}{2\pi^2\hbar^3 c^3} \tag{3.123}$$

We will be concerned here with the K-capture only. The matrix element

$$H_{if} = \int \Psi_f^* \mathrm{H}\Psi_i d\tau = \int \Psi_{fn}^* \Psi_{\nu*} g \Psi_{iN} \Psi_e d\tau$$

Calling the nuclear matrix element

$$M_{if} = \int \Psi_{fN}^* \Psi_{iN} d\tau \tag{3.124}$$

and using the K-shell electron wave function $\Psi_K(0)$ at the origin, i.e. $\Psi_e = \Psi_K(0)$ the neutrino wave function $\Psi_\nu(0) = \frac{1}{\sqrt{V}}$ through plane wave approximation, we can write

$$H_{if} = \frac{g}{\sqrt{V}} M_{if} \Psi_K(0) \tag{3.125}$$

Now

$$\Psi_K(r) = \frac{1}{\sqrt{\pi}} \left[\frac{Z}{a_0}\right]^{3/2} \exp(-Zr/a_0)$$

$$\Psi_K(0) = \frac{1}{\sqrt{\pi}} \left[\frac{Zme^2}{\hbar}\right]^{3/2} \tag{3.126}$$

since

$$a_0 = \frac{\hbar^2}{me^2} \tag{3.127}$$

Using (3.123), (3.125), (3.126) in (3.83) and multiplying the resulting expression by a factor 2 to account for two K-shell electrons, we get

$$\lambda_K = \frac{2g^2 Z^3 m^3 e^6 |M_{if}|^2 (\varepsilon_0 - \varepsilon_B)^2}{\pi^2 \hbar^{10} c^3} \tag{3.128}$$

Fig. 3.19 Plot of $\log(\lambda_K/\lambda_{\beta^+})$ vs ε_0 for various Z

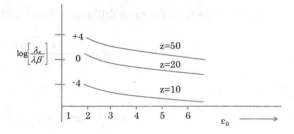

with $\varepsilon_B = \frac{1}{2}(\alpha Z)^2$ and α = fine structure constant. As the nuclear matrix element for electron capture is the same as for β^+ decay, we can find the branching ratio $\lambda_K/\lambda_{\beta^+}$ which is independent of M_{if}.

Combining (3.109) with (3.128) and neglecting the binding energy of electron, we get

$$\frac{\lambda_K}{\lambda_{\beta^+}} = \frac{4\pi^3 z^3 e^6 E_o^2}{\hbar^3 c^7 m^2 f(Z, \varepsilon_{0-})} = 4\pi \left[\frac{Z}{137}\right]^3 \frac{\varepsilon_0^2}{f(Z, \varepsilon_0)} \tag{3.129}$$

where we have used $e^2/\hbar c = 1/137$ and $E_0/mc^2 = \varepsilon_0$. Now, for low Z and large ε_0 as in Eq. (3.110)

$$f(0, \varepsilon_0) \sim \frac{\varepsilon_0^5}{30}$$

From (3.124) and (3.110)

$$\frac{\lambda_K}{\lambda_{\beta^+}} = 120\pi \left[\frac{Z}{137}\right]^3 \frac{1}{\varepsilon_0^3} \tag{3.130}$$

Figure 3.19 is a plot of the branching ratio $\lambda_K/\lambda_{\beta^+}$ (on the logarithmic scale) vs the energy ε_0 for various values of Z. In the range of ε_0 where only electron capture is possible, the branching ratio zooms to infinity. Formula (3.130) shows that for light elements and reasonably large end point energies, K-capture is less probable than for positron emission. As ε_0 approaches the positron threshold or as Z increases, the K-capture becomes more probable. Experimental confirmation of the theory of electron capture relies mainly on measurements of the ratio $\lambda_K/\lambda_{\beta^+}$. The relative probability of L-capture to K-capture is directly related to the relative probability density at the nucleus of the L- and K-electrons. The experimental ratio of L- to K-capture for ^{37}A is 0.087 which is in agreement with the ratio 0.082.

The rate of electron capture depends to a small extent on the chemical environment in which the nucleus is placed, as the electron wave function is slightly modified if the atom is in a compound. This effect will be the largest for light elements. A difference of about 0.08 per cent has been observed between the decay rate of ^{7}Be in BeF$_2$ and in Be metal. As the threshold difference between the electron capture process and positron emission is $(2mc^2 - E_B)$, a nuclide which is unstable against positron emission will also decay by K-capture. If a nuclide is sufficiently heavy, all the three processes, β^-, β^+ and K-capture may occur, as in ^{36}Cl and ^{76}As.

3.5.2 Detection

Neutrino is unobserved for all practical purposes and the nuclear recoil is also too small to be detected. Electron capture is detected by observing atomic process following the consumption of K-shell or L-shell electron, by observing the K-lines and L-lines in the X-ray spectrum. Alternatively, the competitive process of the Auger electron emission (invariably from L-shell) may be observed. They result when another electron from the same shell falls to fill a vacancy in the K-shell and receives enough kinetic energy to be ejected. Obviously, the Auger electrons are of discrete energy and can be conveniently observed experimentally.

3.6 Gamma Decay

Gamma decay of an excited nucleus may occur competitively with α or β decay. The half-lives, however, are usually small. This is the reason why pure γ-ray emitters (except isomers) are not to be found. Ellis and Meitner [1] discovered that nuclei have quantized energy levels, similar to atoms, but with much larger spacing. The γ decay occurs due to radioactive transition between various nuclear levels, resulting in the emission of γ-rays of discrete energy.

3.6.1 Multipole Order of Radiation

The quantum theory of radiation considers the radiation source as an oscillating electric or magnetic moment. The complicated spatial distribution of the corresponding electric charges and currents is represented by spherical harmonies. The multipole order of γ radiation is 2^l, where l is the angular momentum carried by radiation. $l = 1$ corresponds to dipole radiation, $l = 2$ to quadrupole radiation, $l = 3$ to octupole radiation, etc. One consequence of the transverse nature of e.m. wave is that the order $l = 0$ is absent. For multipole order two different waves are possible (i) electric (ii) magnetic multipole radiation. For each value of l, electric and magnetic waves have the same angular momentum but different parity

$$\text{Parity of electric multipole} = (-1)^l$$

$$\text{Parity of magnetic multipole} = -(-1)^l$$

Consider a pair of energy levels of a nucleus with spin I_A and I_B. The angular momentum carried by the radiation is given by the change in nuclear spin. Thus,

$$l = |I_A - I_B| \tag{3.131}$$

where l is non-zero. We can then write

$$\Delta l = |I_A - I_B| \leq l \leq I_A + I_B \tag{3.132}$$

In practice $l = \Delta I$. The competitive case is $l = \Delta I + 1$. If $I_A \neq 0$ and $I_B = 0$ or $I_A = 0$ and $I_B \neq 0$, then $1 = \Delta I$ is only possible. This is the case for transitions to the ground level of even-Z and even-N nuclei for which the ground level has spin zero.

The transition $I_A = 0 \rightarrow I_B = 0$, is absolutely forbidden.

3.6.2 Selection Rules for γ-Emission (or Absorption)

Transition probability $P = \int \Psi_B^*$ (electric moment) $\Psi_A d\tau$. For electric dipole, the relevent moment is the dipole moment $\Sigma e_i X_i$, whose parity is -1

$$P = \int \Psi_B^*(\Sigma e_i X_i)\Psi_A d\tau; \qquad \Delta\pi = \text{Yes} \tag{3.133}$$

For quadrupole moment, $\Sigma e_i X_i^2$ has parity $+1$

$$P = \int \Psi_B^*(\Sigma e_i X_i^2)\Psi_A d\tau; \qquad \Delta\pi = \text{No} \tag{3.134}$$

The value of a definite integral cannot possibly change by the reflection of coordinates through the origin, i.e. parity operation. The integrand must be positive if $P \neq 0$. On the other hand, if the integrand changes its sign then $P = 0$. Thus

$$\text{if } \Psi_A = \text{even}, \quad \Psi_B \left(\text{or } \Psi_B^*\right) = \text{odd}$$
$$\text{if } \Psi_A = \text{odd}, \quad \Psi_B \left(\text{or } \Psi_B^*\right) = \text{even}$$

i.e. parity of final state of the nucleus must be opposite to the initial state.

Conservation of parity in the system as a whole (nucleus + quantum of radiation) then requires that for electric dipole radiation. The photon must have odd parity with respect to the system, in the process of emission or absorption.

Similar reasoning shows that for both electric quadrupole and magnetic radiation, the emission is possible if the parity of final state is the same as that of initial state. Table 3.2 gives the Selection Rules for γ radiation.

3.6.3 γ-Ray Emission Probability

It can be shown that the decay constant is given by the expression:

$$\lambda\gamma = \frac{1}{\tau_{el}} = S \times \frac{2\pi v}{137}\left(\frac{R}{\lambda}\right)^{2l} \tag{3.135}$$

Table 3.2 Selection rules for γ radiation

Type of radiation	Symbol	l	$\Delta \pi$
Electric dipole	$E1$	1	Yes
Magnetic dipole	$M1$	1	No
Electric quadrupole	$E2$	2	No
Magnetic quadrupole	$M2$	2	Yes
Electric octupole	$E3$	3	Yes
Magnetic octupole	$M3$	3	No
Electric 2^l-pole	El	l	No: for l even Yes: for l odd
Magnetic 2^l-pole	Ml	l	Yes: for l even No: for l odd

Fig. 3.20 Plot of $T_{1/2}$ vs E_γ for light and heavy γ-ray emitters

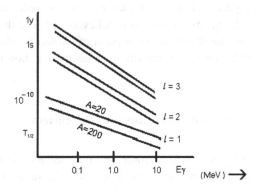

where R is the nuclear radius and S is the statistical factor given by

$$S = \frac{2(2l+1)}{l[1 \times 3 \times 5 \times \cdots \times (2l+1)]^2} \left(\frac{3}{l+3}\right)^2 \tag{3.136}$$

S decreases considerably with the increase in l. Figure 3.20 shows graphs of $T_{1/2}$ vs E_γ, the γ-ray energy for light and heavy nuclei. Lighter nuclei have greater life times for γ-decay.

3.6.4 Internal Conversion

As the $0 \rightarrow 0$ radiative transitions are absolutely forbidden, they may be replaced by the internal process which results in the ejection of a bound atomic electron from the same atom. The energy transfer is caused by a direct interaction between the bound atomic electron and the same multipole field which would have resulted in the photon emission. The energy carried by the ejected electron E_i is given by

$$E_i = W - B_i \tag{3.137}$$

Fig. 3.21 Conversion
electron spectrum

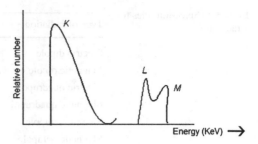

where W is the transition energy and B_i is the binding energy. Observe the similarity of (3.137) with Einstein's photoelectric effect equation. Following the ejection of electron, the atom emits the characteristic X-rays or Auger electrons with energy B_i. Figure 3.21 shows a typical conversion electron spectrum. The mechanism for internal conversion process is believed to be the direct interaction with nuclear volume and not due to the absorption of photon by atomic electron (internal photoelectric effect) as $0 \rightarrow 0$ transition cannot emit a photon.

3.6.4.1   Internal Conversion Coefficient

If $\lambda\gamma$ = probability/unit time for photon emission by radiative multipole transition and λ_e = probability/unit time that the same multipole field would take to transfer its energy W to any bound electron in its own atom, then the total internal coefficient α is defined as

$$\alpha = \frac{\lambda_e}{\lambda_\gamma} = \frac{N_e}{N_\gamma} \tag{3.138}$$

with

$$0 \leq \alpha \leq \infty \tag{3.139}$$

where N_e is the number of conversion electrons and N_γ is the number of photons emitted in the same time interval in the same sample. Total transition probability λ is defined as

$$\lambda = \lambda_\gamma + \lambda e = \lambda_\gamma (1 + \alpha) \tag{3.140}$$

and the total number of nuclei transforming is $N_\gamma + Ne$. Also

$$\alpha = \alpha_K + \alpha_L + \alpha_M \tag{3.141}$$

where α_K refers to both the K-shell electrons and, α_L to all the L-electrons etc.

Fig. 3.22 Plot of α_K
against W

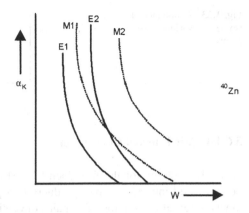

3.6.4.2 K-Shell Conversion Coefficient

The theory gives

$$(\alpha_K)_{e1} \simeq \frac{l}{l+1} Z^3 \left(\frac{1}{137}\right)^4 \left(\frac{2mc^2}{W}\right)^{l+5/2} \tag{3.142}$$

for two electrons, with the condition

$$mc^2 \gg W \gg B_K \tag{3.143}$$

$$(\alpha_K)_{mag} \simeq Z^3 \left(\frac{1}{137}\right)^4 \left(\frac{2mc^2}{W}\right)^{l+3/2} \tag{3.144}$$

α_K increases with l and hence with increasing spin change ΔI in nuclear transition. It increases strongly as Z increases and as W decreases. Usually, $(\alpha_K)_{mag} > (\alpha_K)_{elec}$. Figure 3.22 shows the variation of α_K with W.

3.6.4.3 L-Shell Conversion Coefficient

If $W > B_K$, then $\alpha_K > \alpha_L$, as K-shell electrons have a greater probability for being near the nucleus.

(a) The ratio α_K/a_L decreases as ΔI increases.
(b) As $l = \Delta I$ increases, the decrease in the ratio a_K/α_L is more pronounced for electric 2^l-pole transition than for magnetic 2^l-pole transitions. Thus, for the same W, Z and ΔI

$$\left(\frac{\alpha_K}{\alpha_L}\right)_{el} < \left(\frac{\alpha_K}{\alpha_L}\right)_{mag}$$

Experimental values for α_K/α_L range between 10 (for large W, small ΔI, small Z) and 0.1 (for small W, large ΔI, large Z).

Fig. 3.23 Abundance of
isomers at various mass
numbers

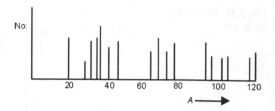

3.6.4.4 Pair Internal Conversion

When $W > 2mc^2$ for the forbidden $0 \to 0$ transitions, the available energy may be converted into e^+, e^- pair production, a process which competes with the ordinary internal conversion. The pair conversion coefficient is almost independent of Z, actually decreasing slightly with increasing Z. The e^+, e^- pair is produced within the nuclear volume as there is no multifold field outside. The angle between the produced e^+ and e^- is small. The energy is also similar, but due to Coulomb interaction, the energy of e^+ is pushed out and that of e^- is pulled in.

3.6.5 Isomers

Certain nuclides are capable of existing in excited state for sufficiently long time to be observed. Such nuclides are called isomers. Their half-lives range from 1 s to 8 months. The phenomenon of isomerism is frequently found in odd A nuclides (odd Z or odd N). Figure 3.23 shows the abundance of isomers at various mass numbers and the isomers clustering into various groups. The existence of islands of isomers is explained within the frame-work of the shell model of nucleus (Chap. 6).

3.6.6 Angular Correlation of Successive Radiation

When γ-rays are emitted from radioactive transitions between two specific levels of identical nuclei, their angular distributions will be isotropic in the lab system. As the atomic nuclei are oriented at random, there is no preferred direction of emission for the photon from the individual transition $I_A \xrightarrow{\gamma} I_B$. The same result holds good for α, β and conversion electron emission. If the transition $I_A \xrightarrow{\gamma_1} I_B$ is followed by a second transition $I_B \xrightarrow{\gamma_2} I_C$, the individual radiations from the second transitions are γ_1, γ_2 also isotropic. However, in two successive cascade transitions, $I_A \xrightarrow{\gamma_1} I_B \xrightarrow{\gamma_2} I_C$, there is usually angular correlation between the direction of emission of two successive photons γ_1 and γ_2, which are emitted from the same nucleus. Similar angular correlations may arise for other pairs of successive radiations like $\alpha-\gamma$, $\beta-\gamma$, $\beta-e^-$, $\gamma-e^-$, ... (where e^- refers to conversion electron).

The angular correlation arises because the direction of first radiation is related to the orientation of the angular momentum I_B of the intermediate level. This orientation is expressible in terms of magnetic angular momentum quantum number m_B with respect to some fixed direction, say that of first radiation. If I_B is not zero and if the intermediate level has a short life time so that I_B persists in orientation, the direction of emission of the second radiation will be related to the direction of I_B and hence of that to the first radiation.

3.6.6.1 γ–γ Angular Correlation

The theory of γ–γ angular correlation can be applied to any γ–γ cascade involving arbitrary multipole orders notwithstanding the mathematical complications introduced in the calculations. Using group theoretical methods Yang first obtained the form of the general angular function. For the generalized γ–γ cascade $I_A(l_1)I_B(l_2)I_C$, the angular correlation function $W(\theta)$ for the angle θ between the successive γ-rays takes the form

$$W(\theta)d\Omega = \sum_{K=0}^{K=l} A_{2K} P_{2K}(\cos\theta)d\Omega \tag{3.145}$$

where A_{2K} are the coefficients that depend on l_1, and l_2 and $P_{2K}(\cos\theta)$ are the even Legendre polynomials. One can also express (3.145) as a power series in even powers of $\cos\theta$ and normalized to $W(90) = 1$ as follows. Therefore

$$W(\theta)d\Omega = 1 + \sum_{K=1}^{l} a_{2K}\left(\cos^{2K}\theta\right) \tag{3.146}$$

where the coefficients $a_2, a_4, \ldots$ are the functions of angular momenta I_A, I_B, I_C, l_1 and l_2 but not the relative parity of the levels.

Equations (3.145) and (3.146) are quite general and with appropriate values of $a_2, a_4, \ldots$ apply to all two-step cascades, α–γ, β–γ, γ–γ, γ–e^-, e^-–e^-, i.e. … as well as to nuclear scattering and nuclear reactions.

Derivations of Eqs. (3.145) and (3.146) are based on the following assumptions:

(a) The magnetic sublevels m_A, of the initial level I_A are equally populated. This is usually true for ordinary radioactive sources at room temperature.
(b) Each nuclear level I_A, I_B, and I_C must be single level with well defined parity and angular momentum, otherwise overlap of broad nuclear levels at high excitation energies may produce interference effects and introduce odd powers of $\cos\theta$ in $W(\theta)$.
(c) Each of the radiation l_1 and l_2 must correspond to a pure multipole. Otherwise radiations of opposite parity may produce interference effects and introduce odd powers of $\cos(\theta)$.
(d) Equation (3.146) is valid for detectors that are insensitive to plane of polarization of radiation.

Table 3.3 Angular correlation coefficients a_2 and a_4	$\gamma-\gamma$ cascades $I_A(l_1)I_B(l_2)I_c$	$W(\theta)d\Omega = (1 + a_2 \cos^2\theta + a_4 \cos^4\theta)d\Omega$	
		a_2	a_4
	0(1)1(1)0	1	0
	1(1)1(1)0	$-1/3$	0
	1(2)1(1)0	$-1/3$	0
	2(1)1(1)0	$-1/3$	0
	3(2)1(1)0	$-3/29$	0
	0(2)2(2)0	-3	$+4$
	1(1)2(2)0	$-1/3$	0
	2(1)2(2)0	$+3/7$	0
	2(2)2(2)0	$-15/13$	$+16/13$
	3(1)2(2)0	$-3/29$	0
	4(2)2(2)0	$+1/8$	$+1/24$

Fig. 3.24 The angular distribution of γ_2 is measured relative to the direction of γ_1

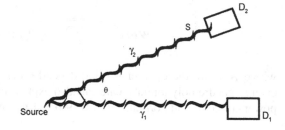

(e) The half-life of intermediate level I_B, must be short enough to permit the orientation of I_B, to be retained.

3.6.6.2 $\gamma-\gamma$ Angular Correlation Coefficients

In Table 3.3 are given the angular correlation coefficients a_2 and a_4 for some dipole and quadrupole $\gamma-\gamma$ cascades of interest, for the special case $I_C = 0$ appropriate for even $-Z$ even N nuclei.

3.6.6.3 Experimental

γ-rays (γ_1) arising from the source are accepted by the fixed detector D_1 (scintillation counter). The second ray γ_2 is accepted by the second movable detector (scintillation counter D_2). The counting rate ratio between D_1 and D_2 is measured in coincidence at different angles θ between the γ_1 and γ_2 rays (Fig. 3.24).

As an example, consider the decay of ^{60}Co which emits two γ-rays in cascade (Fig. 3.25).

Fig. 3.25 Decay scheme
^{60}Co → ^{60}Ni

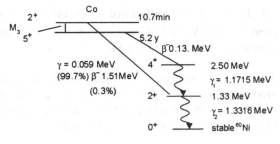

Fig. 3.26 Coincidence
counting rate proportional to
$W(\theta)$ for the $\gamma-\gamma$ cascade in
^{60}Ni following the β decay of
^{60}Co. The *curve* can be fitted
for the angular correlation
distribution for 4(2)2(2)0
cascade (data from Brady and
Deutch)

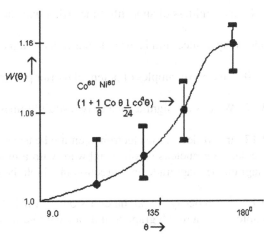

Figure 3.26 shows the measured dependence of coincidence rate on θ for the
$\gamma-\gamma$ cascade in ^{60}Ni, and is seen to be in agreement with the quadrupole transitions
4(2)2(2)0. The observation fixes the angular momenta of the excited levels at 1.33
and 2.5 MeV in ^{60}Ni as $I_B = 2$ and $I_A = 4$.

3.7 Questions

3.1 Explain how the unstable Radon which has only a half-life of 3.8 D can occur
naturally?

3.2 The unit of radioactivity $1\mu c = 3.7 \times 10^4$ disintegrationsls. What is the significance of this number?

3.3 From which naturally occurring radioactive series is the isotope ^{204}Pb produced?

3.4 Radioactive series represented by $4n + 1$ does not exist. Why?

3.5 Neutron or proton emission does not occur in natural radioactivity. Why?

3.6 In the uranium radioactive series, the initial nucleus is $^{238}_{92}U$ and the final nucleus is $^{206}_{82}Pb$. What is the number of α-particles and β-particles emitted when the uranium nucleus decays to lead?

3.7 A radioactive nuclide is capable of decaying via three competitive decay modes, β^- decay with half-life T^-, β^+ decay with half life T^+ and electron capture with T_e. If the observed half-life irrespective of decay is T, write down the equation to show how T is related to T^-, T^+ and T_e.

3.8 In α-emitters of short life times what happens to the potential barrier width?

3.9 Why α-spectrum is discrete but β-spectrum continuous?

3.10 Give two examples of potential barrier penetration in areas other than α-decay.

3.11 What is the significance of two straight lines in the Sargent diagram?

3.12 In a β-decay, the electron is emitted eastward with momentum of 3 units and the daughter nucleus recoils southward with a momentum of 4 units. Indicate on a diagram the magnitude and direction in which the neutrino is emitted.

3.13 ^{210}Po decays by alpha emission with half-life of 138 D into ^{206}Pb which is stable. Draw a rough graph to indicate the rate of formation of ^{206}Pb with time.

3.14 Classify the following β-transitions as Fermi or G-T or both.

(i) $^3H \rightarrow {}^3He$
(ii) $^6He \rightarrow {}^6Li$
(iii) $^{14}O \rightarrow {}^{14}N$
(iv) $^{60}Co \rightarrow {}^{60}Ni$

3.15 Classify the following as super-allowed, allowed or forbidden transitions: $^{14}O \rightarrow {}^{14}N^*$, with $ft = 3103$ s.

3.16 How can the mass of neutrino be estimated from the study of β spectrum?

3.17 The radioactive nuclide of ^{64}Cu can decay either by β^- emission or β^+ emission. The lower side of energy spectrum shows fewer positrons than electrons. Explain.

3.18 In a given type of β^- decay, the β^- spectrum looks as shown in Fig. 3.27. How does The υ-spectrum look like for the same decay scheme?

3.19 What quantities are not conserved if it is assumed that free neutron decays through the scheme $n \rightarrow p + \beta^-$?

Fig. 3.27 β-decay energy
spectrum

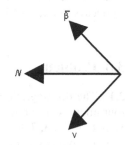

Fig. 3.28 The direction of
decay products in the β-decay

3.20 In the β-decay, should the decay products, β-particle, neutrino and the residual nucleus be coplanar?

3.21 Is the β-decay shown in Fig. 3.28 feasible? If not, why?

3.22 Show that Fermi's factor $F(Z, E)$ given by the expression (3.100) reduces to unity for small Z and large v.

3.23 Explain qualitatively why in high Z atoms, e^- capture is more probable than β^+ decay.

3.24 Explain qualitatively why in the decay by electron capture, K-capture is more probable than L-capture.

3.25 State Laporte rule for radiation transitions in atoms.

3.26 When $0 \rightarrow 0$ radiative transition can not take place, an alternative process of internal conversion is possible. Can one call this process photoelectric effect?

3.27 Explain why pair internal conversion cannot take place outside the nuclear volume.

3.28 A sample of radioactive substance has mass m, decay constant λ and molecular weight M. If Avogadro's number is N_A, show that the activity of the sample is $\lambda \frac{m N_A}{M}$.

3.29 If A_1 is the activity of a sample of radioactive substance at time t_1, and A_2 at time t_2 then show that $A_2 = A_1 \exp(t_1 - t_2)/T$.

3.30 What fraction of initial number of radioactive nuclei in a sample will decay during one mean life?

3.31 What will be the half-life of radioactive nuclei in a sample if three-fourths decay in $3/4$ s?

3.32 A fraction f_1 of a radioactive sample decays in one mean life and a fraction f_2 decays in one half-life. Is $f_1 < f_2$ or $f_1 = f_2$ or $f_1 > f_2$?

3.8 Problems

3.1 The disintegration rate of a radioactive source was measured at intervals of four minutes. The rate was found to be (in arbitrary units) 18.59, 13.27, 10.68, 9.34, 8.55, 8.03, 7.63, 7.30, 6.99, 6.71. and 6.44. Assuming that the source contained only one or two types of radio nucleus, calculate the disintegration constant involved.
[Ans. 0.26 min^{-1}, 0.03 min^{-1}]

3.2 100 millicuries of radon which emits 5.5 MeV α-particles are contained in a glass capillary tube which is 5 cm long with internal and external diameters 2 mm and 6 mm respectively. Neglecting the end effects and assuming that the inside of the tube is uniformly irradicated by the α-particles which are stopped at the surface, calculate the temperature difference between the walls of a tube when steady thermal conditions have been reached. Thermal conductivity of glass $= 0.025$Cal cm^{-2}s^{-1}C^{-1}. Curie $= 3.7 \times 10^{10}$ disintegrations per sec $J = 4.18$ joule cal^{-1}.
[Ans. 4.5×10^{-3} °C]

3.3 Radium being a member of the uranium series occurs in uranium ores. If the half-lives of uranium and radium are 4.5×10^9 and 1620 years respectively, calculate the relative proportions of these elements in a uranium ore which has attained equilibrium and from which none of the radioactive products have escaped.
[Ans. $2.78 \times 10^6 : 1$]

3.4 A sealed box was stated to have contained an alloy composed of equal parts by weight of two metals A and B. These metals are radioactive with half-lives of 12 years and 18 years respectively. When the container was opened it was found to contain 0.53 kg of A and 2.20 kg of B. Deduce the age of the alloy.
[Ans. 73.94 y]

3.5 Determine the amount of $^{210}_{84}$Po necessary to provide a source of α-particles of 5 milli curies strength. Half-life of Polonium $= 138$ D.
[Ans. 1.11 µg]

3.6 A radioactive substance of half-life 100 days which emits β-particles of average energy 5×10^{-7} ergs is used to drive a thermoelectric cell. Assuming the cell to have an efficiency 10 %, calculate the amount (in gram-molecules) of radioactive substance required to generate 5 watts of electricity.
[Ans. 0.02]

3.7 The radioactive isotope, $^{14}_{6}C$ does not occur naturally but it is found at constant rate by the action of cosmic rays on the atmosphere. It is taken up by plants and animals and deposited in the body structure along with natural Carbon, but this process stops at death. The charcoal from the fire pit of an ancient camp has an activity due to $^{14}_{6}C$ of 12.9 disintegrations per minute, per gram of Carbon. If the percentage of $^{14}_{6}C$ compared with normal carbon in living trees is 1.35×10^{-10} %, the decay constant is 3.92×10^{-10} s^{-1} and the atomic weight $= 12.00$, what is the age of the campsite?
[Ans. 1676 y]

3.8 Consider the decay scheme RaE $\xrightarrow{\beta}$ RaF $\xrightarrow{\beta}$ RaG (stable). A freshly purified sample of RaE weighs 2×10^{-10} g at time $t = 0$. If the sample is undisturbed, calculate the time at which the greatest number of atoms of RaF will be present and find this number. Derive any necessary formula. (Half-life of RaE $(^{210}_{83}Bi) = 5.0$ D; Half life of RaF $(^{210}_{84}Po) = 138$ D.)
[Ans. 24.8 D; 1.836×10^{10}]

3.9 It is found that a solution containing 1 g of the α-emitter radium (^{226}Ra) never accumulates more than 6.4×10^{-6} g of its daughter element radon which has a half-life of 3.825 days. Explain how the half life of radium may be deduced from this information and calculate its value.
[Ans. 1637 y]

3.10 An atom of $^{6}_{2}He$ is 0.067 % heavier than another atom $^{6}_{3}Li$. What is the maximum energy of the β-particles emitted by $^{6}_{2}He$?
[Ans. 3.74 MeV]

3.11 A parent nuclide decays with decay constant λ_1 into a daughter of decay constant λ_2 and hence to a stable nuclide. The decays are recorded by detecting equipment which cannot discriminate between the emitted particles. Show that when $\lambda_1 = 2\lambda_2$, the activity indicated by the detector at a time t is $2\lambda_2 N_0 \exp(-\lambda_2 t)$, where N_0 is the number of parent atoms present at time $t = 0$. Comment on the implications of this result.

3.12 Derive an expression for the activity at time t of a nuclide A, given that the members of nuclei of A and B at $t = 0$ are N_0 and 0, respectively. Show that under the condition of secular equilibrium, the total activity of A and B is given by $\lambda_A N_A [2 - \exp(-\lambda_B t)]$, where λ_A and λ_B are the radioactive constants for the nuclides, and N_A is the number of nuclei A at time t.

3.13 Find the mean-life of ^{55}Co radionuclide if its activity is known to decrease 4.0 % per hour. The decay product is non-radioactive.
[Ans. 24.5 h]

3.14 A radioactive specimen emitting β-rays of 2.6 MeV maximum energy is investigated in a sample with a β-ray spectrometer, using a magnetic field of 0.2 weber/m^2. The maximum blackening occurs at a distance of 7.5 cm from the line source. Calculate the energy of the most abundant β-particles and the corresponding momentum of the neutrinos.
[Ans. 1.796 MeV; 2.25 MeV/c]

3.15 What proportion of ^{235}U was present in a rock formed 3000×10^6 y ago, given that the present proportion of ^{235}U to ^{238}U is 1/140?
[Ans. 1/12]

3.16 A source consisting of 1 µg of ^{242}Pu is spread thinly over a plate of an ionization chamber. α-particle pulses are observed at the rate of 80 per second, and spontaneous fission pulses at the rate of 3 per hour. Calculate the half life of ^{242}Pu and the partial decay constants for the two modes of decay.
[Ans. 6.8×10^5 y; 3.23×10^{-14} s^{-1}, 3.36×10^{-19} s^{-1}]

3.17 Samarium emits low-energy α particles at the rate of 90 particles/g s for the element. If Sm-47 (abundance 15 %) is responsible for this activity, calculate its half-life.
[Ans. 4.68×10^{11} y]

3.18 The charcoal in an ancient fire pit shows a beta activity of 25.8 disintegrations per minute per g due to ^{14}C. If the specific activity of ^{14}C in the contemporary charcoal from wood of living trees is 30.6 disintegrations per minute per g, estimate the age of the charcoal sample. (Mean life time of ^{14}C against beta decay = 8035 years.)
[Ans. 1366 y]

3.19 Given the decay scheme $A \xrightarrow{\lambda_A} B \xrightarrow{\lambda_B} C$ (stable), find the number of atoms of B at any time t, if at time $t = 0$ the population in the states A, B and C is respectively, A_0, 0 and 0. Show that the time for the maximum activity of B is given by $t(\max) = \sqrt{\tau_A \tau_B}$ where τ_A and τ_B are the mean life times of radioactive samples A and B, respectively.

3.20 Estimate the amount of cobalt-60 ($Z = 27$) in grams, corresponding to an activity of 1 Curie. (Half-life of Cobalt-60 is 5.3 years.)
[Ans. 8.87×10^{-4} g]

3.21 Show that if $\lambda_A = \lambda_B = \lambda$ Eq. (3.13) reduces to $N_B = \lambda N_A^0 t \exp(-\lambda t)$ and Eq. (3.14) to $t_{\max} = \frac{1}{\lambda}$.

3.22 Find the amount of heat generated by 5.3 MeV α's from 1 mg of ^{210}Po in time equal to mean life time.
[Ans. 1.53×10^6 J]

3.23 If the half-lives of Uranium 235 and 238 are 8.8×10^8 and 4.5×10^9 y, calculate the total number of α-particles emitted per second from one gram of natural Uranium. Both isotopes emit α-particles and the abundance of Uranium 235 is 0.7 %. Assume the atomic weight of Uranium to be 238.
[Ans. 7.53×10^4 s^{-1}]

3.24 ^{90}Sn decays to ^{90}Yn by β-decay with a half-life of 28 years. ^{90}Yn decays by β decay to ^{90}Zn with a half-life of 64 hours. A pure sample of ^{90}Sn is allowed to decay. What is its composition after (a) one hour, (b) after ten years?
[Ans. (a) 3.54×10^5 : 1, (b) 3832 : 1]

3.25 Estimate the transmission coefficient for a rectangular potential barrier of width 10^{-12} cm and height 10 cm for an α-particle of 5 MeV energy.
[Ans. 1.275×10^{-8}]

3.26 $^{194}_{79}$Au is β-unstable and has a mean life of 56 hours. One mode of decay is $^{194}_{79}$Au $\rightarrow$ $^{194}_{78}$Pt $+ \beta^+ + \nu + 1.5$ MeV. The positron is created inside the nucleus and must tunnel through the Coulomb barrier to escape. Applying Gamow's theory of α-decay, show that the barrier factor suppresses the decay rate by a factor of about 4–5 in the case of positron of energy 1 MeV.

3.27 The half-lives of isotopes classified as α-emitters range from 0.3×10^{-6} s ($^{212}_{84}$Po with disintegration energy 8.95 MeV) to 0.16×10^{24} s ($^{142}_{58}$Ce with disintegration energy 1.45 MeV). What deductions can you make from these figures about the heights and thickness of the potential barriers in the two cases? Assuming that the potential outside the nucleus is given by the Coulomb law, calculate a value for the radius of the nucleus $^{212}_{84}$Po.

3.28 A certain preparation includes two β-active components with different half-lives. Using the following data on log (activity) vs time (t)

t (hours)	0	1	2	3	5	7	10	14	20
$\log A$	4.10	3.60	3.10	2.60	2.06	1.82	1.60	1.32	0.90

(i) find the half-lives of both the components and (ii) the ratio of radioactive nuclei of these components at the time $t = 0$.
[Ans. (i) 4.3 h, 1.116 h, (ii) 12.2 h]

3.29 A thin foil of certain stable isotope is irradiated by thermal neutrons falling normally on its surface. Due to capture of neutrons a radio-nuclide with decay constant λ appears. Find the law for accumulation of that nuclide $N(t)$ per unit area of the foil's surface. The neutron flux density is J, the number of nuclei per unit area of the foil's surface is n, and the effective cross-section for the formation of radioactive nuclei is σ.

[Ans. $N(t) = 1 - \exp(-\lambda t)\frac{Jn\sigma}{\lambda}$]

3.30 The isotope ^{226}Th$(Z = 90)$ is α-radioactive with a half-life of 30 min. The energy of the emitted α-particle is 6.5 MeV. Comment on these data in the light of the uncertainty principle assuming that the radius of the nucleus of ^{226}Th is 8.5 fm.

3.31 A cyclotron produced radioactive sample is a mixture of ^{64}Cu (half-life time = 12.8 h) and normal copper. The sample mass and activity are 100 mg and 28 mCi respectively. What is the ratio of the number of stable to the radioactive copper atoms in the sample?

[Ans. $1.37 \times 10^7 : 1$]

3.32 The following results are obtained for the absorption of 1.5 MeV β-rays by aluminium.

Absorber density (mg cm^{-2})	1200	1100	1000	800	600	400	200	100
Counts/minute		22	22	24	25	64	320	1430	2730

The background count was 18 counts per minute. Estimate the range of 1.5 MeV β-rays in aluminium if the counter paralysis time was 400 µS.

[Ans. 800 mg cm^{-2}]

References

1. Ellis, Meitner, (1922)
2. A.A. Kamal, *Particle Physics* (2014)
3. Pauli, (1931)

Chapter 4
General Properties of Nuclei

4.1 Nuclear Sizes

From the alpha scattering experiments of Geiger and Marsden, Rutherford concluded that the nuclear sizes are smaller than atoms by a factor of 10^4, that is the nuclear size is of the order of 10^{-12} cm. It is important to know about nuclear radii as they enter the nuclear reactions quite frequently. Nuclear radii have been determined by a variety of experiments, on the basis of constant density model. The nuclear radius R is defined by

$$R = r_0 A^{1/3} \tag{4.1}$$

where r_0 is a constant and A is the mass number (the number of neutrons and protons). The value of r_0 depends on the type of phenomenon studied. r_0 obtained in those experiments in which nuclear forces are effective, is termed as nuclear radius and that in which electromagnetic forces are involved is known as electromagnetic radius. The value of r_0 ranges from 1.1–1.5 fm.

Various methods that are available for the determination of nuclear radii are based on the study of:

(a) Rutherford scattering with α's of energy 20 to 36 MeV
(b) Coulomb energy term in Weisacker's formula
(c) β-transition energies of mirror nuclei
(d) High energy electron scattering
(e) X-ray energy from mesic atoms
(f) Half life times of α emitters
(g) High energy neutron scattering

4.1.1 Scattering of α Particles

If the α particles penetrate the target nucleus then the Rutherford scattering law breaks down for three reasons: (a) The Coulomb's potential no longer conforms to

A. Kamal, *Nuclear Physics*, Graduate Texts in Physics,
DOI 10.1007/978-3-642-38655-8_4, © Springer-Verlag Berlin Heidelberg 2014

that of a point charge when the incident particle penetrates the nucleus; (b) Nuclear scattering is superimposed on the Coulomb scattering; (c) Incident particle approaches close enough to induce nuclear reactions. The resultant scattering is the so called anomalous scattering. At fixed bombarding energy, the small angle scattering will be adequately described by Rutherford's scattering formula, since the impact parameters would be so large that the particles would stay well outside the nucleus. However, as the scattering angle is progressively increased then at some scattering angle θ' the Rutherford scattering law will break down, the experimental differential cross sections being in disagreement with the theoretical values for all angles $\theta > \theta'$. The angle θ', therefore, signifies that the particles have just started grazing the nuclear boundary, in order that the particles be able to approach the nuclear boundary in the classical sense, it is necessary that their kinetic energy be at least equal to $z_1 z_2 e^2 / 4\pi \varepsilon_0 R$. By Eq. (1.55)

$$r(\min) = \frac{R_0}{2} \left[1 + \sqrt{1 + 4b^2/R_0^2} \right] \tag{4.2}$$

with

$$R_0 = \frac{z_1 z_2 e^2}{4\pi \varepsilon_0 T_0} = \frac{1.44 z_1 z_2}{T_0 \ (\text{MeV})} \ \text{fm} \tag{4.3}$$

The impact parameter b is related to the scattering angle θ' by

$$\tan \frac{\theta'}{2} = \frac{R_0}{2b} \tag{4.4}$$

Combining Eqs. (4.2) and (4.4)

$$R = r(\min) = \frac{R_0}{2} \left(1 + \operatorname{cosec} \frac{\theta'}{2} \right) \tag{4.5}$$

Since R_0 is known, the determination of θ' yields the values of R, the nuclear radius. Alternatively, observations on scattering may be made at a fixed angle by varying the bombarding energy. The energy at which Rutherford scattering starts breaking down is an indication that the incident particles are just grazing the nuclear boundary. Example 1.11 illustrates how the method works for α scattering on the gold nuclei at $\theta = 180°$, the value of R refers to the sum of radii of gold nucleus and α particle. If we subtract a value of ~ 1.6 fm for the radius of α particle, then the actual radius of gold nucleus is deduced as 7.15 fm. Assuming the validity of the constant density model for the nucleus, $R = r_0 A^{1/3}$, we obtain

$$r_0 = \frac{7.15}{(197)^{1/3}} = 1.22 \ \text{fm}$$

4.1.2 Coulomb Energy Term in Weisacker's Mass Formula

The Coulomb energy term which occurs in Weisacker's semi-empirical mass formula (Chap. 6) has the form, $a_c Z^2/A^{1/3}$ with

$$a_c = \frac{3}{5} \cdot \frac{e^2}{4\pi\varepsilon_0 r_0} = \frac{3}{5} \times \frac{1.44}{r_0} \tag{4.6}$$

Inserting the best value of 0.714 MeV for a_c, r_0 is found to be 1.21 fm.

4.1.3 β Transition Energies in Mirror Nuclei

Assuming that the nuclear charge is uniformly distributed in the nucleus, the Coulomb energy is shown to be (Chap. 6)

$$E_c = \frac{3Z^2 e^2}{20\pi\varepsilon_0 R} = \left(\frac{3}{5}\right)\frac{1.44 Z^2}{R} \tag{4.7}$$

for continuous charge distribution.

If protons are regarded as point charges then in (4.7) Z^2 must be replaced by $Z(Z-1)$ as a given proton cannot interact with itself. Consequently (4.7) becomes

$$E_c = 0.864\frac{Z(Z-1)}{R}\text{ MeV} \tag{4.8}$$

for discrete distribution.

Consider β transitions between the mirror nuclei of charge $(Z+1)e$ and Ze. The dominant contribution to the transition energy comes from the Coulomb energy difference and neutron and proton mass difference. It is only in the case of mirror nuclei that nuclear binding energy is substantially the same between the parent and the product. The difference in Coulomb energy

$$\Delta E_c = \frac{3}{5} \times \frac{1.44}{R}\left[Z(Z+1) - Z(Z-1)\right]$$
$$= 1.728\frac{Z}{R} \tag{4.9}$$

In a radioactive decay, when a pair of mirror nuclides are involved, energy conservation gives

$$(M_{Z+1} - M_Z)c^2 = \Delta E_c - (M_n - M_p)c^2 \tag{4.10}$$

Combining (4.1) and (4.10)

$$R = \frac{1.728 Z}{(M_{Z+1} - M_Z + M_n - M_p)c^2} \tag{4.11}$$

Example 4.1 ^{27}Si and ^{27}Al are mirror nuclei. The former is a positron emitter with $E_{max} = 3.48$ MeV. Determine r_0.

Solution

$$^{27}\text{Si} \rightarrow {}^{27}\text{Al} + \beta^+ + \nu$$

$$\Delta E_c = \frac{3}{20\pi\,\varepsilon_0}\frac{e^2}{R}\left[Z^2 - (Z-1)^2\right] = \frac{3e^2(2Z-1)}{20\pi\,\varepsilon R} = 0.6 \times 1.44\frac{A^{2/3}}{r_0}$$

$$\left(\because \quad 2Z - 1 = A \text{ and } R = r_0 A^{1.3}\right)$$

$$E_{max} + m_e c^2 = \Delta E_c - (m_n - m_p)c^2$$

Taking $(m_n - m_p)c^2 = 1.29$ MeV, $m_e c^2 = 0.51$ MeV and $A = 27$, we find $r_0 = 1.47$ fm.

4.1.4 High Energy Electron Scattering

At low energies, electrons are scattered by a nucleus which may be treated as a point charge. However the Rutherford formula or its relativistic generalization, the Mott formula will satisfactorily describe the scattering

$$\sigma_M(\theta) = \frac{d\sigma}{d\Omega} = \left(\frac{Ze^2}{2mc^2}\right)^2\left(\frac{1-\beta^2}{\beta^4}\right)\frac{1}{\sin^4\frac{\theta}{2}}\left(1 - \beta^2\sin^2\frac{\theta}{2}\right) \qquad (4.12)$$

where mc^2 is the electron rest mass energy, $\beta = (v/c)$, and Z is the nuclear charge.

However, at high energies the electrons can no longer be regarded as scattered from a point charge nucleus. Electrons of 20 MeV have the rationalized de Broglie wavelength $\hbar \sim 10$ fm, and at 200 MeV it is 1 fm. Thus, the scattering must be treated by an extended charge of the nucleus. Finite size effects start showing up for $E_0 > 20$ MeV, since the rationalized de Broglie wavelength of electrons becomes comparable with the nuclear sizes. Diffraction effects set in similar to those encountered in optics when the photon wavelength is comparable with the size of the obstacle. In contrast, low energy electrons will not reveal any nuclear structure. In the other extreme, very high energy electrons are likely to be scattered from individual protons in the nucleus corresponding to inelastic scattering by the nucleus as a whole. Therefore, electrons of energies of the order of 100 MeV or so would he suitable for exploring the details of nuclear structure.

The interaction of electrons with the nucleus is almost entirely electromagnetic, the scattering being free from complications arising due to nuclear forces. Also the electron-neutron interaction is believed to be very weak, of the order of few kilo volts. Moderately high energy electrons elastically scattered from target nuclei, are expected to provide the same type of information about nuclear charge distribution

Fig. 4.1 Elastic collision in
the lab system

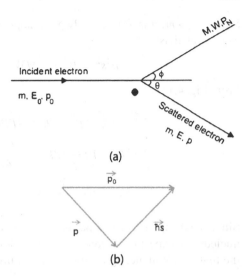

(a)

(b)

as the X-ray diffraction studies yield about the electron distribution in molecules. There is no simple formula for the scattering due to arbitrary charge distribution of the nucleus. Under an assumed charge distribution or the equivalent electrostatic potential, expressions for the differential cross sections can be derived which can be matched with the observed values. The expressions which involve form factors contain all the information concerning the detailed nuclear structure similar to the structure factors in X-ray diffraction scattering.

4.1.4.1 Kinematics of Elastic Scattering

Figure 4.1 shows the essential features at an elastic collision in lab system. Let the electron of mass m, energy E_0 and momentum p_0 suffer an elastic collision with target nucleus of mass M initially at rest. After the collision, the electron is scattered at angle θ, and proceeds with momentum p and energy E, the nucleons recoils with momentum P_N and receives energy W. Assume that the electrons are relativistic so that we may approximate

$$E_0 = cp_0; \qquad E = cp \tag{4.13}$$

Conservation of momentum gives

$$P_N = \hbar s = p_0 - p \tag{4.14}$$

Conservation of energy gives

$$W = E_0 - E = \Delta E = c(p_0 - p) \tag{4.15}$$

Since $M \gg m$, it is a sufficiently good approximation to consider the recoil energy as non-relativistic

$$\hbar^2 s^2 = P_N^2 = 2MW = 2Mc(p_0 - p) \tag{4.16}$$

From Fig. 4.1

$$\hbar^2 s^2 = P_N^2 = p^2 + p_0^2 - 2pp_0 \cos\theta = 2Mc(p_0 - p) \tag{4.17}$$

$$2Mc(p_0 - p) = 4pp_0 \sin^2\frac{\theta}{2} + (p_0 - p)^2 \quad \text{or}$$

$$2Mc(p_0 - p) - \frac{W^2}{c^2} = 4pp_0 \sin^2\frac{\theta}{2} \tag{4.18}$$

Since the recoil energy will be small compared to the rest mass energy of the target nucleus, we expect the second term on the left side will be quite small compared to the first one. With negligible error, we can unite

$$Mc(p_0 - p) = 2pp_0 \sin^2\frac{\theta}{2} \quad \text{or}$$

$$p = \frac{p_0}{1 + \frac{2E_0}{Mc^2} \sin^2\frac{\theta}{2}} \tag{4.19}$$

We introduce the quantity $\hbar q$, the four momentum transfer

$$q^2 = s^2 - \left(\frac{\Delta E}{\hbar c}\right)^2 = \frac{1}{\hbar^2}|p_0 - p|^2 - \frac{1}{\hbar^2}(p_0 - p)^2$$

$$= \frac{4pp_0}{\hbar^2} \sin^2\frac{\theta}{2} \tag{4.20}$$

Using (4.19) in (4.20), gives

$$|q| = \frac{\frac{2p_0}{\hbar} \sin\frac{\theta}{2}}{\sqrt{1 + \frac{2E_0}{Mc^2} \sin^2\frac{\theta}{2}}} \tag{4.21}$$

From (4.16), (4.18) and (4.21)

$$\frac{|s|}{|q|} = \sqrt{1 + \frac{(p_0 - p)^2}{4pp_0 \sin^2\frac{\theta}{2}}} \tag{4.22}$$

The difference in s and q is greater, the larger is the angle of scattering. Even for 1 GeV incident electrons scattered against proton, $|q|$ and $|s|$ differ almost by 25 percent for $\theta = \pi$. With heavier targets and smaller incident energies the difference is still smaller. Of the two quantities, s and q, the latter is more fundamental in that it appears as an invariant in the form factors.

It will be useful to derive a formula which connects the differential cross sections in the lab and C-system

$$\gamma_C = \frac{\gamma + M/m}{\sqrt{1 + 2\gamma \frac{M}{m} + \frac{M^2}{m^2}}} \tag{4.23}$$

(see [9], Appendix A), where γ_C and γ are the Lorentz factors for the C-system and the incident electron in the lab system. Since

$$\frac{M}{m} \gg \gamma$$

$$\gamma_C \simeq \frac{m}{M}\gamma + 1 \simeq 1$$

Use the transformation of longitudinal momentum

$$p\cos\theta = \gamma_C\left(p^*\cos\theta^* + \beta_c u^*\right) \simeq p^*\cos\theta^* + \beta_c u^* \tag{4.24}$$

Differentiate:

$$-p\sin\theta d\theta + \cos\theta dp = -p^*\sin\theta^* d\theta^* \tag{4.25}$$

where p^* and u^* are constant. Further

$$p\left(1 + \frac{2E_0}{Mc^2}\sin^2\frac{\theta}{2}\right) = p_0$$

Differentiate holding E_0 and p_0 as constants, and simplify using the approximation $E_0 \simeq p_0$

$$dp = -\frac{p^2}{Mc}\sin\theta d\theta \tag{4.26}$$

Substitute (4.26) in (4.25)

$$p\sin\theta d\theta\left(1 + \frac{p}{Mc}\cos\theta\right) = p^*\sin\theta^* d\theta^* \tag{4.27}$$

But $p \ll Mc$

$$p\sin\theta d\theta \simeq p^*\sin\theta^* d\theta^* \quad \text{or} \tag{4.28}$$

$$\frac{\sin\theta^* d\theta^*}{\sin\theta d\theta} = \frac{p}{p^*} \tag{4.29}$$

Also

$$\gamma^* = \frac{\gamma + \frac{m}{M}}{\sqrt{1 + \frac{2m}{M}\gamma + \frac{m^2}{M^2}}} \simeq \gamma \tag{4.30}$$

$$\therefore \quad p^* \simeq p_0$$

Fig. 4.2 Born's
approximation

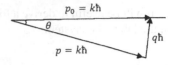

Now

$$\left(\frac{d\sigma}{d\Omega}\right)_{lab} = \left(\frac{d\sigma}{d\Omega}\right)_{cm} \frac{\sin\theta^* d\theta^*}{\sin\theta d\theta} = \left(\frac{d\sigma}{d\Omega}\right)_{cm} \frac{p}{p_0} \qquad (4.31)$$

Using, (4.19), we find

$$\left(\frac{d\sigma}{d\Omega}\right)_{lab} = \frac{1}{1 + \frac{2E_0}{Mc^2}\sin^2\frac{\theta}{2}}\left(\frac{d\sigma}{d\Omega}\right)_{cm} \qquad (4.32)$$

4.1.4.2 Born's Approximation

The scattering problem is considered in the momentum representation. The scattering potential is regarded as something which causes transitions from one state in momentum space to another. The entire potential energy of interaction between the colliding particles is regarded as a perturbation and carry the calculation only to first order. The Born approximation is best applied when the kinetic energy of the colliding particles is large in comparison with the interaction energy. It therefore supplements the method of partial waves (Chap. 5) which is most useful when the bombarding energy is small. Born's approximation gives the scattering amplitude

$$f(\theta) = -\frac{2\mu}{\hbar^2}\int_0^\infty V(r)\frac{r\sin qr dr}{q} \qquad (4.33)$$

The differential cross-section is given by

$$\frac{d\sigma}{d\Omega} = |f(\theta)|^2 = \frac{4\mu^2}{\hbar^4}\left|\int_0^\infty \frac{V(r)r\sin qr dr}{q}\right|^2 \qquad (4.34)$$

where μ is the reduced mass, $V(r)$ is the scattering potential and as shown in Fig. 4.2,

$$q\hbar = 2k\hbar\sin\frac{\theta}{2} \qquad (4.35)$$

What we have is Rutherford scattering (which is true for point charge nucleus) modified partly by relativistic and spin effects, and partly by the finite size of the nucleus, there is no simple scattering formula for an arbitrary charge distribution in the nucleus. Starting with various assumed charge distribution or electrostatic potential, one can calculate the scattering differential cross-section for a given Z and energy and match the calculated and observed angular distributions.

4.1.4.3 Form Factors

Consider the scattering amplitude as given by Born's approximation (4.35)

$$f(\theta) = -\frac{2\mu}{q\hbar^2} \int_0^\infty V(r) \sin(qr) r \, dr \tag{4.36}$$

Integrate by parts to obtain,

$$\int_0^\infty V(r) \sin qr r \, dr = V(r) \left[\frac{1}{q^2} \sin qr - \frac{r}{q} \cos qr \right]_0^\infty$$
$$- \int_0^\infty \frac{dV}{dr} \left(\frac{1}{q^2} \sin qr - \frac{r}{q} \cos qr \right) dr \tag{4.37}$$

The first term on the right side vanishes at both ends since $V(\infty) = 0$

$$\therefore \quad \int_0^\infty V(r) \sin qr r \, dr = -\frac{1}{q^2} \int_0^\infty \frac{dV}{dr} \sin qr \, dr + \frac{1}{q} \int_0^\infty \frac{dV}{dr} r \cos qr \, dr \tag{4.38}$$

Evaluate the second integral on the right side, again by parts to obtain

$$\frac{1}{q} \int_0^\infty \frac{dV}{dr} r \cos qr \, dr = \frac{1}{q} \left[\frac{dV}{dr} \left(\frac{r}{q} \sin qr + \frac{\cos qr}{q^2} \right) \right]_0^\infty$$
$$- \frac{1}{q} \int_0^\infty \left(\frac{r}{q} \sin qr + \frac{\cos qr}{q^2} \right) \frac{d^2 V}{dr^2} dr \tag{4.39}$$

The term $\frac{1}{q^2} r \frac{dV}{dr} \sin qr |_0^\infty$ vanishes at both the limits since we can expect $(\frac{dV}{dr})_{r=\infty} = 0$. Further, integrate by parts to obtain

$$\frac{1}{q^3} \int \cos qr \frac{d^2 V}{dr^2} dr = \frac{1}{q^3} \cos qr \frac{dV}{dr} \Big|_0^\infty + \frac{1}{q^2} \int_0^\infty \frac{dV}{dr} \sin qr \, dr \tag{4.40}$$

$$\frac{1}{q} \int_0^\infty \frac{dV}{dr} r \cos qr \, dr = \frac{1}{q^3} \frac{dV}{dr} \cos qr \Big|_0^\infty - \frac{1}{q^2} \int_0^\infty \frac{d^2 V}{dr^2} r \sin qr \, dr$$
$$- \frac{1}{q^3} \frac{dV}{dr} \cos qr \Big|_0^\infty - \frac{1}{q^2} \int_0^\infty \frac{dV}{d} r \sin qr \, dr \tag{4.41}$$

The first and third terms on the right side cancel out. Therefore, (4.38) becomes

$$\int_0^\infty V(r) \sin qr \, dr = -\frac{1}{q^2} \int \left(\frac{d^2 V}{dr^2} + \frac{2 dV}{r dr} \right) \sin qr r \, dr \tag{4.42}$$

Now, for spherically symmetric potential

$$\nabla^2 V = \frac{d^2 V}{dr^2} + \frac{2dV}{rdr}$$

(4.43)

Also by Poisson's equation

$$\nabla^2 V = -4\pi z e^2 \rho$$

(4.44)

where $\rho = \rho_0/Ze$ is the charge density and Ze is the nuclear charge. Combining (4.36), (4.42), (4.43) and (4.44)

$$f(\theta) = \frac{-8\pi \mu Ze^2}{q^3 \hbar^2} \int_0^\infty \rho(r) \sin(qr) r dr$$

(4.45)

The quantity

$$F(q) = \frac{4\pi}{q} \int_0^\infty \rho(r) \sin(qr) r dr$$

(4.46)

is called the form factor. We can then write

$$\frac{d\sigma}{d\Omega} = \left(\frac{2\mu Ze^2}{q^2 \hbar^2}\right)^2 |F(q)|^2$$

(4.47)

But

$$q^2 \hbar^2 = 4k^2 \hbar^2 \sin^2 \frac{\theta}{2} = 4\mu^2 v^2 \sin^2 \frac{\theta}{2}$$

(4.48)

$$\frac{d\sigma}{d\Omega} = \left(\frac{Ze^2}{2\mu v^2 \sin^2 \frac{\theta}{2}}\right)^2 |F(q)|^2$$

(4.49)

$$\left(\frac{d\sigma}{d\Omega}\right)_{finite\ size} = \left(\frac{d\sigma}{d\Omega}\right)_{point\ charge} |F(q)|^2$$

(4.50)

Since the quantity F^2 multiplies the point charge cross-section in analogy with X-ray diffraction, it is called form factor or structure factor.

4.1.4.4 Scattering from the Shielded Coulomb Potential for a Point Charge Nucleus

$$V = \frac{Z_1 Z_2 e^2}{r} e^{-r/r_0}$$

(4.51)

where r_0 is the shielding radius and is of the order of atomic dimension. For distances $r \gg r_0$, the potential dies off rapidly. Setting $1/r_0 = a$, and inserting (4.51)

Fig. 4.3 Plot showing the
general appearance of the
cross-section

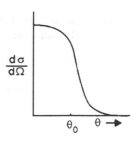

in (4.45)

$$f(\theta) = -\frac{2\mu Z_1 Z_2 e^2}{\hbar^2 q} \int_0^\infty e^{-ar} \sin qr\, dr \qquad (4.52)$$

The value of the particular integral is known to be equal to $q/(q^2 + a^2)$. We therefore, obtain

$$f(\theta) = -\frac{2\mu Z_1 Z_2 e^2}{\hbar^2 [q^2 + a^2]} \quad \text{and} \qquad (4.53)$$

$$\frac{d\sigma}{d\Omega} = |f(\theta)|^2 = \frac{4\mu^2 (Z_1 Z_2 e^2)^2}{\hbar^4 [4k^2 \sin^2 \frac{\theta}{2} + \frac{1}{r_0^2}]^2} \qquad (4.54)$$

Figure 4.3 shows the general appearance of the cross section. With decreasing θ, the curve rises which is reminiscent of Rutherford scattering. However for angles smaller then θ_0, where

$$\sin \frac{\theta_0}{2} \simeq \frac{1}{2kr_0} \qquad (4.55)$$

the curve tends to flatten out. This is because when $q \ll a$, the angular dependence of the cross-section contained is the denominator of (4.54) is damped out. The angle θ_0 may be considered as the limiting angle below which Rutherford scattering ceases because of the shielding by the electron cloud.

We can derive Rutherford scattering formula by setting $a = 0$. This amounts to extending the screening radius r_0 to ∞. In other words, the scattering takes place with bare nucleus. In the limit $a \to 0$, the shielded potential $(Z_1 Z_2 e^2/r)e^{-ar}$ reduces to the ordinary Coulomb potential $Z_1 Z_2 e^2/r$ corresponding to the point charge nucleus. Formula (4.54) then reduces to

$$\frac{d\sigma}{d\Omega} = \frac{1}{4} \frac{\mu^2 (Z_1 Z_2 e^2)^2}{\hbar^4 k^4 \sin^4 \frac{\theta}{2}} = \frac{1}{4} \left(\frac{Z_1 Z_2 e^2}{\mu v^2} \right)^2 \frac{1}{\sin^4 \frac{\theta}{2}} \quad \text{(Rutherford formula)} \quad (4.56)$$

where we have used $k\hbar = \mu v$. This is identical with the Rutherford formula obtained classically.

4.1.4.5 Electron Scattering from an Extended Nucleus of Radius R with Constant Charge Density

$$V(r) = -\frac{Ze^2}{R}\left(\frac{3}{2} - \frac{r^2}{2R^2}\right); \quad 0 < r < R \tag{4.57}$$

$$= -\frac{Ze^2}{r}e^{-ar}; \quad R < r < \infty \tag{4.58}$$

Inside the nucleus the electron sees the potential as given by (4.57) while outside it sees the shielded potential due to point charge nucleus as given by (4.58). Insert (4.57) and (4.58) in (4.33) to obtain

$$f(\theta) = \frac{2\mu Ze^2}{q\hbar^2}\left[\frac{1}{2R}\int_0^R\left(3 - \frac{r^2}{R^2}\right)r\sin qr\,dr + \int_R^\infty \sin qr\,e^{-ar}\,dr\right] \tag{4.59}$$

The second integral in (4.59) can be evaluated as follows

$$\int_R^\infty \sin qr\,e^{-ar}\,dr = \int_0^\infty \sin qr\,e^{-ar}\,dr - \int_0^R \sin qr\,e^{-ar}\,dr$$

$(\mathrm{Lim}\, a \to 0)$

$$= \frac{q}{q^2 + a^2} - \int_0^R \sin qr\,e^{-ar}\,dr$$

$$= \frac{1}{q} - \int_0^R \sin qr\,dr$$

$$= \frac{1}{q}\cos qR \tag{4.60}$$

The first integral in (4.59) can be easily evaluated, we finally obtain

$$f(\theta) = \frac{2\mu Ze^2}{q^2\hbar^2} \times \frac{3}{q^2R^2}\left[\frac{\sin qR}{qR} - \cos qR\right] \tag{4.61}$$

$$\therefore \quad \left(\frac{d\sigma}{d\Omega}\right)_{finite\ size} = \left(\frac{d\sigma}{d\Omega}\right)_{point\ charge}|F(q)|^2 \tag{4.62}$$

where

$$F(q) = \frac{3}{q^2R^2}\left[\frac{\sin qR}{qR} - \cos qR\right] \tag{4.63}$$

The cross-section no longer falls off smoothly. In fact it exhibits sharp maxima and minima. The minima occur whenever the form factor $F(q)$ vanishes, that is when the condition

$$\tan qR = qR \tag{4.64}$$

is satisfied. The maxima and minima are expected to be defined because of the sharp
boundary of the charge distribution. An analogous situation exists in optics. If the
obstacles have sharp edges then the maxima or minima in the diffraction pattern will
be sharp. On the other hand if the refractive index changes slowly as in the case of
a diffuse boundary, then the maxima and minima tend to be washed out. Likewise,
for electron scattering from smoothly varying charge distributions, e.g. Gaussian or
exponential type, the cross section falls off more or less monotonically.

For small momentum transfer, (small incident energy or small scattering angle)
it is readily seen that scattering is almost entirely given by the point charge nucleus.
In the limit $q R \to 0$, the form Factor (4.63) reduces to unity

$$F(q) = \frac{3}{q^2 R^2} \left(\frac{\sin q R}{q R} - \cos q R \right)$$

$(\text{Lim} \, q R \to 0)$

$$= \frac{3}{q^2 R^2} \left(\frac{1}{3} q^2 R^2 - \frac{1}{30} q^4 R^4 + \cdots \right)$$

$$= 1 - \frac{q^2 R^2}{10} + \cdots \tag{4.65}$$

Conversely, the finite size effects begin to show up when $q R \sim 1$, i.e.
$2k R \sin(\theta/2) \sim 1$ or $p \sin(\theta/2) \simeq (\hbar/2R)$. For small angles, this limit becomes,
$\theta \sim \frac{1}{kR} = \lambda/R$. When $|k|$ becomes large enough so that $|k|R \simeq 1$, i.e. when the incident wave is so short that it oscillates several times as it crosses the region in which
the potential is strong then the scattered wave sensitively depends or the details of
the shape of the potential. On the other hand, at smaller energies or smaller scattering angles, the scattering is almost entirely independent of the nature of charge
distribution.

The limiting value of $F(q)$, Eq. (4.65), also follows directly from (4.46). For
small momentum $qr < qR < 1$, $\sin qr \to qr$

$$F(q) = \frac{4\pi}{q} \int_0^\infty \rho(r) \sin(qr) r \, dr$$

$(\text{Lim} \, qr \to 0)$

$$= 4\pi \int \rho(r) \frac{\sin qr}{qr} r^2 dr$$

$$= \int p(r) 4\pi r^2 dr$$

$$= \frac{1}{Ze} \int \rho_0 4\pi r^2 dr$$

$$= 1 \quad \text{(by definition)} \tag{4.66}$$

4.1.4.6 Mean Square Radius

For a given charge distribution $\rho(r)$, the mean square radius is defined by

$$\langle r^2 \rangle = \int_0^\infty r^2 \rho(r) 4\pi r^2 dr \qquad (4.67)$$

For a homogeneous charge distribution the charge density $\rho_0(r)$ is constant and is given by

$$\rho_0 = \frac{3Ze}{4\pi R^3} \qquad (4.68)$$

where R is the nuclear radius. Since $\rho(r) = \frac{\rho_0}{Ze}$

$$r^2 = \frac{4\pi \rho_0}{Ze} \int_0^R r^4 dr = \frac{3}{5} R^2 \quad \text{(homogeneous charge distribution)} \qquad (4.69)$$

4.1.4.7 Geometric Interpretation of the Form Factor

We can expand the sine function in (4.46) and integrate term by term to obtain,

$$F(q) = \frac{4\pi}{q} \int_0^\infty \rho(r) \left[qr - \frac{q^3 r^3}{3!} + \frac{q^5 r^5}{5!} + \cdots \right] r dr$$

$$= \int_0^\infty \rho(r) 4\pi r^2 dr - \frac{q^2}{6} \int_0^\infty r^2 \rho(r) 4\pi r^2 dr + \frac{q^4}{120} \int_0^\infty r^4 \rho(r) 4\pi r^2 dr$$

$$\qquad (4.70)$$

or

$$F = 1 - \frac{q^2}{6} \langle r^2 \rangle + \frac{q^4}{120} + \langle r^4 \rangle + \cdots \qquad (4.71)$$

We conclude that in the approximation quadratic in q^2 the scattering results provide information only on nuclear size since the second term gives the mean square radius of the charge distribution. The shape of the distribution is determined by higher moments, the third and higher terms, which are important only for high momentum transfers.

4.1.4.8 Form Factors and Their Fourier Transforms

Scattering experiments determine the form factors. However, it is possible to invert the procedure and obtain the charge distribution from the experimental form factors, by using the Fourier transform. Now, $F(q)$ is given by

$$F(q) = \frac{4\pi}{q} \int_0^\infty \rho(r) \sin(qr) r dr \qquad (4.72)$$

Set

$$\mathcal{F}(q) = \frac{qF(q)}{4\pi} \tag{4.73}$$

$$f(r) = r\rho(r) \tag{4.74}$$

Therefore

$$\mathcal{F}(q) = \int_0^\infty f(r)\sin qr dr \tag{4.75}$$

then the Fourier transform of $\mathcal{F}(q)$ is

$$f(r) = \frac{2}{\pi}\int_0^\infty \mathcal{F}(q)\sin(qr)qdq \tag{4.76}$$

Use (4.73) and (4.74) in (4.76) to obtain,

$$\rho(r) = \frac{1}{2\pi^2 r}\int_0^\infty F(q)\sin(qr)qdq \tag{4.77}$$

The pair of expressions (4.72) and (4.77) are transforms of each other. Given the form factor $F(q)$, we can deduce the charge distribution and vice versa. As an example consider the form factor of the type

$$F(q) = Ae^{-cq^2} \tag{4.78}$$

where A and c are the constants. Insert (4.78) in (4.77) and integrate by parts to obtain

$$\rho(r) = \frac{A}{8(\pi c)^{3/2}}e^{-r^2/4c} \tag{4.79}$$

The normalization condition

$$\int_0^\infty \rho(r)4\pi r^2 dr = 1$$

yields, $A = 1$. Further, the mean square radius is given by inserting (4.79) with $A = 1$, in (4.67) to find

$$\langle r^2 \rangle = 6c \tag{4.80}$$

In Table 4.1 are listed some of the important charge distributions and the corresponding form factors. Also, included are the values of the mean square radius. Figure 4.4 shows the charge distributions as well as curves for the form factors. For the uniform distribution with a sharp boundary, the quantity F^2 becomes zero whenever the condition $qR = 4.5, 7.7$ etc. is satisfied. For similar distributions with not too diffuse boundary, the zeroes will still be present, but F^2 would decrease much more rapidly at wide angles. Should the F^2 values decrease abruptly for large scattering angles, this would suggest that the associated charge distribution has a long tail.

Table 4.1 Charge distributions and the corresponding form factors

Model	Charge distribution, $\rho(r)$	Form factor, $F(q)$	Mean square Radius, $\langle r^2 \rangle$	
1. Uniform	$\dfrac{3}{4\pi R^3}; r < R$	$\dfrac{3}{q^2 R^2}[\dfrac{\sin qR}{qR} - \cos qR]$	$\dfrac{3}{5}R^2$	(4.81)
2. Exponential	$\dfrac{a^3 e^{-ar}}{8\pi} 0; r > R$	$\dfrac{a^4}{(a^2+a^2)^2}$	$\dfrac{12}{a^2}$	(4.82)
3. Gaussian	$\dfrac{1}{\pi^{3/2} b^3}e^{-r^2/b^2}$	$e^{-\frac{q^2 b^2}{4}}$	$\dfrac{3}{2}b^2$	(4.83)
4. Harmonic oscillator (shell model)	$\dfrac{2}{\pi^{3/2}} \dfrac{1}{a_0^3 (2+3)\alpha}$ $\times (1 + \dfrac{\alpha r^2}{a_0^2})e^{-r^2/a_0^2}$ where $\alpha = \frac{1}{3}(Z-2)$	$[1 - \dfrac{\alpha q^2 a_0^2}{2(2+3\alpha)}]e^{\frac{-q^2 a_0^2}{4}}$	$\dfrac{3}{2}a_0^2 \dfrac{(2+5\alpha)}{(2+3\alpha)}$	(4.84)
5. Fermi (Wood-Saxon) $t = 4.4b$, $c = 1.07A^{1/3} \times 10^{-13}$ cm where the skin thickness is t	$\dfrac{\rho(0)}{1+e^{\frac{r-c}{b}}}$	–	–	(4.85)

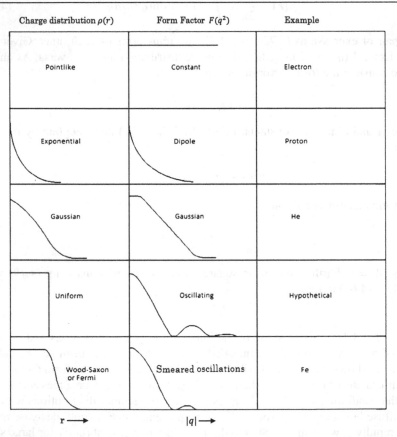

Charge distribution $\rho(r)$	Form Factor $F(q^2)$	Example
Pointlike	Constant	Electron
Exponential	Dipole	Proton
Gaussian	Gaussian	He
Uniform	Oscillating	Hypothetical
Wood-Saxon or Fermi	Smeared oscillations	Fe

$r \longrightarrow$ $|q| \longrightarrow$

Fig. 4.4 Relation between the radial charge distribution and the corresponding form factor on Born approximation

Fig. 4.5 Nuclear charge
density as a function of
distance from the center of
the nucleus found by electron
scattering methods. Ordinate
unit: 10^{19} C cm^{-3} [8]

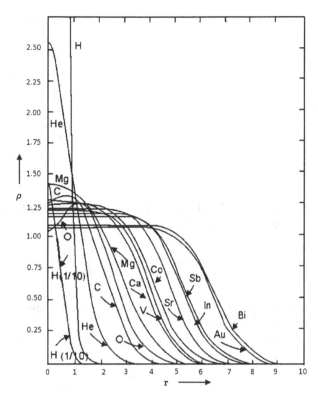

4.1.4.9 Differential Cross-Section

For electron scattering we set $Z_1 = -1$, $Z_2 = Z$. Listed below is the scattering
formula in the order of increasing accuracy.

All nuclei except the light ones exhibit oscillating form factor and the half density radius is c. Figure 4.4 shows form factors for typical charge distributions with
examples.

A practical procedure is to assume a model for the charge distribution and fit the
calculated cross sections with the observed ones, for example, Uniform, Exponential, Gaussian, Yukawa, Wine bottle, Fermi type, Harmonic well, etc. But the choice
is usually narrowed down to two or three types of distributions. It turns out that
data on light elements, such as ^{4}He, ^{12}C, ^{16}O can be fitted well with the assumption
of Gaussian or Harmonic well distribution. Figure 4.5 shows the theoretical curves
based on Born approximation. It is seen that except in the regions of diffraction
minima, the Born approximation applied to low Z elements gives surprisingly good
accuracy. For, ^{4}He the charge distribution can be described well by the function

$$\rho(r) = \frac{1}{(\sqrt{\pi}b)^3} e^{-r^2/b^2} \quad \text{(Gaussian)} \tag{4.86}$$

Fig. 4.6 Fermi model

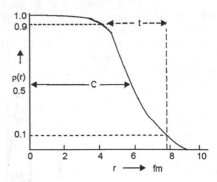

The data for ^{12}C and ^{16}O are in agreement with the Harmonic well, with the two parameter charge distribution

$$\rho(r) = \frac{2}{\pi^{\frac{3}{2}} a_0^3 (2 + 3\alpha)} \left(1 + \frac{\alpha r^2}{a_0^2} \right) e^{-r^2/a_0^2} \quad \text{(Harmonic well)} \qquad (4.87)$$

Note that for ^4He, since $Z = 2$, α is zero, and the Harmonic well reduces to Gaussian charge distribution.

Medium and heavy nuclei are well represented by the Fermi model (Fig. 4.6)

$$\rho(r) = \frac{\rho(0)}{1 + e^{\frac{r-c}{b}}} \qquad (4.88)$$

where c is the half density radius, i.e. the distance from the centre of the nucleus at which charge density falls to half the value, and is given by the relation

$$c = 1.07 \times A^{1/3} \times 10^{-13} \text{ cm} \qquad (4.89)$$

The parameter b is given by

$$t = 4.4b \qquad (4.90)$$

where t, the skin thickness is defined as the distance through which the charge density falls from 90 to 10 % of the value at the centre (see Fig. 4.6). It is found that t is almost constant at a value of about 2.4 fm. The central charge density $\rho(0)$ is given in units of 10^{19} Coulomb per cm^3. It reaches maximum in proton, and falls down to relatively smaller values in heavier elements.

From the form (4.88) we conclude that it is as if the heavier nuclei could be manufactured from the lighter ones simply by stuffing more nucleons in the centre of the nucleus and pushing the thickness outward.

Prior to the high energy electron scattering experiments, it was assumed that the charge is uniformly distributed throughout the nucleus. Results based on the study of α decay, neutron scattering, mirror nuclei etc indicated a value of r_0 between 1.4 and 1.5 fm. However, electron scattering experiments favour the tapering charge distribution (Fermi) for the heavy nuclei, and Harmonic well or Gaussian for the light nuclei, and are in complete disagreement with the constant density model. Since the nuclear boundary can no longer be assumed to be sharp, the concept of nuclear radius must be modified. We have seen that for heavy nuclei, two parameters c and t are necessary in order to specify the entire shape of the distribution. In other words, electron scattering experiments provide information not only on the nuclear sizes but also on the detailed shape of distributions. One can still talk about the nucleus radius r_0 corresponding to an equivalent radius of a spherical nucleus with uniform charge distribution. We can equate the mean square radius for the actual charge distribution to that of equivalent uniform charge distribution

$$\langle r^2 \rangle = a^2 = a_{eq}^2 = \frac{3}{5}R^2 \tag{4.91}$$

where we have used (4.69). It follows that

$$r_0 = \sqrt{\frac{5}{3}}aA^{-1/3} \tag{4.92}$$

When r_0 is calculated from (4.92), it is seen that r_0 is no longer a constant but is a variable, its value ranging from 1.3 or 1.35 fm for light nuclei, to about 1.2 fm for heavy nuclei. The accepted value of r_0 is therefore about 20 % smaller than that derived from the older methods. The redetermination of r_0 based on the study of mirror nuclei is found to be in good agreement with the present value. Further the value of r_0 deduced from experiments on the absorption and diffraction of high energy negative pions (0.6–1.4 GeV) in complex nuclei are also consistent with these results.

Electron scattering experiments are particularly suited for comparing the difference in charge distribution in the neighbouring nuclei, e.g in the pair $^{58}_{28}\text{Ni}$ and $^{60}_{28}\text{Ni}$ or ^{58}Ni and ^{56}Fe. The comparisons are based on the measurement of elastic cross-section for the two nuclei at the same angle. Since the ratio of the cross-sections can be determined more reliably than the individual cross sections and because the theoretical ratio of the cross sections depend only slightly on the exact analytical form of the charge distribution, the method has distinct merits. From such experiments, it has been possible to conclude that in the pairs ^{58}Ni and ^{6}Ni and ^{56}Fe and ^{58}Ni the charge distribution is vastly different. Thus, two extra neutrons in ^{6}Ni have a noticeable influence of the closed shell proton structure in Nickel.

Most of the nuclei studied are spherical in that they possess low values of quadruple moments. Ellipsoidal nuclei are expected to give smooth scattering cross sections, because their random orientations tend to reduce the angular dependence of scattering, and further their extension along the axis amounts to an increase in skin thickness, leading to the smearing of the diffraction pattern.

4.1.5 Mesic Atoms

In passing through a condensed medium a μ^- meson (muon) rapidly loses its energy through excitation and ionization. When the energy is degraded to thermal level the muon due to the Coulomb attraction of a nucleus will be captured in a Bohr orbit of a large principal quantum number to form a mesic atom, then by either radiative or non-radiative (Auger) transitions, the muon cascades down to lower orbits and finally reaches the K-shell in an estimated time of 10^{-13} to 10^{-14} s [3]. According to the calculations of Wheeler, the transition probabilities per unit time for radiative transitions, $2s \rightarrow 2p$ and $2s \rightarrow 1s$ go as Z^4, while for Auger transitions, $2s \rightarrow 1s$, they are independent of Z. For low Z, Auger transitions from the $2s$ level dominate while for high Z, radiative transitions are more important. The ultimate fate of the muon is decided by the competition between the natural β-decay and nuclear capture ($\mu^- + p \rightarrow n + \nu$). Because of its weak nuclear interaction the muon will be able to reach the K-shell with appreciable probability in all but the heavy elements. In vacuum the μ^- have mean decay times of 2.2 μs. However in the presence of nuclear matter, this value is substantially altered. It is shown that the mean lifetime for nuclear capture goes as Z^{-4} for low Z elements and becomes saturated at around 7×10^{-8} s for $Z = 82$. Therefore in light elements, the radioactive decay of μ^- strongly competes with nuclear captures, while in heavy elements nuclear capture is the dominant process. In contrast, the π^- meson owing to its strong interaction with nuclear matter will rarely reach the K-shell except in very light elements ($Z \leq 9$).

The most outstanding peculiarity of the mesic atom is that the orbits for a given principal quantum number are shrunk by a factor approximately equal to the ratio of meson mass to the electron mass as compared to the ordinary atom. Consequently an appreciable part of the wave function corresponding to the lowest levels lies within the nucleus itself for intermediate atomic numbers. For a nucleus with $Z = 47$ (silver) the K-shell orbit of the muon already grazes the nuclear surface (assuming $r_0 = 1.2$ fm). If for simplicity, we assume a uniform charge distribution for the nucleus then the muon experiences a simple harmonic oscillator potential which is quite different from that of a point charge nucleus. This leads to a shift in energy levels for the low lying orbits by an amount that depends on the nuclear radius. Because of the small magnitude of nuclear interaction compared with the Coulomb force, exact calculations for the energy levels are still possible. On the other hand, in light nuclei and for higher orbits the mesic orbits will be outside the range of nuclear forces, and therefore the energy levels of the mesic atom are hydrogen-like to a very good approximation. In this case Dirac's theory which is applicable to muon allows the energy levels to be calculated with sufficient accuracy.

The other peculiarity of the mesic atoms is that the radiative transitions between various levels yield X-rays or even γ rays rather than ordinary photons. The measurements of transition energies for the low lying orbits in heavy elements (for example, $2p \rightarrow 1s$ transitions) which are sensitive to the nuclear radius allow r_0 to be determined. On the other hand corresponding measurements in light nuclei which depend on the mass of the μ^- meson but are independent of the nuclear radius yield a fairly accurate mass determination of the meson.

4.1.5.1 Energy Levels

Treating the nucleus as a point charge, Bohr's simple theory gives

$$E_n = \frac{-\mu c^2 (Z\alpha)^2}{2n^2} \tag{4.93}$$

$$r_n = \frac{\hbar}{\mu e^2} \frac{n^2}{Z} \tag{4.94}$$

$$v_n = \frac{\alpha c Z}{n} \tag{4.95}$$

where E_n is the energy of the level characterised by the principle quantum number n, r_n is the radius of the corresponding orbit, and v_n the classical orbital velocity, $\alpha = 1/137.04$ is the fine structure constant, and $\mu = A m_\mu /(m_\mu + A)$ is the reduced mass of the meson-nucleus system. The dependence of E_n and r_n on μ is clearly borne out by the formulae (4.93) and (4.94), Further, v_n is independent of μ. Relativistic effects for mesons will be as important as for electron.

For pions (π mesons) with zero spin, the Klein-Gordon's relativistic equation is appropriate; it has the solution

$$E_{n,l} = -\frac{\mu c^2}{2n^2} (Z\alpha)^2 \left\{ 1 + \frac{(Z\alpha)^2}{n^2} \left(\frac{n}{l+\frac{1}{2}} - \frac{3}{4} \right) \cdots \right\}$$

(Klein-Gordon relativistic) $\tag{4.96}$

where higher order terms involving $(Z\alpha)$ have been neglected. The formula gives the fine structure which arises due to the relativistic splitting of the states of different l for a given n.

For μ mesons solution of Dirac's equation for spin $(1/2)$ particles gives the expression for the energy levels

$$E_{n,j} = -\frac{\mu c^2}{2n^2} (Z\alpha)^2 \left\{ 1 + \frac{(Z\alpha)^2}{n^2} \left(\frac{n}{j+\frac{1}{2}} - \frac{3}{4} \right) \cdots \right\} \quad \text{(Dirac relativistic)} \quad (4.97)$$

Formula (4.97) has the same form as (4.96) on replacing j with l. Thus spin splitting goes as Z^4 and is therefore, small for low atomic numbers. The relative splitting for the levels $j = l + 1/2$ and $j = l - 1/2$ is the same as that for the electron.

The transition energy $\Delta E = E_2 - E_1$, as calculated by Bohr's formula (4.93) for the μ mesic atom of lead is equal to 14.25 MeV. The more accurate formula (4.97) based on Dirac's theory predicts for the transition $2p^{1/2} \rightarrow 1s$, a value of 15.39 MeV. The splitting of $2p$ state into $2p^{3/2}$ and $2p^{1/2}$ is calculated as 0.425 MeV for the lead μ mesic atom. Formula (4.97) predicts that the $2p$ level is higher than the $2s$ level, since in the former E is less negative than in the latter.

Fig. 4.7 Variation of
Coulomb potential with the
radial distance

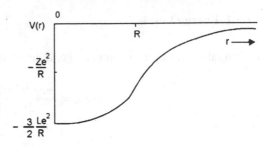

4.1.5.2 Nuclear Size Effects

The finite size effects are most readily seen if it is assumed that the orbit is completely immersed within the nucleus, so that the meson spends negligible time outside. Under the assumption of uniform charge distribution, the potential is that of harmonic oscillator and is given by

$$V(r) = -\frac{Ze^2}{2R}\left(3 - \frac{r^2}{R^2}\right); \quad r \leq R$$

$$= -\frac{Ze^2}{r}; \quad r > R \tag{4.98}$$

Figure 4.7 shows the variation of the Coulomb potential with the radial distance r. Assuming, $R = 1.2 \times 10^{-13} A^{1/3}$, for lead the potential at the surface is $V(R) = -16.6$ MeV, and that at the centre is $V(0) = -25$ MeV. The potential resulting from the finite extension of the nucleus not only grossly alters the energy levels but also leads to the interchange of $2s$ and $2p$ levels. In other words the $2p$ level lies below the $2s$ level. Because of the inversion of levels, radiation transitions do take place from $2p$ to $1s$ level, further, the $2s$ level will be metastable.

The Coulomb force is given by

$$F = -\frac{dV}{dr} = -\frac{Ze^2}{R^3}r; \quad r \leq R \tag{4.99}$$

Balancing the electric force and the centripetal force

$$\frac{Ze^2 r}{R^3} = \frac{\mu v^2}{r} \tag{4.100}$$

Bohr's condition for the quantization of angular momentum gives the relation

$$\mu v r = n\hbar \tag{4.101}$$

Combining (4.100) and (4.101), we find

$$r_n^4 = \frac{n^2 \hbar^2 R^3}{\mu Z e^2} \tag{4.102}$$

$$v^4 = \frac{n^2 \hbar^2 Z e^2}{\mu^3 R^3} \tag{4.103}$$

The total energy is given by

$$E_n = T + V = \frac{1}{2}\mu v^2 - \frac{Z e^2}{2R}\left(3 - \frac{r^2}{R^2}\right) \tag{4.104}$$

Inserting the values of r^2 and v^2 from (4.102) and (4.103)

$$E_n = -\frac{3}{2}\frac{Z e^2}{R} + \frac{n\hbar e}{R^{3/2}}\sqrt{\frac{Z}{\mu}} \tag{4.105}$$

Therefore, the transition energy for successive levels for the case μ meson orbits lying inside the nucleus is given by

$$\Delta E = \frac{\hbar e}{R^{3/2}}\sqrt{\frac{Z}{\mu}} \tag{4.106}$$

Formula (4.106) clearly shows that the transition energies are sensitively dependent on the nuclear radius. For lead mesic atom, with the choice of $R = 1.2 \times 10^{-13} A^{1/3}$ cm we find $\Delta E = 10.1$ MeV, a value which is much smaller than that given by Bohr's formula or Dirac's theory, for the point charge nucleus.

From (4.94) and (4.102), the ratio

$$\frac{r_n (\text{finite size})}{r_n (\text{point charge})} = \left(\frac{\mu R Z e^2}{n^2 \hbar^2}\right)^{3/4} \tag{4.107}$$

which reduce to $\frac{2}{47 n^{\frac{3}{2}}}$ for $A = 2Z$, $r_0 = 1.2 \times 10^{-3}$ cm. Note that for $Z > 47$ for K-shell, the above ratio is larger than unity, but for higher orbits it is less than unity. The reason is that for finite size. $r_n \propto \sqrt{n}$, but for point charge nucleus $r_n \propto n^2$.

It can be shown that the ratio of absorption probabilities from the $2p$ and $1s$ states

$$\frac{(P_{abs})_{2p}}{(P_{abs})_{1s}} = 2 \times 10^{-11} Z^{2.67} \tag{4.108}$$

for the μ mesic atom with $R = 1.2 A^{1/3}$ fm. We therefore conclude that the capture takes place almost exclusively from the $1s$ state.

4.1.5.3 Finite Size Effects on Energy Levels—Quantum Mechanical Treatment

The finite extension of the nucleus modifies the Coulomb potential as compared to that for a point charge. This leads to a shift in energy levels which in case of light nuclei can be calculated by applying the first order perturbation theory. The energy shift ΔE_R can be written as

$$\Delta E_R = -\int \psi^* \delta V \psi \, d\tau \tag{4.109}$$

where ψ is the unperturbed wave function and

$$\delta V = V - V_p \tag{4.110}$$

$$V_p = -\frac{Ze^2}{r} \quad \text{(point charge)} \tag{4.111}$$

$$V = -\frac{Ze^2}{2R}\left(3 - \frac{r^2}{R^2}\right) \quad \text{(uniform charge distribution)} \tag{4.112}$$

The sign convention in (4.109) is that negative ΔE_R implies an increase in binding energy and hence a larger transition energy to a given level. The correction term E_R due to finite size effect is unimportant for all but 1s level. For the 1s level, the hydrogen-like wave function is

$$\psi_0 = \frac{1}{\sqrt{\pi a_0^3}} e^{-r/a_0} \tag{4.113}$$

Combining (4.109) and (4.113), we find

$$\Delta E_R = \int_0^R 4\pi r^2 dr \frac{e^{-2r/a_0}}{\pi a_0^3}\left\{-\frac{Ze^2}{R}\left(3 - \frac{r^2}{R^2}\right) + \frac{Ze^2}{r}\right\} \tag{4.114}$$

As $a_0 \gg R$, we may set the exponential to unity as a good approximation. Then a direct integration yields

$$\Delta E_R = \frac{2}{5}\frac{Ze^2 R^2}{a_0^3} \tag{4.115}$$

Now, according to Bohr's formula, the total energy of the 1s level (binding energy) is given by

$$E_B = \frac{Ze^2}{2a_0} \tag{4.116}$$

The relative shift is therefore given by

$$\frac{\Delta E_R}{E_B} = \frac{4}{5}\left(\frac{R}{a_0}\right)^2 \tag{4.117}$$

For ($Z = 10$), the relative shift amounts to ~0.5 % only. The above formula is valid for the 1s state for $Z \leq 10$. Similer calculations for 2p state give the relative shift of 10^{-6} which is negligible.

Table 4.2 Results for the
$2p_{3/2}$ and $1s$ energy levels
for the nuclear radius
$R = 1.3 \times 10^{-13} A^{1/3}$

Z	$R = 0$		$R = 1.3 \times 10^{-13} A^{1/3}$	
	$1s$	$2p_{3/2}$	$1s$	$2p_{3/2}$
22	1.392	0.346	1.282	0.346
51	7.707	1.874	5.22	1.81
82	22.328	4.914	10.11	4.63

For heavier elements ($Z > 10$) the energy shift of the ls level rapidly becomes large and the first order corrections are no longer adequate. It becomes necessary to solve the wave equation exactly with a suitable potential inside and outside the nucleus. The calculations have to be done numerically and have been carried out for a variety of elements by Fitch and Rainwater [4]. In Table 4.2 are included their results for the $2p_{3/2}$ and $1s$ energy levels for the nuclear radius $R = 1.3 \times 10^{-13} A^{1/3}$. The calculations refer to μ meson with mass $210 m_e$. Also for comparison are given the values by the Dirac equation neglecting the nuclear size ($R = 0$).

The $2p$ doublet splitting ($2p_{1/2} - 2p_{3/2}$) caused by the magnetic moment of μ meson can be roughly estimated by the formula

$$\Delta E = \frac{3\hbar^2}{4\mu^2 c^2} \frac{1}{r} \frac{dV}{dr} \tag{4.118}$$

which is applicable to an ideally heavy nucleus. We find,

$$\frac{1}{r} \frac{dV}{dr} = \frac{Ze^2}{R^3} \tag{4.119}$$

$$\Delta E = \frac{3\hbar^2 Ze^2}{4\mu^2 c^2 R^3}$$

$$= \frac{3}{4} \times \frac{1}{137} \left(\frac{Z}{A}\right) \frac{h^3}{r_0^3 \mu^2 c} \tag{4.120}$$

For lead μ mesic atom, insert $(Z/A) = (82/208)$ and $\mu c^2 = 106$ MeV to obtain $\Delta E \sim 0.87$ MeV. This value is considerably larger than the corresponding result for the point charge nucleus.

By comparing the observed and calculated transition energies, Fitch and Rain water concluded that data are consistent only with $r_0 \sim 1.3$ fm.

4.1.6 Half Lifetimes of α Emitters

If the half lifetime of the α emitter and the α energy be known then the nuclear radius can be found out from Eq. (3.70) which involves the Gamow factor. The value of R thus determined will be the sum of the radii of the daughter nucleus and

α particle. After the correction for the finite radius of α particle, the value of r_0 is found to be little less than 1.48 fm, in agreement with values obtained from constant density model.

4.1.7 High Energy Neutron Scattering

High energy neutron cross-sections can be used to estimate nuclear radii. When a neutron beam is incident on target nuclei each nucleus would cast a shadow in the manner of an opaque disk intercepting a beam of light. This shadow results from the interference of waves scattered from the edge of the opaque sphere (Chap. 5).

It can be shown that exactly the same amount of incident energy is diffracted as is absorbed by the opaque sphere. For fast neutrons with $\lambda \ll R$ this diffraction of "shadow scattering" corresponds to a small angle elastic-scattering for which the cross-section σ_{sc} is the same as σ_{abs} where

$$\sigma_{sc} \simeq (R + \lambda)^2 \tag{4.121}$$

Then the total nuclear cross-section σ_t is

$$\sigma_t = \sigma_{abs} + \sigma_{sc} \simeq 2\pi (R + \lambda)^2 \tag{4.122}$$

which is double the effective geometrical area of the nucleus. When the measured total attenuations cross-sections are used in (4.122), the value of R is obtained. On the constant-density model $R = r_0 A^{1/3}$, whence a value of $r_0 = 1.4$ fm is obtained.

4.2 Constituents of the Atomic Nucleus

An atomic nucleus contains Z protons and N neutrons in a small space of size 10^{-12} cm. Z electrons orbit around the nucleus at distance of the order of 10^{-8} cm. The total positive charge of the nucleus is compensated for by an equal negative charge of electrons so that the atom as a whole appears neutral. There are three reasons which exclude the electrons to stay inside the nucleus.

(a) Suppose the electrons were to be contained inside a nucleus, then their de Broglie wavelength must be equal to the dimension of the nucleus and the minimum momentum corresponding to this wavelength would be $p = h/\lambda$. This is found to be equal to 120 MeV/c when we set $\lambda = 10^{-12}$ cm. This implies a kinetic energy of about 120 MeV—a value which is unreasonably large. With the discovery of neutron (1932), it was natural to assume that protons and neutrons were actually the constituents of the nucleus and electrons do not exist inside the nucleus. With this assumption, for a de Broglie wavelength $\lambda = 10^{-12}$ cm momentum of the neutron would still be 120 MeV/c, but the corresponding kinetic energy would be only 8 MeV—a value which is reasonably low as it is comparable with the binding energy of the particle. Same conclusion is reached from the uncertainty principle.

(b) The nuclear spin (intrinsic angular momentum) arises from the angular momentum of its constituents. The spins of the constituents may be parallel or antiparallel to each other. The spin of each of the particles-proton, neutron and electron, is $(1/2)\hbar$, where $\hbar = $ Planck's constant divided by 2π. Consider the example of nitrogen nucleus under the proton-electron hypothesis. It will have 14 protons and 7 electrons, the total number of particles being 21, i.e. an odd number. The spin of the nitrogen nucleus from odd number of particles is expected to be an odd multiple of $(1/2)\hbar$. But experiments had revealed that the spin of the nitrogen nucleus is $\hbar$, i.e. an even integral multiple of $\hbar/2$. On the other hand, under the proton-neutron hypothesis this difficulty is removed at once. As the nitrogen nucleus will contain 7 protons and 7 neutrons, total number of particles is even, so that the spin of the nucleus will be an even integral multiple of $\hbar/2$ (in agreement with experiment).

(c) If electrons were inside the nucleus, then one could not explain long lifetimes of beta emitters.

4.3 Definitions

Nuclide is a nuclear species.

Nucleon (N) When proton and neutron are not to be distinguished, they are jointly called nucleon.

Atomic number (Z) is the total number of protons in a nucleus. It is also equal to total number of electrons in a neutral atom.

Mass number (A) is the total number of protons (p) and neutrons (N) in a nucleus. $A = Z + N$. In a nucleus X, the values of Z and A are specified by writing them as subscripts and superscripts, respectively $^A_Z X$.

Isotopes are atoms with the same Z but different A, e.g. $[^{16}_8 O, {}^{17}_8 O]$; $[^1_1 H, {}^2_1 H, {}^3_1 H]$. Chemically, isotopes of a given atom are indistinguishable. Isotopes may be stable or unstable against radioactive decay.

Isobars are atoms with the same A but different Z, e.g. $[^3_2 He, {}^3_1 H]$; $[^{14}_7 N, {}^{14}_6 C]$.

Isotones are the atoms with the same N but different Z, e.g. $[^{16}_8 O, {}^{14}_6 C]$.

Isomers are atoms with the same Z and same A but are capable of existing in different nuclear energy states for sufficiently long time to be observed, e.g. $^{80}_{35} Br$ (4.5 hr) and $^{80}_{35} Br$ (18 min).

The 18 min lifetime is associated with beta activity while 4.5 h lifetime goes into the regime of delayed gamma emission. Isomers are metastable excited levels. More than hundred examples of them are known. Gamma emission is usually associated with short mean lifetimes, 10^{-10} s or less. In the case of isomers, the gamma activity can range from 10^{-10} s to few years. It exists only because direct gamma emission is not possible.

Isodiaphes are characterised by constant difference of neutron and proton number, e.g. [$^{12}_{6}$C, $^{14}_{7}$N, $^{16}_{8}$O].

In α decay, the parent and product are isodiaphes.

Mirror nuclei are a pair of nuclei, in which proton and neutron numbers are interchanged, e.g. [$^{27}_{13}$Al, $^{27}_{14}$Si], [$^{3}_{1}$H, $^{3}_{2}$He].

4.4 Atomic Mass Unit

Atomic masses are measured accurately by mass spectrographs. Formerly, atomic mass (weight) was referred to the hydrogen atom and later to the oxygen-16 isotope. Since 1960, the atomic mass unit (amu) has been defined as one-twelfth of the mass of carbon-12 atom. The new standard was chosen because carbon has only two stable isotopes and their proportions are constant in the naturally occurring carbon. Further, the fact that carbon-12 forms numerous compounds is an advantage in modern mass spectroscopic work. 1 amu $= 1.66 \times 10^{-24}$ g. On the carbon scale, the masses of electron, proton and neutron are $m_e = 0.000548$ amu, $m_p = 1.0073$ amu, $m_n = 1.0087$ amu respectively.

4.5 Nuclear Force

In a nucleus, the nucleons are held together by nuclear force. The characteristics of nuclear force are:

(a) They are attractive.
(b) They have short range, of the order of 1 fermi, i.e. their sphere of influence is limited to very small distances. Nuclear forces are said to have *saturation property*. This behaviour is in marked contrast with other types of fundamental forces like Coulomb forces and gravitational forces which obey inverse square law and are therefore of long range.
(c) At small distance, of the order of a fermi, nuclear forces are stronger by a factor 100 than electric forces, by a factor 10^{12} than the weak force such as the one associated with beta decay of a radioactive nucleus and by a factor 10^{36} than the gravitational force.
(d) They are charge independent, i.e. the nuclear force between proton and proton is identical with that between a proton and neutron or a neutron and a neutron, $p-p = p-n = n-n$.
(e) Nuclear forces are spin dependent, i.e. they depend on the orientation of the nuclear spins.

4.6 Mass Defect, Packing Fraction and Binding Energy

The mass of an atom is nearly, but not exactly, integral multiple of mass of hydrogen atom. The departure from the integral number is due to two reasons:

Fig. 4.8 Binding energy per nucleon as a function of mass number A for stable and long lived nuclei

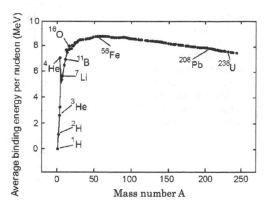

(a) Nuclei contain both protons and neutrons and neutron is slightly heavier than proton.

(b) The mass of a nucleus is not equal to the sum of the masses of neutrons and protons, but is actually smaller by a few tenths of a per cent. We may write

$$M(\text{nucleus}) = Zm_p + (A - Z)m_n - \Delta \qquad (4.123)$$

where Δ is called the mass defect. The quantity Δ arises because some energy is required to break the nucleus into its constituents. In effect, it represents the binding energy $B = \Delta mc^2$ for the nucleus. Another quantity which is frequently used is the separation energy. It is the nuclear analogue of the first ionization potential of atom.

The quantity $(M - A)/A$ is called the *packing fraction*. By definition, this quantity is zero for C^{12}. For other nuclei it can be either positive or negative. The heavier is the nucleus, the larger will be the binding energy since more nucleons are involved. A related quantity of interest is the binding energy per nucleon defined by $f = B/A$. Figure 4.8 shows the variation of f with the mass number A. For small values of A, there are some irregular fluctuations in f, after which it slowly increases up to $A = 50$, then it settles down practically to a constant value around 8.5 MeV up to $A = 150$. Beyond this, it decreases to a value of 7.4 MeV for uranium-238. It follows that the total binding energy of a nucleus is roughly proportional to A. This is a direct consequence of the short range character of nuclear forces. Each nucleon added increases the total binding energy by an equal amount. We can then assume the nuclear density, i.e. the number of nucleons in a given volume to be independent of the size of nucleus

$$A \propto (4/3)\pi R^3, \quad \text{or}$$

$$R = r_0 A^{1/3} \qquad (4.124)$$

where A is the mass number, R the nuclear radius and $r_0 = 1.3$ fm, a constant. For the gold nucleus $R = 1.3 \times (198)^{1/3} \simeq 8$ fm. We can qualitatively understand the

Fig. 4.9 Mass excess
$(\Delta = M - A)$ and
packing-fraction
$P = (M - A)/A$ based on
mass-spectrographic and
nuclear data for beta-stable
nuclei. The smooth curves are
based upon the mass formula.
Ordinates are in millimass
units [5]

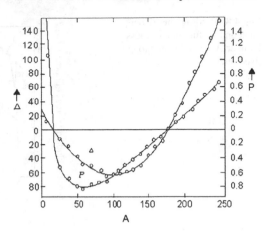

$f–A$ curve given in Fig. 4.8. While the nucleons are held by short range attractive forces, protons repel one another by Coulomb's long range forces. Although the electrostatic forces are generally weaker than the nuclear forces by two orders of magnitude, they become quite important in heavy nuclei as the repulsive Coulomb forces which are proportional to Z^2 tend to counteract the attractive nuclear forces which are only linearly proportional to A. This effect becomes clearly important in heavy nuclei and has the effect of reducing the binding energy per nucleon.

For very low values of A, a smaller value of f arises due to surface tension effect. The nucleons at the surface are less strongly bound than those in the interior. The number of nucleons; lying on the surface of a nucleus of radius R, is proportional to surface area $4\pi R^2$, while number of nucleons in a nucleus is proportional to the nuclear volume $(4/3)\pi R^3$. Hence the fraction of nucleons on the surface is proportional to $4\pi R^2/(4/3)\pi R^3$, or $1/R$. The smaller is the nucleus, the greater will be the fraction of nucleons at the surface. This then explains the lowering of f for very low mass numbers.

The packing fraction P is defined as

$$P = \frac{M - A}{A}$$

where $M - A$ is the mass excess, P can be positive, zero or negative. Note that P has a minimum value of about -8×10^{-4} in the vicinity of iron, cobalt and nickel (Fig. 4.9). This corresponds to maximum value of binding energy per nucleon (f). Fluctuations in f for small A are attributed to shell structure of nuclei.

4.7 Mass and Energy Equivalence

According to Einstein, the energy equivalent to mass m is given by $E = mc^2$. In nuclear physics the basic mass unit is amu, the atomic mass unit which is $1/12$ of the rest mass of carbon-12 atom

$$E = m_0 c^2 = \left(1.66 \times 10^{-27}\ \text{kg}\right) \times \left(3 \times 10^8\ \text{m/s}\right)^2$$

$$= 1.49 \times 10^{-10}\ \text{J} = \left(1.49 \times 10^{-10}/1.6 \times 10^{-13}\right)\ \text{MeV} = 931\ \text{MeV}$$

$$1\ \text{amu} = (1/12)\text{C}^{12}\ \text{mass} = 1.66 \times 10^{-27}\ \text{kg} = 1.49 \times 10^{-10}\ \text{J} = 931\ \text{MeV}/c^2$$

Thus, a unit of energy may be considered as a unit of mass. It is usual to express the masses of fundamental particle in MeV. For example, the mass of electron is 0.51 MeV and that of proton is 938 MeV (actually MeV/c^2).

The mass energy equivalence was first verified in the experiment of *Cockroft* and *Walton*, which was concerned with the nuclear reaction induced with protons accelerated to 300 keV

$$\underset{\text{7.0160 amu}}{^{7}_{3}\text{Li}} \quad + \quad \underset{\text{1.0078 amu}}{^{1}_{1}\text{H}} \quad \rightarrow \quad \underset{\text{4.0026 amu}}{^{4}_{2}\text{He}} \quad + \quad \underset{\text{4.0026 amu}}{^{4}_{2}\text{He}}$$

Total mass of initial particles = 8.0238 amu

Total mass of final particles = 8.0052 amu

(Total initial mass) − (Total final mass)

$$= (8.0238 - 8.0052)\ \text{amu}$$

$$= 0.0186\ \text{amu}$$

$$= (0.0186\ \text{amu}) \times (931\ \text{MeV/amu}) = 17.3\ \text{MeV}$$

If the kinetic energy of 300 or 0.3 MeV of the bombarding protons be added, we find the total energy released as 17.6 MeV. From the range measurements in air, each alpha particle was found to have energy equal to 8.6 MeV so that the energy carried by the two particles is 17.2 MeV, which is in agreement with the calculated value (within experimental errors).

Observe that the masses used in the above problem are atomic masses rather than nuclear masses. But this does not affect the result since the total number of electrons are identical for the initial as well as the final products, and the ionisation potential of electrons is too small to be of any consequence. In nuclear reactions, what is conserved is not mass alone but *mass + energy*. On the other hand, in chemical reactions the energies absorbed or evolved are so small that the conservation of mass alone is sufficiently accurate.

4.8 Nuclear Instability

Stable nuclei contain only a certain combination of protons and neutrons. Figure 4.10 is the plot of N (neutron number) versus Z (proton number). In light stable nuclei $Z = N = A/2$, e.g. $^{4}_{2}\text{He}$, $^{12}_{6}\text{C}$, $^{16}_{8}\text{O}$ are stable. But in heavy nuclei there is significant departure from the $N = Z$ line, stable nuclei having neutron excess.

There are two opposite tendencies in a nucleus. First, the tendency is for N to be equal to Z, which is due to the application of Pauli's principle to protons and

Fig. 4.10 Chart of all
beta-stable nuclei in a ZN
plane

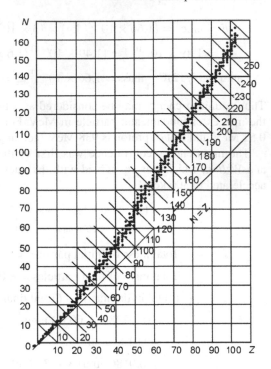

neutrons. In analogy with the electrons in an atom, nucleons in a nucleus can occupy
certain discrete energy levels. Owing to Pauli's principle, not more than two protons
or neutrons can occupy the given energy state. Various energy levels are filled up in
sequence, to give rise to maximum stability for the nucleus. Thus, in absence of the
Pauli's principle, a stable nucleus should have contained neutrons only.

The second tendency for neutrons to exceed protons is due to the repulsive
Coulomb forces between various protons, which tend to weaken the nuclear binding. In order to compensate for this effect, which is more important in heavy nuclei,
a nucleus must be supplied with extra neutrons. The binding energy is maximised in
light and medium elements for $N = Z$. As A increases, the neutron number N exceeds the proton number to overcome coulomb repulsion. Those nuclei which have
proton number in excess of that corresponding to the stability curve, tend to lose
their charge by emitting positrons. Effectively one of the protons becomes a neutron
through

$$p \rightarrow n + \beta^+ + \nu_e \qquad (4.125)$$

Alternatively, the proton captures an orbital electron of the atom, usually the K-shell
electron and transforms itself into a neutron through

$$p + e^- \rightarrow n + \nu_e \qquad (4.126)$$

In both the cases, the neutron remains lodged within the resulting nucleus. On the
other hand, those nuclei which have neutrons in excess of the number corresponding

to the stability curve tend to gain positive charge by β-decay. This is accomplished through the process

$$n \rightarrow p + \beta^- + \bar{\nu}_e \qquad (4.127)$$

Thus, the unstable nuclei transform via β decay along lines of constant A, i.e. diagonally towards the centre of region of stability. Here $\bar{\nu}$ is the antiparticle of ν.

4.9 Stability Against β Decay

The criterion that a nuclide (Z, A) is stable against negative beta decay is

$$M(Z, A) = M_{Nuc} + (Z, A) + ZM(e) \leq M_{Nuc}(Z + 1, A) + ZM(e) + M(\beta) \qquad (4.128)$$

where M_{Nuc} refers to the mass of the nucleus alone, $M(e)$ to the mass of an atomic electron and $M(\beta)$ to the mass of β particle. Since the negative β particle is an electron, the right-hand side represents $M(Z + 1, A)$, the atomic mass. The condition for the stability against β decay is

$$M(Z, A) \leq M(Z + 1, A) \qquad (4.129)$$

where M is the atomic mass.

For the β^+ decay, the corresponding condition for stability is

$$M(Z + 1, A) \leq M(Z, A) + 2M(e) \qquad (4.130)$$

In the right-hand side, the mass of electron appears twice because one electron will be less in the daughter atom and also positron mass which is equal to electron mass must be provided.

A third type of β process occurs in the decay by the capture of an atomic electron by the nucleus, accompanied by neutrino emission. Considering that the neutrino may have zero energy, the condition for stability becomes

$$M(Z + 1, A) \leq M(Z, A) \qquad (4.131)$$

In all the three expressions for the stability of nucleus in the beta process, the change in the binding energy of atomic electrons has been ignored. This is justifiable, since the binding energy of electrons is quite small to affect the results. The corresponding equations for the Q of the decays are

$$Q_{\beta^-} = \left[M(Z, A) - M(Z + 1, A) \right] c^2 = T_{\max} + T_\gamma = T_0 \qquad (4.132)$$

$$Q_{\beta^+} = \left[M(Z + 1, A) - M(Z, A) \right] c^2 = 2m_e c^2 + T_{\max} + T_\gamma$$

$$= 2m_e c^2 + T_0 \qquad (4.133)$$

$$Q_{Ec} = \left[M(Z + 1, A) - M(Z, A) \right] c^2 = T_\nu + T_\gamma = T_0 \qquad (4.134)$$

In case γ-rays accompany β^- and β^+ decays and precede the electron capture, T_0 is the total kinetic energy of the decay products. Note that in (4.131) m_e does not appear on the left hand side. This becomes quite obvious when we consider nuclear masses rather than atomic masses, as was alone for β^- decay in Eq. (4.128)

$$[(Z+1)M' + m_e]c^2 = ZM'c^2 + m_\nu c^2 + T_\nu + T_{M'} + T_\gamma = ZM'c^2 + T_\nu + T_\gamma$$

(as $m_\nu = 0$ and $T_{M'} = $ nuclear mass)

Add Z electrons to both the sides. Then nuclear masses are converted into atomic masses as in (4.128).

4.10 Stability Against Neutron and α Decay and Fission

The criterion for the stability of a nucleus, against disintegration via neutron emission is that the binding energy of the neutron in the nucleus shall be positive. The binding energy of a neutron in a nucleus

$$B(n) = [M(A-1, Z) + M(n) - M(A, Z)]c^2 \qquad (4.135)$$

This quantity is found to be positive for the stable elements, indicating thereby the stability against spontaneous emission of neutron. Same conclusions hold good for proton emission. However, in the case of fission fragments (nuclei of the fission products) which are invariably rich in neutrons, decay can occur via neutron emission. Fission is a special type of nuclear disintegration in which a nucleus is split up into two large fragments and sometimes three.

The binding energy of α particle in a nucleus is

$$B(\alpha) - [M(A-4, Z-2) + M(\alpha) - M(A, Z)]c^2 \qquad (4.136)$$

This quantity becomes negative in the middle of the periodic table long before the natural α emitters are reached. The intervening elements are stable against α decay only because the α energies are so small that their lifetimes are prohibitively long (Gieger-Nuttal law). The periodic table ends beyond $Z = 92$ because of the increasingly negative values of the binding energy for α particles and fission fragments.

4.11 Charge Independence of Nuclear Forces

It is found that in low-energy nucleon-nucleon scattering experiments the $n-p$ and $p-p$ nuclear forces are virtually identical (Chap. 5). This aspect is also evident in the properties of mirror nuclei, such a ^{3}H, ^{3}He; ^{7}Li, ^{7}Be; ^{27}Al, ^{27}Si, etc. which are obtained from the other by transforming all neutrons into protons, and vice versa. Owing to difference in proton number the Coulomb forces in the mirror nuclei are

Fig. 4.11 Excited energy levels for the mirror pair Li^7 and Be^7. The crosshatched areas represent broad levels

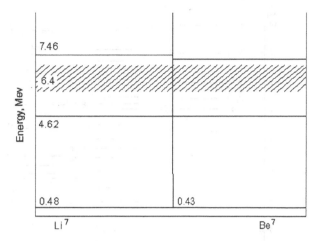

necessarily different. When corrected for electrostatic energy, the nuclei have the same mass. For the mirror nuclei with atomic number $Z + 1$ and Z

$$\Delta M = \frac{3e^2}{5R}\left[(Z+1)^2 - Z^2\right] = \frac{3e^2(2Z+1)}{5r_0 A^{1/3}} \tag{4.137}$$

where R is the nuclear radius. After correcting for the coulomb forces, the energy levels of mirror nuclei bear a remarkable similarity, Fig. 4.11. The similarity of the energy levels in mirror nuclei of the type $N = Z \pm 1$ shows the equality of n–n and p–p forces. This is the principle of charge symmetry which says nothing about the n–p force. The study of energy levels in even A nuclei with neighbours differing by one unit of Z gives evidence for the equality of n–n, n–p, p–p forces, that is the principle of charge independence which is much more stringent than the principle of charge symmetry. Examples for this equality are provided in Figs. 4.12 and 4.13 for (^{14}O, ^{14}N, ^{14}C) and (^{6}He, ^{6}Li, ^{6}Be). The correspondence between levels is clear and the differences in energy are mainly explained by Coulomb effects. Thus, the experiments or N–N scattering and the similarity of energy levels in the isobaric triads strongly support the principle of charge independence, enunciated by Heisenberg et al. [7]. However, the charge independence is only approximate as it does not take into account electromagnetic effects or the neutron-proton mass difference. The principle of charge independence was later generalized to pions by Kemmer [10], and also to the strange particles.

4.11.1 Iso-spin

The principle of charge independence is mathematically expressed by isospin formalism. The nucleon is endowed with another degree of freedom apart from the spatial coordinates and spin, known as isospin or i-spin, designated by T. The T-spin

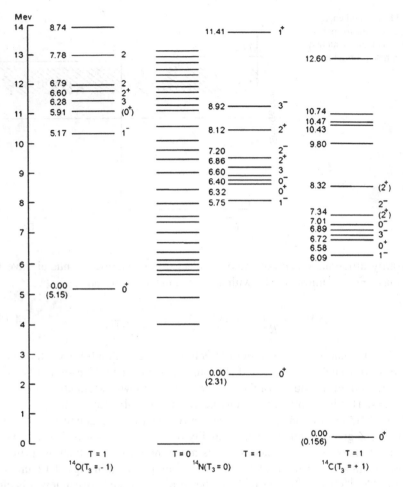

Fig. 4.12 The level schemes for the nuclei with $A = 14$. The relative energies represent atomic masses

operates in a fictitious space and has no relation with ordinary spin. For nucleon, $T = 1/2$. It is a dichotomic variable in that it can take two values. Along the third axis, $T_3 = +1/2$ for proton and $T_3 = -1/2$ for neutron. Neutron and proton are the two aspects of the same particle, nucleon. The isospin operators are the Pauli matries, just as they are for ordinary spin (S) of electron although there is no connection between S and T. They obey the same commutation relations. Furthermore, the algebra for the addition of isospins is identical with that for ordinary spins. This subject is discussed in great detail in [9], Chap. 4.

Pauli's principle is more generalized by incorporating the isospin. It now states that the overall Eigen function for a system of two nucleons is antisymmetric with the exchange of the particles in spatial, spin and isospin coordinates. For a system of particles we shall use I in place of T and the third component I_3 in place of T_3.

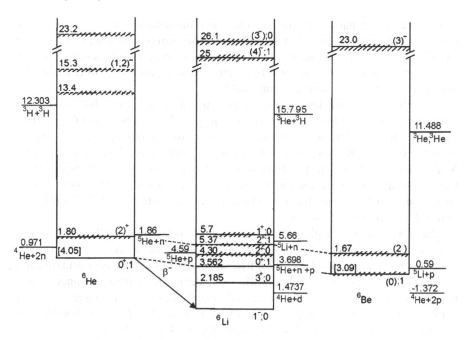

Fig. 4.13 Nuclear levels in ^{6}He, ^{6}Li, and ^{6}Be

It can be shown that for a two-nucleon system

$$l + S + I = \text{odd integer} \tag{4.138}$$

where l is the orbital angular momentum, S the spin and I the isospin for the two-nucleon system. Calling the total isospin of a system of particles by I and its third component by I_3, the system of two protons will have $I_3 = +1$ and $I = 1$, a system of two neutrons will have $I_3 = -1$ and $I = 1$, while neutron-protons system will have $I_3 = 0$, but $I = 1$ or 0.

The charge of a system of nucleons is related to the third component I_3 by

$$\frac{Q}{e} = I_3 + \frac{A}{2} \tag{4.139}$$

where A is the mass number or the number of nucleons and Q/e is the charge in units electron charge. As an example, it is easily verified that $I_3 = -1, 0, +1$ for $^6_2$He, $^6_3$Li and $^6_4$Be, respectively. The ground level of ^{6}Li has $T = 0$ and occurs only in this nucleus, but the excited level at 3.56 MeV has $T = 1$ and occurs in three nuclei corresponding to the ground levels of ^{6}He, ^{6}Li and ^{6}Be. All corresponding levels with the same I also have the same angular momentum and parity, irrespective of T_3 as expected from the postulate of charge independence of nuclear forces. They form an isospin multiplet.

When the states belonging to a given multiplet are corrected for electrostatic interaction, all of them will have the same wave function.

In so far as the nuclear forces are concerned, the total isospin is a constant of motion, in the same way that the total angular momentum is a constant of motion for an isolated system. The foundation of this principle is empirical and its conservation is only approximate as it is violated by electro-magnetic and weak interactions, its use, specially in particle physics is invaluable.

From the nucleon-nucleon scattering experiments it is known that the nuclear forces are identical in the $^1S_0$ state for both $p-p$ and $n-p$ systems both of which correspond to $I = 1$ but each of which corresponds to different I_3 (1 and 0, respectively). It is, therefore, postulated that the nuclear forces depend on I but not on I_3. Mathematically, charge independence is equivalent to invariance with respect to rotation in isospin space because rotation leaves I invariant and changes only I_3. This also amounts to the statement that the Hamiltonian commutes with I; $IH - HI = 0$, or that I is a constant of motion.

The conservation of I gives rise to approximate rules that forbid transitions between states of different isospin under the action of nuclear forces. Some examples are given in [9], Chap. 4, others follow here. Consider the collisions of deuteron and proton with the light nuclei. In the collision with deuteron the isospin cannot change because deuteron has $T = 0$. In proton collision the isospin can change by $1/2$. It is impossible to form ^{10}B in a state at 1.74 MeV which has $T = 1$ by bombarding ^{12}C with deuterons because $T = 0$ for both d and ^{12}C. But the same level can be realized by bombarding ^{13}C.

A stringent test of isospin conservation in nuclear reactions is provided by the angular distributions of a reaction of the type $A + B \rightarrow C + C'$ where B has $T = 0$ and C and C' are the members of the same multiplet [1]. Zero isospin for B implies that the system is in an isospin state with $T = T_A$. In the isospin formalism particles C and C' are the same except for different T_3. Particles C and C' are both fermions or bosons according to their mass number (fermion if A = odd and boson, if A = even). In a given I state, the scattered amplitude must contain only waves with l even or l odd, but not l odd at the same time. Only even powers of $\cos\theta$ appear in the intensity (square of the amplitude). Thus the angular distribution is expected to be symmetric with respect to a plane at 90° to the initial direction in the centre of mass system. This prediction is verified with fair accuracy in Fig. 4.14, for the reaction ^{4}He $+ d \rightarrow ^3$He $+ ^3$H.

4.12 Ground and Excited States of Nuclei

A nucleus is capable of existing in discrete energy levels—a property of a bound quantum mechanical system. Figure 4.15 shows for example the low lying energy levels of ^{224}Ra. The energy value (in MeV) is indicated at the left side of each energy level. The lowest state of energy is called the ground state. Under normal conditions a nucleus like an atom is found in the ground state. If the nucleus is brought into an excited state then it cascades down to lower energy states and ultimately to the ground state via γ, β or even particle emission under various circumstances.

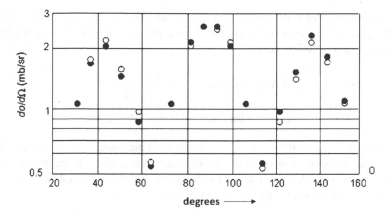

Fig. 4.14 Angular distribution of the ratio of ^{3}He to ^{3}H yields. The differential cross-sections for the process ^{4}He $+$ ^{2}H $\rightarrow$ ^{3}He $+$ ^{3}H at 82 MeV beam energy. *Open circle* represent ^{3}He yields, while *closed circles* ^{3}H yields. According to the Barchay-Temmer theorem, this ratio ought to be 1.0 at all angles [6]

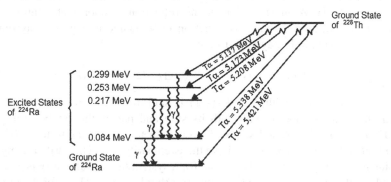

Fig. 4.15 α-particle transitions observed in the decay of ^{228}Th

4.12.1 Nuclear Spin

Protons and neutrons, the constituents of nuclei both have spin quantum number (1/2) (i.e. $s_p = s_n = (1/2)\hbar$). The spin of the proton is represented by a vector operator s_p such that the eigenvalue of s_p^2 is $(1/2)[(1/2) + 1]\hbar$ and of $s_{pz} = (1/2)\hbar$ or $-(1/2)\hbar$ and similarly for S_n. Further, the nucleons may also have orbital angular momentum by virtue of their motion in the nucleus. This is represented by an angular momentum quantum number l $(= 0, 1, 2 \ldots)$ for each nucleon.

The sum total of the spin and the orbital angular momenta of the nucleons combine to give rise to the total intrinsic angular momentum of the nucleus, which is referred to as the nuclear spin and the associated quantum number is denoted by J.

Since an odd number of spin (1/2) particles always combine quantum mechanically to give a half-integer total spin and even number of spin (1/2) particles to integer total spin, it follows that

$$\text{odd } A \text{ nuclei have} \quad J = \frac{1}{2}, \frac{3}{2}, \frac{5}{2}, \ldots$$

$$\text{even } A \text{ nuclei have} \quad J = 0, 1, 2, \ldots$$

These facts are in agreement with the experimental measurements of nuclear spins based on the studies of atomic hyper-fine structure, molecular spectra, nuclear magnetic resonance and other techniques. Furthermore, for even-even (Z even, N even) nuclei the nuclear ground state always has $J = 0$.

4.12.2 Nuclear Parity

The wave function for a nuclear energy state is a function of the coordinates $r_1, r_2, \ldots, r_A$ of the A nucleons. Under the reflection of coordinates through the origin (Parity operation P) the wavefunction may change its sign or may remain unchanged, its magnitude staying constant

$$P\psi(r_1, r_2, \ldots, r_A) = \pm\psi(-r_1, -r_2, \ldots, -r_A) \qquad (4.140)$$

where $P = \pm 1$. A state with $P = +1(-1)$ is said to have even (odd) parity. The nuclear state can be labeled with both its spin and parity, the symbol being J^P (e.g. 0^+, 1^-, 2^+). The ground states of even-even nuclei are found to be 0^+. By convention both neutron and proton have the even parity $(+)$. The parity arising due to orbital motion is given by $(-1)^l$ so that for a system the overall parity is given by the product of intrinsic parity and that due to orbital motion. Thus, for deuteron in the ground state, the overall parity will be

$$(+1)(+1)(-1)^0 = +1$$

If the nuclear Hamiltonian satisfies

$$H(r_1, r_2, \ldots, r_A) = H(-r_1, -r_2, \ldots, -r_A)$$

$$PH = HP \qquad (4.141)$$

then the parity operator P commutes with H, i.e.

$$[P, H] = 0 \qquad (4.142)$$

This means that parity is a constant of the motion, parity is conserved in nuclear processes (as well as electromagnetic processes). However, in the weak nuclear force ($\sim 10^{-7}$ times weaker than the strong force) it changes sign invalidating (4.141), parity is not conserved. The parities of nuclear states is obtained, by studying the angular distribution of particles and photons in nuclear processes, especially in β- and γ-decay (see Sect. 3.6.6).

4.13   Determination of Nuclear Spin

4.13.1   Nuclear Spin from Statistics

Identical particles obey either Fermi statistics or Bose statistics, that is a wave function $\psi(X_1, X_2)$ of particles 1 and 2 will be either symmetrical or antisymmetrical. Under exchange of X_1 and X_2, where X_1 and X_2 are space and spin coordinates

$$\psi(X_2, X_1) = +\psi(X_1, X_2) \quad \text{(Bose)}$$
$$= -\psi(X_1, X_2) \quad \text{(Fermi)} \tag{4.143}$$

Indeed all the particles without exception having even values of spin obey Bose statistics and odd values of spin obey Fermi.

To determine the statistics we shall investigate how an exchange of identical nuclei will effect the wave function of a molecule. Consider a diatomic molecule with identical nuclei. Its wave function may be written as

$$\psi = \psi_{elec}\xi_{vib}\rho_{rot}\sigma_{nuc\ spin} \tag{4.144}$$

Let P be the operator to exchange space and spin coordinates. Then $P\psi_{elec} = \pm\psi_{elec}$. It is known from molecules spectroscopy, usually in the ground state it is positive. Also $\xi_{vib} = +\xi_{vib}$, because ξ_{vib} depends on r, the internuclear distance (specifically, the solution of radial wave equation is of the form $R(r) =$ const $\cdot \exp(-\text{const} \cdot r) \times r^l \times L(r)$, where $L(r)$ is the Laguerre function). Now

$$\rho \sim P_l^m(\cos\theta)e^{im\phi} \tag{4.145}$$

where $P_l^m(\cos\theta)$ is an associated Legendre polynomial, θ, the polar angle and ϕ the azimuth angle are the polar coordinates of the two nuclei. Exchange of $x \to -x$, $y \to -y$, $z \to -z$, implies

$$\theta \to \pi - \theta, \qquad \phi \to \pi + \phi \tag{4.146}$$

Now

$$P_l^m\big(\cos(\pi - \theta)\big) = (-1)^{l+m}P_l^m(\cos\theta) \tag{4.147}$$

and

$$e^{im(\phi+\pi)} = (-1)^m e^{im\phi} \tag{4.148}$$

so that

$$Pp = (-1)^{l+m} P_l^m (\cos\theta)(-1)^m e^{im\theta}$$
$$= (-1)^{2m}(-1)^l p = (-1)^l p \qquad (4.149)$$

where m is an integer. Thus p is symmetrical for even l and antisymmetrical for odd l. Analysis of $P\sigma$

(i) Spin $=$ zero

The total wave function ψ is antisymmetrical for odd l and symmetrical for even l. Now the nuclei must certainly obey either Fermi or Bose statistics. It follows therefore either only the states with even l or only those with odd l can exist. Evidence for this conclusion is obtained from the band spectra of diatomic molecules. These show that if the nuclei have spin zero, every second rotational state of the molecule is absent. Indeed it is found that in every instance only the even rotational states exist, indicating that all the nuclei of zero spin (which have been found previously to have even A) obey Bose statistics. Similarly it has been found that all nuclei of even A, including those with non-zero spin, obey Bose statistics and all those of odd A obey Fermi statistics. The result has been of significance in deciding the model of the nucleus, that is favoring the neutron-proton model and discarding the electron-proton hypothesis (see Sect. 4.1). In case of an even number of particles the exchange of nuclei is equivalent to an even number of changes of sign and ψ must be symmetrical to an interchange of nuclei (Bose statistics). If each nucleus contains an odd number of particles, the exchange of nuclei is equivalent to an odd number of changes of sign, that is ψ is antisymmetrical to nuclear exchange (Fermi statistics).

(ii) Nuclei of non-zero spin

A nucleus of total angular momentum I can have a component M in any prescribed direction taking $2I + 1$ values in all $(I, I - 1, \ldots, -I)$, that is $2I + 1$ states exist. For two identical nuclei $(2I + 1)^2$ wave functions of the form $\psi_{M1}(A)\psi_{M2}(B)$ can be constructed. It the two nuclei are identical, these simple products must be replaced by linear combination of these products which are symmetric or antisymmetric for interchange of nuclei. If $M_1 = M_2$, the products themselves are $(2I + 1)$ symmetric wave functions. The remaining $2I(2I + 1)$ functions with $M_1 \neq M_2$ have the form $\psi_{M1}(A)\psi_{M2}(B)$ and $\psi_{M2}(A)\psi_{M1}(B)$. Each such pair can be replaced by one symmetric and one antisymmetric wave function of the form $\psi_{M1}(A)\psi_{M2}(B) \pm \psi_{M2}(A)\psi_{M1}(B)$. Thus half of $2I(2I + 1)$ functions that is $I(2I + 1)$ are antisymmetric and an equal number symmetric. The total number of symmetric wave functions $= (2I + 1) + I(2I + 1) = (2I + 1)(I + 1)$. Number of antisymmetric functions $= I(2I + 1)$. Therefore, the ratio of the number of symmetric and antisymmetric functions is $(I + 1)/I$.

If the electronic wave function for the molecule is symmetric it was shown that interchange of nuclei produces a factor $(-1)^l$ in the total molecular wave function, where l is the rotational quantum number. Thus if the nuclei obey

Fig. 4.16 Intensity alteration in band spectra of homonuclear diatomic molecules

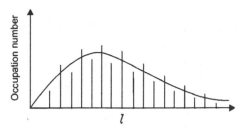

Bose statistics, symmetric nuclear spin functions must be combined with even l rotational states and antisymmetric with odd l. Because of statistical weight attached to spin states, the intensity of even rotational lines will be $(I+1)/I$ as great as that of neighboring odd rotational lines.

For Fermi statistics of the nuclei the spin and rotational states combine in a manner opposite to the previously stated and the odd rotational lines are more intense in the ratio $(I+1/I)$. Thus by determining which lines are more intense, even or odd, the nuclear statistics is determined and by measuring the ratio of intensities of adjacent lines the nuclear spin is obtained. The reason why adjacent lines must be compared is that the rotational lines vary in intensity with l in accordance with the occupation number of rotational state according to the Bultzman distribution, $(2l+1)\exp[-\text{const} \cdot l(l+1)/kT]$ (Fig. 4.16).

4.13.2 Nuclear Spin from Hyperfine Structure

Fine structure of spectral lines is explained by the electron spin while the hyperfine structure can be accounted for by assigning a spin to the nucleus. The nucleus behaves as if it was a miniature magnet, and interacts with the magnetic field just like a magnet. It may be seen that the magnetic effects produced by nuclei are very much small compared to those produced by electrons, even though the spin values are of the same order of magnitude. A many-electron atom can be replaced, from the point of view of magnetism, by three magnetic dipoles.

(1) μ'_L: resulting from orbital angular momentum of all electrons

(2) μ'_S: resulting from spin of all electron

(3) μ'_I: resulting from nuclear angular momentum

μ'_I is very small compared with μ'_S or μ'_L. Magnetic moment always refers not to the magnitude but to maximum projected value the moment can have in any direction. The orbital angular momentum is $L\hbar$ where L is an integer and magnetic moment of electron is $(e/2mc)l\hbar$, where m is the mass of electron. The quantity $(e\hbar/2mc)$ is called one Bohr magneton. In analogy with the Bohr magneton for electron, we define a nuclear magneton, $(eh/2Mc)$ where M is the mass of the nucleon, the nuclear magnet being less by a factor of 1836 compared to the Bohr magneton.

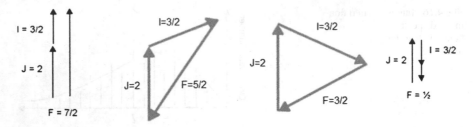

Fig. 4.17 Vector addition

4.13.2.1 Vector Addition

The quantities L and S and also the magnetic moments are space quantized that is their values take up discrete values in definite directions. The total angular momentum of electron, $J = L + S$, where L is the orbital part and S is the spin of electron. The nuclear angular momentum I may be combined with J to yield the resultant $F = J + I$. F takes values

$$J + I, J + I - 1, \ldots, J - I \quad (J > I) \quad (2I + 1 \text{ values})$$

or

$$I + J, I + J - 1, \ldots, I - J \quad (I > J) \quad (2J + 1 \text{ values})$$

As an example consider $J = 2$, $I = \frac{3}{2}$. $F = \frac{7}{2}, \frac{5}{2}, \frac{3}{2}, \frac{1}{2}$, as in Fig. 4.17. It is well known that the interaction between μ'_L and μ'_S (spin-orbit coupling) splits the atomic levels and leads to fine structure in spectrum. In just the same way each of those split levels splits further by $J - I$ coupling leading to hyperfine splitting in spectral lines as in Fig. 4.18. But the hyper-fine splitting is much smaller than the fine structure splitting. The value of nuclear spin can be ascertained directly from the number of hyperfine components of the spectral terms provided the angular momentum in the electron system is large enough and is known.

Example 4.2 In praseodymium (Pr) spectrum the energy level $^5K_7$ corresponding to $J = 7$ is known to be split into 6 components. Find the nuclear spin.

Solution As the number of sub-levels is less that $2J + 1$, that is 15, $2I + 1 = 6$; $I = 5/2$.

Another example is the hyperfine structure of sodium atom.

4.13.2.2 Selection Rule for F

$$\Delta F = \pm 1, 0$$

$$F = 0 \quad \nrightarrow \quad F = 0$$

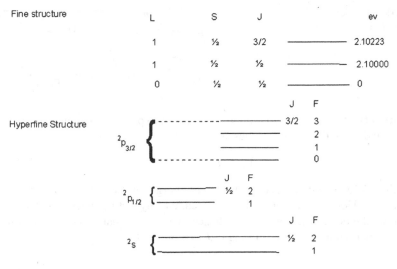

Fig. 4.18 The lowest states of the sodium atoms as modified by hyperfine splitting ($I = 3/2$)

Consider states with the same I and the same J but different F. Let θ be the angle between the unprojected magnetic moment μ'_J and μ'_I. These quantities lie in the direction of the angular momentum vectors. If either μ'_J or μ'_I were zero then there would be no splitting and all the quantum states in question would have the same energy E_0.

In general

$$E_F = E_0 + k\mu'_J \mu'_I \cos\theta \qquad (4.150)$$

where k is a constant

$$\mu'_I = g_I \sqrt{I(I+1)} \left(\frac{e\hbar}{2Mc}\right) \qquad (4.151)$$

where g_I is the nuclear g-factor. Also

$$\mu'_J = g_J \sqrt{J(J+1)} \left(\frac{e\hbar}{2mc}\right) \qquad (4.152)$$

$$E_F - E_0 = k'\sqrt{J(J+1)}\sqrt{I(I+1)} \cos\theta \qquad (4.153)$$

From Fig. 4.19

$$F(F+1) = J(J+1) + I(I+1) + 2\sqrt{J(J+1)}\sqrt{I(I+1)} \cos\theta \quad (4.154)$$

$$E_F - E_0 = \frac{k'}{2}\left[F(F+1) - J(J+1) - I(I+1)\right] \qquad (4.155)$$

where $k' = $ const. It is readily seen that

$$k' = E_F + E_{F-2} - 2E_{F-1} \qquad (4.156)$$

Fig. 4.19 The addition of J
and I to form F

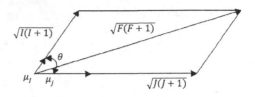

Also from (4.156) we have prediction of the energy spacing between states with the same J and I but different F values as follows

$$E_F - E_{F-1} = k'F \qquad (4.157)$$

Note that (4.157) predicts unequal spacing of energy levels. In fact it is in arithmetic progression. Hence the experimental knowledge of hyperfine spacing of energy levels allows the determination of F and hence that of I. Equation (4.153) also shows that there is no fine structure splitting if $J = 0$ or $I = 0$.

4.13.3 Nuclear Spin from Zeeman Effect

When the magnetic field is so great that the velocity of precession of F about the field direction becomes greater than that of J and I about F, a Paschen-Back effect takes place as for multiplet structure. In the case of hyperfine structure on account of weak coupling between J and I they are independently space quantized in the field direction, the components being M_J and M_I. The space quantization of J gives the ordinary Zeeman effect, see Fig. 4.20. Each term with a given M_J is, however, once again split up into a number of components corresponding to different values of M_I that is $2I + 1$ components. This number of components is the same for all the terms of an atom, since I is constant for a given nucleus. For a transition which without field gives rise to one hyper-multiplet, the selection rules in a strong field are $\Delta M_J = 0, \pm 1$ and $\Delta M_I = 0$. The first of these rules gives the ordinary anomalous Zeeman effect if at first we disregard the nuclear spin. Because of nuclear spin, however, each of the magnetic levels with a certain M_J value has $2I + 1$ equidistant component, the separation being different in the upper and lower states. Therefore, considering $\Delta M_I = 0$, each anomalous Zeeman component is split into $2I + 1$ lines. The splitting dues not depend upon the field strength so long the latter is sufficiently great to produce an uncoupling of J and I. Thus simply by counting up the number of line components the nuclear spin I can be determined. For example, in a strong magnetic field each of Zeeman components of Bi consists 10 components due to nuclear spin so that, $2I + 1 = 10$ or $I = 9/2$.

4.14 Nuclear Magnetic Dipole Moment

Any charged particle moving in a closed path produces a magnetic field which at a large distance can be described as due to a magnetic dipole located at the current

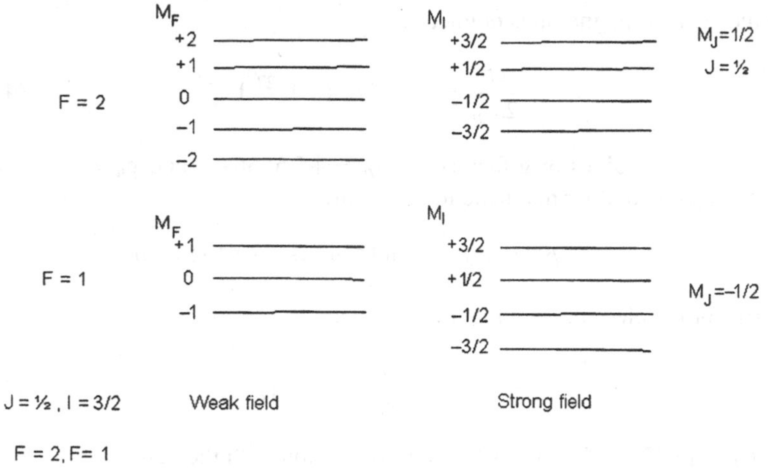

M_F		M_I	
+2	———	+3/2	——————— M_J=1/2
+1	———	+1/2	——————— J = ½
F = 2 0	———	−1/2	———————
−1	———	−3/2	———————
−2	———		

M_F		M_I	
+1	———	+3/2	———————
F = 1 0	———	+1/2	———————
−1	———	−1/2	——————— M_J=−1/2
		−3/2	———————

J = ½ , I = 3/2   Weak field   Strong field

F = 2, F= 1

Fig. 4.20 Splitting of spectral lines due to Zeeman effect

loop. For an electron of charge $-e$ and mass m an orbital magnetic moment is associated with it and is given by

$$\mu_L = -\frac{e}{2m}L \tag{4.158}$$

Similarly, an intrinsic magnetic moment is associated with the spin S

$$\mu_S = -\frac{ge}{2m}S \tag{4.159}$$

where g is known as the g-factor and Dirac's equation gives precisely $g = 2.0000$. However, quantum-electro-dynamical effects modify this value slightly to $g = 2.0023192$. It is remarkable that the experiments are able to match the theoretical value to one part in 10^7.

The value of the proton magnetic moment μ_p and that of neutron magnetic moments μ_n are defined as

$$\mu_p = g_p \frac{e}{2m_p} s_p \tag{4.160}$$

$$\mu_n = g_n \frac{e}{2m_n} s_n \tag{4.161}$$

where m_p and m_n are the masses of proton and neutron, respectively, with different g-factors for each particle. We may rewrite (4.160) and (4.161) as

$$\mu_p = \frac{1}{2}g_p \mu_N \tag{4.162}$$

$$\mu_n = \frac{1}{2}g_n \mu_N \tag{4.163}$$

where the nuclear magneton is defined as

$$\mu_N = \frac{e\hbar}{2m_p} = 5.05078 \times 10^{-27} \text{ JT}^{-1} \tag{4.164}$$

The experimental values of g-factors are $g_p = +5.5856\cdots$ and $g_n = -3.8262\cdots$ so that the corresponding magnetic moments are

$$\mu_p = +2.792S\cdots\mu_N \quad \text{and} \quad \mu_n = -1.9131\cdots\mu_N$$

Similarly, for the electron spin magnetic moment

$$\mu_S = -\frac{1}{2}g\frac{e\hbar}{2m} = -\frac{1}{2}g\mu_B \tag{4.165}$$

where $\mu_B = e\hbar/2m$ is known as the Bohr magneton with the value

$$\mu_B = -9.274\cdots \times 10^{-24} \text{ JT}^{-1} \tag{4.166}$$

Because of the inverse dependence on the mass of the particle, the nucleon magnetic moments are three orders of magnitude smaller than the magnetic moment of electron. Furthermore, they differ significantly from the value predicted for a spin (1/2) particle ($\mu_p = 1\mu_N$, $\mu_n = 0$). This is in contrast with the electron for which μ is close to the Dirac value. This is explained by the fact that nucleons are of finite size and have a complicated structure.

In a complex nucleus, the intrinsic magnetic moments of the constituent neutrons and protons will contribute to the total magnetic moment and there will be an additional contribution from orbital motion of charged protons. The total magnetic moment μ_J is given by

$$\mu_J = g_J\mu_N J \tag{4.167}$$

where g_J is the nuclear g-factor. It is found that the nuclear magnetic moments are spread approximately in the range $-2\mu_N$ to $6\mu_N$.

The techniques used for the determination of magnetic moments of nuclear ground state as well as excited states include hyperfine structure studies, microwave spectroscopy, nuclear magnetic resonance, use of nuclear alignment and atomic beams.

We shall consider the magnetic resonance method which uses molecular beam. It depends essentially upon the resonance between the precession frequency of the nuclear magnet about a constant magnetic field direction and the frequency of an impressed high frequency alternating magnetic field. The magnetic moment

$$\mu = gI\left(\frac{e\hbar}{2m_pc}\right) \tag{4.168}$$

where g is the nuclear g-factor and I the spin. When a nucleus of magnetic moment μ is in a constant magnetic field of intensity B, it will precess about the field

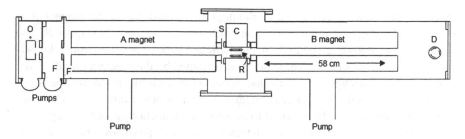

Fig. 4.21 Schematic diagram of the apparatus used by Rabi et al. [11] for molecular beam magnetic resonance experiments

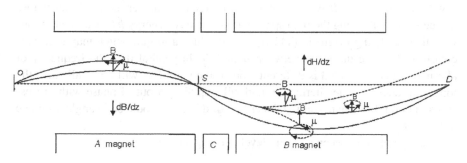

Fig. 4.22 The paths of four molecules in a molecular-beam magnetic resonance apparatus [12]

direction with a frequency ν given by Larmor's theorem

$$\nu = \frac{\mu B}{I h} \tag{4.169}$$

Thus the magnetic moment μ of a nucleus can be found by determining ν which the nucleus of spin I acquires in a known constant magnetic field B. Instead of working with nuclei alone, Rabi used beams of neutral molecules whose electronic angular momentum is zero. A narrow stream of molecular beam issues from the source, which is an oven O, Fig. 4.21. A very small fraction of these molecules will pass through the collimating slit S and reach the detector at D. The beam passes through three magnets, A, C and B. In both A and B the field is strong but inhomogeneous, while in C the field is uniform.

In the absence of any inhomogeneous magnetic deflecting fields the molecules will pass through the collimating slits and traverse the straight line paths OSD and form the direct beam (Fig. 4.22). The magnetic fields of A and B are in the same direction but their gradient dB/dz are in opposite directions. A molecule with magnetic moment μ will be deflected in the direction of the gradient if μ_Z, the projection of μ in the field direction is positive and will be deflected in the opposite direction if μ_Z is negative. Molecules which leave o at some angle with the line OSD will follow paths indicated by solid lines and reach the detector D. The force experienced by any such molecule in the inhomogeneous field due to the magnet A is $F = \mu_Z (\partial B / \partial Z)_A$. A similar expression holds for the force due to magnet B. The actual deflection produced by each magnetic field can be established from a knowledge of the velocity of molecule which is determined by the temperature of source and geometry of arrangement. It no change occurs in μ_Z as the molecule goes from A to B field, the deflections in these fields will be in opposite directions. The magnetic field gradients can be adjusted to make these deflections equal in magnitude and thus to refocus the beam at the detector. Magnet C produces a homogeneous field of intensity B. In the same region there is a high frequency alternating magnetic field B_1 (not shown in the figures) at right angles to the homogeneous field B produced by magnet C. When a molecule of magnetic moment μ enters this region it will precess around B with Larmour frequency ν. Consider the oscillator whose frequency is f flooding the field space with photons of energy hf. The interactions with the oscillating magnetic field B_1 will produce a torque which may either increase or decrease the angle between μ and B. In general if f the frequency of alternating magnetic field is different from ν the net effect will be small since the torque produced by the alternating field will rapidly get out of phase with precessional motion. But when $f = \nu$ the increase or decrease produced in angle might be quite large.

The separation between adjacent levels in magnet C is

$$\Delta E = g \Delta M_J = \left(\frac{e\hbar}{2 M_p c} \right) B = \frac{g e \hbar B}{2 M_p c} \tag{4.170}$$

$$(\because \quad \Delta M_I = \pm 1)$$

Therefore

$$hf = \frac{\mu B}{I} \quad \text{or} \tag{4.171}$$

$$f = \nu = \frac{\mu B}{I h} \tag{4.172}$$

The transition probability is maximum when $f = \nu$.

The molecule will then follow one of the dotted paths while it enters region B and escape from the detector. In some experiments f is kept constant and B (homogeneous magnetic field) is varied. Resonance occurs at a definite value of B as shown in Fig. 4.23

$$g = \frac{\mu}{I} \frac{2Mc}{e\hbar} = \frac{\nu h}{B} \frac{2Mc}{\hbar} = \frac{4\pi Mc}{e} \frac{f}{B} \tag{4.173}$$

If I is also known from some other experiment then μ can be found out from $\mu = g I$, where μ is expressed in nuclear magneton.

Fig. 4.23 A curve showing
the occurrence of magnetic
resonance for ^{7}Li [13]

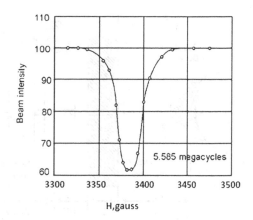

4.14.1 Magnetic Moment of the Neutron

Classically, neutron is not expected to have a magnetic moment as it has no charge. However, magnetic moment depends on the current rather than charge. It is possible for the charge to cancel but not the current. Such is the case of the hydrogen atom. which is electrically neutral but does have a magnetic moment. It turns out that the neutron has a magnetic moment, $\mu_n = -(1.91354 \pm 0.0006)$ nm. The negative sign means that the magnetic moment is oppositely directed to the spin.

The value of μ_n was determined by Block et al. using the magnetic resonance beam method at Stanford. The apparatus used by them is schematically shown in Fig. 4.24. In the place of the magnets A and C they used steel plates as polarizer and analyzer. Deuterons accelerated in a cyclotron on hitting a Be target produce neutrons which are thermalized in paraffin. The thermal neutrons are collimated along the axis of the polarizer and are detected by BF_3 proportional counter on the right.

The microcrystals of magnetized steel plate act as polarizer as the emerging neutrons tend to be aligned with their spins opposed to the direction of magnetization. If the analyzer is magnetized in a parallel fashion then the transmission of the neutrons will be greater. However, if the polarization of neutrons between the steel plates is spoilt then there will be a dip in the neutron intensity at the detector. Magnetic resonance transitions induced in this region can cause a neutron spin to 'flip' from either spin orientation to the other. However the spin orientation that is more heavily populated initially will have more transitions just in proportion to its population. Thus if the neutrons undergo magnetic resonance transitions in the region between the plates, there will be a drop is the intensity of the beam emerging from the second plate. The depolarization is accomplished by employing RF fields. The field strength in the central magnet was measured by observing the proton resonance. Figure 4.25 shows the neutron resonance curve observed by Block et al.

NMR technique has been invaluable in the studies of magnetic properties of materials, the chemical shifts and the structures of certain organic molecules and diagnosis.

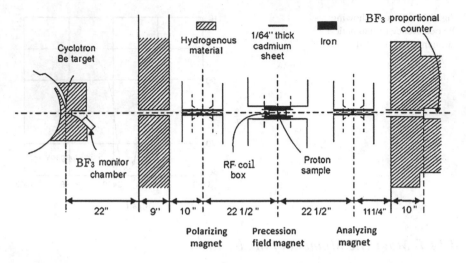

Fig. 4.24 Apparatus used by Block et al. to measure the magnetic moment of the neutron [2]

Fig. 4.25 A typical neutron resonance curve observed by Block et al.

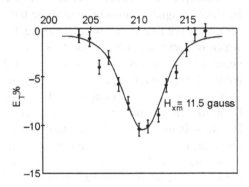

4.15 Electric Quadrupole Moment

The concept of quadrupole comes from the classical electrostatic potential theory. Assume that the nuclear charge is rotating about the nuclear spin I. Then on time average it would appear cylindrically symmetric about I, regardless of the distribution. Because $\nabla^2 \phi = 0$ outside the nucleus, we can expand ϕ in terms of Legendre polynomials

$$\phi(r, \theta) = \frac{1}{r} \sum_{n=0}^{\infty} \frac{a_n}{r^n} P_n(\cos \theta) \tag{4.174}$$

The quantity $(2a_2/e)$ is called the quadrupole. It is given by

$$Q = Q_0 = \frac{1}{e} \int (3z'^2 - r'^2) \rho(r') d\tau' \text{ cm}^2 \tag{4.175}$$

Fig. 4.26 Nuclear
quadrupole shapes for which
$\langle Z^2 \rangle = 1/3 \langle r^2 \rangle$

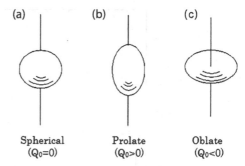

The first term of (4.174) is ordinary coulomb term ϕ. The second term the permanent dipole is not generated by nuclei in the static case because of parity considerations. The third giving a measure of the departure from spherical symmetry is the quadrupole term. A nucleus of the shape illustrated in Fig. 4.26(a) would have $Q \neq 0$. This is of more common occurrence.

Expression (4.175) is simply the average of $3z^2 - r^2$ taken over the charge density distribution. That is, it could be written as

$$Q_0 = Z \left(3 \langle z^2 \rangle - \langle r^2 \rangle \right) \qquad (4.176)$$

where Z is the total nuclear charge measured units of e. The dimensions of Q_0 are (length)^2 and is expressed in units of (meter)^2 or Barn. In the quantum mechanical approach the charge density $\rho(r)$ must be replaced by probability density $\psi^* \psi$. Expression (4.175) then can be written as

$$\frac{Q}{e} = \int \psi^* \left(3z^2 - r^2 \right) \psi \, d\tau \qquad (4.177)$$

This leads to the result

$$Q = \frac{(2I - 1) Q_0}{2(I + 1)} \qquad (4.178)$$

where I is the nuclear spin. Now

$$\langle r^2 \rangle = \langle x^2 \rangle + \langle y^2 \rangle + \langle z^2 \rangle \qquad (4.179)$$

If the nucleus is spherical, $\langle x^2 \rangle = \langle y^2 \rangle = \langle z^2 \rangle$ so that $\langle z^2 \rangle = (1/3)\langle r^2 \rangle$, and for spherical charge distribution, $Q_0 = 0$. The quantity $\langle r^2 \rangle$ is the mean square radius. It the nucleus is non-spherical then $\langle z^2 \rangle \neq (1/3)\langle r^2 \rangle$. There are two possibilities

$$\langle z^2 \rangle > \frac{1}{3} \langle r^2 \rangle \quad \text{(Prolate ellipsoid)} \quad \text{for which } Q_0 > 0 \qquad (4.180)$$

$$\langle z^2 \rangle < \frac{1}{3} \langle r^2 \rangle \quad \text{(Oblate ellipsoid)} \quad \text{for which } Q_0 < 0 \qquad (4.181)$$

Nuclear quadrupole shapes for which $\langle Z^2 \rangle \neq 1/3 \langle r^2 \rangle$ are shown along with the spherical shape in Fig. 4.26.

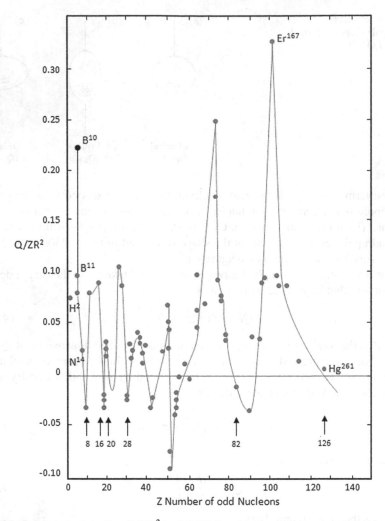

Fig. 4.27 Quadrupole distortion $Q/ZR^2 (\approx \Delta R/R)$ for odd-A nuclei plotted vs. the odd nucleon number N [14]

For an orbiting neutron $Q = 0$

If $|\psi|^2$ is spherically symmetric, $Q = 0$
If $|\psi|^2$ is concentrated in the xy-plane ($z = 0$), then $Q = -\langle r^2 \rangle$
If $|\psi|^2$ is concentrated along z-axis, $Q \simeq +2\langle r^2 \rangle$

For $I = 0$, Q vanishes as no symmetric axis is defined. Also, Q vanishes for $I = 1/2$. Notice that for large values of I, much greater than $\hbar$, Q approaches Q_0, as it should in accordance with Bohr's correspondence principle.

Quadrupole moments for numerous nuclei have been measured using optical hype-fine structure and atomic beam techniques, their values are found to range from -1×10^{-28}, to $+8 \times 10^{-28}$ m^2, the values being particularly high for the rare earths Fig. 4.27. The quadrupole moment values are of great importance for the nuclear models.

Example 4.3   Use the uncertainty relation to estimate the kinetic energy of the nucleons, the nuclear radius is about 8×10^{-13} cm and the mass of a nucleon is about 940 MeV/c^2.

Solution

$$\Delta p_x \sim \frac{\hbar}{\Delta x}$$

but

$$c\Delta p_x = cp = \frac{\hbar c}{\Delta x} = \frac{197\ (\text{MeV fm})}{8\ (\text{fm})} = 24.6\ \text{MeV}$$

$$T \simeq \frac{c^2 p^2}{2Mc^2} = \frac{(24.6)^2}{2 \times 940} = 0.32\ \text{MeV}$$

Example 4.4   Singly-charged lithium ions, liberated from a heated anode, are accelerated by a difference of 625 volts between anode and cathode. They then pass through a hole in the cathode into a uniform magnetic field perpendicular to their direction of motion. The magnetic flux density is 0.1 Wb/m^2 and the radii of the paths of the ions are 8.83 cm and 9.54 cm, respectively. Calculate the mass numbers of the lithium isotopes.

Solution

$$p = q\,Br = \sqrt{2MT} = \sqrt{2MqV}$$

$$M = \frac{q B^2 r^2}{2V}$$

$$M_1 = \frac{1.6 \times 10^{-19} \times (0.1)^2 \times (8.83 \times 10^{-2})^2}{2 \times 625 \times 1.66 \times 10^{-27}} = 6.012 = 6$$

$$M_2 = \frac{1.6 \times 10^{-19} \times (0.1)^2 \times (9.54 \times 10^{-2})^2}{2 \times 625 \times 1.66 \times 10^{-27}} = 7.017 = 7$$

Example 4.5   A narrow beam of singly charged ^{10}B and ^{11}B ions of energy 3.2 keV passes through a slit of width 1mm into a uniform magnetic field of 1200 gauss and after a deviation of 180° the ions are recorded on an photographic plate. (a) What is the spatial separation of the images? (b) What is the mass resolution of the system?

Solution

$$r = \left(\frac{2MV}{q B^2}\right)^{1/2}$$

$$r_{11} = \left(\frac{2 \times 11 \times 1.66 \times 10^{-2.7} \times 3200}{16 \times 10^{-19} \times 0.12^2}\right)^{1/2} = 0.2252 \text{ m}$$

$$r_{10} = \sqrt{\frac{10}{11}} r_{11} = \sqrt{\frac{10}{11}} \times 0.2252 \text{ m} = 0.2147 \text{ m}$$

(a) Spatial separative of images

$$= (r_{11} - r_{10}) \times 2$$

$$= 2(0.2252 - 0.2147) \text{ m} = 0.021 \text{ m}$$

$$= 2.1 \text{ cm}$$

(b) Mass resolution

$$\delta = \frac{M}{\Delta M}$$

$$M \propto r^2$$

$$\frac{\Delta M}{M} = \frac{2\Delta r}{r}$$

$$\therefore \quad \delta = \frac{r}{2\Delta r} = \frac{22 \text{ cm}}{2.1 \text{ cm}} = 10.5$$

Example 4.6 By considering the general conditions for nuclear stability show that the nucleus $^{229}_{90}$Th will decay and decide whether the decay will take place by α or β emission.

The atomic mass excesses of the relevant nuclei are

Element	$^4_2$He	$^{225}_{88}$Ra	$^{229}_{89}$Ac	$^{229}_{90}$Th	$^{229}_{91}$Pa
Mass excess amu $\times 10^{-6}$	2603	23528	32800	31652	32022

Solution When ^{229}Th decays via α emission the daughter nucleus is ^{225}Ra. A body of mass M_1 will decay into $M_2 + m_\alpha$ if $M_1 > M_2 + m_\alpha$. Now, $M - A = \Delta$, where Δ is the mass excess. $M_1 = A + \Delta = 229 + 0.031652 = 229.031652$ amu. For α decay, $M_2 = 225 + 0.02352S = 225.02352S$ amu

$$M_\alpha = 4 + 0.002603 = 4.002603$$

$$M_2 + m_\alpha = 229.0249558$$

Since $M_1 > M_2 + m_\alpha$, $^{229}_{90}$Th will decay via α emission. If ^{229}Th decays via β^- emission then the daughter nucleus would be ^{229}Pa. The criterion for β^- decay is, $M_1 > M_2$. Now, $M_2 = A + \Delta = 229 + 0.032022 = 229.032022$ amu but this value is greater than $M_1 = 229.031625$ amu for ^{224}Th. Therefore β^- decay is not possible.

If ^{229}Th decays via β^+ emission then the criterion for decay is, $M_1 > M_2 + 2m_e$. Here, $M_1 = 229.031652$ amu, and M_2 is ^{228}Ac for which the mass $M_2 = 229 + 0.032800 = 229.032800$ and $M_2 + 2m_e = 229.032800 + 2 \times 0.000548 = 229.03389$, a quantity which is larger than the mass of ^{229}Th atom. Therefore β^+ emission is not possible.

Example 4.7 Consider the β^+ decays

$$^{127}_{51}\text{Sb} \rightarrow {}^{127}_{52}\text{Te} + \beta^+ + 1.60 \text{ MeV}$$

$$^{127}_{55}\text{Cs} \rightarrow {}^{127}_{54}\text{Xe} + \beta^+ + 1.06 \text{ MeV}$$

Using liquid drop model state which of the isobars $^{127}_{53}$I or $^{127}_{54}$Xe is stable against β-decay.

Solution The liquid drop model gives the value of Z_0 for the most stable isobar of mass number A (Chap. 6) by

$$Z_0 = \frac{A}{2 + 0.015A^{2/3}}$$

For $A = 127$, $Z_0 = 53.38$, the nearest Z is 53. Hence $^{127}_{53}$I is stable. On the other hand, $^{127}_{54}$Xe will be unstable against β^+ decay or e^- capture.

Example 4.8 Estimate the ratios of the major to minor axes of $^{181}_{73}$Ta and $^{123}_{51}$Sb. The quadrupole moments are $+6 \times 10^{-24}$ cm^2 for Ta and -1.2×10^{-24} cm^2 for Sb. (Take $R = 1.5A^{1/3}$ fm.)

Solution

$$Q = \frac{4}{5}\eta ZR^2$$

where

$$\eta = \frac{a - b}{R}$$

$$R = 1.5 \times (181)^{1/3} = 8.48 \text{ fm} \quad \text{for Ta}$$

$$R = 1.5 \times (123)^{1/3} = 7.458 \text{ fm} \quad \text{for Sb}$$

Tantalum

$$\eta = \frac{5Q}{4ZR^2} = \frac{5 \times 6 \times 10^{-24}}{4 \times 73 \times (8.48 \times 10^{-13})^2} = 0.143$$

$$\frac{a}{b} \simeq 1 + \eta = 1.143$$

Antimony

$$\eta = \frac{5}{4} \times \frac{(-1.2 \times 10^{-24})}{51 \times (7.458 \times 10^{-13})^2} = -0.053$$

$$\frac{a}{b} = 1 - 0.053 = 0.947$$

Example 4.9 Use the fact that the form factor is the Fourier transform of the charge density distribution to find an expression for $F(q)$ for scattering from a particle whose charge density is given by $\rho(r) = (A/r)\cos(\pi r/2R)$ for $r \le R$ and 0 otherwise, where A is a constant. The values of q which produce zeros in the differential cross-section can be used to find the size of the particle. Find the condition for the occurrence of minima in the differential cross-section.

Solution

$$F(q) = \int_0^\infty \rho(r) \frac{\sin(qr/\hbar)}{qr/\hbar} 4\pi r^2 dr$$

$$= \int_0^R \frac{A}{r} \frac{\cos(\pi r/2R)\sin}{qr/\hbar}(qr/\hbar)4\pi r^2 dr$$

$$= \frac{4\pi A\hbar}{q}\frac{1}{2}\int_0^R \left[\sin\left(\frac{qr}{\hbar} + \frac{\pi r}{2R}\right) + \sin\left(\frac{qr}{\hbar} - \frac{\pi r}{2R}\right)\right]dr$$

$$F(q) = \frac{8\pi^2 A\hbar^3 R}{q(4q^2 R^2 - \pi^2\hbar^2)}\left(\frac{2qR}{\pi\hbar} - \sin\frac{qR}{\hbar}\right)$$

$$F(q) = 0 \quad \text{when} \quad \sin\frac{qR}{\hbar} = \frac{2qR}{\pi\hbar}$$

Example 4.10 The charge distribution in proton may be written as $\rho(r) = A\exp(-r/a)$ where A is a constant and 'a' is known as the characteristic radius. Show that the form factor is proportional to $(1 + q^2 a^2/\hbar^2)$.

Solution

$$F(q) \sim \int_0^\infty re^{-r/a}\sin\left(\frac{qr}{\hbar}\right)rdr$$

Integrate by parts twice. Following definite integrals will be useful to get the desired result

$$\int_0^\infty e^{-ax}\cos bx\,dx = \frac{a}{a^2 + b^2}, \qquad \int_0^\infty e^{-ax}\sin bx\,dx = \frac{b}{a^2 + b^2}$$

Example 4.11 Show that $^{226}_{88}\text{Ra}$ is unstable against α-decay. Use the masses $^{226}_{88}\text{Ra} = 226.025360$ amu, $^{226}_{86}\text{Rn} = 222.017531$ amu, $^4_2\text{He} = 4.002603$ amu.

Fig. 4.28 Decay of $^{28}\text{Al}_{13}$ to $^{28}\text{Si}_{14}$ via β^- emission which in turn decays to the ground state via γ-emission

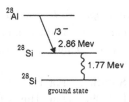

Solution   The decay is

$$^{226}\text{Ra} \rightarrow \, ^{222}\text{Rn} + \, ^4\text{He} + Q$$

The decay is feasible if the mass of ^{226}Ra is larger than the sum of the masses of ^{222}Rn and ^4He

$$^{222}\text{Rn} + \, ^4\text{He} = 222.017531 + 4.002603$$
$$= 226.020134 \text{ amu}$$

Thus, the decay is feasible.

Example 4.12 $^{28}_{13}\text{Al}$ decays to $^{28}_{14}\text{Si}$ via β^- emission with $T_{\text{max}} = 2.865$ MeV. $^{28}_{14}\text{Si}$ is in the excited state which in turn decays to the ground state via γ-emission, see Fig. 4.28. Find the γ-ray energy. Take the masses $^{28}\text{Al} = 27.981908$ amu, $^{28}\text{Si} = 27.976929$ amu.

Solution

$$Q = (27.981908 - 27.976929) \times 931.5 = E_{\text{max}} + E_\gamma$$
$$= 4.638 \text{ MeV}$$
$$E_\gamma = Q - T_\beta = 4.638 - 2.865$$
$$= 1.773 \text{ MeV}$$

Example 4.13 $^{22}_{11}\text{Na}$ decays to $^{22}_{10}\text{Ne}$ via β^+ with $T_{\text{max}} = 0.542$ MeV, followed by γ decay with energy 1.277 MeV. If the mass of ^{22}Ne is 21.991385 amu, determine the mass of ^{22}Na in amu.

Solution

$$^{22}\text{Na} = \, ^{22}\text{Ne} + 2m_c + \frac{T_{\text{max}} + T_\gamma}{931.5}$$
$$21.991385 + 2 \times (0.000548) + \frac{0.542 + 1.277}{931.5}$$
$$= 21.994434 \text{ amu}$$

Example 4.14 $^{7}_{4}$Be undergoes electron capture and decays to $^{7}_{3}$Li. Investigate if it can decay by the competitive decay mode of β^{+} emission. Take the masses $^{7}_{4}$Be $=$ 7.016929 amu, $^{7}_{3}$Li $= 7.016004$ amu.

Solution For E.C. the difference in atomic masses is $7.016929 - 7.016004 =$ 0.000925 amu. This difference is short of $2m_e = 2 \times 0.000549 = 0.001098$ amu necessary for β^{+} decay. Hence minimum energy needed for β^{+} decay is not available.

Example 4.15 Find the energy shift of the ground state of the hydrogen atom due to the finite size of the proton, assuming that the proton is a uniformly charged sphere of radius 1 fm.

Solution Using Eqs. (4.116) and (4.117) and putting $z = 1$

$$\Delta E_R = \frac{4}{5} \cdot \frac{e^2}{2a_0} \left(\frac{R}{a_0} \right)^2 = \frac{4}{5} \times (13.6) \left(\frac{10^{-13}}{0.53 \times 10^{-8}} \right)^2$$

$$= 3.9 \times 10^{-9} \text{ eV}$$

Example 4.16 A $D_{5/2}$ term in the optical spectrum of $^{39}_{19}$K has a hyperfine structure with four components. Find the spin of the nucleus.

Solution Let J be the electronic angular momentum and I the nuclear spin. The multiplicity is $(2J + 1)$ or $(2I + 1)$, which ever is smaller.
   Now $2J + 1 = 2 \times \frac{5}{2} + 1 = 6$. But only four terms are found, $2I + 1 = 4 \rightarrow I = 3/2$.

Example 4.17 In Example 4.16 what interval ratios in the hyperfine quadruplet are expected?

Solution The energy shift in hyperfine structure arises because of the interaction of the nuclear magnetic moment with the magnetic field produced by the electron.

$$\Delta E \sim 2\boldsymbol{I} \cdot \boldsymbol{J} = F(F + 1) - I(I + 1) - J(J + 1)$$

where $F = I + J$ takes on integral values from 4 to 1

$$F = 4, \qquad \Delta E = 20 - \frac{25}{2}$$

$$F = 3, \qquad \Delta E = 12 - \frac{25}{2}$$

$$F = 2, \qquad \Delta E = 6 - \frac{25}{2}$$

$$F = 1, \qquad \Delta E = 2 - \frac{25}{2}$$

The intervals are 8, 6 and 4, the ratios being $4 : 3 : 2$.

Example 4.18   Obtain an expression for the potential at a distance r from the centre of a sphere of radius R in which charge q is homogeneously distributed.

Solution

Region I $(r > R)$

$$V(r) = \frac{q}{4\pi\varepsilon_0 r}$$

The charge is assumed to be concentrated at the centre.

Region II $(r < R)$

Let q' be the charge within the sphere of radius r. Then $q' = q(\frac{r}{R})^3$. The electric field will be

$$E = \frac{q'}{4\pi\varepsilon_0 r^2} = \frac{qr}{4\pi\varepsilon_0 R^3}$$

$$V = -\int E dr = -\int \frac{1}{4\pi}\frac{qrdr}{\varepsilon_0 R^3} + c$$

$$= -\frac{qr^2}{8\pi\varepsilon_0 R^3} + c$$

At

$$r = R, \qquad V(R) = \frac{q}{4\pi\varepsilon_0 R}$$

$$\therefore \quad \frac{q}{4\pi\varepsilon_0 R} = -\frac{q}{8\pi\varepsilon_0 R} + C \quad \text{or}$$

$$C = \frac{3}{2}\frac{q}{4\pi\varepsilon_0 R}$$

$$V = \frac{q}{8\pi\varepsilon_0 R}\left(3 - \frac{r^2}{R^2}\right)$$

Example 4.19   Given that the proton has a magnetic moment of 2.79 magnetons and a spin quantum number of one half, what magnetic field strength would be required to produce proton resonance at a frequency of 50 MHz in a nuclear magnetic resonance spectrometer?

Solution

$$\nu = \frac{\mu B}{Ih}$$

$$B = \frac{vIh}{\mu} = \frac{50 \times 10^6 \times 1/2 \times 6.62 \times 10^{-34}}{2.79 \times 5.05 \times 10^{-27}} = 1.167 \text{ T}$$

4.16 Questions

4.1 Why are there no mirror nuclei with $A \gtrsim 40$?

4.2 Give four characteristics of nuclear forces.

4.3 In what region of A is the B/A value maximum?

4.4 How do you account for the drop of B/A for low A and larger A?

4.5 Distinguish between charge symmetry and charge independence.

4.6 Give two examples of mirror nuclei.

4.7 Give two examples of isobaric triplets to justify charge independence of nuclear forces.

4.8 List four methods for the determination of nuclear radii. Which method would you rank as most important?

4.9 One method to determine nuclear radius is to measure E_{max} for β energy in the decay of mirror nuclei. What is the merit of choosing mirror nuclei?

4.10 What information is obtained from the sign and magnitude of quadrupole moment?

4.11 If the quadrupole moment of a nucleus is zero, what do you infer?

4.12 What value of quadrupole moment is expected for nuclide with $I = 0$? with $I = \frac{1}{2}$?

4.13 What is the cause of alternating intensities of rotational lines in the band spectrum of homonuclear diatomic molecules? When will be the alternating lines missing?

4.14 The kinetic energy of the two nuclei produced in the fission of ^{235}U is about 200 MeV. Approximately what fraction of the original mass appears as kinetic energy?

4.17 Problems

4.1 Calculate the binding energy of the last neutron in ^{13}C, given the atomic masses based on ^{12}C

$$^{1}_{0}n = 1.008665 \text{ amu}, \qquad ^{13}_{6}C = 13.003354 \text{ amu}$$

[Ans. 4.95 MeV]

4.2 The radius of ^{165}Ho is 7.731 fm. Find the radius of ^{4}He.
[Ans. 2.238 fm]

4.3 What is the mass number A of an element whose nuclear radius is 2.71 fm ($r_0 = 1.3$ fm).
[Ans. 9]

4.4 Calculate the total binding energy of $^{4}_{2}$He and $^{5}_{2}$He. Which one of these nuclei is more stable? Use the following atomic masses

$$M(^{4}_{2}\text{He}) = 4.003873 \text{ amu}; \qquad M(^{4}_{2}\text{He}) = 5.013888 \text{ amu}$$

$$M(^{1}_{1}\text{H}) = 1.008145 \text{ amu}; \qquad M(^{1}_{0}n) = 1.008986 \text{ amu}$$

[Ans. 28.3, 27.4 MeV, ^{4}He more stable]

4.5 Determine the density of a nucleus. Given that the mass of proton/neutron $= 1.67 \times 10^{-27}$ kg. Radius of the nucleus $= 1.3A^{1/3}$ fm.
[Ans. 1.8×10^{17} kg m^{-3}]

4.6 Show that when a nucleus of rest mass M absorbs a photon of energy $h\nu$, the excitation energy of the nucleus is given by, $E_{ex} = Mc^2(\sqrt{1 + \frac{2h\nu}{Mc^2}} - 1)$.

4.7 Show that if $E(\gg mc^2)$ is the laboratory energy of electrons incident on a nucleus of mass M, the nucleus will acquire kinetic energy

$$E_N = \frac{E^2}{Mc^2} \frac{(1 - \cos\theta)}{[1 + \frac{E}{Mc^2}(1 - \cos\theta)]}$$

4.8 Using the uncertainty principle and the fact that the maximum energy of a β-particle is of the order of 1 MeV, show that a free electron is not likely to be found inside a nucleus, whose dimension δx is of the order of 10^{-12} cm, by using the uncertainty principle.

4.9 Singly charged chlorine ions are accelerated, through a fixed potential difference and then caused to travel in circular paths by means of a uniform field of

magnetic induction of 1500 gauss. What increase in induction is necessary to cause the mass 37 ion to follow the path previously taken by the mass 35 ion?
[Ans. 42.26 G]

4.10 Two isotopes of silver ^{107}Ag and ^{109}Ag are to be separated electromagnetically. The singly charged ions are first accelerated through an electrostatic potential of 10 kV and then deflected in a uniform magnetic field through a semi circular path of radius 1 m.

(i) What magnetic field intensity is required?
(ii) If the entrance and exit slits have the same size, calculate the maximum slit width for which the two isotopes will be completely separated.

[Ans. (i) 1.497 kg, (ii) 1.86 cm]

4.11 The scattering amplitude by a spherically symmetric potential $V(r)$ with a momentum transfer q is given by

$$A = \int_0^\infty \frac{\sin(qr/\hbar)}{qr/\hbar} V(r) 4\pi r^2 dr$$

Assuming a Yukawa type potential show that the scattering amplitude is proportional to $(q^2 + m^2 c^2)^{-1}$.

4.12 From the β^+ decay of ^{14}O an excited state of ^{14}N is formed. The ^{14}N γ-rays have an energy of 2.313 MeV and the maximum energy of the positrons is 1.835 MeV. The masses of ^{14}O and electron are 14.008623 amu and 0.000548 amu, respectively. Calculate the mass of ^{14}N.
[Ans. 14.003074 amu]

4.13 Estimate the ratio of the major and mirror axis of $^{176}_{71}$Lu and $^{127}_{53}$I. The quadrupole moments are 7×10^{-24} and -2.86×10^{-24} cm^2 for Lutecium and Iodine. (Take $R = 1.5 A^{1/3}$ fm.)
[Ans. $(a/b)_{Lu} = 1.18$, $(a/b)_I = 0.97$]

4.14 The empirical mass formula is

$$^A_Z M = 0.99198A - 0.000841Z + 0.01968A^{2/3}$$

$$+ 0.0007668Z^2 A^{-1/3} + 0.09966\left(Z - \frac{A}{2}\right)^2 A^{-1} - \delta$$

in atomic mass units, where $\delta = \pm 0.01204 A^{-1/2}$ or 0. Investigate the β-decay stability of the nuclide $^{27}_{12}$Mg.
[Ans. Unstable]

4.15 Verify that the Coulomb repulsive energy in ^{238}Pu is about 70 % of the total binding energy.

4.16 Use the uncertainty principle to show that a nucleus which is confined to a region of the order of $\Delta x \sim 1$ fm must have a kinetic energy of the order of 20 MeV.

4.17 By what factor must the mass number be increased in order to double the nuclear radius?
[Ans. 8]

4.18 In a magnetic resonance experiment using water as sample what would be the magnetic field in glass if the resonance frequency is 30 megacycles per second (magnetic moment of the proton = 2.85 nuclear magnetons).
[Ans. 6.9 kg]

References

1. Barshay, Temmer (1964)
2. F. Block, D. Nicodemus, H.H. Staub, Phys. Rev. **74**, 1025 (1948)
3. Fermi, Teller (1947)
4. Fitch, Rainwater (1953)
5. A.S. Green, Rev. Mod. Phys. **30**, 569 (1958)
6. E.E. Gross, E. Newman, W. Roberts, R.W. Rutrowsky, A. Zucker, Phys. Rev. Lett. **24**, 43 (1970)
7. Heisenberg, Condon, Cassen (1932)
8. R. Hofstadter, Annu. Rev. Nucl. Sci. **7**, 231 (1957)
9. A.A. Kamal, *Particle Physics* (Springer, Berlin, 2014)
10. Kemmer (1938)
11. I.I. Rabi, J.M.B. Kellogg, J.R. Zacharias, Phys. Rev. **46**, 157 (1934)
12. I.I. Rabi, J.R. Zacharias, S. Millman, P. Kusch, Phys. Rev. **53**, 392 (1938)
13. Rabi, Zacharias, Millman, Kusch
14. E. Segre, *Nuclei and Particles* (Benjamin, New York, 1964). Reproduced by permission

Chapter 5
The Nuclear Two-Body Problem

5.1 Deuteron

Deuteron consists of a proton and a neutron. It is essentially a loose structure as implied by its low binding energy at around 2 MeV, i.e. 1 MeV per particle, in contrast with a value of 8 MeV for the average binding energy per nucleon in medium and heavy nuclei. The binding energy of deuteron has been determined by various methods.

5.1.1 Binding Energy of Deuteron

(a) Photo-Disintegration of Deuteron ($h\nu + d \rightarrow p + n$) The threshold energy for this reaction is equal to W, the binding energy of deuteron. For γ-ray energy and E_γ greater than W, the surplus energy, $E_\gamma - W$ appears as the kinetic energy shared between the product particles. For E_γ not much above the threshold energy, the deuteron would receive negligible momentum. Consequently, proton and neutron would be projected with equal and opposite momentum, and they would share nearly equal energy. The binding energy is then given by $W = h\nu - 2E_p$, where E_p is the kinetic energy of proton which can be measured either from ionization measurements or range measurements. Chadwick and Goldhaber [4] using 2.62 MeV γ-rays from Th C'' obtained a value of 2.14 MeV for the binding energy, while Stetter and Jentschke determined it to be 2.19 ± 0.03 MeV.

(b) Threshold Energy of the Photo-Disintegration Reaction The neutron yield from the photo-disintegration reaction is measured for various γ-ray energies. In practice, electrons accelerated from a Van de Graff generator, on bombarding a heavy target produce the γ-rays. The γ-ray energy was calibrated from the known electron energy. By extrapolating the neutron yield to zero value corresponding to the threshold of the reaction, a value of $W = 2.226 \pm 0.003$ MeV is obtained.

A. Kamal, *Nuclear Physics*, Graduate Texts in Physics,
DOI 10.1007/978-3-642-38655-8_5, © Springer-Verlag Berlin Heidelberg 2014

(c) **Inverse Reaction** $(n + p \rightarrow h\nu + d)$ When slow neutrons are absorbed in hydrogen, deuteron is formed and a γ-ray is emitted. This reaction corresponds to the inverse of photo-disintegration of deuteron and is also called radiative capture. The energy that is released is equal to W. This is shared between the photon and the deuteron. The latter carries negligible amount of energy. From the conservation of momentum and energy, it is easily shown that

$$W = E_\gamma + \frac{E_\gamma^2}{2Mc^2} \tag{5.1}$$

The second term on the right side corresponding to the deuteron energy, is small, being about 0.0013 MeV. A precise measurement of γ-ray energy for example by a γ-ray spectrometer, yields the binding energy. From the observations of the radiative capture of slow neutrons from reactors in hydrogen, Bell et al. [1] obtained a value of 2.23 ± 0.007 MeV.

(d) **Mass Spectrography**   The most accurate method for the estimation of binding energy is based on the direct mass determination of deuteron from mass spectrographs. Since the proton mass is accurately known and that of neutron is fairly well-known from the measurements of end-point energy of β-rays in the decay of free neutron, the binding energy of deuteron is simply given by

$$W = (m_p + m_n - m_d)c^2 = 2.225 \pm 0.015 \text{ MeV}$$

Collecting the results of various recent measurements, the accepted value of W is 2.2246 ± 0.0002 MeV.

5.1.2 The Ground State of Deuteron

By convention, the intrinsic parity of both proton and neutron is positive. It follows that the intrinsic parity (π_d) of deuteron must also be positive. Now, the nucleons by virtue of their motion can have the relative angular momentum values $l = 0, 1, 2$.
  The parity of deuteron is then given by

$$\pi_d = \pi_p \cdot \pi_n (-1)^l \tag{5.2}$$

But the intrinsic parities, $\pi_d = \pi_p = \pi_n = +1$. Consequently, only even angular momentum states can exist. In particular, the two lowest possible states for deuteron are the S- and D-states, corresponding to $l = 0$ and $l = 2$, respectively. We shall see that the experimentally measured value of J, the total angular momentum of the deuteron, is effectively contributed by the spins of neutrons and protons lined up parallel, i.e. the total spin of the particles $s = s_n + s_s = 1$ (triplet state), so that the ground state of deuteron is predominantly the S-state ($l = 0$) and the corresponding wave function is spherically symmetrical.

The nuclear forces that provide the necessary binding of the nucleons have a short range character, the characteristic range of interaction being of the order of $\hbar/m_\pi c \simeq 1.4$ fm. Corresponding to the force we may introduce a potential V given by $F = -(\partial V/\partial r)$. Because the nuclear forces are attractive, the sign of the potential must be negative. We assume that the forces between the two particles are essentially central forces, i.e. the force is directed along the line joining the two particles. We start with the time-independent Schrödinger equation in spherical polar coordinates in the CM system of proton and neutron

$$\nabla^2 \psi(r, \theta, \phi) + \frac{2\mu}{\hbar^2}\left[E - V(r)\right]\psi(r, \theta, \phi) = 0 \qquad (5.3)$$

where r is the distance between neutron and proton, and μ is the reduced mass given by

$$\mu = \frac{M_n M_p}{M_n + M_p} \simeq \frac{M}{2} \qquad (5.4)$$

with M as the proton or neutron mass.

The total energy E is negative and is numerically equal to the binding energy W. Under the assumption of central forces, the $\psi(r, \theta, \phi)$ function is spherically symmetric and consequently the angular derivatives in the Laplacian vanish. The Schrodinger equation (5.3) is then reduced to

$$\frac{1}{r^2}\frac{d}{dr}\left(r^2 \frac{d}{dr}\right)\psi(r) + \frac{M}{\hbar^2}\left[E - V(r)\right]\psi(r) = 0 \qquad (5.5)$$

With the introduction of the radial wave function $u(r)$ defined by

$$\psi(r) = \frac{u(r)}{r} \qquad (5.6)$$

Equation (5.5) is simplified to

$$\frac{d^2 u}{dr^2} + \frac{M}{\hbar^2}\left[E - V(r)\right]u = 0 \qquad (5.7)$$

Because of the short range nature of nuclear forces, the potential is assumed to vanish for all practical purposes beyond a distance $r > R$, where R is of the order of 3 fm. The simplest choice of the potential is the square well potential defined by

$$V = -V_0, \quad r < R$$

$$= 0, \quad r > R \qquad (5.8)$$

The parameters V_0 and R are called the depth and range of the potential respectively. Writing $E = -W$ and $V = -V_0$ for the region $r < R$ and $V = 0$ for the region, $r > R$, with both W and V_0 positive, the equations become

Fig. 5.1 Wave functions u_I
and u_{II} in the square well
potential

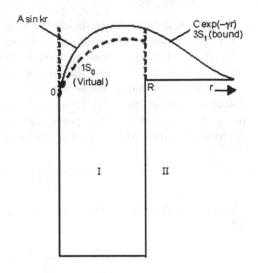

$$\frac{d^2u}{dr^2} + K^2u = 0, \quad r < R \tag{5.9}$$

$$\text{with } K^2 = \frac{M(V_0 - W)}{\hbar^2} \quad \text{and} \tag{5.10}$$

$$\frac{d^2u}{dr^2} - \gamma^2U = 0, \quad r > R \tag{5.11}$$

$$\text{with } \gamma^2 = \frac{MW}{\hbar^2} \tag{5.12}$$

Equation (5.9) has the solution

$$u_I = A\sin Kr + B\cos Kr, \quad r < R \tag{5.13}$$

and Eq. (5.11) has the solution:

$$u_{II} = ue^{-\gamma r} + De^{+\gamma r}, \quad r > R \tag{5.14}$$

where A, B, C, and D are the constants of integration. The boundary condition imposed on the solutions is that u_I must vanish as $r \to 0$ and that u_{II} must vanish as $r \to \infty$. The first condition when used in (5.13) yields $B = 0$; and the second condition when applied to (5.14) gives us $D = 0$. The physically acceptable solutions are then

$$u_I = A\sin kr, \quad r < R \tag{5.15}$$

$$u_{II} = Ce^{-\gamma r}, \quad r > R \tag{5.16}$$

The wave functions u_I and u_{II} in the square well potential are shown in Fig. 5.1. Equation (5.16) shows that the wave function outside the range of nuclear forces is completely determined by W. The quantity $R_0 = (1/\gamma) = (\hbar/\sqrt{MW}) = 4.36$ fm is loosely called the size of deuteron.

Since the wave function must be single valued and continuous throughout, we require the amplitude and first derivative of the functions u_I and u_{II} to be equal at the boundary $r = R$, i.e.

$$(u_I)_{r=R} = (u_{II})_{r=R}$$

$$\left(\frac{du_I}{dr}\right)_{r=R} = \left(\frac{du_{II}}{dr}\right)_{r=R}$$

Therefore

$$A \sin KR = C \exp(-\gamma R) \tag{5.17}$$

$$AK \cos KR = -\gamma C \exp(-\gamma R) \tag{5.18}$$

On dividing (5.18) by (5.17), we obtain the relation

$$K \cot KR = -\gamma \tag{5.19}$$

which does not contain the unknown constants A or C. Figure 5.1 shows the radial wave function u in regions $r < R$ and $r > R$. Inserting the values of K and γ from (5.10) and (5.12) in (5.19) we obtain

$$\cot KR = \frac{-\gamma}{K} = -\sqrt{W/(V_0 - W)} \tag{5.20}$$

It is plausible to assume that $W \ll V_0$, in which case we conclude that $\cot KR$ is negative and small numerically. This means that KR is only slightly larger than $(n + 1/2)\pi$, with $n = 0, 1, 2, \ldots$. The correct solution for the ground state is $KR \simeq (\pi/2)$. For, if we accept the second solution $KR \simeq (3\pi/2)$, the corresponding wave function in the region $r < R$ would produce a node at $kr = \pi$, and would not be consistent with the ground state. Equation (5.19) is essentially a relation between the depth V_0 and the width R of the potential. We can approximate Eq. (5.20) to obtain a more direct relation between K and R. Multiplying both sides of (5.19) by R, we get

$$KR \cot KR = -\gamma R \tag{5.21}$$

we have already seen that KR is slightly larger than $(\pi/2)$. Put the value

$$KR = \frac{\pi}{2} + \varepsilon \tag{5.22}$$

where ε is a small quantity

$$\left(\frac{\pi}{2} + \varepsilon\right) \cot\left(\frac{\pi}{2} + \varepsilon\right) = -\gamma R$$

But $\cot(\pi/2+\varepsilon) = -\varepsilon$. It, therefore, follows that $\varepsilon \simeq 2\gamma R/\pi$. This value of ε when used in (5.22) yields the formula which is a good approximation:

$$K \simeq \frac{\pi}{2R} + \frac{2\gamma}{\pi} \tag{5.23}$$

A very simple relation between V_0 and R follows if we use the following approximation:

$$KR \simeq \frac{\pi}{2} \quad \text{or}$$

$$K^2 R^2 \simeq \left(\frac{\pi}{2}\right)^2 \quad \text{or}$$

$$R^2 \frac{M(V_0 - W)}{\hbar^2} = \left(\frac{\pi}{2}\right)^2 \tag{5.24}$$

Further, using the approximation $W \ll V_0$, we get

$$V_0 R^2 = \frac{h^2}{16M}$$

$$= 103 \text{ MeV fm}^2 \tag{5.25}$$

With an arbitrary value of $R = 2$ fm, we find $V_0 = 103/(2)^2$ or 26 MeV. The calculated value of V_0 is quite sensitive to the choice of R. In general, the smaller the width of the well, the deeper is the potential depth and vice versa. For an assumed value of R, a more precise value of V_0 is obtained by the use of Eq. (5.23).

5.1.3 The Probability that the Neutron and Proton Are Found Outside the Range of Nuclear Forces

The expression $4\pi |\psi|^2 r^2 dr$ or $4\pi |u|^2 dr$ represents the probability that the neutron and proton are at a distance r and $r + dr$ apart. We shall first determine the constants from the normalization condition

$$\int_0^\infty |\psi|^2 d\tau = \int_0^\infty |\psi|^2 4\pi r^2 dr = 4\pi \int_0^\infty |u|^2 dr = 1 \tag{5.26}$$

using the relevant wave functions for the region $r < R$ and $r > R$, we can re-write the above condition as:

$$4\pi \int_0^R |u_I|^2 dr + 4\pi \int_R^\infty |u_{II}|^2 dr = 1$$

Inserting the wave functions u_I and u_{II} from (5.15) and (5.16) we obtain

$$A^2 \int_0^R \sin^2 kr\,dr + C^2 \int_R^\infty e^{-2\gamma r}dr = \frac{1}{4\pi} \tag{5.27}$$

A straightforward integration yields the result

$$A^2\left(R - \frac{\sin 2Kr}{2K}\right) + \frac{C^2}{\gamma}e^{-2\gamma R} = \frac{1}{2\pi} \tag{5.28}$$

We also have another relation between A and C either from (5.17) or (5.18). Therefore

$$A\sin KR - Ce^{-\gamma R} = 0 \tag{5.29}$$

Solving (5.28) and (5.29) for A^2 and C^2, we get

$$A^2 = \frac{\gamma}{2\pi(1+\gamma R)} \tag{5.30}$$

$$C^2 = \frac{\gamma \sin^2 K Re^{2\gamma R}}{2\pi(1+\gamma R)} \tag{5.31}$$

where we have used (5.19) in simplifying the expressions. The probability P of finding the neutron and proton separated by a distance larger than R is given by

$$P = \int_R^\infty 4\pi|u_{II}|^2 dr = 4\pi C^2\int_R^\infty e^{-2\gamma r}dr = \frac{2\pi C^2 e^{-2\gamma R}}{\gamma}$$

where the value of u_{II} has been used from (5.16). Inserting the value of C^2 from (5.31), we obtain

$$P = \frac{\sin^2 KR}{1+\gamma R} \tag{5.32}$$

Since $KR \simeq \pi/2$, we find the approximate value

$$P \simeq \frac{1}{1+\gamma R} \tag{5.33}$$

From the experimentally measured value of $W = 2.225$ MeV, we find

$$\gamma = \frac{\sqrt{MW}}{\hbar} = \frac{\sqrt{Mc^2W}}{\hbar c} = \frac{\sqrt{939\times 2.225}}{197} = 0.232 \text{ fm}^{-1}$$

Therefore, with the assumed value of $R = 2$ fm, we find $P = 1/(1+0.232\times 2) = 0.68$. This means that 70 percent of time the neutrons and protons are found outside the range of nuclear forces. This is caused by the quantum mechanical 'tunnel effect', the wave function leaks into the forbidden region. The size of the deuteron, represented by $R = (1/\gamma) = 4.31$ fm is comparable with a nucleus containing 20 nucleons.

5.1.4 Excited States of Deuteron

For $l = 0$ there are no bound excited states. This is because if we had accepted the solution $KR \simeq (3\pi/2)$ corresponding to the possible first excited state, then we would have got the result analogous to (5.25)

$$K^2 R^2 = \frac{9\pi^2}{4} \quad \text{or} \tag{5.34}$$

$$V_0 - W_1 = \frac{9\pi^2}{4} \frac{\hbar^2}{MR^2} \tag{5.35}$$

where W_1 is the binding energy of the first excited state. But the corresponding relation for the ground state given by (5.26) can be rewritten as:

$$V_0 - W = \frac{\pi^2}{4} \frac{\hbar^2}{MR^2} \tag{5.36}$$

Comparing (5.35) with (5.36)

$$W_1 = 9W - 8V_0$$

$$\simeq 20 - 8V_0 \tag{5.37}$$

where we have used the value $W = 2.225$ MeV.

For any reasonable value of V_0, the binding energy W_1 is certainly negative, indicating thereby that the first excited state is not a bound state. Higher excited states would have increasingly large negative values of binding energy.

We can also prove the above result from a different angle. We may set $W_1 = 0$ as a criterion for giving the minimum well depth. Comparison of (5.36) (wherein W is neglected compared to V_0) with (5.35) shows that the required well depth for the first excited state is about nine times that for the ground state, a result which is untenable.

We shall now prove that deuteron has no bound states for higher values of l. It will be assumed that the nuclear forces between neutron and proton for the $l \neq 0$ states are identical with that for the ground state $l = 0$. The Schrodinger's equation for the $l \neq 0$ state is modified as:

$$\frac{d^2 u}{dr^2} + \frac{M}{\hbar^2}(E - V)u - \frac{l(l+1)u}{r^2} = 0 \tag{5.38}$$

This is equivalent to the s-wave radial equation (5.7) with the effective potential given by

$$V_{eff}(r) = V(r) + \frac{\hbar^2 l(l+1)}{Mr^2} \tag{5.39}$$

The second term on the right side represents the potential arising from the centrifugal force which is essentially positive and therefore repulsive. Physically, the centrifugal barrier has the effect of making the particles run away from each other. Since the binding of neutron and proton is provided by the negative potential $V(r)$, the $l \neq 0$ states would cause a decrease in the binding energy of the lowest bound state compared to that for the $l = 0$ state. For $l = 1$, we obtain the result:

$$KR \sin KR = 0$$

but $KR \neq 0$ since neither K nor R is zero. It follows that $\sin KR - 0$. The smallest positive root of this equation is $KR = \pi$ or $K^2 R^2 = \pi^2$. Using the value of K^2 from (5.10) and setting $W = 0$, we find the relation

$$V_0 R^2 = \frac{\pi^2 \hbar^2}{M} = \frac{h^2}{4m} \tag{5.40}$$

Comparison of (5.40) with (5.27) shows that the minimum well depth required for the $l = 1$ state is four times the actual value for the ground state. Therefore, we conclude that the state cannot be a bound state. Similarly, it can be shown that higher values of l would require larger values of KR and hence still deeper well depths. The result that the deuteron cannot have bound excited states follows simply from the fact that the binding energy for the ground state itself is already very small.

It must be conceded that the square well potential chosen in the deuteron problem is unrealistic, and has been adopted for mathematical simplicity. However, the results obtained from the assumed potential are atleast qualitatively correct. Calculations have also been made for other forms of potentials, for example the exponential function e^{-r}, the error function e^{-r^2} and the Yukawa's potential (e^{-r}/r). But, the results are insensitive to the choice of the potential.

5.1.5 Root Mean Square Radius

It might appear that any arbitrary well depth with an appropriate width can be choosen to satisfy the relation (5.19) or (5.23), or the more approximate version (5.25). But, other considerations such as the root mean square radius of deuteron and the results on neutron proton scattering, remove this arbitrariness and allow the values of V_0 and R to be defined within a narrow range. Since the distance of proton or neutron from the centre of mass is half the distance separating the two particles, the mean square radius of the deuteron is given by

$$\langle r_d^2 \rangle = \frac{1}{4} \langle r^2 \rangle \tag{5.41}$$

Thus

$$\langle r_d^2 \rangle = \frac{A^2}{4} \int_0^R 4\pi r^2 \sin^2 Kr \, dr + \frac{C^2}{4} \int_R^\infty 4\pi r^2 e^{-2\gamma r} \, dr \tag{5.42}$$

Using the values of A^2 and C^2 from (5.30) and (5.31) and simplifying the trigono-metric functions through the application of (5.19), we find the following after some algebra:

$$\langle r_d^2 \rangle = \frac{1}{8\gamma^2} - \frac{1}{8K^2} + \frac{R(1+\gamma R)}{8\gamma} - \frac{R^3\gamma}{24(1+\gamma R)} \qquad (5.43)$$

So far we have considered proton to be point charge. On the other hand, if the charge distribution of proton is also taken into account then under the assumption that the charge is spherically symmetrical, the actual mean square radius of deuteron is then given by

$$\langle r_d^2 \rangle = \frac{1}{8\gamma^2} - \frac{1}{8K^2} + \frac{R(1+\gamma R)}{8\gamma} - \frac{R^3\gamma}{24(1+\gamma R)} + \langle r_p^2 \rangle \qquad (5.44)$$

The high energy electron scattering experiments have determined the root mean square radius of proton, $\sqrt{\langle r_p^2 \rangle} = 0.8$ fm. Inserting the measured value of $\sqrt{\langle r_d^2 \rangle} = 2.1$ fm (MC Intgre et al.) and the accepted value of $\gamma = 0.232$ fm^{-1} Eq. (5.44) yields a relation between K and R. Another relation between K and R is provided from (5.19) or its approximate version (5.23). On solving the equations, we find $R = 2.08$ fm, and $K = 0.902$. This value of K and the experimentally known value of $W = 2.225$ MeV when used in (5.10) yield $V_0 = 35.8$.

5.1.6 The Inclusion of Hard Core Potential in the Square Well

The square well potential may be modified to include the 'hard core' potential in-finitely high for $r < b$, which effectively prevents the particles from approaching each other within distance b (region I). Evidence on the existence of hard core is furnished by high energy scattering experiments. The solution to this problem is illustrated in the following example.

Example 5.1 A 'hard core' potential for the interaction between two nucleons is:

$$V(r) = \infty, \qquad 0 < r < b$$
$$= 73 \text{ MeV}, \quad b < r < R$$
$$= 0, \qquad\qquad r > R$$

(a) If $b = 0.4$ fm is the radius of the hard core and the binding energy of deuteron $W = 2.225$ MeV, calculate the value of R. (b) What is the most probable distance between the proton and neutron in the deuteron ground state? (c) Calculate the root mean square radius of deuteron.

Fig. 5.2 Square well
potential with a "hard core"
potential

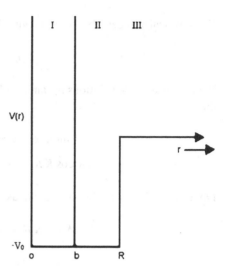

Solution   (a) Figure 5.2 shows the potential as a function of r for the three regions. We write down Schrodinger equation

$$\frac{d^2u}{dr^2} + \frac{M}{\hbar^2}[E - V(r)]u = 0$$

For region II, this equation takes the form

$$\frac{d^2u}{dr^2} + K^2u = 0, \quad \text{with } K^2 = \frac{m}{\hbar^2}(V_0 - W)$$

The general solution is:

$$u = A \sin Kr + B \cos Kr \tag{5.45}$$

We now impose the boundary condition, $u \to 0$ at $r = b$

$$0 = A \sin Kb + B \cos Kb \quad \text{or}$$

$$B = -A\frac{\sin Kb}{\cos Kb} \tag{5.46}$$

Using the value of B from (5.46) in (5.45), the solution simplifies to

$$u_{II} = A \sin K(r - b) \tag{5.47}$$

where the factor $\frac{1}{\cos Kb}$ has been included in the constant A.
   For region III, the equation has the form

$$\frac{d^2u}{dr^2} - \gamma^2u = 0, \quad \text{with } \gamma^2 = \frac{MW}{\hbar^2}$$

The physically acceptable solution with the boundary condition $u \to 0$ as $r \to \infty$ is

$$u_{III} = Ce^{-\gamma r} \tag{5.48}$$

We now match the solutions u_{II} and u_{III} both in amplitude and the first derivative at the boundary $r = b + R$

$$A \sin KR = Ce^{-\gamma(b+R)} \tag{5.49}$$

$$AK \cos KR = -\gamma C \exp\left[-\gamma(b+R)\right] \tag{5.50}$$

Dividing (5.71) by (5.70), we obtain, as before, the relation

$$K \cot KR = -\gamma \quad \text{or}$$

$$R = \frac{1}{K} \text{arc cot}\left(\frac{-\gamma}{K}\right) \tag{5.51}$$

Now

$$\frac{\gamma}{K} = \sqrt{\frac{W}{V_0 - W}} = \sqrt{\frac{2.225}{73 - 2.225}} = 0.177 \quad \text{and}$$

$$K = \sqrt{\frac{M(V_0 - W)}{\hbar^2}} = \frac{1}{\hbar c}\sqrt{Mc^2(V_0 - W)} = \frac{\sqrt{939(73 - 2.225)}}{1.9732 \times 10^{-11}} = 1.31 \text{ fm}^{-1}$$

$$R = \frac{1}{1.31} \text{arc cot}(-0.177) = 1.33 \text{ fm}$$

where we have chosen the angle in the second quadrant corresponding to the bound ground state.

(b) Since the probability that the neutron and proton are between r and $r + dr$ apart is given by $4\pi |u|^2 dr$, the most probable value of r is the one for which the wave function is maximum. From Fig. 5.2 it is seen that this condition is determined by finding the maximum value of $u_{II} = \sin k(r - b)$. Obviously, u_{II} will be maximum when $K(r - b) = (\pi/2)$ or

$$r_{most\ probable} = \frac{T}{2K} + b = \frac{\pi}{2 \times 1.31} + 0.4 = 1.24 \text{ fm}$$

(c) With the inclusion of hard core potential of radius b in the square well potential, the mean square radius of deuteron can be calculated similar to the treatment given in Sect. 5.5

$$\langle r_d^2 \rangle = \frac{A^2}{4} \int_0^{b+R} r^2 \sin^2 K(r - b) dr + \frac{C^2}{4} \int_{b+R}^{\infty} r^2 e^{-2\gamma r} dr \tag{5.52}$$

where the radial functions u_{II} and u_{III} defined by (5.47) and (5.48) have been used. The constants A^2 and C^2 are determined from the normalization condition:

$$A^2 \int_a^{a+b} \sin^2 K(r-b)dr + C^2 \int_{b+R}^{\infty} e^{-2\gamma r}dr = 1 \qquad (5.53)$$

together with the relation (5.49) or (5.50). We obtain

$$A^2 = \frac{2\gamma}{1+\gamma R} \qquad (5.54)$$

$$C^2 = \frac{2\gamma \sin^2 KR}{1+\gamma R} \exp\left[2\gamma(b+R)\right] \qquad (5.55)$$

The integrals in (5.52) can be readily evaluated by partial integrations. The values of A^2 and C^2 are substituted from (5.54) and (5.55) and the trigonometric functions are removed by the use of (5.51). After some algebraic manipulations we finally obtain the expression

$$\langle r_d^2 \rangle = \frac{1}{8\gamma^2} - \frac{1}{8K^2} + \frac{(2+R)(1+\gamma R)}{8\gamma} + \frac{b^2}{4} - \frac{\gamma R^3}{24(1+\gamma R)} + \langle r_p^2 \rangle \qquad (5.56)$$

As before, the quantity $\langle r_p^2 \rangle$ has been added to correct for the charge distribution of proton. In the limit $b \to 0$ expression (5.56) reduces as it should, to (5.44) which was derived without the inclusion of 'hard core'. As before, if we use the experimental values, we find

$$\sqrt{\langle r_d^2 \rangle} = 2.1 \text{ fm}, \qquad \sqrt{\langle r_p^2 \rangle} = 0.8 \text{ fm}, \qquad b = 0.4 \text{ fm}, \qquad \gamma = 0.232 \text{ fm}^{-1}$$

Equation (5.51) provides another relation between R and K. Solution of Eqs. (5.56) and (5.51), yields $R = 1.337$ fm and $K = 1.31$ fm which implies $V_0 = 73$ MeV.

5.1.7 Use of the Exponential Wave Function in the Solution of a Square Well Potential Problem

If we make the drastic approximation that the outside wave function of the form (5.16) may be used in the inside region as well then calculations are considerably simplified. The approximate wave function together with the actual wave functions are shown in Fig. 5.3. It is seen that most of the area under the curve is contributed from the region II, $r > R$. Although, the wave function $e^{-\gamma r}$ when extended into region I does not satisfy the boundary condition $U_I \to 0$ as $r \to 0$, but a normalized wave function of the exponential form adopted for the entire region $0 < r < \infty$, would atleast give results of a qualitative nature. Its use in crude calculations is found justified regardless of the details of potential provided the forces decrease rapidly with distance.

Fig. 5.3 Exponential wave function

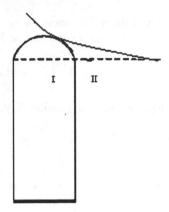

Example 5.2 Find the root mean square distance of separation of neutron and proton in deuteron under the assumption that the ground state can be described by the approximate wave function $\psi = (1/r)\sqrt{\gamma/2\pi}\,e^{-\gamma r}$, with $(\frac{1}{\gamma}) = 4.3$ fm. Further estimate the error by the use of this approximation.

Solution   Since the wave function is normalized the mean square distance of separation of the neutron and proton is given by

$$\langle r^2 \rangle = \int_0^\infty 4\pi r^2 |\psi|^2 r^2 dr = 2\gamma \int_0^\infty r^2 e^{-2\gamma r} dr = \frac{1}{2\gamma^2}$$

The root mean square distance of separation is given by

$$\sqrt{\langle r^2 \rangle} = \frac{1}{\sqrt{2}\gamma} = \frac{4.3}{\sqrt{2}} = 3.0 \text{ fm} \tag{5.57}$$

We can estimate the fractional error introduced through this approximation. The wave function can be written as:

$$u = Be^{-\gamma r} \tag{5.58}$$

$$\text{with } B = \sqrt{\frac{\gamma}{2\pi}} \tag{5.59}$$

This is to be compared with the more exact constant for the square well potential (5.31)

$$C = \sqrt{\frac{\gamma}{2\pi}}\,\frac{e^{\gamma R}}{\sqrt{1+\gamma R}}\sin KR \simeq \sqrt{\frac{\gamma}{2\pi}}\,\frac{e^{\gamma r}}{\sqrt{1+\gamma R}} \tag{5.60}$$

where the approximation, $KR \simeq \pi/2$ has been used in the simplification; if $\gamma R \ll 1$, then approximately we have the result:

$$\frac{C}{B} = \frac{e^{\gamma R}}{\sqrt{1+\gamma R}} = (1+\gamma R+\cdots)\left(1-\frac{\gamma R}{2}+\cdots\right) \simeq 1+\frac{\gamma R}{2}$$

This shows that the constant B used in this approximation is smaller than C used for the square well potential. The fractional error incurred is,

$$\frac{C - B}{B} \simeq \frac{\gamma R}{2} \simeq \frac{0.232 \times 2.0}{2} = 0.23 \simeq 23 \%$$

Inspite of a large error on B, the qualitative conclusions deduced earlier by the use of more exact wave function, remain unaltered. For example, the root mean square distance of separation given by (5.57) implies a value of about 3 fm (for $\gamma = 0.232$), which is much larger than the range of nuclear forces ($R \sim 2$ fm). Furthermore, when the value of γ given by (5.12) is used in (5.57), we find that $\sqrt{\langle r^2 \rangle} \propto (1/\sqrt{W})$. Thus, a low value of the binding energy implies a loose structure expressed by the large magnitude of the root mean-square distance of separation.

5.1.8 Magnetic Dipole Moment of Deuteron

Hitherto, the ground state has taken no account of the fact that the total angular momentum has the value 1. It was supposed that the ground state is an S-state, i.e. the orbital angular momentum $l = 0$. The total angular momentum then arises exclusively from the combined intrinsic spins of the nucleons aligned parallel (triplet). In the spectroscopic notation the ground state is designated as $^3S_1$ state. The potential which has been studied in the preceding sections is the triplet potential. The magnetic moment of deuteron is contributed partly by the intrinsic moments of the nucleons and partly by the orbital motion. Then in the S-state for the parallel spins the magnetic moment of deuteron is expected to be the sum of intrinsic magnet moments of the nucleons, $\mu_p + \mu_n$, since in the S-state proton does not contribute to the magnetic moment from its orbital angular momentum and neutron because of absence of charge does not contribute to the magnetic moment from any orbital state for that matter. Now, the measured magnetic moments are, $\mu_p = 2.792716$ and $\mu_n = -1.913148$ nuclear magnetons, so that $\mu_p + \mu_n = 0.87961$ nm which is to be compared with the experimentally measured value, $\mu_d = 0.85739$ nm.

The discrepancy of about 0.022 nm, is completely outside the experimental errors. We, therefore, conclude that the deuteron magnetic moment is not exactly given by the simple addition of the neutron and proton intrinsic magnetic moments in the $^3S_1$ state.

Because the discrepancy between the expected and measured value of μ_d is small, it follows that to a good approximation, the ground state of deuteron is essentially the, $^3S_1$ state with a small admixture of other states. Since the total spin of the neutron and proton can take on values 0 (singlet) or 1 (triplet), and the total angular momentum $J = 1$, then in accordance with the rule of vector addition of angular momentum, the orbital angular momentum values are limited to $l = 0, 1$ and 2. Apart from the $^3S_1$ state, the other conceivable states are $^1P_1$ and $^3P_1$ and $^3D_1$ states. However, since the P states have odd parity their mixture with even parity states cannot be considered. We, are therefore left with the mixture of $^3S_1$, and

$^3D_1$, states. The wave function can be written as

$$\psi = a\psi_S + b\psi_D \tag{5.61}$$

where ψ_S and ψ_D are the $^3S_1$ and $^3D_1$ functions respectively. We shall now estimate the contribution to the magnetic moment from the orbital motion of the proton in the D state in the presence of non-central forces (with central forces, states of different values of l cannot be mixed).

The intrinsic magnetic moment operator of proton is given by $M_p = \mu_p \sigma_p$, where σ_p is the Pauli spin operator. Similarly, $M_n = \mu_n \sigma_n$.

Assuming additivity of moments the deuteron moment operator is given by

$$M = \mu_0 \sigma_p + \mu_n \sigma_n + L_p \tag{5.62}$$

Where $L_p = (1/2)L$, is the angular momentum of the proton relative to the centre of mass of the system. As already mentioned, neutron does not contribute to the magnetic moment by virtue of orbital motion. We can consequently rewrite (5.62) as follows:

$$M = \frac{1}{2}(\mu_n + \mu_p)(\sigma_n + \sigma_p) + \frac{1}{2}(\mu_n - \mu_p)(\sigma_n - \sigma_p) + \frac{L}{2} \tag{5.63}$$

where $S = (1/2)(\sigma_p + \sigma_n)$ is the total spin angular momentum operator. Now, the anti-symmetric operator $\sigma_p + \sigma_n$ transforms a singlet spin state (anti-symmetric) into a triplet (symmetric) spin state, and a triplet into a singlet spin state. Therefore, both in the triplet and singlet spin states, the eigen value of $\sigma_p + \sigma_n$ is zero. We can then write

$$M = (\mu_n + \mu_p)S + \frac{1}{2}L$$

$$= (\mu_n + \mu_p)J - \left(\mu_n + \mu_p - \frac{1}{2}\right)L \tag{5.64}$$

since the total angular momentum $\mathbf{J} = \mathbf{L} + \mathbf{S}$. The magnetic moment of deuteron is given by the expectation value of the operator M in the state for which the total angular momentum has maximum projection along the z-axis

$$\langle M_z \rangle = \frac{\langle (\mathbf{M} \cdot \mathbf{J}) J_z \rangle}{J(J+1)} \tag{5.65}$$

since $J_z = J = 1$ and $J^2 = j(j+1)$ and $\mathbf{L} \cdot \mathbf{J} = \frac{j(j+1)+l(l+1)-s(s+1)}{2}$

$$\langle M_z \rangle = (\mu_n + \mu_p) - \left(\mu_n + \mu_p - \frac{1}{2}\right)\left\{\frac{j(j+1) + l(l+1) - s(s+1)}{2j(j+1)}\right\} \tag{5.66}$$

We give below the value of the expectation value of the deuteron magnetic moment in nuclear magnetons for various states calculated from (5.66). Using the spectroscopic notation $2S + 1_{L_J}$

$$^3S_1; \quad (j = 1, l = 0, s = 1); \quad \langle M_z \rangle = \mu_p + \mu_n = 0.879$$

$$^1P_1; \quad (j = 1, l = 1, s = 0); \quad \langle M_z \rangle = 0.5$$

$$^3P_1; \quad (j = 1, l = 1, s = 1); \quad \langle M_z \rangle = \frac{1}{2}\left(\mu_p + \mu_n + \frac{1}{2}\right) = 0.689$$

$$^3D_1; \quad (j = 1, l = 2, s = 1); \quad \langle M_z \rangle = \frac{3}{4} - \frac{1}{2}(\mu_p + \mu_n) = 0.310$$

Note that each of the last three states if considered as pure states would make the discrepancy in the expected and observed magnetic moment of deuteron still worse. On the other hand, if we consider the deuteron moment to arise from the mixture of S and D states, then

$$\langle M_z \rangle = |a|^2 \langle M_z \rangle_s + |b|^2 \langle M_z \rangle_D \quad \text{or}$$

$$0.85739 = 0.879|a|^2 + 0.310|b|^2 \tag{5.67}$$

The quantities $|a|^2$ and $|b|^2$ are the probabilities of finding the system in the S and D state, respectively. We have another relation provided by the normalization condition:

$$|a|^2 + |b|^2 = 1 \tag{5.68}$$

Solving (5.67) and (5.68), we find, $p_D = |b|^2 = 0.039$, i.e. there is about 4 % probability for the system to be found in the D state. It must be pointed out that relativistic corrections have been ignored in the foregoing analysis. Furthermore, it is plausible that the intrinsic magnetic moments of the nucleons may be altered by the meson field with which they interact. Considering various, uncertainties, it is reasonable to conclude that p_D lies between 2 percent and 6 percent.

5.1.9 Tensor Force

We have seen that the binding energy and angular momentum are consistent with the assumption of central forces and that the deuteron is essentially in the $^3S_1$ state. But an S state implies a spherically symmetric wave function which gives uniform density distribution and has no angular dependence. Such a state cannot account for the quadrupole moment of the deuteron. In 1939, the observed fine structure of radio-frequency magnetic resonance spectrum of deuterium revealed that it could be explained only by ascribing a non-spherical charge distribution to the deuteron (Rabi and Nordsieck). The quadrupole moment corresponding to this charge distribution results in an additional energy $E = -(1/4)(\partial E/\partial z)Q(\psi)$, to the deuteron in the inhomogeneous electric field of the molecule. A value of the quadrupole moment

$Q = (2.74 \pm 0.02) \times 10^{-27}$ cm^2 was found necessary to explain the experimental results. The existence of quadrupole moment of deuteron shows that the ground state of deuteron is not a pure $^3S_1$ state. On the other hand, the fact that the magnitude of the quadrupole moment is small implies only a small admixture of higher l-states. This can be seen by comparing the magnitude of the quadrupole moment with the mean square radius of the deuteron which is of the order of 2×10^{-25} cm^2. Thus Q is two orders of magnitude smaller than the mean square radius, implying thereby that the ground state is essentially spherically symmetrical and only slightly distorted by higher angular momentum states. It does not of course follow that the non-central part of the force is also very small. For, a very appreciable, non-central part of the force will, in general, cause a relatively small asymmetry of the wave function corresponding to the ground state.

While the central force depends only on the distance between the particles and the spin alignment, the tensor (non-central) force depends on the angles between the spin directions and the line joining the particles. Under the central forces, the magnitude of angular momentum L is a constant of motion, since the orbital angular momentum is conserved if the potential is solely dependent on r. On the other hand, with non-central forces the total angular momentum is a constant of motion but L is not. Since parity is a good quantum number, states of the same parity belonging to different values of l (e.g. $^3S_1\,^3D_1$) but the same value of total angular momentum J may be combined together. Thus the tensor force can explain the quadrupole moment.

We now consider the general form of the potential. We shall again assume that the nuclear forces are derivable from a potential and that they are velocity independent. The potential must include apart from the relative position vector $\mathbf{r}$ the spin coordinates σ_n and σ_p in order to account for the quadrupole moment. The choice of the potential has a restricted form because of the requirement that it must be invariant under rotations and reflections of the coordinate system. In other words, it must be a scalar. Thus, the number of conceivable potentials is limited due to the following conditions.

1. The vector $\mathbf{r}$ changes sign under inversion (reflection followed by rotation), and hence can occur only in even powers.
2. The vectors σ_n and σ_p remain unchanged under reflection since they transform like angular momentum under reflection $\mathbf{r} \times \mathbf{p} \rightarrow (-\mathbf{r}) \times (-\mathbf{p})$.
3. The components of σ_n and σ_p are not invariant under rotation but $\sigma_n \cdot \sigma_p$ is.
4. Higher powers of σ_n and σ_p can always be reduced to the first power by applying the commutation rules for the Pauli operators.
5. Derivatives of $\mathbf{r}$ must not occur since the velocity dependent forces have been excluded.
6. $(\sigma \cdot \mathbf{r})$ is invariant under rotation but not under inversion. We must therefore have even moments of $\sigma \cdot \mathbf{r}$ such as $(\sigma_n \cdot \mathbf{r})(\sigma_p \cdot \mathbf{r})$.

The number of scalars satisfying all these conditions are limited to the terms:

$$V(r); \quad \sigma_n \cdot \sigma_p; \quad (\sigma_n \cdot \mathbf{r})(\sigma_p \cdot \mathbf{r}); \quad (\sigma_n \times \mathbf{r}) \cdot (\sigma_p \times \mathbf{r}) \qquad (5.69)$$

or their products.

The last one can be simplified by the vector identity:

$$(\mathbf{a} \times \mathbf{b}) \cdot (\mathbf{c} \times \mathbf{d}) = (\mathbf{b} \cdot \mathbf{d})(\mathbf{a} \cdot \mathbf{c}) - (\mathbf{b} \cdot \mathbf{c})(\mathbf{a} \cdot \mathbf{d}) \quad \text{or}$$

$$(\boldsymbol{\sigma}_n \times \mathbf{r}) \cdot (\boldsymbol{\sigma}_p \times \mathbf{r}) = r^2 \boldsymbol{\sigma}_n \cdot \boldsymbol{\sigma}_p \rightarrow -(\boldsymbol{\sigma}_n \cdot \mathbf{r})(\boldsymbol{\sigma}_p \cdot \mathbf{r})$$

and can, therefore, be represented in terms of the remaining three.

It can be readily shown that terms with higher powers like $(\boldsymbol{\sigma} \cdot \mathbf{r})^2$ can be reduced to those already assumed. Thus

$$(\boldsymbol{\sigma} \cdot \mathbf{r})^2 = (\sigma_x x + \sigma_y y + \sigma_z z)^2$$

$$= \sigma_x^2 x^2 + \sigma_y^2 y^2 + \sigma_z^2 z^2 + (\sigma_x \sigma_y + \sigma_y \sigma_x)xy$$

$$+ (\sigma_y \sigma_z + \sigma_z \sigma_y)yz + (\sigma_z \sigma_x + \sigma_x \sigma_z)zx$$

But

$$\sigma_x^2 = \sigma_y^2 = \sigma_z^2 = 1;$$

$$(\sigma_x \sigma_y + \sigma_y \sigma_x) = (\sigma_y \sigma_z + \sigma_z \sigma_y) = (\sigma_z \sigma_x + \sigma_x \sigma_z) = 0$$

$$\therefore \quad (\boldsymbol{\sigma} \cdot \mathbf{r})^2 = x^2 + y^2 + z^2 = r^2$$

The first two terms of the potential (5.69) are invariant not only under combined rotation of space and spin coordinates but also under separate rotations of these coordinates. Such potentials are called central potentials. However, the third potential is different in that it couples the space and spin coordinates of the particles belonging to the two-body system, and hence to the orbital and spin angular momenta, with the result, the orbital angular momentum is no longer a constant of motion, although the total angular momentum is a constant of motion. Such a potential is called tensor or non-central potential.

It is convenient to define the non-central potential in such a way that it vanishes when averaged over all directions. Now

$$\frac{1}{4\pi} \int (\boldsymbol{\sigma}_n \cdot \mathbf{r})(\boldsymbol{\sigma}_p \cdot \mathbf{r}) d\Omega = \frac{1}{4\pi} \int [\sigma_{nx} x + \sigma_{ny} y + \sigma_{nz} z][\sigma_{px} x + \sigma_{py} y + \sigma_{pz} z] d\Omega$$

$$= \frac{1}{4\pi} \sigma_{nx} \sigma_{px} \int x^2 d\Omega + \frac{1}{4\pi} \sigma_{ny} \sigma_{py} \int y^2 d\Omega$$

$$+ \frac{1}{4\pi} \sigma_{nz} \sigma_{pz} \int z^2 d\Omega$$

The cross products terms xy, yz, zx vanish upon integration. Using the substitutions $x = r \cos\phi \sin\theta$, $y = \sin\phi \sin\theta$, $z = r \cos\theta$; $d\Omega = d(\cos\theta)d\phi$, each of the integrals gives the result $(4\pi/3)r^2$

$$\frac{1}{4\pi} \int (\boldsymbol{\sigma}_n \cdot \mathbf{r})(\boldsymbol{\sigma}_p \cdot \mathbf{r}) d\Omega = \frac{1}{3} r^2 \boldsymbol{\sigma}_n \cdot \boldsymbol{\sigma}_p$$

Hence we define the tensor operator

$$S_{np} = \frac{3}{r^2}(\boldsymbol{\sigma}_n \cdot \mathbf{r})(\boldsymbol{\sigma}_p \cdot \mathbf{r}) - \boldsymbol{\sigma}_n \cdot \boldsymbol{\sigma}_p \tag{5.70}$$

We may rewrite (5.70) as:

$$S_{np} = 3(\boldsymbol{\sigma}_n \cdot \mathbf{e}_r)(\boldsymbol{\sigma}_p \cdot \mathbf{e}_r) - \boldsymbol{\sigma}_n \cdot \boldsymbol{\sigma}_p \tag{5.71}$$

where e_r is a unit vector along the vector r. The non-central potential then has the form $V = V_T(r)S_{np}$. The complete potential is given by

$$V = V_R(r) + V_\sigma(r)\sigma_1\sigma_2 + V_T(r)S_{np} \tag{5.72}$$

The subscript T has been used to denote the tensor interaction, which is actually a scalar product of two second-rank tensors. The first two terms comprise the central potential V_C; the second term is the spin-dependent part so that it allows for the fact that the central potential for triplet and singlet states is different as in low energy neutron-proton scattering. We can then write

$$V = V_C + V_T(r)S_{np} \tag{5.73}$$

$$\text{with } V_C = V_R(r) + V_\sigma(r)\sigma_n\sigma_p \tag{5.74}$$

Now

$$\sigma_n\sigma_p = +1 \quad \text{for the triplet state}$$
$$= -3 \quad \text{for the singlet state}$$

Consequently

$$V_C(\text{trip}) = V_R(r) + V_\sigma(r) \tag{5.75}$$

$$V_C(\text{sing}) = V_R(r) - 3V_\sigma(r) \tag{5.76}$$

It is seen that the triplet state is lower than the singlet.

The operator S_{12} has been defined in such a way that its average over all directions is zero. In the singlet state, the spins have no preferential direction and so we expect S_{np} to be zero. That this is so can be easily verified

$$S = \frac{1}{2}(\sigma_n + \sigma_p) = 0, \quad \text{implies that } \sigma_n = -\sigma_p$$

$$\therefore \quad S = -3(\sigma_p\sigma_r)^2 + \sigma_p\sigma_p = -3 + 3 = 0$$

5.1.10 Constants of Motion for the Two-Body System

Since central forces are invariant under rotation of space and spin coordinates separately, L^2, S^2, L_3, S_z are constants of motion. The non-central forces are invariant only under coupled rotation of space and spin so that the total angular momentum is a constant of motion. That L and S separately are not constants of motion can be demonstrated from the fact that the potential $V_T(r)S_{np}$ does not remain invariant under a rotation of space coordinates or spin coordinates separately. Another constant of motion is the parity. The operator S_{np} has been constructed in such a way as to remain invariant under space inversion. In the special case of two particles of spin $(1/2)$ it turns out that S^2 is a constant of motion even under non-central forces. To demonstrate this, examine the behavior of the potential $V_T(r)S_{np}$ under the exchange of σ_n and σ_p. Clearly it remains unchanged. This then means that the states associated with the system must be either symmetric or anti-symmetric with respect to the spin exchange. But in our special case of two $(1/2)$ spin particles, only two states are possible, the symmetric triplet state and the anti symmetric singlet state. Hence, the classification into symmetric and anti-symmetric states under the exchange of spin coordinates of the two particles is actually a classification into triplet and singlet states, proving thereby S^2 is a constant of motion. Thus I^2, I_3, P, S^2 are constants of motion.

We may classify the states according to whether they are singlet or triplet. States with even (odd) parity will consist of a linear combination of even (odd) l states. It is apparent that the non central forces do not affect the singlet states.

We now use the normalized "Spin angle" wave function $y_{I_{l,s}}^{I_3}$ belonging to state of total angular momentum I whose z-component is I_z, and which results from the combination of orbital angular momentum l and a spin s. Expanding $y_{I_{l,s}}^{I_z}$ by the use of Clebsch-Gordon coefficients:

$$\int_{I_{l,s}}^z = \sum_{m_l+m_s=I_z} C_{ls}(I, I_z; m_l, m_s)Y_{l,m_l}\chi_{sm_s} \tag{5.77}$$

From the tables of Clebsch-Gordon coefficients, for $I = I_z = s = 1$, and $l = 0$ and 2, we find respectively.

Normalized wave function for

$$^3S_1 \text{ state} \quad \Rightarrow \quad y_{101}^1 = Y_{0,0}\chi_{11} = \frac{1}{\sqrt{4\pi}}\chi_{11} \tag{5.78}$$

Normalized wave function for

$$^3D_1 \text{ state} \quad \Rightarrow \quad y_{121}^1 = \sqrt{\frac{6}{10}}Y_{2,2}\chi_{1,-1} - \sqrt{\frac{3}{10}}Y_{2,1}\chi_{1,0} + \sqrt{\frac{1}{10}}Y_{2,0}\chi_{1,1} \tag{5.79}$$

We have arbitrarily assumed $I_z = 1$, since I_z is a constant of motion and there is no preferred direction. Table 5.1 shows that the application of the tensor force operator S_{np} on the $^3S_1$ wave function can lead only to a linear combination of $^3S_1$ and $^3D_1$ wave functions. Thus

$$S_{np}y_{101}^1 = Ay_{101}^1 + By_{121}^1 \tag{5.80}$$

Table 5.1 Application of the tensor force operator S_{np} on the $^3S_1$ wave function

I	Singlet ($s = 0$)		Triplet ($s = 1$)	
	Even parity	Odd parity	Even parity	Odd parity
0	$^1S_0$	–	–	$^3p_0$
1	–	$^1p_1$	$^3S_1, {}^3D_1$	$^3p_1$
2	$^1D_2$	–	$^3D_2$	$^3p_2, {}^3F_2$

Table 5.2 First few spherical harmonics

$$Y_{0,0} = \frac{1}{\sqrt{4\pi}} \qquad\qquad Y_{2,2} = \sqrt{\tfrac{15}{32\pi}}\, \sin^2\theta e^{2i\phi}$$

$$Y_{1,1} = -\sqrt{\tfrac{3}{8\pi}}\, \sin\theta e^{i\phi} \qquad Y_{2,1} = -\sqrt{\tfrac{15}{8\pi}}\, \sin\theta \cos\theta e^{i\phi}$$

$$Y_{1,0} = \sqrt{\tfrac{3}{4\pi}}\, \cos\theta \qquad\qquad Y_{2,0} = \sqrt{\tfrac{5}{4\pi}}(\tfrac{3}{2}\cos^2\theta - \tfrac{1}{2})$$

$$Y_{1,-1} = \sqrt{\tfrac{3}{8\pi}}\, \sin\theta e^{-i\phi} \qquad Y_{2,-1} = \sqrt{\tfrac{15}{8\pi}}\, \sin\theta \cos\theta e^{-i\phi}$$

$$Y_{2,-2} = \sqrt{\tfrac{15}{32\pi}}\, \sin^2\theta e^{-2i\phi}$$

where A and B are constants. Now, on averaging over all directions in space, we find using the orthonormal property of spherical harmonics

$$\text{av}\quad S_{12}y^1_{101} = 0, \qquad \text{av}\quad y^1_{101} \neq 0, \qquad \text{av}\quad y^1_{121} = 0$$

It follows that $A = 0$. One way of evaluating B is to consider a special case in which r points in the z direction so that $\theta = 0$. First few spherical harmonics are listed in Table 5.2.

Since r is pointing in the z-direction $S_{np} = 3\sigma_{nz}\sigma_{pz} - \sigma_n\sigma_p = 3 - 1 = 2$, where we have used the fact that in the triplet state

$$\sigma_{nz} = \sigma_{pz} \quad \text{so that } \sigma_{nz}\sigma_{pz} = \sigma_{nz}^2 = 1 \text{ and } \sigma_n\sigma_p = +1$$

Direct evaluation of (5.80) with $A = 0$, yields

$$\frac{2}{\sqrt{4\pi}}\chi_{11} = By^1_{121}(\theta = 0) = B\sqrt{\frac{1}{10}}\sqrt{\frac{5}{4\pi}}\chi_{11}$$

or $B = \sqrt{8}$

$$S_{np}y^1_{101} = \sqrt{8}y^1_{121} \tag{5.81}$$

Similar to (5.80) we must have

$$S_{np}y^1_{121} = By^1_{101} + Cy^1_{121} \tag{5.82}$$

where B is the same B appearing in (5.80) since the tensor operator S_{12} is Hermetian. Using the values, $B = \sqrt{8}$, $S_{12} = 2$, $y^1_{12}(\theta = 0) = \sqrt{(1/8\pi)}$, $y^1_{101} = \sqrt{(1/4\pi)}$

in (5.82)

$$2\sqrt{\frac{1}{8\pi}}\chi_{11} = \sqrt{8}\sqrt{\frac{1}{4\pi}}\chi_{11} + c\sqrt{\frac{1}{8\pi}}\chi_{11}$$

or $c = -2$

$$S_{12}y_{121}^1 = \sqrt{8}y_{101}^1 - 2y_{121}^1 \tag{5.83}$$

We shall now write the complete wave functions for the S and D states with space and spin dependence. Corresponding to the S-state

$$\phi_S = \frac{u(r)}{r}y_{101}^1 \tag{5.84}$$

where $u^2(r)dr$ is the probability of finding neutron and proton in the S-state between r and $r + dr$ apart

$$p_S = \int_0^\infty u^2(r)dr \tag{5.85}$$

Similarly

$$\phi_D = \frac{\omega(r)}{r}y_{121}^1 \tag{5.86}$$

Corresponding to the D-state, where $\omega^2(r)dr$ represents the probability of finding neutron and proton at a distance between r and $r + dr$ apart

$$p_D = \int_0^\infty \omega^2(r)dr$$

$$p_S + p_D = \int_0^\infty [u^2(r) + \omega^2(r)]dr = 1 \tag{5.87}$$

The mixed S and D state, corresponding to the deuteron ground state can then be written as:

$$\phi = \phi_S + \phi_D \tag{5.88}$$

$$\left\{\frac{\hbar^2}{M}\nabla^2 + E - V_C(r) - V_T(r)S_{np}\right\}\phi = 0 \tag{5.89}$$

Now, the Laplacian has the value

$$\nabla^2\phi = \frac{1}{r}\frac{d^2}{dr^2}(r\phi) - \frac{l(l+1)\phi}{r^2} \tag{5.90}$$

with $l(l+1) = 0$ for the s-state

$$= 6 \quad \text{for the } D\text{-state} \tag{5.91}$$

Using (5.88), (5.90), (5.91) and (5.84) and (5.86) in (5.89), we find

$$\left\{\frac{\hbar^2}{Mr}\frac{d^2}{dr^2} + \frac{E}{r} - \frac{V_C(r)}{r}\right\}u(r)y_{101}^1 - V_T(r)S_{np}\left\{\frac{u(r)y_{101}^1}{r} + \frac{\omega(r)y_{121}^1}{r}\right\}$$

$$+ \left\{\frac{\hbar^2}{Mr}\left(\frac{d^2}{dr^2} - \frac{6}{r^2}\right) + \frac{E}{r} - \frac{V_C}{r}\right\}\omega(r)y_{121}^1 = 0 \qquad (5.92)$$

Upon using (5.81) and (5.83) in (5.92), we find

$$\left\{\frac{\hbar^2}{M}\frac{d^2}{dr^2} + E - V_C(r)\right\}u(r)y_{101}^1 - \sqrt{8}y_{121}^1 u(r) - V_T(r)\left\{\sqrt{8}y_{101}^1 - 2y_{121}^1\right\}\omega(r)$$

$$+ \left\{\frac{\hbar^2}{M}\left(\frac{d^2}{dr^2} - \frac{6}{r^2}\right) + E - V_C(r)\right\}\omega(r)y_{121}^1 = 0 \qquad (5.93)$$

This equation must be valid for all the angles θ and ϕ, and therefore the coefficients of y_{101}^1 and y_{121}^1 must vanish separately. The following system of two coupled differential equations in $u(r)$ and $\omega(r)$ are obtained:

$$\frac{\hbar^2}{M}\frac{d^2u}{dr^2} - V_C(r)u + Eu = \sqrt{8}V_T(r)\omega \qquad (5.94)$$

$$\frac{\hbar^2}{M}\left(\frac{d^2\omega}{dr^2} - \frac{6\omega}{r^2}\right) - \left[V_C(r) - 2V_T(r)\right]\omega + E\omega = \sqrt{8}V_T(r)u \qquad (5.95)$$

These equations were first derived by Schwinger and Rarita. The total energy E is negative and is equal to $-W$, corresponding to the ground state. The above coupled equations do not lend themselves to exact solutions. Detailed numerical solutions exist only for the cases where both the potentials V_C and V_T have the square well shape or both of them have Yukawa well shape. Calculations are rendered difficult not only because of the nature of the coupled differential equations, but also owing to the increased number of parameters. In fact, there are four parameters, the depth and range for each of the potentials $V_C(r)$ and $V_T(r)$ which are to be adjusted. In principle, the four sources of experimental data viz., the binding energy of deuteron, the quadrupole moment, the magnetic moment, and the effective range of neutron-proton scattering in the triplet state, enable the four parameters to be fixed up. But in practice, these parameters cannot be determined uniquely. Indeed, there are a large sets of well parameters which are in agreement with the known experimental data on deuteron. Thus, for example, the strength of the tensor force can be increased with a corresponding decrease in the strength of the central force, without contradicting the known properties of the deuteron.

5.1.11 Quadrupole Moment

The electric quadrupole moment of deuteron can be explained by invoking for the tensor forces which depend on the angle θ between the line joining the particles and the axis of the total spin. Pure central forces offer no solution as states with different L cannot mix. By definition

$$Q = \int_0^\infty \phi_{I,I_z}^* r^2 (3\cos^2\theta - 1)\phi_{I,I_z}d\tau$$

$$= 2\sqrt{\frac{4\pi}{5}} \int_0^\infty \phi_{I,I_z}^* r^2 Y_{2,0}(\theta)\phi_{I,I_z}d\tau \tag{5.96}$$

where r is the distance of the proton from the centre of mass of the deuteron. However, it is desirable to measure the distance relative to neutron, so that the wave functions which have been previously derived can be readily used in the calculations. Thus, replacing r by $r/2$ in (5.96)

$$Q = \sqrt{\frac{\pi}{5}} \int_0^\infty \phi_{I,I_z}^* r^2 Y_{2,0}(\theta)\phi_{I,I_z}d\tau \tag{5.97}$$

As the ground state of the deuteron is assumed to be the mixture of 3 $^3S_1$ and $^3D_1$ states, the expectation value of Q will be given by,

$$(\theta, Q\phi) = (\phi_S, Q\theta_S) + 2(\phi_S, Q\phi_D) + (\phi_D, Q\phi_D)$$

The first term is zero, since the s-state is spherically symmetrical and cannot give rise to a quadrupole moment. This can also be checked by the direct evaluation of the integral:

$$(\phi_S, Q\phi_S) = \sqrt{\frac{\pi}{5}} \int_0^\infty \phi_S^* r^2 Y_{2,0}(\phi)\tau \tag{5.98}$$

writing $d\tau = d\Omega r^2 dr$ and using the value

$$\phi_S = \frac{u(r)}{r} y_{101}^1 = \frac{u(r)}{r} \frac{\chi_{11}}{\sqrt{4\pi}},$$

$$(\phi_S, Q\phi_S) = \sqrt{\frac{1}{80\pi}} \int_0^\infty r^2 u^2(r)dr \chi_{11}^2 \int_0^\infty Y_{2,0}(\theta)d\Omega = 0 \tag{5.99}$$

since $\int_0^\infty Y_{2,0}(\theta)d\Omega$ vanishes.

The second term can be evaluated as follows:

$$2(\phi_S, Q\phi_D) = 2\int_0^\infty r^2 dr \int_0^\infty d\Omega \frac{u(r)}{r} y_{101}^1 Y_{2,0}(\theta)\frac{\omega(r)}{r} Y_{121}^1 r^2$$

$$= 2\sqrt{\frac{\pi}{5}} \int_0^\infty r^2 dr \int_0^\infty \frac{\chi_{11}}{\sqrt{4\pi}} \frac{u(r)}{r} Y_{2,0}(\theta)\frac{\omega(r)}{r} r^2 \sqrt{\frac{1}{10}} Y_{2,0}(\theta)\chi_{11}d\Omega$$

where ϕ_S and ϕ_D are inserted from (5.84) and (5.85), respectively. Terms involving cross products of spin eigen functions are dropped off since they are orthonormal.

The only terms which survive are those which involve square of the spin eigen function χ_{11}. But integration over spin variables yields unity because of normalization. Further

$$\int_0^\infty Y_{2,0}^*(\theta) Y_{2,0}(\theta) d\Omega = 1$$

Therefore

$$2(\phi_S, Q\phi_D) = \frac{1}{\sqrt{50}} \int_0^\infty u(r)\omega(r)dr \qquad (5.100)$$

We shall now evaluate the third term:

$$(\phi_D, Q\phi_D) = \sqrt{\frac{\pi}{5}} \int_0^\infty r^2 dr \int_0^\infty d\Omega \omega^2(r) y_{121}^{*1} Y_{2,0}(\theta) y_{121}^1$$

$$= \sqrt{\frac{\pi}{5}} \int_0^\infty r^2 \omega^2(r) dr \left\{ \frac{6}{10} \int_0^\infty Y_{2,2}^* Y_{2,0} Y_{2,2} d\Omega \right.$$

$$\left. + \frac{3}{10} \int_0^\infty Y_{2,1}^* Y_{2,0} Y_{2,1} d\Omega + \frac{1}{10} \int_0^\infty Y_{2,0}^* Y_{2,0} Y_{2,0} d\Omega \right\} \qquad (5.101)$$

Using the general result:

$$\int Y_{l_3 m_3}^* Y_{l_2 m_2} Y_{l_1 m_1} d\Omega = \sqrt{\frac{(2l_1+1)(2l_2+1)}{4\pi(2l_3+1)}} C(l_1, l_2, l_3 : m_1, m_s) C(l_1, l_2, l_3 : 0, 0)$$

$$(5.102)$$

and the Clebsch-Gordon coefficients:

$$C(2,2,2,0,0) = -\sqrt{\frac{2}{7}}; \qquad C(2,2,2,2,2) = \sqrt{\frac{2}{7}}; \qquad C(2,2,2,1,1) = -\sqrt{\frac{1}{14}}$$

$$(\phi_D, Q\phi_D) = \sqrt{\frac{\pi}{5}} \int_0^\infty r^2 \omega(r) d\Omega \sqrt{\frac{5}{4\pi}} \left\{ \frac{6}{10}\left(-\frac{2}{7}\right)\left(\frac{2}{7}\right) + \frac{3}{10}\left(-\sqrt{\frac{2}{7}}\right)\left(-\sqrt{\frac{1}{14}}\right) \right.$$

$$\left. + \frac{1}{10}\left(-\sqrt{\frac{2}{7}}\right)\left(-\sqrt{\frac{2}{7}}\right) \right\}$$

$$= -\frac{1}{20} \int_0^\infty r^2 \omega^2(r) dr \qquad (5.103)$$

Note that a pure D-state (5.103) would give a negative quadrupole moment contrary to the experiment. Adding (5.100) and (5.103), we find

$$(\phi, Q\phi) = \frac{1}{\sqrt{50}} \int_0^\infty r^2 u(r)\omega(r)dr - \frac{1}{20} \int_0^\infty r^2 \omega^2(r) dr \qquad (5.104)$$

We have seen that the deuteron ground state is predominantly an S-state. Therefore, in (5.104) the first term which is linear in $w(r)$ predominates over the second term.

Further, the contribution to the integral $\int_0^\infty r^2 u(r)\omega(r)dr$ comes mainly from the region outside the nuclear forces. Now for the outside region, both the potentials V_c and V_T vanish, and consequently, the coupled equations (5.94) and (5.95) become independent and have the solutions:

$$u(r) = N_S \exp(-\gamma r) \tag{5.105}$$

$$\omega(r) = N_D \exp(-\gamma r)\left[\frac{3}{\gamma^2 r^2} + \frac{3}{\gamma r} + 1\right] \tag{5.106}$$

substituting (5.105) and (5.106) in the first term of (5.104) we obtain

$$Q \simeq \frac{N_S N_D}{\sqrt{50}} \int_0^\infty e^{2\gamma r}\left[\frac{3}{\gamma^2 r^2} + \frac{3}{\gamma r} + 1\right] r^2 dr$$

$$= \frac{N_S N_D}{\sqrt{50}\gamma^3} \times \frac{5}{2}$$

$$= \sqrt{\frac{1}{8}}\frac{N_S N_D}{\gamma^3} \tag{5.107}$$

we can make a rough estimate of N_S by neglecting the D-state probability compared to unity, and by using the asymptotic form of $u(r)$, (5.105), throughout the region. The normalization condition is approximately given by

$$\int_0^\infty u^2(r)dr = N_S^2 \int_0^\infty e^{-2\gamma r} dr = \frac{N_S^2}{2\gamma} = 1 \quad \text{or} \tag{5.108}$$

$$N_S \simeq \sqrt{2\gamma}$$

on using (5.106) in (5.105), we find,

$$N_D \simeq 2Q\gamma^{5/2} = \frac{2Q}{R_0^{5/2}} \tag{5.109}$$

with $(1/\gamma) = R_0$, the size of deuteron. Thus, the wave function, $\omega(r)$ outside the range of nuclear forces is to the first approximation, is completely determined by the quadrupole moment.

For small values of r, $\omega(r)$ behaves roughly as the D-state function, and goes like r^3 near the origin, whilst $u(r)$ behaves like an S-state function and goes like r. However, outside the range of forces $\omega(r)$ goes roughly like r^{-2} as indicated by Eq. (5.106) (Fig. 5.4). Therefore, there will be a sharp maxima in $\omega(r)$ just outside the region $r \simeq R_T$ beyond which it goes into the asymptotic form (5.106). It is seen that most of the contribution to P_D, the D-state probability (5.87) comes from the region near $\simeq r = R_T$. On the other hand, the contribution to P_S the S-state probability, comes mainly for values of $r \simeq R_0$, the range of nuclear forces. In Fig. 5.4 is plotted the function $\omega^2(r)$ against r for two different values of R_T. The behaviour of

Fig. 5.4 Plot of function $\omega^2(r)$ and r, for two different values of R_T

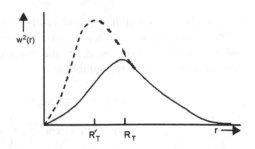

the curve for $r > R_T$ is identical. For a smaller value of R_T the maximum becomes larger. Hence, the D-state probability P_D becomes larger for a smaller range R_T. In what follows, we shall assume that $R_T \ll R_0$, where R_T is approximately the position of the maximum and R_0 is the size of the deuteron. Neglecting all the terms save the first one in the parenthesis of (5.106), setting the exponential equal to unity, and putting $R_0 = (1/\gamma)$

$$\int_{R_T}^{\infty} \omega^2(r)dr \simeq \int_{R_T}^{\infty} 9N_D^2 \left(\frac{R_0}{r}\right)^4 dr = 3N_D^2 \frac{R_0^4}{R_T^3}$$

$$\simeq \frac{12Q^2}{R_0 R_T^3} \tag{5.110}$$

where we have inserted the value of N_D from (5.109). Multiplying the integral $\int_{R_T}^{\infty} \omega^2(r)dr$ by a factor of 2, we find very approximately the value of P_D with say 20 % uncertainty from

$$P_D = \int_0^{\infty} \omega^2(r)dr \simeq 2\int_{R_T}^{\infty} \omega^2(r)dr \simeq \frac{24Q^2}{R_0 R_T^3} \tag{5.111}$$

Inserting the value of $P_D = 0.04$ obtained from the analysis of magnetic moment of deuteron, $R_0 = 4.3$ fm obtained from the binding energy of the deuteron, and from the experimental value of the quadrupole moment $Q = 0.274$ fm^2, we find from (5.111) an approximate value, $R_T = 2.2$ fm, a value which is much smaller than the size of the deuteron (4.3 fm) but is somewhat larger than the range of central forces. The fact that the measured quadrupole moment is positive implies that a stretched-out configuration like, cigar-shaped, is more favored than the prolate configuration in Fig. 5.5a. Both the figures correspond to the triplet state. In Fig. 5.5a the spins are aligned, one behind the other while in Fig. 5.5b, they are aligned side by side perpendicular to r. Now the tensor operator S_{np} in Fig. 5.5a has the value

$$S_{np} = 3(\sigma_n e)(\sigma_p e) - (\sigma_n \sigma_p) = 3\sigma_{nz}^2 - 1 = 3 - 1 = +2$$

since the spins point out in the z-direction, and the value of $\sigma_n \sigma_p = 1$ for the triplet state.

In Fig. 5.5b the spins are perpendicular. Therefore, the term $3(\sigma_n e)(\sigma_p e)$ vanishes, and consequently $S_{np} = -1$. We, therefore, conclude that S_{np} is positive in

Fig. 5.5 In the deuteron the
tensor force favours the
prolate shape
configuration (**a**) over the
oblate shape configuration (**b**)

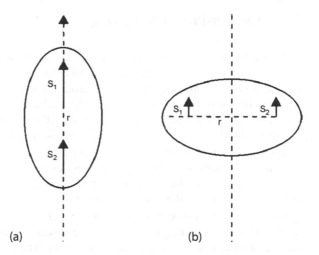

(a)   (b)

the stretched out configuration (Fig. 5.5a) and negative in the prolate configuration
(Fig. 5.5b). Therefore an attractive (negative) $V_T(r)$ will deform the deuteron wave
function from a spherical shape into a cigar-shape and because the observed quadruple moment is indeed positive, we conclude that $V_T(r)$ is attractive (negative).

To a first approximation the function $\omega(r)$ outside the range of the forces is determined completely by the value of the quadruple moment. This then implies that the
D-state probability P_D depends strongly on the tensor force range R_T and increases
rapidly as R_T is shortened.

For $r > R_T$, the curves coincide. Near the origin, $r = 0$, both the curves are proportional to $(r^3)^2 = r^6$, but with different constants. Thus there is a very sharp maximum of $\omega^2(r)$ just inside the range of the tensor force, and most of the contribution
to the integral $\int_0^\infty \omega^2(r)dr$ comes from the neighborhood of this maximum. Figure 5.4 shows that the maximum becomes rapidly raised as R_T is decreased. Thus
the D-state probability P_D for a given Q increases sharply with decreasing R_T.
Equation (5.111) implies that the tensor force cannot have an arbitrarily small range,
otherwise the ground state would become a predominantly D-state rather than a predominantly S-state. This argument was first advanced by Schwinger.

5.2 Nucleon-Nucleon Scattering: Phase Shift Analysis

5.2.1 Introduction

Study of nucleon-nucleon scattering provides valuable information on the nature
of nuclear forces. It is easier to handle the problem of neutron-proton scattering
rather than that of proton-proton scattering. The latter is complicated due to the
superposition of coulomb scattering on nuclear scattering and secondly due to the
application of the Pauli's principle on the system of two identical particles. We shall
first study the neutron-proton scattering.

5.2.2 Neutron-Proton Scattering

In the scattering problem we abandon the classical concept of well defined trajectories of particles, since according to the uncertainty principle, particles with well
defined momenta cannot be precisely localized. In quantum mechanical description
of scattering, we consider a beam of particles incident along the positive z-direction
represented by a wave packet which is essentially a plane wave which we take as
of unit amplitude, $\psi_i = e^{ikz}$, where $k = (p/\hbar)$ is the wave number and is the reciprocal of λ, the rationalized de Broglie wavelength. Here, the time dependent factor
$e^{-i\omega t}$ has been omitted for brevity. We consider the scattering in the centre of mass
of neutron and proton so that the two-body problem is essentially reduced to the
scattering of the reduced mass μ by a fixed scattering centre. The scattering caused
by the nuclear forces is represented by any suitable choice of potential $V(r)$ which
guarantees their short range character. The forces are assumed to be central, and
the scattering to possess an azimuthal symmetry. The origin is taken at the scattering centre and the scattered wave is assumed to be a spherical wave of the form
$\psi_s = (1/r)e^{ikr} f(\theta, \phi)$ where the factor $(1/r)$ accounts for the fact that the intensity of the scattered particles diminishes inversely with the square of radial distance
from the scattering centre. The factor $f(\theta, \phi)$ is in general some function of the
scattering angle θ and the azimuthal angle ϕ and is called the scattering amplitude.
By virtue of the assumption of azimuthal symmetry f will be a function of θ only.
The total wave function ψ which describes the scattering is then the superposition
of the incident wave and the scattered wave (Fig. 5.6), i.e.

$$\psi = \psi_i + \psi_s$$

$$= e^{ikz} + f(\theta)\frac{e^{ikr}}{r} \tag{5.112}$$

Note that because the scattering is assumed to be elastic, the momentum in the
C-system before and after the scattering is unchanged, and consequently the wave
number k appearing in the scattered wave ψ_s is identical with that for the incident
wave.

We shall now derive a formula for the differential cross section. The probability
of finding the scattered particles in the volume element $d\tau_s$ is equal to $|\psi_s|^2 d\tau_s$ or
$|(f(\theta)/r)|^2 \times 2\pi r^2 \sin\theta d\theta dr$, the last expression can be rewritten as $|f(\theta)|^2 d\Omega dr$,
where $d\Omega$ is the element of solid angle.

The probability current of the scattered particles, which is the probability for the
particles to be scattered in the element of solid angle $d\Omega$ per second is given by
$|f(\theta)|^2 d\Omega (dr/dt)$ or $v|f(\theta)|^2 d\Omega$ where $v = (dr/dt)$ is the velocity of the scattered as well as incident particles in the C-system, since in the elastic scattering the
velocity does not change. The rate at which the particles are scattered per unit solid
angle is then given by $v|f(\theta)|^2$. Now, the probability for finding the incident parti-
cles in the volume element $d\tau_i$ is given by $|\psi_i|^2 d\tau_i$ or simply $dx\,dy\,dz$. The proba-
bility current of incident particles, which is the rate at which the particles cross unit
area perpendicular to the incident direction, per second is given by (dz/dt) or v.

Fig. 5.6 Superposition of the incident wave and the scattered wave

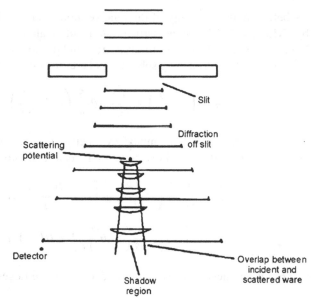

By definition, the differential cross-section is given by

$$\frac{d\sigma}{d\Omega} = \frac{\text{Number of particles scattered per unit solid angle per second}}{\text{Number of incident particles crossing unit area per second}}$$

$$= \frac{v|f(\theta)|^2}{v} = |f(\theta)|^2 \tag{5.113}$$

5.2.3 Phase-Shift Analysis

We start by writing down the Schrodinger's equation in the C-system

$$\nabla^2\psi + \frac{2\mu}{\hbar^2}[E - V(r)]\psi = 0 \tag{5.114}$$

where E is the energy of the relative motion and μ is the reduced mass. Equation (5.114) may be interpreted to represent the collision of a fictitious particle of mass μ by a scattering centre through a potential $V(r)$. The radial distance r is then the distance of separation of the particle of mass μ and the origin located at the centre of the scattering potential.

When the incident particles are well outside the range of nuclear forces, the potential $V(r)$ is zero, and Eq. (5.114) in that case essentially represents the incident wave and reduces to

$$\nabla^2\psi + k^2\psi = 0 \quad \text{with } k = \frac{\sqrt{2\mu E}}{\hbar} \tag{5.115}$$

As before E is the energy of the relative motion and is related to the Lab energy by $E(\text{Lab}) = 2E$. Since the particles are incident along the positive z-direction, the Laplacian in (5.115) reduces to $(d^2\psi/dz^2)$ and the solution as expected, is the plane wave $\psi_i = e^{ikz}$. Introducing spherical polar co-ordinates (5.115) becomes

$$\frac{1}{r^2}\frac{\partial}{\partial r}\left(r^2\frac{\partial\psi}{\partial r}\right) + \frac{1}{r^2}\sin\theta\frac{\partial}{\partial\theta}\left(\sin\theta\frac{\partial\psi}{\partial\theta}\right) + k^2\psi = 0$$

which by virtue of an azimuthal symmetry does not contain the ϕ term. Let $\psi(r,\theta) = R(r)F(\theta)$; the above equation is separated into radial and angular parts

$$\frac{1}{r^2}\frac{d}{dr}\left(r^2\frac{dR}{dr}\right) + \left[k^2 - \frac{l(l+1)}{r^2}\right]R = 0 \tag{5.116}$$

$$\frac{1}{\sin\theta}\frac{\partial}{\partial\theta}\left(\sin\theta\frac{\partial F}{\partial\theta}\right) + l(l+1)F = 0 \tag{5.117}$$

where l is an integer. The solution of (5.117) are the Legendre functions, i.e.

$$F(\theta) = P_l(\cos\theta) \tag{5.118}$$

Further, with the change of variable $x = kr$ (5.116) is easily brought in the form

$$x^2\frac{d^2R}{dx^2} + 2x\frac{dR}{dx} + \left[x^2 - l(l+1)\right]R = 0 \tag{5.119}$$

The physically acceptable solution of (5.119) are the spherical Bessel functions $j_l(x)$

$$R(x) = j_l(x) = \sqrt{\frac{\pi}{2x}}J_{l+\frac{1}{2}}(x) \tag{5.120}$$

where $J_{l+1/2}(x)$ is the ordinary Bessel functions. The other solution is the spherical Neumann function $n_l(x)$ which at small r starts as $x^{-(l+1/2)}$ and is therefore inadmissible.

The most general solution of (5.115) is then

$$\psi_i = \sum_{l=0}^{\infty} A_l j_l(kr)P_l(\cos\theta) \tag{5.121}$$

which has axial symmetry, i.e. does not involve the azimuthal angle ϕ and is finite at the origin. The A_l's are constants which need not be real. Since $\psi_i = e^{ikz}$, we have the expansion for the plane wave

$$e^{ikr\cos\theta} = \sum_{l=0}^{\infty} A_l j_l(kr)P_l(\cos\theta) \tag{5.122}$$

where we have put $z = r\cos\theta$. We shall now calculate the constants A_l's. Multiplying both sides of (5.122) by $P_l(\cos\theta)d(\cos\theta)$ and integrating between the limits -1 and $+1$

$$\int_{-1}^{+1} e^{ikr\cos\theta} P_l(\cos\theta)d(\cos\theta) = \int_{-1}^{+1} P_l(\cos\theta)d(\cos\theta)\sum_{l=0}^{\infty} A_l j_l(kr)P_l(\cos\theta) \tag{5.123}$$

We now make use of the orthonormal properties of Legendre polynomials ([6], Appendix C)

$$\int_{-1}^{+1} P_l(\cos\theta)P_{l'}(\cos\theta)d\cos(\theta) = \frac{2}{2l+1}\delta_{ll'} \tag{5.124}$$

where $\delta_{ll'}$ is the Kronecker delta. On the right side of (5.123) only one term survives upon integration

$$\int_{-1}^{+1} e^{ikr\cos\theta} P_l(\cos\theta)d(\cos\theta) = \frac{2}{2l+1}A_l j_l(kr) \tag{5.125}$$

Integrating the left side by parts

$$\frac{1}{ikr}e^{ikr\cos\theta} P_l(\cos\theta)\Big|_{-1}^{+1} - \frac{1}{ikr}\int_{-1}^{+1} e^{ikr\cos\theta} P_l'(\cos\theta)d(\cos\theta) \tag{5.126}$$

where $P_l'(\cos\theta)$ means differentiation of $P_l(\cos\theta)$ with respect to $\cos\theta$. Since we are interested in the behaviour of the wave function at large distances r from the scattering centre, the second term upon integration involves $(1/r^2)$ which may be neglected compared to the first term

$$\frac{1}{ikr}\left[e^{ikr} P_l(1) - e^{-ikr} P_l(-1)\right] = \frac{2}{2l+1}A_l j_l(kr) \tag{5.127}$$

But $P_l(1) = 1$ and $P_l(-1) = (-1)^l$

$$\therefore \quad \frac{1}{ikr}\left[e^{ikr} - (-1)^l e^{-ikr}\right] = \frac{2}{2l+1}A_l j_l(kr)$$

Further, we use the identity

$$e^{\frac{i\pi l}{2}} = i^l \tag{5.128}$$

$$A_l j_l(kr) = \frac{(2l+1)}{2ikr}i^l\left[e^{i(kr-\frac{\pi l}{2})} - e^{-i(kr-\frac{\pi l}{2})}\right]$$

$$= (2l+1)i^l\frac{\sin(kr - \frac{\pi l}{2})}{kr} \tag{5.129}$$

$$\therefore \quad \psi_i = \sum_{l=0}^{\infty}(2l+1)i^l\frac{\sin(kr - \frac{\pi l}{2})}{kr}P_l(\cos\theta) \tag{5.130}$$

This may be rewritten as

$$e^{ikz} = \frac{1}{2ikr} \sum_{l=0}^{\infty} (2l+1) P_l(\cos\theta) \left[e^{ikr} - (-1)^l e^{-ikr} \right] \tag{5.131}$$

This expresses the unscattered wave as a superposition of outgoing and ingoing spherical waves. Now, the total wave function ψ is the solution of (5.114), which may be rewritten as

$$\nabla^2 \psi + \left[k^2 - \frac{2\mu V(r)}{\hbar^2} \right] \psi = 0 \tag{5.132}$$

Upon introducing the spherical polar coordinates, the solution is found to be

$$\psi = \sum_{l=0}^{\infty} B_l g_l(kr) P_l(\cos\theta) \tag{5.133}$$

where $g(kr)$ satisfies the differential equation

$$\frac{1}{r^2} \frac{d}{dr} \left(r^2 \frac{dg}{dr} \right) + \left\{ k^2 - \frac{2\mu V(r)}{\hbar^2} - \frac{l(l+1)}{r^2} \right\} g = 0 \tag{5.134}$$

and B_l's are arbitrary constants. Setting, $g_l(r) = (1/r)u(r)$, Eq. (5.134) reduces to

$$\frac{d^2 u}{dr^2} + \left[k^2 - \frac{2\mu V(r)}{\hbar^2} - \frac{l(l+1)}{r^2} \right] u = 0 \tag{5.135}$$

For large values of r, the last two terms in the parenthesis tend to zero, and the asymptotic solution is expected to be of the form

$$u \sim A \sin(kr + \delta) \tag{5.136}$$

where A and δ are constants. That this is so may be tested by setting $u = F(r)e^{ikr}$ and substituting it in (5.135), we obtain

$$\frac{d^2 F}{dr^2} + 2ik \frac{dF}{dr} - \left[\frac{2\mu V(r)}{\hbar^2} + \frac{l(l+1)}{r^2} \right] F = 0 \tag{5.137}$$

For large r, F may be assumed to be nearly constant and $(d^2 F/dr^2) \ll k(dF/dr)$. Neglecting the first term in (5.137) and integrating, we obtain

$$2ik \ln F = \int_0^{\infty} \left[\frac{2\mu V(r)}{\hbar^2} + \frac{l(l+1)}{r^2} \right] dr \tag{5.138}$$

The integral on the right side approaches a constant value if $rV(r) \to 0$ as $r \to \infty$. Clearly, this condition can be satisfied if the potential varies faster than $(1/r)$. Thus for fields which fall off more rapidly than the Coulomb potential, the asymptotic solution of u has the form (5.136). Since $V(r) \to 0$ as $r \to \infty$, Eq. (5.134) reduces to (5.116) appropriate for free particle. We, therefore, expect the asymptotic solution of (5.133), to be similar to (5.121) except for a difference in phase and a multiplying constant. Accordingly, the asymptotic solution for ψ which is finite at the origin may be written in view of (5.130) as

$$\psi_{r \to \infty} = \sum_{l=0}^{\infty} \frac{B_l}{kr} \sin\left(kr - \frac{\pi l}{2} + \delta_l\right) P_l(\cos\theta) \qquad (5.139)$$

where δ_l is a constant which is real and depends on k and on the shape of the potential.

Inserting (5.131) and (5.130) in (5.112), and rearranging the terms, we obtain,

$$f(\theta)e^{ikr} = r(\psi - \psi_i)$$

$$= \frac{1}{k}\left[\sum_{l=0}^{\infty} B_l \sin\left(kr - \frac{\pi l}{2} + \delta_l\right) P_l(\cos\theta)\right.$$

$$\left. - \sum_{l=0}^{\infty} i^l (2l+1) \sin\left(kr - \frac{\pi l}{2}\right) P_l(\cos\theta)\right] \quad \text{or}$$

$$f(\theta)e^{ikr} = \frac{1}{2i}\sum_{l=0}^{\infty} P_l(\cos\theta)\left\{B_l\left[e^{i(kr - \frac{\pi l}{2} + \delta l)} - e^{-i(kr - \frac{\pi l}{2} + \delta l)}\right]\right.$$

$$\left. - i^l(2l+1)\left[e^{i(kr - \frac{\pi l}{2})} - e^{-i(kr - \frac{\pi l}{2})}\right]\right\} \qquad (5.140)$$

where the sine functions have been expressed as exponentials. Equating the coefficients of e^{-ikr} in (5.140)

$$0 = \frac{1}{2ik}\sum_{l=0}^{\infty} P_l(\cos\theta)\left\{i^l(2l+1)e^{\frac{i\pi l}{2}} - B_l e^{i\left(\frac{\pi l}{2} - \delta l\right)}\right\}$$

It follows that

$$B_l = i^l(2l+1)e^{i\delta l} \qquad (5.141)$$

Inserting the value of B_l from (5.141) in (5.140) and equating the coefficients of e^{ikr}, we find

$$f(\theta) = \frac{1}{2ik}\sum_{l=0}^{\infty} P_l(\cos\theta)i^l(2l+1)e^{-\frac{i\pi l}{2}}\left(e^{2i\delta l} - 1\right)$$

$$= \frac{1}{2ik}\sum_{l=0}^{\infty}(2l+1)P_l(\cos\theta)\left(e^{2i\delta l} - 1\right) \qquad (5.142)$$

where we have used (5.128) in simplifying the above expression. We can rewrite (5.142) as

$$f(\theta) = \frac{1}{k} \sum_{l=0}^{\infty} (2l+1) e^{i\delta l} \sin \delta_l P_l(\cos \theta)$$

The scattering amplitude $f(\theta)$ is in general a complex function

$$\frac{d\sigma}{d\Omega} = |f(\theta)|^2 = \frac{1}{k^2} \left| \sum_{l=0}^{\infty} (2l+1) e^{i\delta l} \sin \delta_l P_l(\cos \theta) \right|^2 \qquad (5.143)$$

Upon using (5.142), (5.112) and (5.131), we can write an expression for ψ in the form

$$\psi = \frac{1}{2ikr} \sum_{l=0}^{\infty} (2l+1) P_l(\cos \theta) \left[e^{2i\delta l} e^{ikr} - (-1)^l e^{-ikr} \right] \qquad (5.144)$$

Comparison of (5.144) with (5.131) shows that for large r the ingoing spherical waves are unaffected by the potential. The amplitude of the outgoing wave is also unaffected. However, a phase factor has been introduced for each wave. The total cross section σ is given by

$$
\begin{aligned}
\sigma &= \int \left(\frac{d\sigma}{d\Omega} \right) d\Omega = 2\pi \int_{-1}^{+1} \left(\frac{d\sigma}{d\Omega} \right) d(\cos \theta) \\
&= \frac{2\pi}{k^2} \int_{-1}^{+1} \left| \sum_{l=0}^{\infty} (2l+1) e^{i\delta l} \sin \delta_l P_l(\cos \theta) \right|^2 d(\cos \theta) \\
&= \frac{4\pi}{k^2} \sum_{l=0}^{\infty} (2l+1) \sin^2 \delta_l \\
&= 4\pi \lambda^2 \sum_{l=0}^{\infty} (2l+1) \sin^2 \delta_l \qquad (5.145)
\end{aligned}
$$

where by virtue of the orthonormal property of Legendre polynomials ([6], Appendix C) all the cross products drop off.

The above method is called the method of partial waves and was originally employed by Rayleigh, for the analysis of scattering of sound waves.

5.2.4 Physical Interpretation of Partial Waves and Phase-Shifts

In view of the similarity of the angular dependent factor $P_l(\cos \theta)$ occurring in (5.117) with the one which arises in the bound state central field problem, it is reasonable to associate an orbital angular momentum $\sqrt{l(l+1)}$ with a vanishing z-component (because of azimuthal symmetry, the azimuthal quantum number $m = 0$) with the lth partial wave in expression (5.121) and (5.131). Using the spectroscopic notation, the wave function with $l = 0$ is called s-wave, that with $l = 1$ is called p-wave, that with $l = 2$ is d-wave, etc.

The angle δ_l is called the phase shift of the lth partial wave, since it represents the phase difference between the asymptotic form (5.144) with the potential and (5.131) without the potential. It can be determined by imposing the boundary conditions on the solution of differential equations in particular problems of interest. The set of δ_l completely determine the scattering.

It is seen from (4.143) and (4.145) that both the differential and total cross-section would vanish when all the δ_l's are zero or $180°$. If the potential vanishes, i.e. no force acts between the particle then all δ_l's must identically vanish, and there will be no scattering. That this is so is readily seen from the integral expression for the phase shift which relates the phase shift and the potential. This expression is derived in the next section.

At low incident energies only a few waves contribute to the cross-section. This can be seen by a semi-classical argument. If p is the momentum and b is the impact parameter, then the angular momentum is given by

$$bp = l\hbar \quad \text{or} \quad l = \frac{bp}{h} = \frac{b}{\lambdabar} \tag{5.146}$$

Now, the interaction will take place if b is smaller than the range of nuclear forces, i.e. if

$$l < \frac{R}{\lambdabar} \tag{5.147}$$

Thus, at a given incident energy and hence at a definite wavelength only a limited number of l's contribute to the scattering cross-section. The said criterion is equivalent to the statement that a classical particle is not scattered if its angular momentum is too high to penetrate the potential region $r < R$. A useful numerical relation between $\lambdabar$ (the rationalized de Broglie wavelength in the C-system) and E, the neutron kinetic energy in the Lab-system, can be readily obtained

$$\lambdabar = \frac{1}{k} = \frac{\hbar}{p^*} = \frac{\hbar}{\sqrt{2ME^*}} = \frac{\hbar}{\sqrt{\frac{2ME_0}{4}}} = \frac{\hbar c}{\sqrt{\frac{1}{2}Mc^2 E_0}}$$

Inserting $\hbar c = 197 \, \text{fm MeV}$ and $Mc^2 = 939 \, \text{MeV}$

$$\lambdabar \, (\text{fm}) = \frac{9}{\sqrt{E_0}} \, (\text{MeV}) \tag{5.148}$$

Another useful relation is

$$k^2 \left(\text{in } 10^{24} \, \text{cm}^2\right) = 1.206 E_0 \, (\text{in MeV}) \tag{5.149}$$

We can find the energy up to which the s-wave alone is important. It is shown in example (5.4) that the p-wave begins to show up only beyond 20 MeV neutron energy. Similarly, the d-wave would contribute for energy greater than about 40 MeV, and so on. The method of phase shift analysis finds useful application only at low energies at which the partial waves are limited in number, and consequently only a few phase- shifts contribute to the cross-section.

5.2.5 Integral Expression for Phase Shift

The same result will now be proved from quantum mechanical considerations. The radial equation (5.135) is

$$\left[\frac{d^2}{dr^2} - \frac{l(l+1)}{r^2} - \frac{2\mu V(r)}{\hbar^2} + k^2\right]u_l(r) = 0 \tag{5.150}$$

and has the asymptotic form

$$u_l(r) \underset{r\to\infty}{\to} \sin\left(kr - \frac{\pi l}{2} + \delta l\right) \tag{5.151}$$

We shall now derive a useful expression for the phase shift. The field free equation corresponding to (5.150) can be written as

$$\left[\frac{d^2}{dr^2} - \frac{l(l+1)}{r^2} + k^2\right]v_l(r) = 0 \tag{5.152}$$

and has the asymptotic form

$$v_l(r) = kr j_l(kr) \underset{r\to\infty}{\to} \sin\left(kr - \frac{\pi l}{2}\right) \tag{5.153}$$

Multiplying (5.150) by $v_l(r)$ and (5.152) by $u_l(r)$ subtracting and integrating the resultant expression over r from 0 to ∞

$$\int_0^\infty \left(v_l\frac{d^2u_l}{dr^2} - u_l\frac{d^2v_l}{dr^2}\right)dr = \frac{2\mu}{\hbar^2}\int_0^\infty V(r)u_l(r)v_l(r)dr \tag{5.154}$$

Integrating by parts, the left hand side becomes

$$v_l\frac{du_r}{dr}\bigg|_0^\infty - \int_0^\infty \frac{du_l}{dr}dv_l - u_l\frac{dv_l}{dr}\bigg|_0^\infty + \int_0^\infty \frac{dv_l}{dr}du_l$$

the second and the fourth term get cancelled—we are left with

$$\left(v_l\frac{du_l}{dr} - u_l\frac{dv_l}{dr}\right)\bigg|_0^\infty = \frac{2\mu}{\hbar^2}\int_0^\infty V(r)u_lv_l(r)dr \tag{5.155}$$

where we have used (5.154).

We shall now show that $u_l(0) = v_l(0) = 0$.
If $R \ll \lambda$ then

$$r^2 < \frac{l(l+1)}{k^2} = l(l+1)\lambda^2 \tag{5.156}$$

For example at 4 MeV, $\lambda = 4.5$ fm which is the range of nuclear force. The condition (5.156) is certainly fulfilled if $r < l\lambda$. Neglecting the last two terms in the parenthesis of (5.150), the differential equation reduces to

$$\frac{d^2 u_l}{dr^2} - \frac{l(l+1)}{r^2} u_l = 0; \quad r < l\lambda \tag{5.157}$$

which has the solution

$$u_l = Cr^{l+1}; \quad r < l\lambda \tag{5.158}$$

where C is the constant of integration. Thus for $r \to 0$, $u_l(0) = 0$. Similarly $v_l(0) = 0$. We are therefore concerned only with the upper limit in (5.155). Using the asymptotic expressions (5.151) and (5.153) in (5.155)

$$k\sin\left(kr - \frac{\pi l}{2}\right)\cos\left(kr - \frac{\pi l}{2} + \delta_l\right) - k\sin\left(kr - \frac{\pi l}{2} + \delta_l\right)\cos\left(kr - \frac{\pi l}{2}\right)$$

$$= \frac{2\mu}{\hbar^2}\int_0^\infty V(r)u_l(r)v_l(r)dr$$

Simplifying the L.H.S.

$$-k\sin\delta_l = \frac{2\mu}{\hbar^2}\int_0^\infty V(r)u_l(r)v_l(r)dr \quad \text{or}$$

$$\sin\delta_l = \frac{-2\mu}{k\hbar^2}\int_0^\infty V(r)u_l(r)v_l(r)dr \tag{5.159}$$

Since δ_l is also contained in the function $u_l(r)$ under integral, the above expression does not yield the phase shift explicitly. However, some useful information is provided by the above identity.

We shall now show that at low neutron energy (<10 MeV) all δ_l's, save δ_0, would be small.

Equation (5.158) shows that u_l decreases rapidly with decreasing r where r is smaller than $l\lambda$. Since the range of forces R has been assumed to be smaller than λ the foregoing remarks are clearly valid for $r = R$. In other words, u_l ($l \neq 0$) will have a very small value in the region where the potential is significant. Similar conclusion is reached for the field free function v_l. In view of the identity (5.159), it follows that the integral will be negligible if $R < l\lambda$, i.e. $l > (R/\lambda)$ and consequently the corresponding δ_l will be very small. Thus the interaction corresponding to the lth partial wave will be effective only if $l < \frac{R}{\lambda}$, a result which is identical with the condition (5.147) derived from classical argument.

We shall now investigate the sign dependence of the phase shift on the nature of the potential (i.e. attractive or repulsive). Replacing v_l by the asymptotic form, for a square well of depth V_0 (5.159) becomes,

$$\sin \delta_l = \frac{-2\mu V_0}{\hbar^2} \int_0^\infty r j_l(kr) u_l(r) dr \tag{5.160}$$

where $V = -V_0$ for attractive potential and $V = +V_0$ for repulsive potential, V_0 being positive.

Now, in the limit of very large energies, such that $k^2 \gg (2\mu |V_0|/\hbar^2)$ the solution for (5.150) approaches that for the free particle equation (5.152), i.e. in this limit $u_l \to v_l = k r j_l(kr)$ since $k\hbar = \sqrt{2\mu E}$, and $E = E_0/2$. This condition is equivalent to $E_0 \gg 2|V_0|$. In this case, the corresponding phase shift becomes very small, so that $\sin \delta_l \to \delta_l$ and (5.160) reduces to

$$\delta_l \simeq -\frac{2\mu V_0 k}{\hbar^2} \int_0^R r^2 j_l^2(kr) dr \tag{5.161}$$

Because the integral is positive, in the high energy limit, the phase shift tends to zero as a positive or negative quantity depending on whether the potential is attractive or repulsive, respectively.

If on the other hand when $V_0 \to 0$ then $\delta_l \to 0$, and there is no scattering. The dependence of the phase shift on the potential and energy is quite clear from (5.161). The approximation (5.161) is valid not only in the limit $k \to \infty$ for any value of l, but also in the limit $l \to \infty$ for any value of k. This result follows from the fact that if $(l(l+1)/r^2) \gg (2\mu |V|/\hbar^2)$ for all r, $k r j_l(kr)$ is a sufficiently good approximation to u_l.

Another interesting result which follows from (5.161) is the limiting k dependence of δ_l for the square well potential. When the potential falls off sufficiently rapidly as $r \to \infty$ (this is certainly true of the square well potential), we can approximate j_l in (5.161) by the asymptotic formula

$$j_l(x) \to \frac{x^l}{(2l+1)!!} \tag{5.162}$$

where $(2l+1)!! = (2l+1)(2l-1)\cdots 5\cdot 3\cdot 1$

$$\delta_{l(l \gg kR)} \sim \frac{-2\mu V_0}{\hbar^2} \frac{k^{2l+1}}{[(2l+1)!!]^2} \int_0^R r^{2l+2} dr \quad \text{or}$$

$$\delta_l \sim -\left(\frac{2\mu V_0 R^2}{\hbar^2}\right) \frac{(kR)^{2l+1}}{[(2l+1)!!]^2 (2l+3)} \tag{5.163}$$

It follows that

$$\frac{\delta_{l+1}}{\delta_l} \sim \frac{(kR)^2}{(2l+3)(2l+5)} \tag{5.164}$$

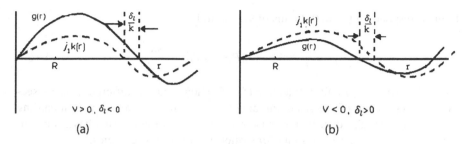

Fig. 5.7 Schematic plots of the effect of (**a**) positive (repulsive) potential and (**b**) negative (attractive) potential, on the force-free radial wave function $j_l(kr)$, the range of the potential is R is each case

As mentioned earlier, the phase shifts corresponding to $l \gg kR$ fall off rapidly and can be ignored in the calculation of cross section. So long as $kR \ll 1$, i.e. $\lambda \gg R$, δ_1 itself will be much smaller than unity, for V_0 of the order of 30 MeV. Higher phase shifts will be still smaller. In a similar way, we can estimate δ_l for Yukawa's potential

$$V(r) = -V_0 \frac{R}{r} e^{-r/R}$$

$$\delta_l \sim -\frac{2\mu V_0 R}{\hbar^2} \frac{k^{2l+1}}{[(2l+1)!!]^2} \int_0^\infty r^{2l+1} e^{-r/R} dr$$

The integral can be evaluated by successive integration by parts. We finally obtain

$$\delta_l \sim -\left(\frac{2\mu V_0 R^2}{\hbar^2}\right) \frac{(2l+1)!}{[(2l+1)!!]^2} (kR)^{2l+1} \qquad (5.165)$$

If the potential decreases as r^{-n} ($n > 2$) for large r, the above limiting energy dependence of δ in general does not apply.

The effect of phase shift on the wave function $g_l(r)$ is shown schematically in Fig. 5.7 for the positive (repulsive) and negative (attractive) potentials, separately, as compared to the function $j_l(kr)$ corresponding to the free particle. In Fig. 5.7a, $g_l(r)$ in the presence of positive potential is pushed out, i.e. has a retarded phase (negative δ_l) compared to $j_l(kr)$ whilst in Fig. 5.7b, $g_l(r)$ in the presence of negative potential is pulled in, i.e. has an advanced phase (positive δ_l) compared to j_l. The scattering is completely determined by the phase shift although the amplitudes have no direct physical significance. Note that both the functions vanish at the origin. The difference between the neighbouring nodes of j_l and g_l when the former has gone through several oscillations, is given by (δ_l / k).

5.2.6 Angular Distribution of Scattered Neutrons at Low Energies

In the very low energy region, $k \to 0$, so that $kR \ll 1$ in which case the s-waves alone contribute to the scattering. In the formula (5.143) we need to accept only one term corresponding to $l = 0$. In this limit, we find

$$\frac{d\sigma}{d\Omega} = \frac{1}{k^2} \left| e^{i\delta_0} \sin \delta_0 P_0(\cos \theta) \right|^2 = \frac{\sin^2 \delta_0}{k^2} \tag{5.166}$$

since $P_0(\cos \theta) = 1$ for all θ. Equation (5.166) shows that the differential cross section is independent of the scattering angle θ. Therefore, at very low bombarding energies, the angular distribution is isotropic in the C-system. This has been confirmed in various experiments on $n-p$ scattering at very low energies.

As the incident energy is raised, say $E_0 > 20$ MeV, the p-wave can no longer be neglected. If we do not go to much higher energies, so that d and higher waves may be ignored, then we can express the differential cross section (5.143) in the presence of s- and p-waves only

$$\begin{aligned}
\frac{d\sigma}{d\Omega} &= \frac{1}{k^2} \left| e^{i\delta_0} \sin \delta_0 + 3e^{i\delta_1} \sin \delta_1 \cos \theta \right|^2 \\
&= \frac{1}{k^2} \left(e^{i\delta_0} \sin \delta_0 + 3e^{i\delta_1} \sin \delta_1 \cos \theta \right) \left(e^{-i\delta_0} \sin \delta_0 + 3e^{-i\delta_1} \sin \delta_1 \cos \theta \right) \\
&= \frac{1}{k^2} \left[\sin^2 \delta_0 + 6 \sin \delta_0 \sin \delta_1 \cos(\delta_0 - \delta_1) \cos \theta + 9 \sin^2 \delta_1 \cos^2 \theta \right] \tag{5.167}
\end{aligned}$$

where we have used the fact that $P_0(\cos \theta) = 1$ and $P_1(\cos \theta) = \cos \theta$. Expression (5.167) is of the form

$$\frac{d\sigma}{d\Omega} = A + B \cos \theta + C \cos^2 \theta \tag{5.168}$$

where A, B and C are constants. The first term in (5.167) arises for s-waves alone and the third one for p-waves alone. On the other hand, the second term which is the cross product term, results from the interference between the s- and p-wave. Suppose that $\delta_0 \gg \delta_1$ so that the $\cos^2 \theta$ term may be neglected. The $\cos \theta$ cross-product term may change the angular distribution in a significant way from spherical symmetry, which is expected for s-wave alone. The question of interference terms does not arise in the total cross section and the total cross section contributed by s- and p-waves only will be given by

$$\sigma = \frac{4\pi}{k^2} \left(\sin^2 \delta_0 + 3 \sin^2 \delta_1 \right) \tag{5.169}$$

If the condition $\delta_0 \gg \delta_1$ is satisfied, then it is seen that the p-waves contribute but little to the total scattering cross-section. On the other hand the presence of p-waves affects the differential cross-section in a marked way at a lower energy, than that at which it becomes significant in the total cross section. For example, if $\delta_0 = 30°$ and $\delta_1 = 3°$ at a particular incident energy, the p-wave contributes only 3 % to the total cross-section while it makes the ratio $\frac{d\sigma(0°)}{d\Omega} / \frac{d\sigma(180°)}{d\Omega} = 3.5$. Figure 5.8 shows the predicted angular distribution for the above choice of phase shifts. In general if L is

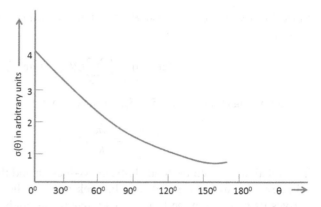

Fig. 5.8 Angular distribution for $\delta_0 = 30°$ and $\delta_1 = 3°$

the largest value of l for which δ_l is appreciably different from zero, then the largest power of $\cos\theta$ appearing in the differential cross section is $(\cos\theta)^{2L}$.

Scattering experiments serve to determine the *phase-shifts* and they in their turn afford the determination of the sign and magnitude of the potential $V(r)$. Since the angular distribution contains interference terms, the relative signs of the phase shifts can be measured. On the other hand, if we reverse the sign of all the phase shifts, i.e. make the substitution $\delta_l \to -\delta_l$ for all l, then the scattering amplitude $f \to -f^*$ which leaves the differential cross-section (5.143) unaltered. Hence, by the use of angular distribution, the overall sign of the set δ_l remains indeterminate. The knowledge of the sign of δ_l is important since it affords the determination of the sign of the potential. If $V(r)$ is attractive everywhere then $\delta_l > 0$ and repulsive, if $\delta_l < 0$, a result which is rigorously correct if $V(r)$ never changes sign as r is varied. The overall sign of δ_l can be determined only by having recourse to further interference experiments such as those which are involved in Coulomb scattering in p–p collisions or coherent scattering of neutrons with molecular hydrogen.

Formula (5.145) imposes restrictions on the value of σ_l for the lth wave; the cross-section attains maximum value for $\delta_l = (\frac{\pi}{2})$. Thus

$$\sigma_{el}^l(\text{max}) = 4\pi \lambdabar^2(2l+1) \tag{5.170}$$

This is called unitary limit.

5.2.7 Optical Theorem

The total scattering cross-section can be related to $f(0)$, the scattering amplitude in the forward direction ($\theta = 0$). Now the imaginary part of the amplitude is given by (5.142) as

$$\text{Im } f(\theta) = -\frac{1}{2k} \sum_{l=0}^{\infty} (2l+1)(\cos 2\delta_l - 1)P_l(\cos\theta) \tag{5.171}$$

Since for $\theta = 0$, $\cos\theta = 1$ and $P_l(1) = 1$ for all values of l, the above expression becomes

$$\text{Im } f(0) = \frac{1}{k} \sum_{l=0}^{\infty} (2l + 1) \sin^2 \delta_l \tag{5.172}$$

On comparing (5.172) with (5.145), it follows that

$$\sigma = \frac{4\pi}{k} \text{Im } f(0) \tag{5.173}$$

This relation between the total elastic cross-section and the forward scattering amplitude is due to Bohr, Peierls and Placzek and is called the optical theorem and derives its name from the fact that a similar result holds good in the scattering of light by spherically symmetrical obstacles. Actually Eq. (5.173) is valid much more generally, when f depends on θ as well as ϕ and when σ includes inelastic scattering, absorption as well as elastic scattering. A physical interpretation may be given to Eq. (5.173). There is interference between the contributions of various partial waves to the differential cross-section but not to the total cross-section. This is to be expected since the total cross-section is just a measure of the total number of partials scattered per unit beam flux. However, the fact that the scattered particles appear at angles $\theta > 0$ means that there is depletion of the beam behind the scattering region ($\theta = 0$) than in front of it. This can occur only by interference between the two terms in the asymptotic expression (5.112), i.e. this redistribution is produced by destructive interference between the incident plane wave and the scattered wave in the forward direction. This close connection between the forward scattering amplitude and the total cross section finds an expression in the optical theorem.

5.2.8 Total Cross Section

In the deuteron problem with a square well the inside and outside radial wave functions $u_I = A \sin \hbar r$ and $u_{II} = ce^{-\gamma r}$ were matched at the boundary $r = R$. The condition that these wave functions must join smoothly at the boundary is equivalent to identify the logarithmic derivatives

$$\left(\frac{u_I'}{u_I} \right)_{r=R} = \left(\frac{u_{II}'}{u_{II}} \right)_{r=R} = -\gamma \tag{5.174}$$

where dash means differentiation with respect to r

$$K = \frac{\sqrt{2\mu(V_0 + E)}}{\hbar} \quad \text{and} \quad \gamma = \frac{\sqrt{2\mu W}}{\hbar}; \qquad E = -W$$

W being the binding energy of deuteron. In the deuteron problem $E = -W$, since a bound state exists, while for $n-p$ scattering at very low energies, $E \sim 0$ or slightly positive. Since the potential well V_0 which is known to be of the order of 30 MeV, is much greater than W, we expect that the value $(u'_I/u_I)_{r=R}$ will not be drastically altered, when the energy E is changed from -2.2 MeV (corresponding to the binding energy of deuteron) to a value zero or even slightly positive. Thus by using the approximation

$$
\left(\frac{u'_I}{u_I}\right)_{r=R} = K \cot kR
$$

$$
= \frac{\sqrt{2\mu(V_0 + E)}}{\hbar} \cot \frac{R}{\hbar}\sqrt{2\mu(V_0 + E)}
$$

$$
= -\gamma \tag{5.175}
$$

γ is altered between 20 percent and 50 percent depending mainly on the choice of R, when E is altered from 2.2 MeV to zero. To the extent that this approximation is correct, we can match the asymptotic form given by (5.139) for $r > R$ with the inside deuteron wave function. Using (5.139) and accepting only the term with $l = 0$ and $B_0 = e^{i\delta_0}$ from (5.141)

$$
\psi_{r\to\infty} = \frac{e^{i\delta_0}}{kr} \sin(kr + \delta_0)
$$

where we have put $P_0(\cos\theta) = 1$. It follows that

$$
u_{II} = r\psi_{r\to\infty} = \frac{e^{i\delta_0}}{k} \sin(kr + \delta_0) \tag{5.176}
$$

It is then sufficient to set the logarithmic derivative of $(e^{i\delta_0}/k) \sin(kr + \delta_0)$ at $r = R$, equal to $-\gamma$. Accordingly, we have

$$
k \cot(kR + \delta_0) = K \cot KR = -\gamma \tag{5.177}
$$

As before, in the low energy limit, we set $kR \to 0$. This is justifiable even at few MeV energy if we choose an arbitrary small value of R, since the main features of the deuteron problem remain unaltered by reducing the width of the well to an arbitrarily small value and at the same time by increasing the depth of the potential to the correspondingly large value such that the quantity $V_0 R^2$ remains a constant. This approximation which consists of ignoring the range R is called *zero range approximation*. We then obtain

$$
\cot\delta_0 \simeq -\frac{\gamma}{k} \quad \text{or} \tag{5.178}
$$

$$
\sin^2\delta_0 = \frac{k^2}{k^2 + \gamma^2} \tag{5.179}
$$

It follows that

$$\sigma = \frac{4\pi}{k^2} \sin^2 \delta_0 = \frac{4\pi}{k^2 + \gamma^2} = \frac{4\pi \hbar^2}{M} \frac{1}{E + W}$$

$$= \frac{4\pi \hbar^2 c^2}{Mc^2} \frac{1}{(\frac{E_0}{2} + W)} = \frac{5.2}{(\frac{E_0}{2} + W)} \text{ b} \qquad (5.180)$$

where we have used the numerical values $\hbar c = 197.3$ MeV fm and $Mc^2 = 940$ MeV.

5.2.9 Comparison of Experimental Cross-Sections with the Theory and Evidence for Spin Dependence of Nuclear Forces

In the early days, because of experimental difficulties in obtaining monoergic beams of neutrons, precise estimations of cross sections were scanty. Nevertheless, it was concluded that for neutrons of few MeV, the experimental cross section (5.180) were in rough agreement with the theoretical estimates. Thus, for example for $E_0 = 4.3$ MeV, Chadwick obtained a cross section between 0.5 and 0.8 b which may be compared with a theoretical value of 1.2 b. At 2.1 MeV, the experimental value was between 1.1 and 1.5 b while the theoretical value was 1.6 b. Because of the uncertainties in the experimental cross-sections and the approximations in the theory, it was believed that a rough agreement was established for neutrons of a few MeV. But at extremely low neutron energies, the formula completely failed to account for the observed cross section. Thus at $E_0 \sim 0$, $\sigma_{theor} = 2.6$ b, while $\sigma_{free} \sim 50$ b for protons bound in molecules. Fermi pointed out that the observed cross sections for bound protons must be divided by a factor of 2.5 in order to reduce them to free neutron cross sections. With this correction, σ_{free} becomes ~ 20 b. Even then a discrepancy of a factor of 8 remained. It may be remarked that the 'zero-energy' cross sections refer to energies $5 < E_0 < 100$ eV where they are constant and are so called because this value is extrapolated to zero energy for any assumed shape and finite range. Theoretical estimates of the cross-section for any assumed shape and finite range could be improved, but it appeared improbable that σ_{theor} would be raised by more than a factor of 2. On the other hand, the zero range approximations was expected to be more valid for very low neutron energies than those of few MeV. This discrepancy was finally resolved by Wigner [9] who pointed out that the observed deuteron binding energy referred to neutron and proton in the triplet state (spins parallel) and provides no information regarding the $n-p$ interaction in singlet state (spin antiparallel). Following this suggestion, it is clear that the interaction in the singlet may be quite different from that in the triplet state, i.e. the nuclear forces are spin dependent. In analogy with the triplet state, we introduce the quantity W_s corresponding to the singlet state. If it is assumed that W_s is small then we can reach agreement with theory since $\sigma \propto 1/(E + W_s)$. Now, if an unpolarized beam of neutrons falls on protons,

Fig. 5.9 Total n–p scattering cross section as a function of the incident neutron energy

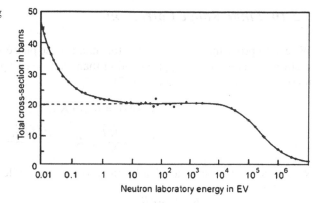

the statistical weight for the triplet interaction must be three times as large as that for the total spin (s) equal to 1 (triplet), $2s + 1 = 2 \times 1 + 1 = 3$, i.e. there are three different quantum states corresponding to the spin components $+1, 0, -1$, in a given direction. On the other hand, for the total spin equal to zero (singlet) $2s + 1 = 2 \times 0 + 1 = 1$, and only one quantum state exists. We can then write the total cross section as

$$\sigma = \frac{1}{4}\sigma_s + \frac{3}{4}\sigma_t \qquad (5.181)$$

Introducing the parameters W_t and W_s in the scattering formula for the triplet and singlet scattering, respectively and using the statistical weights (5.181), we find

$$\sigma = \frac{\pi \hbar^2}{M}\left(\frac{3}{E + W_t} + \frac{1}{E + |W_s|}\right) \qquad (5.182)$$

where $W_t = 2.22$ MeV. The value of W_s can be fixed only by comparing σ_{theor} and σ_{expt}. The best value of $|W_s|$ is found to be 68 keV, which not only gives good agreement with the observed cross-sections at very small neutron energies, but also has the merit of preserving agreement in the MeV region. This is so because for $E \gg W_t$ formula (5.182) approximately reduces to (5.180) which was previously shown to be in accord with the observations. Thus, the entire n–p scattering, from thermal energies up to few MeV can be explained by invoking for the spin dependence of nuclear forces. In Fig. 5.9 is shown the variation of experimental cross section based on comparatively recent data, for neutron energies ranging from 10 MeV, down to thermal energies. It may be pointed out that so far we are free to choose the sign of W_s. A negative value of the binding energy implies a bound state whilst a positive sign means a virtual state. The energy W_s has no direct significance in that nothing extraordinary happens to σ_s when $E \sim W_s$. Therefore, W_s may be taken simply as an adjustable parameter.

5.2.10 Finite Range Correction

We can improve the accuracy of the formula for the phase shift and hence the cross-section through a better approximation than that given by the zero range approximation. We start with relation (5.177)

$$k \cot(kR + \delta_0) = K \cot KR \tag{5.183}$$

$$\text{with } k = \frac{\sqrt{ME}}{\hbar}; \quad K = \frac{\sqrt{M(V_0 + E)}}{\hbar} \tag{5.184}$$

We further make use of fact that for the ground state of deuteron

$$K_0 \cot K_0 R = -\gamma \tag{5.185}$$

$$\text{with } K_0 = \frac{\sqrt{M(V_0 - W)}}{\hbar}; \quad \gamma = \frac{\sqrt{MW}}{\hbar} \tag{5.186}$$

where $W = 2.22$ MeV is the deuteron binding energy. Now

$$K^2 - K_0^2 = \frac{M}{\hbar^2}(V_0 + E) - \frac{M(V_0 - W)}{\hbar^2}$$

$$= \frac{ME}{\hbar^2} + \frac{MW}{\hbar^2} = k^2 + \gamma^2 \tag{5.187}$$

Therefore

$$K - K_0 = \frac{k^2 + \gamma^2}{K + K_0} \simeq \frac{k^2 + \gamma^2}{2K_0} \quad \text{or} \quad K = K_0 + \frac{k^2 + \gamma^2}{2K_0} \tag{5.188}$$

Expanding $K \cot KR$ about K_0 by Tayler's series

$$K \cot KR = K_0 \cot K_0 R + (K - K_0)\left(\cot K_0 R - K_0 R \operatorname{cosec}^2 K_0 R\right)$$

$$= -\gamma + \frac{(k^2 + \gamma^2)}{2K_0}\left[-\frac{\gamma}{K_0} - K_0 R\left(1 + \frac{\gamma^2}{K_0^2}\right)\right] \tag{5.189}$$

where (5.185) and (5.188) have been used. Neglecting the terms $-(\gamma/K_0)$ and $(-R\gamma^2/K_0)$ in comparison with $K_0 R$ in the parenthesis of (5.189), we obtain

$$K \cot KR \simeq -\gamma - (k^2 + \gamma^2)\frac{R}{2} \tag{5.190}$$

On inserting (5.190) in (5.183) we have

$$\cot(kR + \delta_0) = -\frac{\gamma}{R} - \frac{R}{2k}(k^2 + \gamma^2) \tag{5.191}$$

Expanding the left side, we obtain

$$\cot \delta_0 - kR \operatorname{cosec}^2 \delta_0 = -\frac{\gamma}{k} - \frac{R}{2k}(k^2 + \gamma^2) \tag{5.192}$$

using the approximate value for $\csc^2 \delta_0$ from (5.179) in the second term on the left side, we have

$$\cot \delta_0 = -\frac{\gamma}{k} + \frac{R}{k}(k^2 + \gamma^2) - \frac{R}{2k}(k^2 + \gamma^2)$$

$$= -\frac{\gamma}{k} + \frac{R}{2k}(k^2 + \gamma^2) \tag{5.193}$$

Finally

$$\sigma = \frac{4\pi}{k^2} \sin^2 \delta_0 = \frac{4\pi}{k^2(1 + \cot^2 \delta_0)} \simeq \frac{4\pi}{k^2 + \gamma^2 - \gamma R(k^2 + \gamma^2)}$$

$$= \frac{4\pi}{(k^2 + \gamma^2)(1 - \gamma R)} \simeq \frac{4\pi}{k^2 + \gamma^2}(1 + \gamma R) = \frac{4\pi \hbar^2}{M(E + W)}(1 + \gamma R) \tag{5.194}$$

where we have neglected higher powers of γR other than the first one. Comparison of (5.194) and (5.180) shows that the more accurate treatment introduces a correction factor $(1 + \gamma R)$ compared to that given in the zero range approximation. Now, we have seen in the deuteron problem that $(1/\gamma) = 4.31$ fm and with the choice of $R = 2$ fm, the correction factor amounts to 1.5, for the scattering of neutrons by protons with parallel spin; on the other hand, for the opposite spins, the correction factor will be $(1 + \beta R)$ with $\beta^2 = (MW_s/\hbar^2)$. For $W_s = 70$ keV, the correction factor will amount to only 1.05. For very low neutron energies where singlet scattering dominates over triplet scattering, the correction to the estimated cross-section is rather small.

5.2.11 Evidence for Neutron Spin (1/2)

Scattering experiments also provide strong evidence in favour of the assignment of the neutron spin as 1/2 rather than 3/2. For, if it were 3/2, then there would be two states of the n–p system contributing to the scattering, one quintet state with a statistical weight 5 corresponding to $S = 2$ (the neutron spin 3/2 and proton spin 1/2 parallel) and a triplet state with a statistical weight 3 corresponding to $S = 1$ (neutron and proton spins antiparallel). Consequently, the corresponding formula for the cross section would be

$$\sigma = \frac{\pi \hbar^2}{2M} \left(\frac{3}{E + W_t} + \frac{5}{E + W_q} \right) \tag{5.195}$$

If this formula is to be brought in agreement at thermal energies with a suitable choice of W_q then for $E_0 = 2E$ ranging from 400 to 800 keV, it gives results which are larger than a factor of 1.5 or greater compared to the observed values, the discrepancy being far outside the experimental errors. We, therefore, conclude that the results on scattering experiments are only consistent with the neutron spin value 1/2.

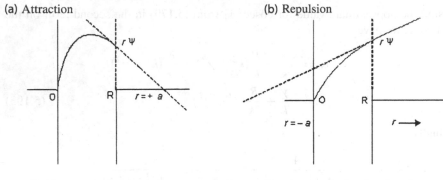

(a) Attraction

(b) Repulsion

Positive scattering length, bound state   Negative scattering length, unbound state

Fig. 5.10   Geometrical interpretation of Fermi scattering length

Furthermore, with the assumption of neutron spin greater than $3/2$, the angular momentum must be a state with $l \neq 0$ in order to account for a total spin of 1 for the deuteron ground state. But it was shown that the choice of $l \neq 0$ is highly improbable on general grounds.

5.2.12 Scattering Length

In the low energy limit $E \to 0$, $k \to 0$, the scattering cross section given by $\sigma = (4\pi/k^2)\sin^2\delta_0$ should remain finite and must not go to zero. In that case, we demand that $(\sin^2\delta_0/k^2) \to (\delta_0/k)^2 \to a^2$, where '$a$' having the dimension of length, in the zero energy limit is called Fermi scattering length. Its sign remains to be determined. A geometrical interpretation of 'a' is given in Fig. 5.10. In the zero energy limit, the asymptotic wave function outside the range of nuclear forces goes as $\sin(kr + \delta_0) \to kr + \delta_0$ which is linear in r and extrapolates to zero at $r = -(\delta_0/k)$. The node may be on either side of the origin. The intercept is then equal to the scattering length. The choice of the sign of 'a' is such that $a = -(\delta_0/k)$. Thus, a positive phase shift implies a negative scattering length and the low energy wave function has the form $k(r - a)$. In accordance with this convention, scattering against an impenetrable sphere of radius 'a' will have a scattering length $+a$.

The magnitude of the Fermi scattering length is determined by the experimental cross-section at low energy. Its sign can be fixed only with additional experiments which are capable of measuring the sign of the phase shifts by exploiting the interference effects. Henceforth, 'a' the low energy scattering length will be referred to as $a(0)$. With a totally repulsive potential, such as the Coulomb's potential for the like charged particles, the Fermi scattering length is always positive. On the other hand, with an attractive potential, $a(0)$ may be either positive or negative, depending on the details of the shape of the potential.

We may extend the definition of scattering length to higher energies. We shall call $a(k)$ the general scattering length for the corresponding wave number k. If we now define

$$\tan \delta_0 = -ka(k)$$

$$a(k) \underset{k \to 0}{\to} a(0) \quad \text{and} \quad \delta_0 \to -ka \tag{5.196}$$

then in the low energy limit, $k \to 0$, the cross-section can be expressed as

$$\sigma = \frac{4\pi}{k^2} \sin^2 \delta_0 = \frac{4\pi}{k^2(1 + \cot^2 \delta_0)} = \frac{4\pi}{k^2 + \frac{1}{a^2(k)}} \tag{5.197}$$

The entire s-wave scattering is completely determined by the length $a(k)$ which is related to the phase shift through (5.197). The cross-section at zero energy becomes

$$\sigma_0 = 4\pi a^2 \tag{5.198}$$

which is identical with the zero energy cross-section of an impenetrable sphere of radius a. The measurement of zero energy cross-section leads directly to the estimation of the magnitude of Fermi scattering length but not its sign. The connection between 'a' and σ_0 can also be seen directly from (5.112) with $k = 0$. In that case we have the relation $\psi = 1 + (f/r)$; the radial function then is given by $u(r) = \psi r = r + f$ which may be compared with the form $k(r - a)$ deduced for the low energy limit. We conclude that $f = -a$ leads to $(d\sigma/d\Omega) = a^2$ which upon integration yields (5.198).

5.3 Effective Range Theory

We shall now proceed to show that regardless of the shape and depth of the potential, the inverse of the general scattering length $a(k)$ is a linear function of energy, and has a slope given by another parameter r_0 called the effective range which has the dimension of length. In the zero energy limit, $a(k)$ of course approaches 'a', the Fermi scattering length. We use the equations

$$\frac{d^2 u}{dr^2} + \frac{2\mu}{\hbar^2}\left[E - V(r)\right]u = 0 \quad (E \text{ finite}) \tag{5.199}$$

where $u(r)$ is the wave function for the energy E and

$$\frac{d^2 u_0}{dr^2} - \frac{2\mu V(r)u_0}{\hbar^2} = 0 \quad (E \text{ zero}) \tag{5.200}$$

where u_0 is the inside function for zero energy. We multiply (5.199) by $u_0(r)$ and (5.200) by $u(r)$ and subtract

$$\frac{d}{dr}\left(uu_0' - u_0 u'\right) = k^2 u u_0 \tag{5.201}$$

Fig. 5.11 (a) Forms of $u(r)$ and $v(r)$. (b) Schematic plot of $u_0(r)$ and the asymptotic form $v_0(r)$ for zero energy

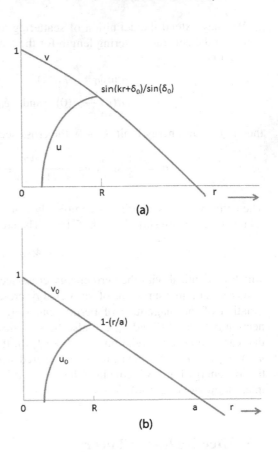

A similar relation holds for the asymptotic forms $v(r)$ of $u(r)$ and $v_0(r)$ of $u_0(r)$, i.e. in the limit $r \rightarrow \infty$

$$\frac{d}{dr}(vv_0' - v_0v') = k^2 v v_0 \tag{5.202}$$

We subtract (5.202) from (5.201) and integrate over r from zero to infinity

$$\left[uu_0' - u_0u' - vv_0' + v_0v'\right]_0^\infty = k^2 \int_0^\infty (uu_0 - vv_0)dr \tag{5.203}$$

The forms of $u(r)$ and $v(r)$ are shown in Fig. 5.11a. Let us examine these functions in some detail. Now, the factor $\sin \delta_0$ in the denominator has been introduced so as to satisfy the normalization condition

$$u(r) \underset{r \rightarrow \infty}{\rightarrow} v(r) = \frac{\sin(kr + \delta_0)}{\sin \delta_0} \tag{5.204}$$

$$v(0) = 1 \tag{5.205}$$

Here the normalization condition is different from the conventional one, and is used here for convenience. It is seen that outside the range of nuclear forces R, the functions $u(r)$ and $v(r)$ coincide. Inside the range of forces, the function $u(r)$, in the presence of potential $V(r)$, bends around and satisfies the condition

$$u(0) = 0 \qquad (5.206)$$

On the other hand $v(r)$ extrapolates to $v(0) = 1$ and the inside function has been plotted as if the potential is absent.

Figure 5.11b is a schematic plot of $u_0(r)$ and the asymptotic form $v_0(r)$ for zero energy. As mentioned earlier, $v_0(r)$ is a straight line and cuts the r-axis at the Fermi scattering length a'. The function $v_0(r)$ is also normalized in the same way as $v(r)$

$$v_0(0) = 1 \qquad (5.207)$$

also

$$u_0(0) = 1 \qquad (5.208)$$

That the asymptotic form of $u_0 \to v_0$ is a straight line follows immediately from the fact that for $E \to 0$ and $V(r)_{r \to \infty} \to 0$, $v_0(r)$ is a solution of (5.200) which is reduced to $(d^2 v_0/dr^2) = 0$. This has the solution

$$v_0(r) = C(a - r) = 1 - \frac{r}{a} \qquad (5.209)$$

which is an equation for a straight line. The constant C has been chosen in such a way that the normalization condition (5.207) is satisfied.

Now, in the limit of zero energy (5.204) also yields and expression for $v_0(r)$

$$\underset{k \to 0}{v(r) \to v_0(r)} = \frac{\sin(kr + \delta_0)}{\sin \delta_0} = \cot \delta_0 \sin kr + \cos kr$$

$$= kr \cot \delta_0 + 1 \qquad (5.210)$$

Comparing (5.210) with (5.209), we obtain

$$\tan \delta_0 = -ka(k) \qquad (5.211)$$

which is identical with (5.196). Going back to (5.203) we note that in the limit $r \to \infty$, $u \to v$; $u_0 \to v_0$. Hence, the expression in the parenthesis vanishes identically at the upper limit. At the lower limit, we use the expressions (5.205) to (5.208), $u(0) = v_0(0) = 0$, $v(0) = v_0(0) = 1$. Further, from (5.210) we note $v'(0) = k \cot \delta_0$ and $v_0'(0) = -(1/a)$. Expression (5.203) is then simplified to

$$k \cot \delta_0 = -\frac{1}{a} + k^2 \int_0^\infty (vv_0 - uu_0) dr \qquad (5.212)$$

This is an exact equation. In order to derive useful information from this equation
we shall proceed to make certain approximations. First we note that the significant
contribution to the integral comes from those regions in which the functions u and
u_0 are appreciably different from their respective asymptotic forms, viz, v and v_0.
This is precisely the inside region where the nuclear forces are effective. But, in
this region the forces are so strong and the corresponding potential $V(r)$ so great
($V_0 \sim 30$ MeV) compared to E, the energy of relative motion in the C-system, is
hardly changed when E is raised from zero to a small value. Consequently, the
wave function $u(r)$ is only slightly changed (compare $u(r)$ and $u_0(r)$). It is then a
sufficiently good approximation to replace $u(r)$ by the zero energy form $u_0(r)$ in
the integral. Similarly, $v(r)$ may be replaced by $v_0(r)$. Under these approximations
we obtain the formula

$$k \cot \delta_0 = \frac{1}{a(k)} = -\frac{1}{a} + \frac{1}{2} r_0 k^2 \tag{5.213}$$

where the quantity r_0 is defined by

$$r_0 = 2 \int_0^\infty \left(v_0^2 - u_0^2 \right) dr \tag{5.214}$$

In the higher approximation, indeed (5.213) is modified to

$$\frac{1}{a(k)} = -\frac{1}{a} + \frac{1}{2} k^2 r_0 + P k^4 r_0^3 \tag{5.215}$$

where P is a small numerical co-efficient which depends on the detailed shape of the
potential and varies between -0.04 and $+0.15$ for typical potentials; for a square
well the curve bends down, while for a Yukawa potential it tends to curl up.

The quantity r_0 has the dimension of length and is a function of $V(r)$ but is
independent of the energy. Outside the range of forces, the integrand is zero and
inside it is of the order of unity. The factor 2 outside the integral tends to make
r_0 assume a value somewhere near the edge of the potential well. The parameter
r_0 is, therefore, called the "effective range" and the second term occurring on the
right side of (5.213) is called the range correction. The effective range, it must be
emphasized, depends not only on the width of the well but also on its depth. The
observed cross-sections at various energies (k^2) allow the scattering length $a(k)$ to
be determined.

Formula (5.213) shows that if the inverse of the scattering length, ($1/a(k)$) be
plotted as a function of E (since k^2 is proportional to energy) then the curve must
be a straight line. The slope of the curve determines the effective range r_0 and its
intercept at $k^2 = 0$ yields 'a', the Fermi scattering length. For E ranging from zero
to 10 MeV, the curve is actually a straight line. But for higher values of E deviations
begin to occur which are attributed to the shape dependent terms involving higher
powers of k which are included in (5.215) in the higher approximation.

The approximation (5.213) is called "shape-independent" since it involves only
two independent parameters, the scattering length 'a', and the range r_0. For any

assumed shape of potential these two experimental parameters can be obtained with a proper choice of range and depth of the well. So long as k is small, observed cross sections determine only the depth and the range of potential and not its detailed shape, since the approximation (5.213) is shape independent.

The s-wave scattering is completely described in terms of two parameters 'a' and r_0

$$\sigma = \frac{4\pi}{k^2 + \frac{1}{a^2(k)}} = \frac{4\pi}{k^2 + [\frac{1}{2}r_0 k^2 - \frac{1}{a}]^2} \tag{5.216}$$

5.3.1 Triplet Scattering

In our previous treatment of triplet scattering there appeared only a single free parameter, whereas here, there are two. This is because previously, the relationship between range and depth provided by the binding energy was used. Here, again it is possible to eliminate one of the parameters for establishing a relationship between a_t and r_{0t}, the triplet scattering length and the triplet effective range, respectively. This is achieved by extending (5.213) to negative energies (imaginary k), i.e. to the ground state, with $E = -W$, by simply replacing the square of outside wave number k^2 by $-\gamma^2$, where $\gamma = \sqrt{MW/\hbar^2} = \frac{1}{R}$, R being the size of deuteron. The asymptotic wave function (normalized) for the ground state has the form $v(r) = e^{-\gamma r}$. Hence $v'(0) = -\gamma$ rather than $k \cot \delta_0$. The shape independent approximation then becomes

$$\gamma = \frac{1}{a_t(0)} + \frac{1}{2}r_{0t}\gamma^2 \quad \text{or} \tag{5.217}$$

$$r_{0t} = \frac{2}{\gamma}\left(1 - \frac{1}{\gamma a_t(0)}\right) \tag{5.218}$$

Eliminating the scattering length between (5.215) and (5.216), the triplet cross section is given as

$$\sigma_t = \frac{4\pi}{(k^2 + \gamma^2)[1 - \gamma r_{0t} + \frac{r_{0t}^2}{4}(k^2 + \gamma^2)]} \tag{5.219}$$

This formula involves only one adjustable parameter r_{0t}, the effective range since $R = (1/\gamma)$ is completely determined by the binding energy of the deuteron. It is interesting to note that in the low energy limit, to the first approximation, (5.219) reduces to

$$\sigma_t \simeq \frac{4\pi}{(k^2 + \gamma^2)(1 - \gamma r_{0t})} = \frac{4\pi(1 + \gamma r_{0t})}{(k^2 + \gamma^2)} \tag{5.220}$$

On identifying r_{0t} with R, the width of the well, this formula becomes identical with (5.194) which was derived by Bethe without using the shape independent approximation.

We shall now investigate into the variation of σ_t with the neutron energy. For this purpose we need to know the values of $a_t(0)$ and r_{0t}.

We can make a rough calculation for $a_t(0)$. For $E = 0$, (5.194) gives the triplet cross section, using square well potential

$$\sigma_t(0) = \frac{4\pi \hbar^2}{MW}(1 + \gamma R) = 4\pi \left(\frac{1}{\gamma^2} + \frac{R}{\gamma}\right) \tag{5.221}$$

But

$$\sigma_t(0) = 4\pi a_t^2(0) \tag{5.222}$$

$$a_t^2(0) = \frac{1}{\gamma^2} + \frac{R}{\gamma} \tag{5.223}$$

With $(1/\gamma) = 4.31$ fm, and $R = 2$ fm

$$a_t(0) = +5.21 \text{ fm} \tag{5.224}$$

The positive sign has been chosen since it is a bound state. For the purpose of calculations, however we shall use the generally accepted value of $+5.38$ fm which corresponds to a cross-section $\sigma_t(0) = 3.63$ b. Inserting this value for $a_t(0)$ in (5.218) or (5.219), we find the effective range

$$r_{0t} = 1.7 \text{ fm} \tag{5.225}$$

Inserting these values of $a_t(0)$ and r_{0t} in (5.105), expressing k^2 in terms of energy E, we obtained

$$\sigma_t \text{ (in barns)} = \frac{4\pi}{0.04E^2 + 1.66E + 3.46} \tag{5.226}$$

At $E = 0$, this formula predicts $\sigma_t(0) = 3.63$ b, as it should, since this is a direct consequence of our estimate of $a_t(0)$. At $E = 2$ MeV, it gives a value of 1.8 b, while at $E = 5$ MeV, it yields a value of about 1 b. We, therefore, conclude that the triplet cross section is a slow decreasing function of energy, in the range that we are concerned.

5.3.2 Singlet Scattering

Since deuteron does not have bound singlet state, a formula analogous to (5.218) for this state cannot be derived. We can, however, obtain an expression for the effective range r_{0s} for the singlet scattering from our knowledge of the wave functions $u(r)$ and $v(r)$ which have the form

$$v = 1 - \frac{r}{a_s} \tag{5.227}$$

Let

$$u = A_s \sin K_s r; \quad r < R_0 \tag{5.228}$$

where

$$K_s = \frac{\sqrt{M(V_s + E)}}{\hbar}$$

the continuity condition yields the relations

$$A_s \sin K_s R_s = 1 - \frac{R_s}{a_s} \tag{5.229}$$

$$A_s K_s \cos K_s R_s = -\frac{1}{a_s} \tag{5.230}$$

Dividing (5.230) by (5.229) gives

$$K_s \cot K_s R_s = \frac{1}{R_s - a_s} \tag{5.231}$$

Then using the functions u and v in (5.214)

$$r_{0s} = 2 \int_0^{R_s} \left[\left(1 - \frac{r}{a_s} \right)^2 - A_s^2 \sin^2 K_s r \right] dr \tag{5.232}$$

A direct calculation of the integral which uses expressions (5.229) and (5.230), yields

$$r_{0s} = R_s - \frac{1}{3} \frac{R_s^3}{a_s^2} - \frac{1}{a_s K_s^2} \tag{5.233}$$

It is interesting to note that for $a_s \to \pm\infty$, i.e. for the singlet binding energy exactly zero, $r_{0s} \to R_s$. Thus, in the case of square well potential which has such a depth that the binding energy just vanishes, the effective range is identical with the actual range.

We can estimate a_s from our knowledge of neutron-proton scattering cross-section σ_0 at very low energies. Accepting the value $\sigma_0 = 20.36 \pm 0.10$ b, and using the relation (5.181), we can write

$$\sigma_0 = 20.36 \times 10^{-24} = \frac{1}{4}\sigma_s + \frac{3}{4}\sigma_t = \pi \left(a_s^2 + 3a_t^2 \right) \tag{5.234}$$

Further, using the known value of the triplet scattering length at zero energy, viz $a_t(0) = +5.38$ fm, we find

$$a_s = -23.6 \text{ fm} \tag{5.235}$$

the corresponding cross-section being $\sigma_s = 4\pi a_s^2 = 70$ b. The negative sign on a_s has been chosen for reasons that will become apparent later. Because of the large magnitude of $|a_s|$, the last two terms in (5.233) are small, and it is a sufficiently good approximations to write

$$r_{0s} = R_s \tag{5.236}$$

in analogy with (5.216), we have the formula for the cross section in the singlet state

$$\sigma_s = \frac{4\pi}{k^2 + (\frac{1}{2}r_{0s}k^2 - \frac{1}{a_s})^2} \tag{5.237}$$

Inserting the values of a_s and r_{0s} from (5.235) and (5.236) and expressing k^2 in terms of energy

$$\sigma_s \text{ (b)} = \frac{4\pi}{0.014R_s^2 E^2 + E(2.4 + R_s) + 0.18} \tag{5.238}$$

where E is in MeV and R_s is in fermis. It will be found that with the choice of $R_s = 2.5$ fm, the experimental cross sections are brought in fair agreement with the theory, throughout the low energy region. The formula then becomes

$$\sigma \text{ (b)} = \frac{4\pi}{0.088E^2 + 4.9E + 0.18} \tag{5.239}$$

At $E = 0$, the formula predicts the large expected value $\sigma_s = 70$ b, at $E = 2$ MeV, it drops to 1.2 b, while at $E = 5$ MeV, it gives 0.5 b. We conclude that the cross-section in the singlet state is a rapidly decreasing function of energy, which is in contrast with the slow varying behaviour of the triplet cross-section. The total cross section, which is the mixture of σ_t and σ_s with the appropriate statistical weights, can then be written as

$$\sigma = \frac{3\pi}{k^2 + (\frac{1}{a_t(0)} - \frac{1}{2}k^2 r_{0t})^2} + \frac{\pi}{k^2 + (\frac{1}{a_s(0)} - \frac{1}{2}k^2 r_{0s})^2} \tag{5.240}$$

A numerical formula based on (5.226) and (5.239) is

$$\sigma = \frac{3\pi}{0.04E^2 + 1.66E + 3.46} + \frac{\pi}{0.088E^2 + 4.9E + 0.18} \tag{5.241}$$

In both the terms, the co-efficient of the quadratic term E^2 is small. The difference in the remaining terms in the denominators of the partial cross-sections of triplet scattering and singlet scattering arises from the difference in signs and magnitudes of the amplitudes $a_t(0)$ and $a_s(0)$. This becomes more transparent on examining the two equivalent terms in (5.240).

Figure 5.12 shows the variation of the total cross section with energy as given by (5.241). The partial contributions, $(3/4)\sigma_t$ and $(1/4)\sigma_s$ corresponding to the first and second terms respectively in (5.241) are also shown separately. It is seen that for very small energies, the singlet scattering dominates. With the increase of energy, both the contributions diminish, the contribution for the singlet state falling much

Fig. 5.12 Variation of the total cross-section with energy

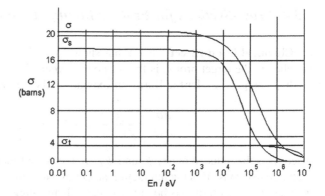

Fig. 5.13 (a) Wave function for triplet $n-p$ scattering at $E_n \sim 200$ keV and a well radius of ~ 1 fm. The scattering length is positive. (b) Wave function for singlet $n-p$ scattering with negative scattering length

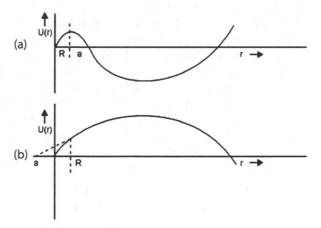

more rapidly than that for the triplet state. At around 1 MeV, the contributions become about equal; and at higher energies the triple scattering becomes much more important.

5.3.3 Nature of the Singlet and Triplet States

Figure 5.13 shows the nature of wave functions for (a) a bound state (b) an unbound system. The sign of the scattering length is closely connected with the existence or non-existence of a bound system. Now, the relation (5.217) has a solution with real γ, only if 'a' is positive, corresponding to bound state. Conversely, if 'a' is negative then γ and R will also be negative, i.e. the system cannot have a bound state. When the system is unbound the inside wave function does not have enough curvature to bend around and consequently the linear extrapolation of the outside function intersects the r-axis at a negative value. Experiments, which we shall consider later, prove that the sign of a_t is positive and that of a_s as negative.

5.3.4 Cross-Sections for Protons Bound in Molecules

(a) **Chemical Bond Effect**   Fermi has shown that, the $n-p$ cross-section depends on whether the target proton is free or bound, when E_0 is much less than chemical binding energies. Under the Born approximation, the differential cross-section is

$$\frac{d\sigma}{d\Omega} = \frac{\mu^2}{4\pi^2\hbar^2}\left|\int \psi_f^* V \psi_i d\tau\right|^2 \tag{5.242}$$

where μ is the reduced mass, ψ_i and ψ_f are the wave functions of the incident and scattered neutrons, V is the interaction potential. Now, the reduced mass depends on whether the target proton is bound or free, while the integral does not depend on this fact. For a free proton, the reduced mass is $\mu = (M/2)$. In the other extreme case when the proton is bound to a very heavy molecule, such as paraffin, $\mu \simeq M$. In that case we should then expect σ (bound) $= 4\sigma$ (free). This result has been confirmed in the experiments of Rainwater et al. Born approximation, it must be pointed out, is not strictly valid for the very low energy neutrons (few eV), since the perturbation potential is of the order of 10 MeV or greater and should therefore grossly distort the wave—a feature which runs counter to the basic assumption implied in the Born approximation. However, the application of Born approximation to the present problem has been shown to be justified by using artificial potentials of much shallower depths and correspondingly larger widths. In general, for scattering against target mass number A

$$\sigma\,(\text{bound}) = \left(\frac{A+1}{A}\right)^2 \sigma\,(\text{free}) \tag{5.243}$$

Following Fermi, the proton is essentially bound to the molecule if $E_0 \ll h\nu$ where ν is the frequency of the proton in the sub-group of the molecule. In the case of CH bound in paraffin $\hbar\nu \sim 0.4$ eV. For $E_0 < \hbar\nu$, no energy will be imparted in the way of the vibration of the molecule. When E_0 coincides with $\hbar\nu$, the probability of losing one quantum of energy to the vibration becomes very high, leading to an abrupt rise in the cross-section. Similar discontinuities in the cross-section occur at $E_0 = 2h\nu$ etc. (Fig. 5.14). At higher energies ($E_0 \gg \hbar\nu$), the proton can be easily knocked off from the molecule and therefore acts as if it were a free particle, and consequently the cross-section approaches σ (free).

It was mentioned that for $E_0 < 0.4$ eV, energy transfer to the CH bond is not possible. However, energy transfer can be effected to the vibration of the whole CH_2 group which have much smaller quantum energies.

(b) **Effect of Target Motion**   For neutrons which have energies comparable with the thermal energies (<0.1 eV) we can no longer ignore the thermal motion of the target nucleus. The scattering depends on the relative velocity of the neutron and the target nucleus, and because the relative velocity is directly related to the thermal motion of the molecules, the scattering cross-sections are increased. This is called the temperature effect, which is essentially similar to the Doppler effect

Fig. 5.14 Discontinuities in
the cross-section at different
energy values

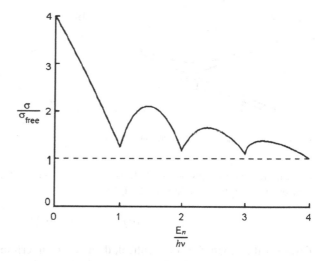

which affects the Breit-Wigner peaks in the absorption and scattering of neutrons in
nuclei, Fig. 5.14. The variation of cross-section with E_0, as measured in water in the
very low energy region $(0.003 < E_0 < 100$ eV) shows that at $E_0 = 0.004$ eV, the
observed cross-section (82 b) is about 4σ (free), which is consistent with the value
expected from the chemical binding effect. But at lower energies the cross-section
is still increasing. This increase is partly due to the temperature effect and partly due
to the absorption of neutrons by protons $(n + p \rightarrow d + \gamma)$, a process which becomes
important at exceedingly low neutron energies.

5.4 Proton-Proton Scattering; Low Energy

Compared to neutron-proton scattering the proton-proton scattering is complicated
by the fact that the indistinguishability of the particles introduces new quantum mechanical effects. Secondly, the Coulomb interaction is superimposed on the nuclear
scattering resulting in interference effects.

The application of Pauli's principle to the two-proton system is that it can exist
only in states of $1\ ^1S, {}^3P, {}^1D$ etc. This can be proved by considering the spherical-
harmonic part of the eigen function. On exchanging the protons, which is identical
with changing θ to $\pi - \theta$, the spherical harmonic is multiplied by $(-1)^l$, where l
is the orbital angular momentum. Now the spin wave function is even for the triplet
$(S = 1)$ state and odd for the singlet $(S = 0)$ state. The overall wave function will be
odd for $S = 0$ and $l =$ even or for $S = 1$ and $l =$ odd, as required by Pauli's principle
$p-p$ scattering in the CM system. The two diagrams are identical for indistinguishable particles.

Differential scattering cross-section is given by

$$\frac{d\sigma}{d\Omega} = |f(\theta)|^2 \tag{5.244}$$

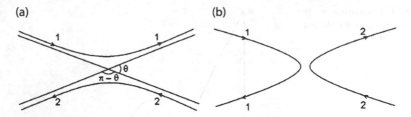

Fig. 5.15 Exchange of the two proton co-ordinates

For distinguishable particles of the same mass, the probability that either will be scattered through an angle θ in the CMS is

$$|f(\theta)|^2 + |f(\pi - \theta)|^2 \tag{5.245}$$

However if the particles are identical, their waves interfere and instead of summing the squares of the amplitudes, we must first sum the amplitudes and then square them. Further, for fermions the eigen function must be antisymmetric with respect to exchange of particles. The portion of the eigen functions containing only spatial coordinates without spins will then be

$$f(\theta) \pm f(\pi - \theta) \tag{5.246}$$

Exchange of the two proton coordinates means exchanging θ with $\pi - \theta$, as in Fig. 5.15. If plus sign is used in (5.246) the expression is symmetric with respect to the exchange of coordinates and if the minus sign is used the expression is antisymmetric to the exchange. First we shall consider pure Coulomb scattering of identical particles. The amplitude for Coulomb scattering is given by

$$f_c(\theta) = -\frac{1}{2ik} \sum_{l=0}^{\infty} (2l + 1)\left(e^{2i\sigma_l} - 1\right) P_l(\cos\theta) \tag{5.247}$$

It can be shown that this is equivalent to

$$f_c(\theta) = -\frac{\eta}{2k\sin^2\frac{\theta}{2}} e^{-i\eta \ln \sin 2(\theta/2) + 2i\sigma_0} \tag{5.248}$$

with $\eta = \frac{e^2}{\hbar v}$; v is the relative velocity of protons.

The part of the eigen functions containing only spatial coordinates without spin, will be

$$f(\theta) \pm f(\pi - \theta) \tag{5.249}$$

It was pointed out that triplet spin state is symmetric while singlet spin state is antisymmetric. For protons the total eigen function must be antisymmetric hence the triplet states will be associated with

$$f(\theta) - f(\pi - \theta) \tag{5.250}$$

and the in singlet states will be associated with

$$f(\theta) + f(\pi - \theta) \tag{5.251}$$

The scattering cross-section in triplet states is thus

$$\frac{d\sigma_t}{d\Omega} = |f(\theta) - f(\pi - \theta)|^2 = |f(\theta)|^2 + |f(\pi - \theta)|^2 - 2R_e[f(\theta)f^*(\pi - \theta)]$$
$$\tag{5.252}$$

and in singlet states it is

$$\frac{d\sigma_s}{d\Omega} = |f(\theta) + f(\pi - \theta)|^2 = |f(\theta)|^2 + |f(\pi - \theta)|^2 + 2R_e[f(\theta)f^*(\pi - \theta)]$$
$$\tag{5.253}$$

At very low energy the Coulomb scattering alone will be important as protons due to repulsion will not be allowed to come close enough to undergo nuclear scattering. The amplitude for Coulomb scattering given by (5.247) can be rewritten as

$$f_c(\theta) = \frac{e^2}{Mv^2 \sin^2(\frac{\theta}{2})} \exp\left(-i\eta \ln \sin^2 \frac{\theta}{2}\right) \tag{5.254}$$

For unpolarized proton beams, the triplet and singlet scattering cross-sections are added with statistical weights 3/4 and 1/4 respectively to obtain

$$\left(\frac{d\sigma}{d\Omega}\right)_c = \frac{3}{4}\left(\frac{d\sigma}{d\Omega}\right)_t + \frac{1}{4}\left(\frac{d\sigma}{d\Omega}\right)_s \tag{5.255a}$$

Using (5.252) and (5.253) in (5.255a) we get

$$\left(\frac{d\sigma}{d\Omega}\right)_c = \frac{1}{4}|f(\theta) + f(\pi - \theta)|^2 + \frac{3}{4}|f(\theta) - f(\pi - \theta)|^2 \quad \text{or} \tag{5.255b}$$

$$\left(\frac{d\sigma}{d\Omega}\right)_c = |f(\theta)|^2 + |f(\pi - \theta)|^2 - R_e f^*(\theta) f(\pi - \theta) \tag{5.256}$$

Using (5.254)

$$\left(\frac{d\sigma}{d\Omega}\right)_c = \left(\frac{1}{4\pi\varepsilon_0}\right)^2 \left(\frac{e^2}{Mv^2}\right)^2 \left[\frac{1}{\sin^4(\frac{\theta}{2})} + \frac{1}{\cos^4(\frac{\theta}{2})} - \frac{\cos[\eta \ln \tan^2(\frac{\theta}{2})]}{\cos^2(\frac{\theta}{2})\sin^2(\frac{\theta}{2})}\right]$$

(Mott scattering)
$$\tag{5.257}$$

The first two terms on the right represent the classical Rutherford scattering, the second term takes care of the target protons which recoil at angle $\pi - \theta$ in the CMS. The detector cannot distinguish between the two possibilities. Thus the first two terms are classical terms. The third term is a quantum mechanical interference term. The three terms constitute the Mott scattering. A term similar to the third term, but of opposite sign occurs for identical bosons such as alpha scattering with helium.

Fig. 5.16 Interference effect
for carbon-carbon elastic
scattering at 5 MeV

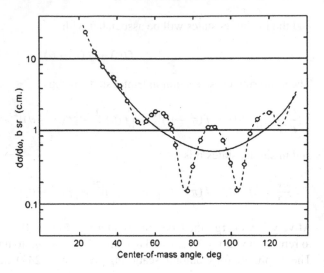

Note that the numerator of the third term is nearly 1 for protons of energy larger
than 1 MeV and for angles not too close to 0° or to 90°. An example of interference
effect is shown in Fig. 5.16 for carbon-carbon elastic scattering at 5 MeV from the
work of [3].

So far we have neglected the presence of nuclear forces. At low energy, the nuclear phase shifts δ_l are all zero except δ_0, and due to the Pauli principle, the two
protons must be in a singlet state. However, the Coulomb phase shifts contribute to
the scattering amplitude also for $l \neq 0$ and the scattering amplitude becomes

$$f(\theta) = f_c(\theta) + f_n(\theta) = \frac{1}{2ik} \sum_l (2l+1)\left(e^{2i\sigma_l} - 1\right) P_l(\cos\theta) + \frac{1}{k} e^{2i\sigma_0} e^{2i\delta_0} \sin\delta_0$$

$$(5.258)$$

For even l, only singlet states contribute to $f_c(\theta)$, for odd 1 only triplet states. We
finally get

$$\frac{d\sigma}{d\Omega} = \left(\frac{1}{4\pi\varepsilon_0}\right)^2 \left(\frac{e^2}{Mv^2}\right)^2 \left\{ \frac{1}{\sin^4\frac{\theta}{2}} + \frac{1}{\cos^4\frac{\theta}{2}} - \frac{\cos[\eta \ln \tan^2(\frac{\theta}{2})]}{\sin^2(\frac{\theta}{2})\cos^2(\frac{\theta}{2})} \right\}$$

← Rutherford scattering → ← Quantum mechanical interference

$$\frac{-2}{\eta} \sin\delta_0 \left[\frac{\cos(\delta_0 + \eta \ln \sin^2\frac{\theta}{2})}{\sin^2\frac{\theta}{2}} + \frac{\cos(\delta_0 + \eta \ln \cos^2\frac{\theta}{2})}{\cos^2\frac{\theta}{2}} \right] + \frac{4\sin^2\delta_0}{\eta^2} \qquad (5.259)$$

← Quantum mechanical interference → ← pure nuclear →

between Coulomb and nuclear scattering potential scattering

δ_0 is the $l = 0$ phase-shift for pure nuclear scattering. The six terms can be identified.
(1) The $\sin^{-4}\theta/2$ is the Rutherford scattering. (2) Since the two protons are identical
one cannot tell the case in which the incident proton comes out at θ and the target

Fig. 5.17 The data are from [7]

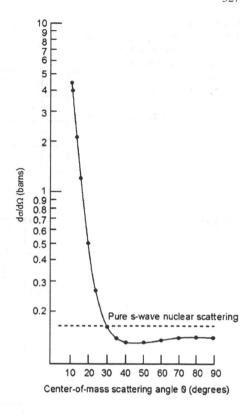

proton at $\pi - \theta$ (in the centre-of-mass system) from the case in which the incident proton comes out at $\pi - \theta$, and the target proton at θ. Thus, the scattering cross-section must contain a term $\sin^{-4}(\pi - \theta)/2 = \cos^{-4}\theta/2$. (3) This term describes the interference between Coulomb scattering at θ and $\pi - \theta$. (4 and 5) These two terms result from the interference between Coulomb and nuclear scattering. (6) The last term is the pure nuclear scattering term.

For a given kinetic energy the differential cross-section can be measured as a function of angle, and δ_0 can be extracted from the best fit of (5.259) Fig. 5.17 shows the differential cross-section for incident energy at 3.037 MeV. Fitting the data the last equation yields $\delta_0 = 50.966°$ at $T = 3.037$ MeV. The cross section for pure nuclear scattering would be $0.0165b$. The observation of smaller value gives evidence for interference between Coulomb and nuclear parts of the wave function.

The dependence of δ_0 on energy as obtained from a number of experiments is shown in Fig. 5.18.

The ^1S wave is the only part of the incoming wave which is appreciably changed by the nuclear effects at energies below 10 MeV, the phase shift δ_0 is the only change from pure Coulomb scattering. All the other partial waves ($^3P, ^1D, ^3F, \ldots$) are effectively subject to the Coulomb force only since the particles do not get close enough to each other to experience nuclear forces. This then means that at a given energy the scattering cross-section is fixed by a single parameter δ_0. Thus it is possi-

Fig. 5.18 Experimental p–p differential scattering cross sections as represented by the nuclear phase shift δ_0, as a function of the incident proton energy. The data are compiled by Jackson and Blatt; and Worthington et al.

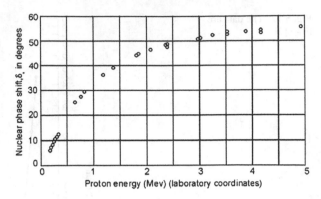

Fig. 5.19 Curtesy from [5]

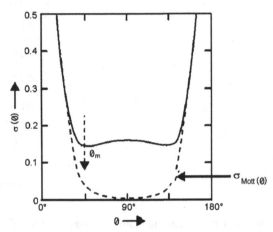

ble to fit the observed differential cross-section at all scattering angles by the use of only one adjustable parameter. A perfect agreement between theory and experiment is a triumph of the theory.

Figure 5.19 shows the differential cross-section for proton-proton scattering at $E_{lab} = 2.4$ MeV. At small angles in CMS the scattering is essentially pure Coulomb (Rutherford) scattering. At larger angles the Coulomb scattering interferes appreciably with the nuclear scattering. At still wider angles the nuclear scattering predominates. In the central region of angles, the cross-section is approximately constant in this region because the nuclear scattering is s-wave scattering ($l = 0$) only. The dip around $\theta = \theta_m$, Fig. 5.19 is caused by interference between nuclear and Coulomb scattering. Since Coulomb scattering is repulsive and nuclear scattering attractive, the interference is destructive. At angles $\theta < \theta_m$ and correspondingly at $\theta > \pi - \theta_m$, the Coulomb scattering predominates.

Figure 5.20 shows the angular distribution of p–p scattering in the CMS for the incident proton energies (Lab) marked on each curve.

The most important result of p–p scattering analysis is that it permits the determination of the sign of the phase shift which occurs as linear in (5.266). Positive values of δ_0 correspond to an attractive interaction, negative values of δ_0 signify

Fig. 5.20 Angular distribution of $p-p$ scattering (in C coordinates), for the incident proton energies (in L coordinates) marked on each curve [2]

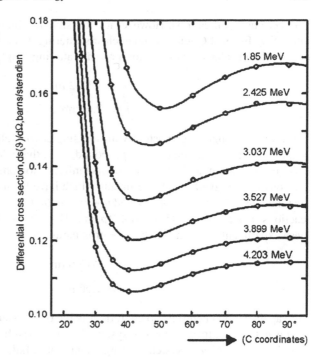

a repulsive force. This is also reflected in the scattering lengths. The comparison between neutron-proton and proton-proton forces in the ^1S state gave rise to the hypothesis of the charge independence of the nuclear forces.

By measuring the differential scattering cross-section as a function of angle at a specific incident kinetic energy, one can extract δ_0 from the best fit of the curve using (5.259). As an example, Fig. 5.20 shows such a fit from which a value of $\delta_0 = 50.966°$ is deduced at $T = 3.037$ MeV. From numerous such experiments the dependence of δ_0 on energy can be found out as in Fig. 5.18.

The neutron-proton scattering differs in two aspects

1. There are two phase shifts available, for the ^3S and ^1S scattering.
2. The differential cross-section is spherically symmetrical in the CMS.

A complicated angular distribution in proton-proton scattering is caused by the interference between Coulomb and nuclear scattering and therefore provides a much more sensitive test of the theory.

The important parameters which are energy independent are scattering length and effective range. Calculations are rendered difficult due to the fact that Coulomb interaction has infinite range and even in the $k \to 0$ limit one cannot neglect the higher order terms of Eq. (5.260)

$$k \cot \delta_0 = \frac{1}{a} + \frac{1}{2} r_0 k^2 + \cdots \qquad (5.260)$$

with certain modifications. However it is possible to obtain an expression incorporating the effects of Coulomb and nuclear scattering in a form similar to (5.253) and thus to obtain values for the proton-proton scattering length and effective range

$$a = -17.2 \text{ fm} \qquad (5.261)$$

$$r_0 = 2.65 \text{ fm} \qquad (5.262)$$

The effective range is consistent with the singlet np values. The fact that 'a' is negative, suggests that there is no pp bound state, that is the nucleus ^{2}He does not exist. For nn parameters, one has to extract information only indirectly as free target neutrons are not available. Experiments which have been used are concerned with the reactions $\pi^- + {}^2\text{H} \rightarrow 2n + \gamma$ and $n + {}^2\text{H} \rightarrow 2n + p$. Also comparison of mirror reactions such as $^3\text{He} + {}^2\text{H} \rightarrow {}^3\text{H} + 2p$ and $^3\text{H} + 2\text{H} \rightarrow {}^3\text{He} + 2n$. The analysis of these experiments give the neutron-neutron parameters

$$a = -16.6 \text{ fm} \qquad (5.263)$$

$$r_0 = 2.66 \text{ fm} \qquad (5.264)$$

Here again the two neutrons do not form a stable bound state just as for $p-p$ system.

It is not correct to say that $p-p$ system does not exist because of Coulomb repulsion. In the case of $n-n$ system this argument would fail.

The correct explanation for two identical fermions is that di-proton and di-neutron systems must have antisymmetric or singlet spin states (for spatially symmetric state $l = 0$) which are unbound.

5.5 High Energy Nucleon-Nucleon Scattering

The low-energy nucleon-nucleon scattering is well described by phase-shift analysis based on quantum mechanics. The analysis gives some information about the strength of nucleon-nucleon interaction but not about the detailed shape of the potential. To this end scattering experiments were continued at higher energies. The meeting ground between experiment and theory is a set of phase shift for the various angular momentum states in which the two particles interact. As the bombarding energy is increased a greater number of phase shifts are involved. In low-energy proton-proton scattering, because of Pauli's exclusion principle we deal with only one such state, and in low energy neutron-proton scattering, we deal with two states, one singlet S-state and one degenerate triplet S-state. For each value of l greater than zero there are four $n-p$ states, one singlet and three triplet states, but in the $p-p$ scattering the exclusion principle allows only one (antisymmetric) if l is even (symmetric) or three states (symmetric) if l is odd (antisymmetric).

At a given energy sufficient experimental information is needed to determine the phase shift for each of these states. In addition, if tensor forces are present then the so-called mixing parameters need to be introduced to connect, for example, the

$^3S_1$-state and the $^3D_1$-state. Thus in the case of $n-p$ scattering at a given energy, sufficient information must be available in order to determine $(5l_{max} + 1)$ parameters (phase shifts and mixing parameters). Here l_{max} is the maximum l-value, taken as odd, that must be considered for the energy used. For $p-p$ scattering, the number of parameters is $(5l_{max} + 3)/2$. As an example, if $l_{max} = 3$, for $n-p$ scattering we need to determine fourteen phase shifts and two mixing parameters while for $p-p$ scattering, eight phase shifts and one mixing parameter need to be determined. The experimental information that is most readily available in the differential cross-section as a function of the angle, that is the angular distribution [11] and [10] can be written as

$$\frac{d\sigma(\theta)}{d\Omega} = \sum_{n=0}^{2l_{max}} A_n Y_{n_0}, \quad (\theta) \tag{5.265}$$

The number of terms in the sum is $2l_{max} + 1$, and so this is the maximum number of parameters, A_n, that can be determined from the $n-p$ angular distribution measurement. In proton-proton scattering, the angular distribution is symmetrical around $90°$ in the CMS, so that only even n spherical harmonics would be present in Eq. (5.265). The number of coefficients A_n that is obtained from $p-p$ angular distribution measurements is therefore only $l_{max} + 1$. It is obvious that in both cases the angular distribution measurements at a given energy do not yield adequate information to permit the determination of the phase shifts and mixing parameters at that energy. Additional information can be obtained from polarization experiments which determine the spin direction of the scattered particles.

5.5.1 Polarization

Normally the spins of beam particles are randomly oriented. But somehow if the spins are oriented in a particular direction then the beam is said to be polarized. The polarization can be achieved in the scattering of particles, similar to the scattering of light. With a purely central interaction potential and an unpolarized target nucleus, the scattered particles would not show up any left-right asymmetry, even with a polarized incident beam. The direction left-right is taken perpendicular to the scattering plane. However, if the interaction has a tensor or non-central component with the coupling of the spin and orbital angular momentum then a polarized beam showing left-right asymmetry, may be produced. The larger is the orbital angular momentum the greater will be the asymmetry in the scattered beam. Consequently, polarization becomes more important at high energy where a number of partial waves associated with high values of l contribute to the phase-shift. In the polarization experiments the beam is first polarized by scattering and the polarization is detected in the second scattering experiment. These two constitute the double scattering experiment. The first one plays the role of polarizer and the second one that of analyzer as for light.

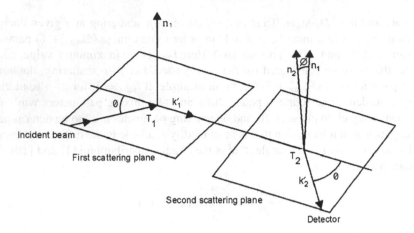

Fig. 5.21   Double-scattering experiment to measure polarization

The spin direction of the scattered particle, that is spin 'up' or 'down' relative to the plane of the scattering event is measured. In order to detect the polarization the particle is allowed to scatter a second time. The geometry of such a double scattering experiment is shown in Fig. 5.21. The apparatus is arranged so that the scattering angle θ is identical in both events. The trajectories of the incident particle and the scattered particle define the scattering plane for that event. The vectors n_1 and n_2 shown is Fig. 5.21 are unit vectors normal to the scattering planes. The second scattering plane subtends an angle ϕ with the first scattering plane. The polarization of the nucleons in a beam (or in a target) is defined as

$$P = \frac{N(\uparrow) - N(\downarrow)}{N(\uparrow) + N(\downarrow)} \tag{5.266}$$

where $N(\uparrow)$ and $N(\downarrow)$ refer to the number of nucleons with their spins pointed up and down respectively. When an unpolarized beam is scattered, values of P range from $+1$, for a 100 % spin-up polarized beam, to -1, for a 100 % spin-down polarized beam. An unpolarized beam, with $P = 0$, has equal numbers of nucleons with spins pointing up and down.

Let there be n particles in the beam which is initially unpolarized. The first scattering takes place with an unpolarized target T_1. Since the beam is unpolarized there will be $n/2$ particles with spins up $(+)$ and $n/2$ particles with spin down $(-)$. After scattering once to the left by T_1, the number of particles reaching the second unpolarized target T_2 with polarization P_1 and spin up is $(n/2) f_1 (1 + P_1)$, where f_1 is the fraction of all particles reaching T_2. The number reaching T_2 with spin down is $(n/2) f_1 (1 - P_1)$. Similarly, those particles scattered once to the right by T_1, $(n/2) f_1 (1 + P_1)$, will reach T_2 with spin down and $(n/2) f_1 (1 - P_1)$ will reach T_2 with spin up. Let P_2 be the polarization with spin up and f_2 the fraction of all particles scattered by the second target into the detector. Thus the number scattered twice to the left with spin up is

$$\left(\frac{n}{2}\right) f_1 f_2 (1 + P_1)(1 + P_2) \qquad (5.267)$$

and with spin down is

$$\left(\frac{n}{2}\right) f_1 f_2 (1 - P_1)(1 - P_2) \qquad (5.268)$$

Number of particles scattered twice, first to the left by T_1 and then to the right by T_2, with spin up, is

$$\left(\frac{n}{2}\right) f_1 f_2 (1 + P_1)(1 - P_2) \qquad (5.269)$$

and with spin down is

$$\left(\frac{n}{2}\right) f_1 f_2 (1 - P_1)(1 + P_2) \qquad (5.270)$$

Hence, the number of particles scattered twice to the left (LL) is

$$LL = \left(\frac{n}{2}\right) f_1 f_2 [(1 + P_1)(1 + P_2) + (1 - P_1)(1 - P_2)]$$

$$= \left(\frac{n}{2}\right) f_1 f_2 (2 + 2 p_1 p_2) \qquad (5.271)$$

and the number scattered first to the left and second to the right (LR) will be

$$LR = \left(\frac{n}{2}\right) f_1 f_2 [(1 + P_1)(1 - P_2) + (1 - P_1)(1 + P_2)]$$

$$= \left(\frac{n}{2}\right) f_1 f_2 (2 - 2 P_1 P_2) \qquad (5.272)$$

Hence

$$\epsilon = \frac{(LL) - (LR)}{(LL) + (LR)} = P_1 P_2 \qquad (5.273)$$

If the second scattering is identical to the first, then $P_1 = P_2 = P$ and (5.273) becomes

$$\epsilon = P^2$$

The quantity ϵ is called the asymmetry. The quantity ϵ is directly measurable while the magnitude of P can be found out, although not the sign of P. The sign may be determined by studying the interference of nuclear scattering with Coulomb scattering.

Figure 5.22 shows the differential cross-sections as a function of the CMS scattering angle, for p–p scattering at various energies. Figure 5.23 shows that the larger is the bombarding energy the greater is the polarization.

Fig. 5.22 Differential
cross-section for p–p elastic
scattering at several incident
energies. The experimental
data are compared with
theoretical predictions [8]

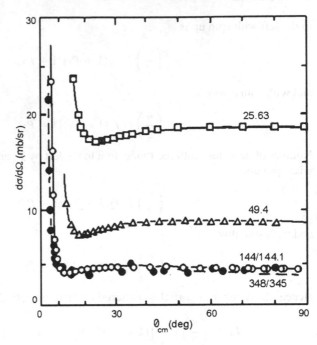

Fig. 5.23 Polarization in
p–p elastic scattering at
various incident energies [8]

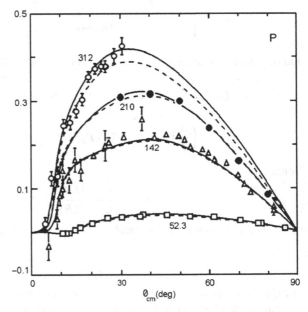

5.5.2 Mechanism of Polarization

Scattering that involves only s-waves will be spherically symmetric and does not
exhibit the phenomenon of polarization. We can now see how the spin-orbit in-

Fig. 5.24 Two nucleons with spin-up incident on spin-up target so that total spin $S = 1$

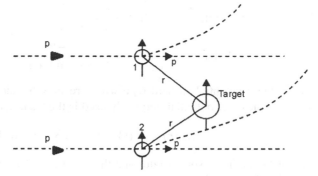

teraction can give rise to polarization for higher order angular momentum waves. Figure 5.24 shows two nucleons with spin-up incident on spin-up target so that total spin $S = 1$. The P-wave ($l = 1$) scattering of identical nucleons has an antisymmetric spatial wave function and hence a symmetric wave function. For incident nucleon 1, $\mathbf{l} = \mathbf{r} \times \mathbf{p}$ points down (into the page) and so $\mathbf{l} \cdot \mathbf{s}$ is negative because $\mathbf{l}$ and $\mathbf{s}$ point in opposite directions. If we assume that the spin-orbit potential $V_{so}(r)$ is negative so that the combination $V_{so}(r)\mathbf{l} \cdot \mathbf{s}$ is positive and so there is a repulsive force between the target nucleus and the incident nucleon 1 which pushes it to move to the left. For nucleon 2, $\mathbf{l}$ points up, $\mathbf{l} \cdot \mathbf{s}$ is positive, and the interaction being attractive, nucleon 2 is pulled in towards the target and is also scattered toward left. Thus the spin-up nucleons are preferentially scattered toward left and by a similar argument spin-down nucleons to the right. Consequently, the spin-orbit interaction can produce polarized scattered beams when unpolarized particles are incident on a target.

At low energies, where s-wave scattering dominates, polarization is not expected. As the bombarding energy increases, the contribution of p-waves increases leading to an increased polarization. These expectations are borne out in Fig. 5.23. The observations on the variation of P with θ and with energy elucidate vital information on the forms of the potential $V_{so}(r)$.

From the study of a large number of experiments at various energies a wealth of information in available on the differential and total cross-sections, spin dependence and polarization. This has permitted to introduce various phenomenological potentials to fit the observed data on nucleon-nucleon scattering.

The most frequently used expressions are:

$$\text{Square well} \quad V(r) = -V_c \qquad (r \leq a)$$

$$= 0 \qquad (r > a) \qquad (5.274)$$

$$\text{Exponential} \quad V(r) = -V_0 e^{-ar} \qquad (5.275)$$

$$\text{Gaussian} \quad V(r) = -V_0 e^{-a^2 r^2} \qquad (5.276)$$

$$\text{Yukawa} \quad V(r) = \frac{-V_0 e^{-ar}}{r} \qquad (5.277)$$

$$\text{Hulthén} \qquad V(r) = \frac{-V_0 e^{-ar}}{1 - e^{-ar}} \tag{5.278}$$

$$\text{Eckart} \qquad V(r) = \frac{-V_0}{(1 + e^{ar})(1 + e^{-ar})} \tag{5.279}$$

To all these potentials a hard repulsive core may be added for $r \le r_c$ with $r_c \simeq$ 0.5 fm. One other potential frequently used is the Gammel Thaler potential

$$V_{GT}(r) = V_c(r) + V_t(r)S_{12} + V_{Ls}(r)\mathbf{L} \cdot \mathbf{S} \tag{5.280}$$

where S_{12} is the tensor operator and the functions $V(r)$ are Yukawa functions for $r \ge r_c$ and go to infinity for $r < r_c$.

With the availability of fast computers, more complicated potentials have been proposed to fit the scattering data more accurately. Among these the Hamada-Johnston, Yale and Reid potentials have been extensively used.

5.6 Properties of the Nucleon-Nucleon Force

1. At short distances (~ 1 fm) nuclear force is stronger than Coulomb force.
2. At long distances ($\sim 10^{-8}$ cm) the nuclear force is negligibly small. The interactions among atoms in a molecule can be understood in terms of Coulomb force alone.
3. Some particles like electrons or muons are unaffected by nuclear force.
4. Nuclear force is charge independent, that is the force between the pairs of $n-n$, $n-p$, $p-p$ is identical.
5. The nucleon-nucleon force depends on the orientation of spins, that is the spins are parallel or antiparallel. Thus the nuclear force is spin dependent.
6. The nucleon-nucleon force has a tensor or non-central component which does not conserve the orbital angular momentum which is a constant of motion under central forces.
7. The nucleon-nucleon force includes a repulsive term which prevents the nucleons to come too close to each other.

5.6.1 Exchange Forces

The fact that nuclear density is constant suggests that nuclear forces have saturation property which can be explained by assuming that these forces are 'exchange forces', similar to the force that binds ordinary chemical molecules.

Regardless of the origin of these forces let us enumerate various types of exchange forces that exist between a pair of nucleons and then investigate the effect of these forces on the properties of the deuteron and on the saturation of the binding.

Four types of interactions may be distinguished.

5.6.1.1 Wigner Force

For an ordinary (non-exchange) central force the Schrodinger equation for two particles in the centre of mass system is

$$\left[\left(\frac{\hbar^2}{M}\right)\nabla^2 + E\right]\psi(r_1, r_2, \sigma_1, \sigma_2) = V(r)\psi(r_1, r_2, \sigma_1, \sigma_2) \tag{5.281}$$

Here the interaction does not cause any exchange between coordinates of the two particles and the spin coordinates remain unaffected.

5.6.1.2 Majorana Force

Here the interaction interchanges the space coordinates, the spin coordinates remaining unaffected, in addition to multiplication of ψ by the potential $V(r)$. For such an interaction the Schrodinger equation is

$$\left[\left(\frac{\hbar^2}{M}\right)\nabla^2 + E\right]\psi(r_1, r_2, \sigma_1, \sigma_2) = V(r)\psi(r_2, r_1, \sigma_1, \sigma_2) \tag{5.282}$$

5.6.1.3 Bartlett Force

Here the spin coordinates are interchanged but the spatial coordinates are unaffected. The Schrodinger equation is

$$\left[\left(\frac{\hbar^2}{M}\right)\nabla^2 + E\right]\psi(r_1, r_2, \sigma_1, \sigma_2) = V(r)\psi(r_1, r_2, \sigma_2, \sigma_1) \tag{5.283}$$

5.6.1.4 Heisenberg Force

Here both the space and spin coordinates are interchanged. The Schrodinger equation is

$$\left[\left(\frac{\hbar^2}{M}\right)\nabla^2 + E\right]\psi(r_1, r_2, \sigma_1, \sigma_2) = V(r)\psi(r_2, r_1, \sigma_2, \sigma_1) \tag{5.284}$$

5.6.2 Effect of Exchange Forces

When exchange forces are central forces with $V(r)$, l's are not mixed. However, if a tensor is used in place of $V(r)$ as the multiplying potential, l's are mixed and the quadrupole moment of the deuteron can be explained. It may be pointed out that the tensor force does not by itself lead to saturation.

5.6.2.1 Majorana Force

The Majorana interaction replaces (r) by $(-r)$ in ψ. This is just the parity operation which replaces $\psi(r)$ by $(-1)^l \psi(r)$. The Schrodinger equation then can be rewritten as

$$\left[\left(\frac{\hbar^2}{M}\right)\nabla^2 + E\right]\psi(r) = (-1)^l V(r)\psi(r) \tag{5.285}$$

This is equivalent to having an ordinary potential that changes sign according to whether l is even or odd, and is independent of spin. Since the potential is attractive for $l = 0$, it would be repulsive for $l = 1$ if the interaction were of pure Majorana type.

5.6.2.2 Bartlett Force

Two nucleons in triplet state (total $S = 1$) will have symmetric spin function and in the singlet state $(S = 0)$ it will be in antisymmetric state. For the Bartlett force, Schrodinger equation may be written as

$$\left[\left(\frac{\hbar^2}{M}\right)\nabla^2 + E\right]\psi(r) = (-1)^{s+1} V(r)\psi(r) \tag{5.286}$$

This is equivalent to an ordinary potential which changes sign between $S = 0$ and $S = 1$. Since we know from neutron-proton scattering data that both the ^3S and ^1S potentials are attractive, the nuclear force cannot be pure Bartlett type.

5.6.2.3 Heisenberg Force

From the discussion of Majorana and Bartlett force the Schrodinger equation for Heisenberg may be written as

$$\left[\left(\frac{\hbar^2}{M}\right)\nabla^2 + E\right]\psi(r) = (-1)^{l+s+1} V(r)\psi(r) \tag{5.287}$$

This is equivalent to an ordinary potential which changes sign according to whether $1 + S$ is even or odd. We give below some examples

$$
\begin{array}{cccc}
^3S & ^1S & ^3P & ^1P \\
V(r) & -V(r) & -V(r) & +V(r)
\end{array}
$$

The reversal of sign between ^3S- and ^1S-states indicates that the nuclear force cannot be purely of Heisenberg type. With the assumption of the interaction of 25 percent Heisenberg or Bartlett and 75 percent Wigner or Majorana the difference between neutron-proton well depths, 21 and 12 MeV respectively for $a = 2.8$ fm can be explained.

5.6.3 Exchange Forces and Saturation

The Bartlett spin exchange force does not lead to saturation of the binding energy per particle. For nuclear force of Bartlett type heavy nuclei should exist with all spins aligned. In such a case the number of interacting pairs is $A(A - 1)/2$, which leads to binding energy proportional to atleast the square of A.

However, the space exchange in the Majorana and the Heisenberg forces does lead to saturation because of the alteration in sign of the potential between odd and even l.

Consider the nucleus ^{4}He. The spatial wave function can be symmetrical in all the four particles without violating Pauli's principle by giving antiparallel spins (antisymmetric spin wave functions) to the two neutrons as well as two protons.

In the next heavier nucleus—^{5}He or ^{5}Li the Pauli principle can no longer be satisfied by spin wave functions alone. Therefore, the spatial wave function must have atleast one node. In other words, only four particles can be in an s-states and will therefore be repelled by the other particles, ^{5}He and ^{5}Li should thus be unstable, in agreement with the experiment. This is a first sign of saturation.

Wigner force does not lead to saturation. Saturation is achieved either by the space-exchange Majorana force or the space-exchange part of the Heisenberg force.

The Heisenberg force gives special stability to the deuteron, and the Majorana force to the alpha particle. The Bartlett forces do not give saturation. It is concluded that the exchange force of the Majorana type does exist. This is proved directly by high-energy neutron-proton scattering which shows maxima in the angular distribution of protons in the CMS.

We can understand the results by the use of Born's approximation.

$$f(\theta) = \frac{\mu}{2\pi \hbar^2} \int \exp(-ik_f r) V(r) \exp(ik_i r) d\tau \tag{5.288}$$

where $r = r_1 - r_2$ and μ is the reduced mass. The integral for a short-range $V(r)$ is appreciable only if $k_i - k_f \cong 0$, otherwise the function $\exp[i(k_i - k_f)r]$ oscillates so rapidly over the region for which $V(r)$ is appreciable so that the result averages to 0. Now $k_i - k_f = 0$ indicates forward scattering. We thus find that the striking neutrons are scattered forward, contrary to experiment. If we now introduce the exchange force, the potential $V(r)$ is replaced by $V(r)P_M$ which gives us

$$f_{ex}(\theta) = \frac{\mu}{2\pi \hbar^2} \int \exp(+ik_f r) V(r) P_M \exp(ik_i r) d\tau \tag{5.288(a)}$$

as the P_M operator changes particles 1 and 2 and hence transforms r into $-r$.

The $f_{ex}(\theta)$ due to exchange force is large when

$$k_f + k_i = 0 \tag{5.289}$$

That is when the neutron is scattered backward and the proton forward. The experimental result of Fig. 5.25 demonstrates clearly two maxima in the forward and

Fig. 5.25 Result of two maxima in the forward and backward at high energies

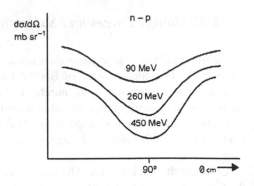

backward at high energies. We conclude that both normal and exchanges forces are present and that they are of comparable intensities. The interaction is assumed to take place due to exchange of field quanta, that is mesons. Since the range of nuclear interaction is finite the mesons were assumed to be massive. This is in contrast with photons which are the quanta of electromagnetic field, with zero rest mass and infinite range. According to quantum field theory, the field itself is quantized, that is object with field quanta and the second object interact only with the field, not directly with the first object, and absorb the field quanta (and reemit them back to the first object) the two objects interact directly with the exchanged field quanta and therefore, indirectly with each other.

Since the nucleons have spin 1/2 and if they have to be transformed into one another the quanta to be exchanged between two nucleons must have spin 0 or 1 and must carry electric charge $+$ or $-$ or 0 charge.

The particle which is exchanged is assumed to represent the nuclear force and is called meson, from the Greek meso, meaning intermediate, because the predicted mass is between that of proton and electron. Suppose a nucleon emits a particle X. A second nucleon absorbs the particle X

$$N_1 \rightarrow N_1 + X$$

$$X + N_2 \rightarrow N_2$$

When the particle of rest mass energy $m_X c^2$ is emitted or absorbed there will be an apparent violation of conservation of energy in these processes. However, if the process takes place within a short time Δt such that $\Delta t < \hbar m_X c^2$ as given by the uncertainty principle, then we will be unaware that the energy $m_\pi c^2$ has been violated. The maximum range of the force is determined by the maximum distance that the particle X can travel in the time Δt. If the exchanged particle travels at a speed of the order of c, then the range R can be at most

$$R = c\Delta t = \frac{c\hbar}{m_X c^2} = \frac{197 \text{ MeV fm}}{m_X c^2} \tag{5.290}$$

Equation (5.290) gives the relationship between the mass energy of the exchanged particle and the range of force, the smaller is the range the more massive will be the particle. For nuclear force with a range of about 1 fm, the exchanged particle must have mass energy of the order of 200 MeV. For 1.0–1.5 fm, pions are responsible for the exchange forces. For exchange between all the possible states of nucleon, pions must exist in all the three charge states +, 0 and −. The pions have zero spin and exist in three charged states with rest mass energies of 139.5 MeV for $\pi^{\pm}$ and 135 MeV for π^0.

The single pion that is exchanged between two identical nucleons must be a π^0

$$n_1 \to n_1 + \pi^0, \qquad \pi^0 + n_2 \to n_2$$
$$p_1 \to p_1 + \pi^0, \qquad \pi^0 + p_2 \to p_2$$

The one pion exchange model is abbreviated as OPE model. The neutron-proton interaction can be carried by charged as well as neutral pions.

$$n_1 \to n_1 + \pi^0, \qquad \pi^0 + p_2 \to p_2$$
$$n_1 \to p_1 + \pi^-, \qquad \pi^- + p_2 \to n_2$$
$$p_1 \to n_1 + \pi^+, \qquad \pi^+ + n_2 \to p_2$$

At shorter ranges (0.5–1.0 fm) two-pion exchange is probably responsible for the nuclear binding. At much shorter ranges (0.25 fm) the exchange of ω mesons ($mc^2 = 783$ MeV) may contribute to the force while the exchange of ρ mesons ($mc^2 = 769$ MeV) may contribute to the spin-orbit part of the interaction.

The exchanged particles are called virtual particles as they are unobserved. However they can be created as real particles at high enough energies and have identical properties to real particles.

5.7 Yukawa's Theory

In 1935, Hideki Yukawa, a Japanese physicist proposed a potential to represent the nucleon-nucleon interaction. The potential was to describe the exchange of particles giving rise to nuclear force, is analogy with the electromagnetic potentials which describe the exchange of photons that give rise to electromagnetic force. The major difference between the electromagnetic interaction and the strong nuclear interaction is the infinite range for the former in contrast with a short range of the order of 1 fm for the latter.

The basic equations for the electromagnetic field are Maxwell's equations which govern the propagation of photons. The relevant equation for spin zero particles is the Klein-Gordon equation. Consider the relativistic equation

$$E^2 = c^2 p^2 + m^2 c^4 \tag{5.291}$$

In quantum mechanics energy $E = i\hbar\partial/\partial t$ and momentum $p = -i\hbar\nabla$. Carrying out their substitutions in the relativistic equation, we obtain the Klein-Gordon equation

$$\left(\nabla^2 - \frac{m^2c^2}{\hbar^2}\right)\phi = \frac{1}{c^2}\frac{\partial^2\emptyset}{\partial t^2} \tag{5.292}$$

where ϕ represents the amplitude of the field. Observe that for $m = 0$, Eq. (5.292) reduces to the familiar wave equation for the electromagnetic field. The time-independent equation is

$$\nabla^2\phi - K^2\phi = 0 \tag{5.293}$$

where $K = mc/\hbar$. The spherically symmetric solution of (5.293) is

$$\phi = g\frac{e^{-kr}}{r} \tag{5.294}$$

where g is a constant that represents the strength of the pion field, in analogy with the electronic charge e which represent the strength of the electromagnetic field.

The nuclear forces should have the range of the order of $K^{-1} = \hbar/mc$, identical with Eq. (5.290) derived from arguments based on uncertainty principle. E is the total energy, and p is the momentum of a free particle of mass m.

Klein-Gordon equation describes the propagation of spinless particles of mass m

$$\nabla^2\phi - \frac{m^2c^2}{\hbar^2}\phi - \frac{1}{c^2}\frac{\partial^2\phi}{\partial t^2} = 0 \tag{5.295}$$

ϕ may be interpreted either as the potential at a point in space and time, or as the wave amplitude. Here we are not so much interested with propagation of particle waves as in static potentials. It we drop the time-dependent term, the resulting equation for the static potential U has the spherically symmetric form

$$\nabla^2U(r) = \frac{1}{r^2}\frac{\partial}{\partial r}\left(\frac{r^2\partial U}{\partial r}\right) = \frac{m^2c^2}{\hbar^2}U(r) \tag{5.296}$$

for values of $r > 0$ from a point source at the origin, $r = 0$. Integration gives

$$U(r) = \frac{g}{4\pi r}e^{-r/R} \tag{5.297}$$

$$\text{where } R = \frac{\hbar}{mc} \tag{5.298}$$

Here, the quantity g is a constant of integration identical with the strength of the point source. The analogous equation in the electromagnetism is $\nabla^2U(r) = 0$ for $r > 0$, with solution $U = Q/4\pi r$, where Q is the charge at the origin. Thus, g in the Yukawa theory plays the same role as charge in electrostatics and measures the "strong nuclear charge".

Example 5.3 A beam of 100 keV neutrons is attenuated to 50 % of its initial intensity in passing through 10 g cm^{-2} of carbon. What can you say about the s-wave phase-shift for the scattering of neutrons from carbon nuclei?

Solution

$$I = I_0 e^{-\Sigma x}$$

$$\frac{I}{I_0} = \frac{50}{100} = e^{-10\Sigma}$$

$$\Sigma = \frac{1}{10} \ln 2 = 0.0693 \text{ cm}^2 \text{ g}^{-1}$$

$$\sigma \frac{N_{av}}{A} = \frac{6 \times 10^{23} \sigma}{12} = 0.0693$$

$$\sigma = 1.38 \times 10^{-24} \text{ cm}^2 = 1.38 \times 10^{-28} \text{ m}^2$$

$$E = 100 \text{ keV} = 0.1 \text{ MeV}$$

$$\mu = \left(\frac{1 \times 12}{1 + 12}\right) M = \frac{12}{13} M$$

$$k^2 h^2 = 2\mu E$$

$$k^2 = 2 \times \frac{12}{13} \times \frac{1.67 \times 10^{-27} \times 0.1 \times 1.6 \times 10^{-13}}{(1.05 \times 10^{-34})^2} = 0.447 \times 10^{28} \text{ m}^2$$

$$\sigma_0 = \frac{4\pi \sin^2 \delta_0}{k^2}$$

$$\sin \delta_0 = \sqrt{\frac{\sigma_0 k^2}{4\pi}} = \sqrt{\frac{1.38 \times 10^{-28} \times 0.447 \times 10^{28}}{4\pi}} = \pm 0.2216 \text{ rad}$$

$$\delta = \pm 12.8°$$

Example 5.4 Consider the photo disintegration of deuteron, $\gamma + d \rightarrow p + n$. Assuming that the proton and neutron are emitted with equal energy, see Fig. 5.26, calculate the angle of emission of proton.

Solution

$$h\nu = W + T_p + T_n = W + 2T_p \quad \text{(energy conservation)} \qquad \text{(i)}$$

$$p_n^2 = p_p^2 + \frac{h^2 \nu^2}{c^2} - 2P_p \frac{h\nu}{c} \cos\theta \quad \text{(momentum conservation)} \qquad \text{(ii)}$$

Given $T_p = T_n$. Hence $P_p = P_n$.

Fig. 5.26 Photo
disintegration of deuteron

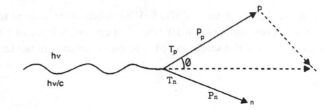

From (ii)

$$\cos\theta = \frac{\frac{h\nu}{c}}{2P_p} = \frac{h\nu}{2c\sqrt{2MT_p}} = \frac{h\nu}{2c\sqrt{M(h\nu - W)}}$$

Example 5.5   Find the root mean square separation of neutron and proton in
deuteron using the normalized exponential wave function for the ground state

$$\psi = \frac{1}{r}\sqrt{\frac{\alpha}{2\pi}}e^{-\alpha r}\left(\frac{1}{\alpha} = 4.3\times10^{-15}\ \text{m}\right)$$

Solution

$$\langle r^2\rangle = \int \psi^* r^2 \psi\, d\tau$$

$$= \int_0^\infty \frac{r^2}{r^2}\frac{\alpha}{2\pi}e^{-2\alpha r}4\pi r^2 dr$$

$$= \frac{1}{4\alpha^2}$$

hence

$$\sqrt{\langle r^2\rangle} = \frac{1}{2\alpha} = \frac{4.3}{2} = 2.15\ \text{fm}$$

Example 5.6   At what neutron energy will *p*-wave be important in *n*–*p* scattering?

Solution   In the CMS for $l = 1$

$$ap_{cm} = \hbar$$

$$E_{cm} = \frac{p_{cm}^2}{2\mu} = \frac{p_{cm}^2}{2\times\frac{M}{2}} = \frac{p_{cm}^2}{M} = \frac{\hbar^2}{Ma^2}$$

$$E_{lab} = 2E_{cm} = \frac{2\hbar^2}{Ma^2} = \frac{2\hbar^2 c^2}{Mc^2 a^2}$$

$$= \frac{2\times(197)^2}{940\times 2^2} = 20.6\ \text{MeV}$$

where we have put $a = 2$ fm. Thus below 20 MeV s-waves ($l = 0$) alone are important.

Example 5.7 Show that at a given energy p-waves affect $d\sigma/d\Omega$ to a larger extent than σ.

Solution

$$\sigma = \frac{4\pi}{k^2} \sum_{l=0}^{1} (2l + 1) \sin^2 \delta_l$$

$$= \frac{4\pi}{k^2} \left(\sin^2 \delta_0 + 3 \sin^2 \delta_1 \right)$$

$$\frac{d\sigma}{d\Omega} = \frac{1}{k^2} \left[\sin^2 \delta_0 + 6 \sin \delta_0 \sin \delta_1 \cos(\delta_1 - \delta_0) \cos \theta + 9 \sin^2 \delta_1 \cos^2 \theta \right]$$

As an example, let $\delta_0 = 20°$ and $\delta_1 = 2°$ at a certain energy. The p-wave ($l = 1$) contributes 3 % to total cross-section. But

$$\frac{\frac{d\sigma(0°)}{d\Omega}}{\frac{d\sigma(180°)}{d\omega}} = 3.5$$

Example 5.8 1 MeV neutrons are scattered on a target. The angular distribution of the neutrons in the centre-of-mass proves to be isotropic. The total cross-section is measured to be 10^{-25} cm^2. Using the partial wave representation, calculate the phase shift of the s-wave.

Solution Only s-waves ($l = 0$) are expected to be involved since scattering is isotropic

$$\sigma = \frac{4\pi}{k^2} \sin^2 \delta_0$$

but

$$k^2 \hbar^2 = p^2 = 2mE$$

$$\sigma = \frac{4\pi \hbar^2}{2mE} \sin^2 \delta_0$$

$$\sin^2 \delta = \frac{2mE\sigma}{4\pi \hbar^2} = \frac{mc^2 E\sigma}{2\pi \hbar^2 c^2} = \frac{940 \times 1 \times 10}{2\pi \times (197)^2} = 0.03857$$

where $\sigma = 10^{-25}$ cm$^2 = 10$ fm^2

$$\delta_0 = \pm 11.3°$$

Example 5.9 In the analysis of scattering of particles of mass m and energy E from a fixed centre with range 'a' the phase shift for the lth partial wave is given by

$$\delta_l = \sin^{-1}\left[\frac{(iak)^l}{\sqrt{(2l+1)(l!)}}\right]$$

show that the total cross-section at a given energy is approximately given by

$$\sigma = \frac{2\pi\hbar^2}{mE}\exp\left(\frac{-2mEa^2}{\hbar^2}\right)$$

Solution

$$\sigma = \frac{4\pi}{k^2}\sum_{l=0}(2l+1)\sin^2\delta_l$$

By problem

$$\sin\delta_l = \frac{(iak)^l}{\sqrt{(2l+1)(l!)}}$$

$$\sin^2\delta_l = \frac{(iak)^{2l}}{(2l+1)l!} = \frac{(-a^2k^2)^l}{(2l+1)l!}$$

$$\sigma = \frac{4\pi\hbar^2}{k^2\hbar^2}\sum_{l=0}\frac{(2l+1)(-a^2k^2)^l}{(2l+1)l!} = \frac{4\pi\hbar^2}{2mE}\sum_{l=0}\frac{(-a^2k^2)^l}{l!}$$

If the summation goes to infinite number of terms then

$$\sigma = \frac{2\pi\hbar^2}{mE}\exp(-a^2k^2) = \frac{2\pi\hbar^2}{mE}\exp\left(-\frac{2mEa^2}{\hbar^2}\right)$$

Example 5.10 Consider the scattering from a hard sphere of radius 'a' such that the D-wave phase-shift is negligible, the potential being

$$V(r) = \infty \quad \text{for } r < a$$
$$\quad\quad = 0 \quad \text{for } r > a$$

Show that

$$\sigma(\theta) = a^2\left[1 - \frac{(ka)^2}{3} + 2(ka)^2\cos\theta + \cdots\right] \quad \text{and}$$

$$\sigma = 4\pi a^2\left[1 - \frac{(ka)^2}{3}\right]$$

Solution

$$\sigma(\theta) = \frac{1}{k^2}\left|\Sigma(2l+1)e^{i\delta_l}\sin\delta_l P_l(\cos\theta)\right|^2$$

If only s- and p-waves are present

$$\sigma(\theta) = \frac{1}{k^2} |e^{i\delta_0} \sin\delta_0 P_0(\cos\theta) + 3e^{i\delta_1} \sin\delta_1 P_1(\cos\theta)|^2$$

but

$$P_0(\cos\theta) = 1; \qquad P_1(\cos\theta) = \cos\theta$$

$$\sigma(\theta) = \frac{1}{k^2} |e^{i\delta_0} \sin\delta_0 + 3e^{i\delta_1} \sin\delta_1 \cos\theta|^2$$

$$= \frac{1}{k^2} [\sin^2\delta_0 + 6\sin\delta_0 \sin\delta_1 \cos(\delta_0 - \delta_1)\cos\theta + 9\sin^2\delta_1 \cos^2\theta]$$

$$= \frac{1}{k^2} \left[\left(\delta_0 - \frac{\delta_0^3}{3!} \right)^2 + 6\delta_0\delta_1 \cos\theta \right]$$

$$= \frac{1}{k^2} \left[\delta_0^2 - \frac{\delta_0^4}{3} + 6\delta_0\delta_1 \cos\theta \right]$$

$$= \frac{1}{k^2} \left[k^2a^2 - \frac{1}{3}k^4a^4 + 6(ka)\frac{(k^3a^3)}{3} \cos\theta \right] \quad \left(\because \delta_0 = ka, \ \delta_1 = \frac{k^3a^3}{3} \right)$$

$$= a^2 - \frac{k^2a^4}{3} + 2k^2a^4 \cos\theta$$

$$\frac{d\sigma}{d\Omega} = a^2 \left[1 - \frac{(ka)^2}{3} + 2(ka)^2 \cos\theta \right]$$

$$\sigma = \int \left(\frac{d\sigma}{d\Omega} \right) d\Omega = 2\pi \int_{-1}^{+1} a^2 \left[1 - \frac{(ka)^2}{3} + 2(ka)^2 \cos\theta \right] d\cos\theta$$

$$= 4\pi a^2 \left[1 - \frac{(ka)^2}{3} \right]$$

Example 5.11   Show that the expectation value of the potential energy of deuteron described by a square well of depth V_0 and width R is given by

$$\langle V \rangle = -V_0 A^2 \left[\frac{1}{2}R - \frac{1}{4K} \sin 2KR \right]$$

Solution   As $V(r) = 0$ for $r > R$, the contribution to $\langle V \rangle$ comes only from within the well

$$\langle V \rangle = \int_0^R u_1^*(-V_0)u_1 dr = -V_0 \int_0^R A^2 \sin^2 kr \, dr$$

Integration gives the desired expression.

5.8 Questions

5.1 What fraction of time is spent by deuteron outside the range of nuclear forces?

5.2 Why deuteron is said to have a loose structure?

5.3 Why deuteron has only one bound state?

5.4 The magnetic dipole moments of proton and neutron do not add up exactly to that of the deuteron. How is this fact explained?

5.5 Deutron has positive electrical quadrupole moment. Does this imply a cigar shaped or prolate structure?

5.6 How is the existence of electrical quadrupole moment of deuteron explained?

5.7 List the requirements for the potential representing the static nuclear force to fulfill?

5.8 Why we cannot have a term like $\mathbf{S} \cdot \mathbf{r}_{12}$ or $\mathbf{S} \cdot \mathbf{P}_{12}$ for the potential?

5.9 Give an example of a static force and velocity dependent force.

5.10 What is the significance of the expression $S_{12} = \frac{3}{r^2}(\sigma_1 r)(\sigma_2 r) - \sigma_1 \sigma_2$.

5.11 Write down the most general static potential as the sum of six terms showing $s = 0/1, l = \text{odd/even}$, force = central/tensor.

5.12 Mention the four types of exchange forces with necessary description.

5.13 Mention various methods for the determination of binding energy of deuteron. Which method would you rate as most accurate?

5.14 Mention some of the widely used nucleon-nucleon potentials.

5.15 Explain why $R = \hbar/\sqrt{MB}$ is called the deuteron radius.

5.16 Given that there are no bound states of the dineutron and the diproton, what can you infer about the force between the two nucleons?

5.17 The deuteron ground state has the following properties (1) spin and parity $j^\pi = 1^+$, (2) magnetic dipole moment $\mu = 0.857\mu_N$, (3) electric quadrupole small and positive.

   What information do these quantities give about the ground state wave function of the deuteron? The proton and neutron dipole magnetic moments are, respectively, 2.793 and $-1.913\mu_N$.

Fig. 5.27 Differential
cross-sections for Coulomb
scattering of carbon isotopes

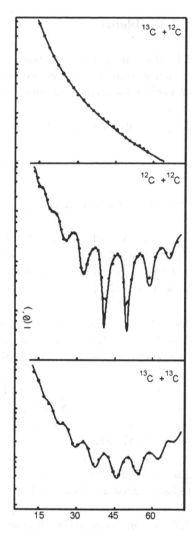

5.18 In the deuteron problem on using square well potential of depth V_0 and
width R, one finds a relation $V_0 R^2 = $ const, find the value of the constant.
[Ans. 103 MeV fm^2]

5.19 Figure 5.27 shows the differential cross-sections for Coulomb scattering of
^{13}C on ^{12}C (top), ^{12}C on ^{12}C (middle) and ^{13}C on ^{13}C (bottom) at an incident lab
energy of 4 MeV. Explain why these cross-sections are so different.

5.20 Describe the evidence for this following properties of the nucleon-nucleon
interaction (1) short range, (2) charge independence, (3) spin dependence.

5.9 Problems

5.1 Using the exponential wave function $u = C \exp(-kr)$, with $k = 0.232$ fm^{-1} for the ground state of deuteron $(0 < r < \infty)$, show that $\langle r \rangle = 1.83$ fm and the probability for neutron and proton to stay outside 2 fm is $p = 0.395$.

5.2 Show that the threshold energy for photo disintegration of deuteron is

$$E_\gamma = E_B + \frac{E_B^2}{2m_d C^2}$$

where E_B is the binding energy.

5.3 A particle of mass m moves in a potential $V(r) = -V_0$ when $r < a$, and $V(r) = 0$ when $r > a$. Find the least value of V_0 such that there is a bound state of zero energy and zero angular momentum.
[Ans. $V_0 = \frac{h^2}{32ma^2}$]

5.4 The $n-p$ interaction in deuteron may be described by square well potential of width a and depth $-V_0$. Assuming that the binding energy of deuteron is much smaller than the potential well depth, show that $V_0 a^2 = $ const.

5.5 When slow neutrons of negligible energy are captured by $_1^1H$ to form $_1^2D$, γ-rays of 2.224 ± 0.005 MeV are observed. Find this mass of neutron in MeV/c^2. Given $M_p = 938278$ MeV/c^2, $M_D = 1875.625$ MeV/c^2.
[Ans. 939.571 MeV/c^2]

5.6 Show that the nucleon-nucleon potential can only contain terms given by a radial function multiplied by 1, σ_1, $\sigma_2 \cdot (\sigma_1 \cdot \mathbf{r})(\sigma_2 \cdot \mathbf{r})$, and $(\sigma_1 + \sigma_2) \cdot (\mathbf{r} \times \mathbf{p})$.

5.7 A more accurate Hamiltonian for deuteron has the form

$$\frac{3(\sigma_1 \cdot \mathbf{r})(\sigma_2 \cdot \mathbf{r})}{r^2} - \sigma_1 \cdot \sigma_2$$

Explain various symbols in the above expression. What is the bearing of this expression on $n-p$ scattering results.

5.8 Show that if deuterons are scattered by protons, the maximum scattering angles in the lab system and the CMS are 30° and 120°, respectively, but that if protons are scattered by deuterons, the maximum angle in both the systems is 180°.

5.9 Show that the expectation value of the electric quadrupole moment for a neutron-proton system in the $^3S_1$ state is zero.

5.10 Calculate the phase shift δ_0 for an impenetrable sphere of radius R. Compare its cross-section to its geometrical area.

[Ans. ka, $\frac{\sigma}{\sigma_g} = 4$]

5.11 The small binding energy of the deuteron indicates that the maximum of $u(r)$ lies only just inside the range R of the well. Use this information to estimate the value of R if $V_0 = 22.7$ MeV.

[Ans.1.5 fm]

5.12 Explain why the following dependences for potentials are not acceptable for the description of nucleon-nucleon potential.

(a) $[(\mathbf{r} \times \mathbf{s})(\mathbf{r} \times \mathbf{s})][\mathbf{s} \cdot \mathbf{s}]$
(b) $(\mathbf{r} \cdot \mathbf{p})(\mathbf{r} \cdot \mathbf{s})$
(c) $(\mathbf{r} \cdot \mathbf{p})(\mathbf{L} \cdot \mathbf{s})$
(d) $(\mathbf{L} \cdot \mathbf{s})(\mathbf{L} \cdot \mathbf{L})$
(e) $(\mathbf{r} \times \mathbf{L}) \cdot \mathbf{P}$

[See Sect. 5.1.9]

References

1. Bell et al. (1950)
2. C. Breit, R.L. Gluckstern, Annu. Rev. Nucl. Sci. **2**, 365 (1953)
3. A.D. Bromley, J.A. Kuhner, E. Almquist, Phys. Rev. Lett. **4**, 365 (1960)
4. Chadwick, Goldhaber (1934)
5. J.D. Jackson, J.M. Blatt, Rev. Mod. Phys. **22**, 77 (1950)
6. A.A. Kamal, *Particle Physics* (Springer, Berlin, 2014)
7. D.J. Knecht et al., Phys. Rev. **148**, 1031 (1966)
8. R. Machleidt, Adv. Nucl. Phys. **19**, 159 (1989)
9. Wigner (1935)
10. L. Wolfenstein, Phys. Rev. **75**, 1664 (1949)
11. C.N. Yang, Phys. Rev. **74**, 764 (1948)

Chapter 6
Nuclear Models

6.1 Need for a Model

The models used in nuclear physics as well as in atomic physics are invented because we do not know how to solve the many-body Schrodinger equation either with
Coulomb forces or nuclear forces. The observed features of light and heavy nuclei
are too complex to be explained by a reliable theory. In the absence of an exact theory, a number of nuclear models have been developed. These are based on different
sets of simplifying assumptions. Each model is capable of explaining only a part of
our experimental knowledge about nuclei. The experimental facts which are to be
explained by a model are:

1. Nuclear Spins I of ground state
2. Magnetic dipole moments μ as summarized in Schmidt diagrams
3. Electrical quadrupole moments Q
4. Existence of isomers and the occurrence of islands of *isomerism*
5. Parity of nuclear levels
6. Discontinuities of nuclear binding energy for certain values of N or Z
7. Substantially constant density of nuclei
8. Dependence of the neutron excess $(N-Z)$ on $A^{5/3}$ for stable nuclides
9. Approximate constancy of the binding energy per nucleon B/A
10. Fission by thermal neutrons of ^{235}U and other odd nuclides
11. Nonexistence in nature of nuclides heavier than ^{238}U
12. Wide spacing of low-lying excited levels in nuclei, in contrast with the close
    spacing of highly excited levels
13. Existence of resonance-capture reactions.

6.2 Type of Nuclear Models

In Chap. 4 we have noted that the nucleons in the nucleus are confined to an approximately spherical volume of radius about $1.2A^{1/3}$ fm and that the system is capable

A. Kamal, *Nuclear Physics*, Graduate Texts in Physics,
DOI 10.1007/978-3-642-38655-8_6, © Springer-Verlag Berlin Heidelberg 2014

of existing in various energy states each with its own distinctive properties. A sound theory would be concerned with the movement of nucleons in the nucleus, the nature of interaction between various nucleons and the properties of the ground and excited levels.

To begin with the problem may be formulated classically but eventually the solution must be found in quantum mechanical terms. In particular we need the total wave function of the nucleus which is possible only for the simplest nuclei. For a large complex nucleus the total wavefunction even if available would be too complicated to be of any practical use.

In this situation we are under the necessity of resorting to various nuclear models. These are simple analogies based on certain similarities with some other physical systems which are mathematically well understood. A given model based on certain assumptions attempts to explain only selected specific features of nuclei. Other features must be explained by some other model. In the past a number of models have been used to explain the whole lot of properties of nuclei based on very strong interacting nucleons to very weakly interacting nucleons. The contrast between the strong and weak interaction based models is resolved by invoking for Pauli's principle.

The earlier nuclear models addressed the problems of α-decay (Gamow) and the reactions of nucleons with nuclei. The models used to explain reactions of nucleons with nuclei assumed very weak interaction corresponding to independent particle motion as in Bethe's potential model and strong interactions as in Bohr's compound nucleus model. Subsequently, the nuclear models have been developed to a high degree of sophistication both semi-classically and quantum mechanically and fall into independent and collective categories. Among the more important models mention should be made of the liquid drop model which involves collective motion of nucleons and the Fermi gas model which treats nucleons as a gas of non-interacting particles. Both the models are semi-classical but are the fore-runners of more important quantum mechanical models like the Shell model and the collective model involving rotation and vibration.

The collective and single-particle aspects of nuclear structure are unified in the form of the Nilsson model which considers the independent motion of nucleons in a deformed potential. Another way is to consider the α-particle model which permits a simple way of calculating some properties of nuclei that are composed only of α-particles.

6.3 Fermi Gas Model

In the Fermi gas model, the nucleons are considered as a gas of non-interacting particles moving around in the nucleus with momenta ranging from zero to maximum value, P_F. The reality of Fermi momentum has been demonstrated by the study of the energy spectrum of electrons scattered off a thin H_2O target.

First consider the elastic scattering of electrons off free protons (neutrons) at rest. For a given beam energy E at a fixed scattering angle, the scattered electrons will

Fig. 6.1 Energy spectrum of the electrons scattered off a thin H_2O target. Data are taken at the linear accelerator MAMI-A at a beam energy of 246 MeV and at a scattering angle of 148.5° by [10]

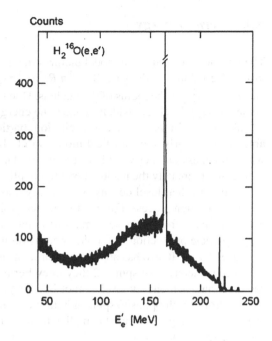

be scattered at energy E' given by [9], Chap. 6

$$E' = \frac{E}{1 + \frac{E}{Mc^2}(1 - \cos\theta)} \tag{6.1}$$

where M is the mass of the target. Repeating the scattering experiment with the same beam energy and at the same scattering angle but with complex nuclei containing several nucleons gives a more complicated spectrum. Figure 6.1 shows the spectrum of electrons scattered off free protons as well as oxygen nuclei.

The narrow peak observed at $E' \simeq 160$ MeV occurs due to elastic scattering off the free protons in hydrogen. On this is superimposed a broad distribution with the maximum shifted towards smaller scattering energies, near $E' \simeq 150$ MeV. This part of the spectrum may be identified with scattering of electrons off individual nucleons within the ^{16}O nucleus—a process known as quasi-elastic scattering. The sharp peaks at high energies are due to scattering off the ^{16}O nucleus as a whole. The left side of the curve is interpreted to be formed from the tail of the Δ-resonance. In the quasi-elastic scattering process the nucleon is assumed to be knocked out of the nucleus. The shift of the maximum in the energy of the scattered electrons towards lower energies is due to the binding energy of the nucleon. From the observations of broadening of the maximum, compared to the elastic scattering off free protons in the hydrogen atom, we conclude that the nucleus is not a static object with fixed nucleons, rather the nucleons move around in the quasi-elastic fashion within the nucleus. Consequently, the kinematics of scattering are altered compared to scattering off a nucleon at rest.

6.3.1 Fermi Energy

The gas model or statistical model pictures the nucleus as a gas of protons and neutrons. The volume of the gas, $\Omega = \frac{4}{3}\pi R^3 = \frac{4}{3}\pi r_0^3 A$, where A is the mass number and $R = r_0 A^{1/3}$ is the radius of the nucleus. Due to restriction to a small volume the nuclear energy levels are widely spaced, the energy level density being proportional to this volume. In the absence of excitation, particles will occupy the lowest available states. Clearly, the statistical model would be applicable for heavy nuclei for which the mass number would be sufficiently large. This model is useful for computing approximately the nuclear potential depth, to explain semi-quantitatively the increase in nuclear level density with energy and to consider emission of particles as an evaporation process. Further, the model explains the odd-even and asymmetry energy terms in the Weisacker's mass formula. The model also explains the lowering of particle production threshold in collision with complex nuclei as opposed to hydrogen target. It also has application for neutron stars.

Since nucleons have spin 1/2 they obey Fermi-Dirac statistics. Pauli's principle requires that each energy level cannot be occupied by more than two protons and two neutrons (with opposite spins in both the cases). The number of quantum states (n) corresponding to momenta smaller than a given value p equals the available phase space divided by h^3

$$n = 2 \times \frac{4}{3}\pi p^3 \frac{\Omega}{h^3} \tag{6.2}$$

The factor 2 arises due to spin 1/2 (multiplicity of states) in a nucleus containing Z protons and $A - Z$ neutrons. The maximum Fermi momenta P_F of protons and neutrons are given by

$$2\left(\frac{4}{3}\pi\right)^2 r_0^3 P_F^3(p) = Z \quad \text{(for proton)} \tag{6.3}$$

$$2\left(\frac{4}{3}\pi\right)^2 r_0^3 P_F^3(n) = A - Z \quad \text{(for neutron)} \tag{6.4}$$

The corresponding kinetic energy

$$E_F(p) = \frac{p_F^2(p)}{2m_p} = \left(\frac{9}{32\pi^2}\right)^{\frac{2}{3}} \frac{h^2}{2m_p r_0^2}\left(\frac{Z}{A}\right)^{2/3} \tag{6.5}$$

$$E_F(n) = \frac{p_F^2(n)}{2m_n} = \left(\frac{9}{32\pi^2}\right)^{\frac{2}{3}} \frac{h^2}{2m_n r_0^2}\left(\frac{A-Z}{A}\right)^{2/3} \tag{6.6}$$

In nuclei containing approximately equal number of neutrons and protons, the values of P_F and F_F are the same for both neutron and proton. Using numerical values $r_0 = 1.2$ fm, and nucleon mass $m_N = 940$ MeV/c^2, we find $P_F = 216$ MeV/c and $E_F = 25$ MeV. This then suggests that in light nuclei ($A = 2Z$), the potential for

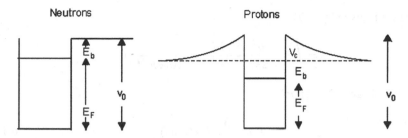

Fig. 6.2 Potential well depths for neutron and proton

neutrons ~33 MeV since the average binding energy ~8 MeV. Notice that since in heavy nuclei E_F is smaller for protons than for neutrons, the total depth of the potential well for protons (nuclear and electric) must be smaller than the depth of the potential well for neutrons (nuclear only), Fig. 6.2. Thus the effect of the Coulomb's repulsion more than compensates for the somewhat greater attractive nuclear forces acting upon the protons.

In the ground state of a nucleus the gas is completely degenerate and so all the states up to the maximum are filled.

Number of possible states with momenta between p and $p + dp$ is

$$dn = \frac{\Omega 4\pi p^2 dp 2}{h^3} = \frac{\Omega p^2 dp}{\pi^2 \hbar^3} \tag{6.7}$$

Integrating

$$n = \int dn = \frac{\Omega 8\pi p_{max}^3}{3h^3} \tag{6.8}$$

where $n = N$ or Z

$$P_n(max) = \frac{\hbar}{r_0} \left(\frac{9\pi}{4}\right)^{\frac{1}{3}} \left(\frac{N}{A}\right)^{1/3} = K \left(\frac{N}{A}\right)^{1/3} \tag{6.9}$$

$$E_N(max) = \frac{p_{max}^2}{2m} = \frac{\hbar^2}{2r_0^2 M} \left(\frac{9\pi}{4}\right)^{\frac{2}{3}} \left(\frac{N}{A}\right)^{2/3} = \frac{K^2}{2M} \left(\frac{N}{A}\right)^{2/3} \tag{6.10}$$

Similarly, for protons

$$E_Z(max) = \frac{K^2}{2M} \left(\frac{Z}{A}\right)^{2/3} \tag{6.11}$$

The total kinetic energy of all the protons in the gas is given by

$$E_Z = \int_0^Z E \, dn$$

where dn is given by (6.7)

$$E_Z = \int_0^{P_{max}} \frac{p^2}{2M} \frac{p^2 \Omega dp}{\pi^2 \hbar^3} = \frac{\Omega}{10 M \pi^2 \hbar^3} (p_{max})^5$$

Using (6.11) and $\Omega = \frac{4}{3} \pi R^3 A$,

$$E_Z = \frac{3}{5} Z E_Z(\text{max}) \tag{6.12}$$

Similarly

$$E_N = \frac{3}{5} N E_N(\text{max}) \tag{6.13}$$

6.3.2 Asymmetric Term (δ) in the Mass Formula

Let the kinetic energy of a nucleus be represented by $E(N, Z)$ and the standard nucleus have $N = Z = A/2$. We shall now show that the standard nucleus is the isobar with least energy. For this purpose we can compute the increase in kinetic energy of an isobar with a neutron or proton excess. This is clearly

$$\delta E = E(N, 0) + E(0, Z) - E\left(\frac{A}{2}, \frac{A}{2}\right)$$

$$= \frac{3}{5} \frac{K^2}{2M A^{2/3}} \left[N^{5/3} + Z^{5/3} - 2\left(\frac{A}{2}\right)^{5/3} \right]$$

$$= \frac{3}{5} \frac{K^2}{2M A^{2/3}} \left[\left(\frac{A}{2} + \Delta\right)^{5/3} + \left(\frac{A}{2} - \Delta\right)^{5/3} - 2\left(\frac{A}{2}\right)^{5/3} \right] \tag{6.14}$$

where $\Delta = N - \frac{A}{2} = \frac{A}{2} - Z$.

For any real nuclei Δ is small. We can therefore expand (6.14) as far as Δ^2 and obtain the result for the increase in kinetic energy

$$\delta E = \frac{3}{5} \frac{K^2}{2M A^{2/3}} \times \left[\left(\frac{A}{2}\right)^{5/3} + \frac{5}{3} \Delta \left(\frac{A}{2}\right)^{2/3} + \left(\frac{A}{2}\right)^{2/3} + \frac{5}{9} \Delta^2 \left(\frac{A}{2}\right)^{-1/3} \right.$$

$$\left. + \left(\frac{A}{2}\right)^{5/3} - \frac{5}{3} \Delta \left(\frac{A}{2}\right)^{2/3} + \frac{5}{9} \Delta^2 \left(\frac{A}{2}\right)^{-1/3} - 2\left(\frac{A}{2}\right)^{5/3} \right]$$

$$= \frac{2^{1/3} K^2}{3M} \frac{\Delta^2}{A}$$

$$\therefore \quad \delta \propto \left(\frac{A}{2} - Z\right)^2 / A \tag{6.15}$$

This is always positive and as the depth of the potential well is independent of N, Z and A, it follows that the standard nucleus is the isobar with the least energy. Note that it is only the exclusion principle which makes it possible for stable nuclei to contain protons at all. Without this it would always be energetically more favourable to add another neutron rather than a proton to the nucleus, so that the most state nuclei would consist entirely of neutrons.

6.3.3 Odd-Even Term in the Mass Formula

If there is an odd number of neutrons in the nucleus then the highest energy state is only half filled and so the next neutron will go into it. This means that the energy of the nucleus does not increase smoothly with N, as had been assumed but rather than in steps, that is compared with the smooth increase an even-even nucleus will have less energy and an odd-odd one. This step-like increase of energy is ensured by the δ-term in the mass formula. A rough form of the δ-term follows from the gas model. Clearly, the energy of an even-even nucleus (N, Z) has been overestimated in comparison with the nucleus $(N - 1, Z)$ by $E_{\max}(N) - E_{\max}(N - 1)$ and this is represented by $f(A)$

$$f(A) = \frac{K^2}{2MA^{2/3}}\left[N^{2/3} - (N - 1)^{2/3}\right] \simeq \frac{\text{const}}{A^{2/3}N^{1/3}}$$

But

$$N \simeq \frac{A}{2}$$

$$f(A) \simeq \frac{\text{const}}{A} \tag{6.16}$$

Actually

$$f(A) \simeq \frac{1}{A^{3/4}} \tag{6.17}$$

6.3.4 Threshold for Particle Production in Complex Nuclei

Consider the pion (π-meson) production in proton collisions in hydrogen

$$p + p \rightarrow n + p + \pi^+$$

The target proton is essentially at rest and the threshold for single pion production is found to be 289.37 MeV (see [9], Chaps. 3, 6 and 7). In case of a complex target nucleus, high energy collisions are expected to take place with individual nucleons, because of small de Broglie wavelength of the incident particle. However, because of the Fermi momentum of the target nucleons the threshold energy becomes much smaller. For a nucleon moving with maximum momentum of 218 MeV/c (E_F = 25 MeV) in the opposite direction to that of incidence, less energy is needed for the incident proton for the reaction to proceed. Consequently the required threshold energy is dramatically lowered to 160 MeV (see [9], Example 3.27).

For antiproton-proton pair production in the reaction

$$p + p \rightarrow p + p + p + p^-$$

the threshold is calculated as $6 \, Mc^2$ or 5.64 GeV for collisions in hydrogen target. But in complex nuclei the threshold is reduced to about 4.0 GeV.

6.3.5 Application to Neutron Stars

For these objects Coulomb energy is not to be considered. Apart from the attractive nuclear force which would lead to a density ρ_0, the gravitational force can cause the resulting density to go up to ten times larger.

Neutron stars are produced in supernova explosions. The burnt out centre of the star whose mass is between one and two solar masses, and is mainly made of iron, collapses under the gravitational force. The high density increases the electron's Fermi energy so much that the inverse of β-decay, $+e^- \rightarrow n + \nu$, takes place, while the beta decay, $n \rightarrow p + e^- + \overline{\nu_c}$ is forbidden by the Pauli's principle. All the protons in the atomic nuclei are eventually converted into neutrons. The Coulomb barrier disappears, the nuclei lose their identity and the interior of the star is solely composed of neutrons

$$^{56}_{26}\text{Fe} + 26e^- \rightarrow 56n + 26\nu_e$$

The implosion is only stopped by the Fermi pressure of the neutrons at a density of 10^{18} kg/m^3. If the mass of the central core is greater than double the solar mass the Fermi pressure can not withstand the gravitational force and the star ends up as a black hole.

The known neutron stars have masses 1.3 to 1.5 solar mass with typical radius R of the order of 10 km. In the simplest model the innermost core is composed of a degenerate neutron liquid with a constant density. To a good approximation the neutron star may be considered as a gigantic nucleus held together by its own gravitational force. We can now estimate the size of a typical neutron star with a mass $M = 3 \times 10^{30}$ kg which is about 1.5 times the solar mass and corresponds to a neutron number $N = 1.8 \times 10^{57}$. Assuming that the neutron star is a cold neutron gas the Fermi momentum is given by (6.9)

$$P_F = \left(\frac{9\pi N}{4}\right)^{1/3} \frac{\hbar}{R} \tag{6.9}$$

The average kinetic energy per neutron is given by (6.13)

$$\langle E_{kin}/N \rangle = \frac{3}{5}\frac{p_F^2}{2M_n} = \frac{C}{R^2} \quad . \tag{6.18}$$

$$\text{where } C = \frac{3\hbar^2}{10M_n}\left(\frac{9\pi N}{4}\right)^{2/3} \tag{6.19}$$

The gravitational energy of a star with constant density implies that the average potential energy per neutron is

$$(E_{pot}/N) = -\frac{3}{5}\frac{GNM_n^2}{R} \tag{6.20}$$

where M_n is the mass of the neutron and G is the gravitational constant. The star is in equilibrium if the total energy per nucleon is minimized

$$\frac{d}{dR}\langle E/N \rangle = \frac{d}{dR}\left[\langle E_{kin}/N \rangle + \langle E_{pot}/N \rangle\right] = 0 \tag{6.21}$$

Inserting (6.18) and (6.20) in (6.21) we find

$$R = \frac{\hbar^2(9\pi/4)^{2/3}}{GM_n^3N^{1/3}} \tag{6.22}$$

Using the numerical values, $R \simeq 12$ km for such a neutron star which is close to the experimental value and an average neutron density of 0.25 nucleons/fm^3, which is about 1.5 times the density $\rho_0 = 0.17$ nucleons/fm^3 for the atomic nucleus. The calculations have ignored the mutual repulsion of neutrons at high densities. In spite of the crudeness of the model the result of calculations are satisfactory.

6.3.6 Energy Levels of Individual Nucleons

The Fermi gas model is generally employed to describe the macroscopic phenomena like conduction of electrons in metal, nucleons in neutron stars, electrons in white dwarf etc., where the quantization of angular momentum may be neglected. By contrast a microscopic system, for example a nucleus is so small that it possesses distinct energy levels with distinct angular momenta. The energy levels in a spherically symmetric potential are calculated to possess orbital angular momentum $l = 0, 1, 2, \ldots$. At zero temperature the lowest lying states are all occupied. The interaction between the nucleons can only cause the nucleons to swap their place in the energy level spectrum. This is unobservable as the total energy of the nucleon is unchanged. This is the reason for associating the individual nucleons in the nucleus with definite energy and angular momentum state. The wave function that describes such a state is the one-particle wave function. The nuclear wave function is the product of all the one-particle wave functions.

It is not at all obvious that nucleons move freely inside the nucleus. This is demonstrated by using Λ-hyperons as probes.

In 1970's an elegant experiment was conducted at CERN to investigate the energy levels of the individual nucleons by employing Λ-hyperons as a probe. Λ-hyperon is produced frequently in the absorption of K^--meson with a nucleon via

$$K^- + p \rightarrow \Lambda + \pi^0 \tag{6.23}$$

$$K^- + n \rightarrow \Lambda + \pi^- \tag{6.24}$$

When the K^- is captured by a complex nucleus the Λ-hyperon thus produced may be attached to the nucleus after knocking out one of the nucleons. The nucleus thus formed is known as hyper-nucleus (see [9], Sect. 3.8, Chap. 7). A Λ particle in the nucleus cannot decay strongly since strangeness is conserved in such an interaction. Its lifetime is therefore approximately that of a free Λ particle; that is $\sim 10^{-10}$ s. This is long enough to permit the analysis to be made.

Since the neutron in the reaction (6.24) is bound and the Λ also remains inside the nucleus the energy difference between the K^- and the π^- yields the difference between the binding energies of the neutron and the Λ:

$$B_\Lambda = B_n + T_\pi - T_K + (M_\Lambda - M_n)c^2 \tag{6.25}$$

where T_π and T_K are pion and kaon kinetic energies.

The Pauli principle does not restrict the states the hyperon occupies. Consequently the hyperon may be captured in any bound state however deep it might be.

Binding energies of Λ-hyperon have also been determined in the interactions of secondary pion at Brookhaven [5]

$$\pi^+ + A \rightarrow \Lambda^A + K^+$$

Results of such experiments have yielded the binding energies of the Is states as well the excited p, d and f states for various nuclei, Fig. 6.3.

This shows the dependence of these binding energies upon the mass number A of the hypernuclei and that the Λ hyperons occupy discrete energy levels. The curves shown are the theoretical curves based or the calculations which assume a potential with uniform depth $V_0 \simeq 30$ MeV and the nuclear radius $R = r_0 A^{1/3}$ [6, 12]. The scale $A^{-2/3}$ corresponds to R^{-2} and is chosen because $B_\Lambda R^2$ is almost constant for states with the same quantum numbers.

6.4 Shell Model

The basic assumption of the shell model is that nucleons move around in a nucleus in an average potential rather freely. There is an apparent contradiction with Bohr's

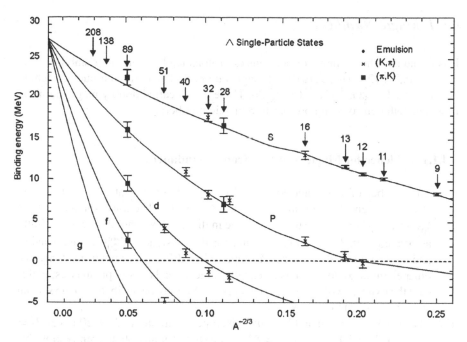

Fig. 6.3 Data on binding energies of s, p, d and f single-particle states of the Λ as a function of $A^{-2/3}$ [4]

idea of liquid drop model in which the particles are supposed to be highly interacting with each other. This paradox is, however, resolved if we consider Pauli's exclusion principle which says that no two nucleons can be in the same state. In other words, even if a nucleon hits another nucleon inside the nucleus no transfer of energy will take place since the other nucleon will have to be raised on the energy level but the nucleus being a degenerate gas, the higher levels are already filled up. Hence transfer of energy is highly suppressed. Thus because of Pauli's exclusion principle nucleons inside nuclear matter may have a very long mean free path. The assumption made is that nucleons more around very much similar to the motion of electrons in an atom. The underlying idea of shell model is that nucleons move in orbits of definite energy and angular momentum and that the outstanding stability of the nuclides in due to the completion of a neutron or proton shell in the same way as the stability of the rare gas atoms is due to completion of an electron shell. The situation in the case of nucleus is complicated by two factors:

(i) The "central" potential is really an average potential and the addition of an extra nucleon modifies this potential far more than the addition of an electron in the atomic case.

(ii) Because of the Coulomb repulsion of the protons, the number of neutrons and protons in a nucleus are not even approximately the same in all but the lightest nuclides. It is, therefore, most unlikely that a nuclide with a closed shell number of neutrons can also have a closed number of protons and vice versa.

6.4.1 Magic Numbers

It is found that the numbers of neutrons or protons lead to particular stability, 2, 8, 20, 28, 50, 82, 126. These are called Magic numbers. The particular nuclei ought to be and are $^4_2$He, $^{16}_8$O, $^{40}_{20}$Ca, $^{48}_{20}$Ca, $^{208}_{82}$Pb. These are so called doubly magic numbers in which both neutrons and protons have magic numbers.

6.4.1.1 Evidence for Magic Numbers from Abundances

(i) The number of stable and long-lived isotopes is greater at $Z = 20$ (six) and at $Z = 50$ (ten) than for any other even Z close by. The mass spread between $^{46}_{20}$Ca and $^{48}_{20}$Ca is double the usual value in the neighbourhood. Special stability associated with $N = 28$ may account for the existence of $^{48}_{20}$Ca. The heaviest and lightest naturally occurring isotopes of tin differ by 12 neutrons. Only one other element, xenon, attains an equally large spread of isotopic masses; in this case the existence of the heaviest isotope $^{134}_{54}$Xe may be attributed to the special stability associated with $N = 82$.

(ii) The number of stable and long-lived isotopes is greater at $N = 20$ (five), $N = 28$ (five), $N = 50$ (six) and $N = 82$ (seven) than for any other even N close by.

(iii) Pairs of stable and long-lived nuclides with different odd Z, but with the same even number of neutrons, occur only at $N = 20(^{37}_{17}\text{Cl}, ^{39}_{19}\text{K})$, $N = 50(^{87}_{37}\text{Rb}, ^{89}_{39}\text{Y})$ and $N = 82(^{139}_{57}\text{La}, ^{141}_{59}\text{Pm})$.

(iv) There are several exceptions to the statement that the relative abundance of an even A isotope is in general much less than 60 %. In these exceptional cases, N is either 50 or 82. Examples of these exceptional cases are $^{88}_{38}$Sr, $^{138}_{56}$Ba, $^{140}_{58}$Ce.

(v) The study of absolute abundances show peaks at Zr (50 neutrons), Sn (50 protons), Ba (82 neutrons) and Pb (82 protons or 126 neutrons). Thus these elements are more abundant than their neighbors.

6.4.1.2 Evidence for Magic Numbers from Stability

(i) Data show a sharp reduction in the binding energy of the last nucleon added to the magic number nuclei, for example for a neutron added to $N = 126$ and a proton added to $Z = 82$ nuclides.

(ii) The plot of the binding energy per nucleon against mass number shows special stability for the higher magic numbers. Specially significant are the changes in B/A at mass number 208 associated with the completion of the 82-neutron shells, at or near mass number $140(\text{Ce}^{140})$ associated with the completion of the 82-neutron shell, at or near mass number $88(\text{Sr}^{88})$ associated with the completion of the 50-neutron shell.

(iii) The known delayed neutron emitters are $^{87}_{36}$Kr, $^{137}_{54}$Xe and $^{17}_{8}$O, the emission occurring in all cases from one or in more excited states. These nuclei emit neutrons when excited by the β-decay of a parent radioactive fission product. These are extreme examples of the below average binding energy of the odd particle when N or Z exceeds the magic value by one. The shell binding energy of the odd particle makes neutron emission possible at relatively low excitation energies.

6.4.1.3 Evidence for Magic Number from Neutron Cross-Sections

Neutron absorption cross-sections are generally small for nuclides containing 50, 82 and 126 neutrons for neutron energies in the range 0.4–1.0 MeV. This can be understood in terms of a relatively low excitation energy of the compound nucleus containing 51, 83 or 127 neutrons. At low excitation energies, the level density is small and consequently also the cross-section. No significant resonant scattering or absorption is found for neutrons of energies below 0.1 MeV in $^{90}_{40}$Zr, $^{119}_{50}$Sn, $^{139}_{57}$La, $^{209}_{83}$Bi.

6.4.2 Theory

Various potential wells have been used for the central potential in which nucleons move around, for example square well, oscillator well, and infinite spherical well. The model was originally suggested as a possible explanation of the fluctuations in the relative abundance and relative masses of nuclei in the periodical table. Such fluctuations were associated with shell filling and shell closures at magic numbers. In the simplest approach Schrodinger's equation in three dimensions was used for the assumed potential.

Infinite Harmonic Oscillator Well   The potential has the form

$$V(r) = -V_0 + \frac{1}{2}M\omega^2 r^2 \tag{6.26}$$

where m is the reduced mass of the nucleon and $\omega = 2\pi \nu$ is the angular frequency of the oscillator. Schrodinger's equation is

$$\left(-\frac{\hbar^2}{2M}\nabla^2 + V(r) - E\right)\psi = 0 \tag{6.27}$$

The solution for a spherically symmetric well has the form

$$\psi_{nlm} = u_{nl}(r)Y_m^l(\theta, \phi) \tag{6.28}$$

where $Y_m^l(\theta, \phi)$ are spherical harmonics whose presence expresses the fact that the angular momentum l and its z-component m are constants of motion. The function

Table 6.1 Isotropic
Harmonic states

Λ	$\frac{E_\Lambda}{\hbar\omega}$	Orbitals				$2\Sigma(2l+1)$	Nor Z
6	15/2	1i	2g	3d	4s	56	168
5	13/2	1h	2f	3p		42	112
4	11/2	1g	2d	3s		30	70
3	9/2	1f	2p			20	40
2	7/2	1d	2s			12	20
1	5/2	1p				6	8
0	3/2	1s				2	2

$u_{nl}(r)$ contains the radial dependence of the function and satisfies the equation

$$\left(-\frac{\hbar^2}{2M}\frac{d^2}{dr^2} + V(r) + \frac{\hbar^2}{2M}\frac{l(l+1)}{r^2} - E_{n,l}\right) r u_{nl} = 0 \tag{6.29}$$

Put

$$q = \sqrt{\frac{M\omega}{\hbar}}\, r \tag{6.30}$$

Then (6.29) reduces to

$$-\frac{d^2}{dq^2} + q^2 + \frac{l(l+1)}{q^2} - \left(\frac{2E_{n,l}}{\hbar\omega}\right)(q u_{nl}) \tag{6.31}$$

The only eigen values of such an equation are shown to be

$$E = E_\Lambda = \left(\Lambda + \frac{3}{2}\right)\hbar\omega, \qquad \Lambda = 0, 1, 2\ldots \tag{6.32}$$

These are a number of degenerate eigen functions classified by their l-values.

For Λ even, $l = 0, 2, 4, \ldots, \Lambda$
For Λ odd, $l = 1, 3, 5, \ldots, \Lambda$
For given Λ

$$u_{nl} \sim q^l \exp\left(-\frac{1}{2}q^2\right) f_{nl}(q) \tag{6.33}$$

where the function $f(q)$ is a polynomial in even powers of q starting with a non-vanishing constant term. Table 6.1 exhibits the structure of the first few levels including nl values, degeneracy and the total number of like particles (neutrons or protons) making up the closed shells.

The states in the same row (like $1h$, $2f$, $3p$) are degenerate, the energy levels being identical. Closed shells occur at 2, 8 and 20 and are in agreement with the empirical indications, but 28, 50, 82 and 126 are not in evidence.

Table 6.2 States for rectangular potential well of infinite depth

Level	ω_{nl}	$2(2l+1)$	$2\Sigma(2l+1)$
$2g$	11.7049	18	156
$1i$	10.5128	26	138
$3p$	10.9041	6	112
$2f$	10.4171	14	106
$1h$	9.3555	22	92
$3s$	9.4248	2	70
$2d$	9.0950	10	68
$1g$	8.1826	18	58
$2p$	7.7253	6	40
$1f$	6.9879	14	34
$2s$	6.2832	2	20
$1d$	5.7635	10	18
$1p$	4.4934	6	8
$1s$	3.1416	2	2

Rectangular Well of Infinite Depth The radial function inside the well can be expressed in terms of a Bessel function of half integral order

$$u_{nl}(r) = r^{-1/2} J_{l+1/2}\left(\frac{\omega r}{R}\right) \tag{6.34}$$

with

$$\omega = \sqrt{\frac{2MER^2}{\hbar^2}} \tag{6.35}$$

The boundary condition at $r = R$ requires

$$J_{l+1/2}(\omega) = 0$$

Thus ω_{nl} is the nth zero of the Bessel function of order $l + \frac{1}{2}$; also

$$E_{nl} = \frac{\hbar^2 \omega_{nl}^2}{2MR^2} \tag{6.36}$$

Numerical results are listed in Table 6.2. The degeneracy with respect to l characteristic of the oscillator potential does not exist in the rectangular well. Otherwise, the order of levels is remarkably similar. The closed shell numbers 2, 8 and 20 occur again and also many others having no apparent connection with the observations. Once more 28, 50, 82 and 126 are missing from the list of closed shell numbers. The third column gives the number of particle $N_{nl} = 2(2l + 1)$ in each orbit and the last column the accumulating number. Some other potentials were also tried with similar results.

6.4.3 LS Coupling

A quite different suggestion was put forward independently by Mayer and by Haxel Jensen and Suess. So far no account has been taken of the possible splitting of each energy level into two according as the spin S and angular momentum l of the particle in the level are in the same or in opposite directions. Such a splitting could be due to a spin-orbit coupling which introduces a term of the form $\mathbf{l} \cdot \mathbf{s}$ into energy. If such a splitting is postulated and if it is further assumed that

(i) the $j = l + \frac{1}{2}$ level is below the $j = l - \frac{1}{2}$ level
(ii) the splitting increases with l which follows naturally from the $\mathbf{l} \cdot \mathbf{s}$ form of the spin-orbit term, then the level scheme is the one given in Table 6.3. The scheme clearly reproduces all the magic numbers

To reproduce the higher magic numbers a spin-orbit coupling term is added to the oscillator potential so that the single-particle potential has the form

$$V(r) = -V_0 + \frac{1}{2}M\omega^2 r^2 - \frac{2\alpha}{\hbar^2}(\mathbf{L} \cdot \mathbf{S}) \tag{6.37}$$

Now

$$\frac{2}{\hbar^2}(\mathbf{L} \cdot \mathbf{S}) = j(j+1) - l(l+1) - s(s+1)$$

For a nucleon, $s = \frac{1}{2}$ and $j = l \pm \frac{1}{2}$

$$\frac{2}{\hbar^2}(\mathbf{L} \cdot \mathbf{S}) = \begin{cases} l & \text{for } j = l + \frac{1}{2} \\ -(l+1) & \text{for } j = l - \frac{1}{2};\ l \neq 0 \end{cases}$$

(for $l = 0$, the spin orbit term vanishes). We can write (6.37) in the form

$$V(r) = -V_0 + \alpha \begin{cases} -l \\ l+1 \end{cases} + \frac{1}{2}M\omega^2 r^2 \quad \text{for } j = l \pm \frac{1}{2} \tag{6.38}$$

Since the spin-orbit contributions are constant, like V_0, the oscillator eigen functions are not altered by the introduction of a spin-orbit coupling.

Now for the oscillator, $V(r) = -V_0 + \frac{1}{2}M\omega^2 r^2$ (without coupling) and the energy eigen values are given by

$$E_{nl} = -V_0 + \left(2n + l - \frac{1}{2}\hbar\omega\right) \quad (l = 0, 1, 2\ldots),\ (n = 1, 2, 3\ldots) \quad \text{or} \tag{6.39a}$$

$$E_{nl} - E_{10} = \Lambda\hbar\omega \quad \Lambda = 2(n-1) + l = 0, 1, 2, \ldots \tag{6.39b}$$

The lowest lying energy state is given by the eigen value

$$E_{10} = -V_0 + \frac{3}{2}\hbar\omega$$

Table 6.3   States with spin-orbit coupling

Spin orbit		Split levels	$E_{n,j,l} - E_{1,1/2,0}$	N_j	ΣN_j
		$1i_{11/2}$			
1i					
		$1i_{13/2}$	$6\hbar\omega - 6\alpha$	14	126
		$1h_{9/2}$	$5\hbar\omega + 6\alpha$	10	
		$2f_{5/2}$	$5\hbar\omega + 4\alpha$	6	
2f		$3p_{1/2}$	$5\hbar\omega + 2\alpha$	2	
3p					
		$3p_{3/2}$	$5\hbar\omega - \alpha$	4	
1h		$2f_{7/2}$	$5\hbar\omega - 3\alpha$	8	
		$1h_{11/2}$	$5\hbar\omega - 5\alpha$	12	82
		$1g_{7/2}$	$4\hbar\omega + 5\alpha$	8	
		$2d_{3/2}$	$4\hbar\omega + 3\alpha$	4	
2d					
3s		$3s_{1/2}$	$4\hbar\omega$	2	
1g		$2d_{5/2}$	$4\hbar\omega + 2\alpha$	6	
		$1g_{9/2}$	$4\hbar\omega - 4\alpha$	10	50
		$2p_{1/2}$	$3\hbar\omega + 4\alpha$	6	
2p		$1f_{5/2}$	$3\hbar\omega + 2\alpha$	2	
1f					
		$2p_{3/2}$	$3\hbar\omega + \alpha$	4	
		$1f_{7/2}$	$3\hbar\omega - 3\alpha$	8	28
		$1d_{3/2}$	$2\hbar\omega + 3\alpha$	4	20
2s					
1d		$2s_{1/2}$	$2\hbar\omega$	2	
		$1d_{5/2}$	$2\hbar\omega - 2\alpha$	6	
		$1p_{1/2}$	$\hbar\omega - 2\alpha$	2	8
1p					
		$1p_{3/2}$	$\hbar\omega - \alpha$	4	
1s		$1s_{1/2}$	0	2	2

For the spin-orbit case, therefore in (6.39a), we must replace $-V_0$ by

$$-V_0 + \alpha \begin{cases} -l \\ l+1 \end{cases}$$

to obtain

$$E_{njl} = \left(2n + l - \frac{1}{2}\right)\hbar\omega - V_0 + \alpha \begin{cases} -l & \text{for } j = l + \frac{1}{2} \\ l+1 & \text{for } j = l - \frac{1}{2} \end{cases}$$

$$E_n, l+\frac{1}{2}, l - E_1, \frac{1}{2}, 0 = \Lambda\hbar\omega - \alpha l$$

$$E_n, l-\frac{1}{2}, l - E_1, \frac{1}{2}, 0 = \Lambda\hbar\omega + \alpha(l+1)$$

where Λ is given by (6.39b).

The degeneracy is therefore partly reduced; the energy eigen values for L and S antiparallel ($j = l - 1/2$) are shifted to higher energies. For L and S parallel the eigen values are shifted to lower energies. This occurs in such a way that the centre of gravity of the energy is not changed: the degeneracy of the state $j = l + 1/2$ or $j = l - 1/2$ is of degree

$$2j+1 = 2\left(l+\frac{1}{2}\right)+1 = 2(l+1)$$

$$2j+1 = 2\left(l-\frac{1}{2}\right)+1 = 2l$$

Hence we have

$$-\alpha l \times 2(l+1) + \alpha(l+1) \times 2l = 0$$

and the centre of gravity of the energy is indeed the same.

In a nucleon shell with fixed Λ, the state with $l = \Lambda$ and $j = l+1/2 = \Lambda + 1/2$ is lowered most strongly by the spin-orbit coupling. If the shift is so great that the lowest level of the Λ shell must be calculated in the next lower shell, while the lowest level of the next higher shell is shifted into the Λ shell; we get the following occupation numbers

$$N_\Lambda = \sum_l^\Lambda 2(2l+1) = 2\big[(2\Lambda+1) + (2\Lambda-3) + (2\Lambda-7) + \cdots + 1\big]$$

Number of terms in the arithmetic series $= \frac{\Lambda+2}{2}$.

Sum $= N_\Lambda = 2(\frac{\Lambda+2}{2})\{\frac{(2\Lambda+1)+1}{2}\} = (\Lambda+1)(\Lambda+2)$

$$N_\Lambda = 2,\ 6,\ 20,\ 30,\ 42,\ 56,\ \ldots$$

The lowest level of the energy shell with the quantum number Λ has the maximal occupation number

$$2\left(\Lambda+\frac{1}{2}\right)+1 = 2(\Lambda+1)$$

Hence the new occupation number N'_Λ of the energy shell Λ is

$$N'_\Lambda = \underset{\text{Depleted}}{N_\Lambda - 2(\Lambda+1)} + \underset{\text{Repleted}}{2(\Lambda+2)}$$

$$= (\Lambda+1)(\Lambda+2) - 2(\Lambda+1) + 2(\Lambda+2)$$

$$= (\Lambda+1)(\Lambda+2) + 2$$

Table 6.4 Reproduction of magic numbers by spin-orbit scheme

Λ	N_s	ΣN_s
5	$N'_\Lambda = 44$	126
4	$N'_\Lambda = 32$	82
3	$N'_\Lambda = 22$	50
3	$2(\Lambda + 1) = 8$	28
2	$N_\Lambda = 12$	20
1	$N_\Lambda = 6$	8
0	$N_\Lambda = 2$	2

For the first three shells ($\Lambda = 0, 1, 2$) the splitting is still so small that the occupation number N_s is given by N_Λ. The deepest level of the shell $\Lambda = 3$ has the occupation number $N_s = 8$. This level turns out to lie, because of the spin-orbit splitting just between the two shell $\Lambda = 2$ and $\Lambda = 3$. It therefore forms its own shell. For $\Lambda \geq 3$ the splitting is so strong that the maximal occupation number is given by $N_s = N'_\Lambda$.

Thus the magic numbers are successfully reproduced with the spin-orbit coupling scheme, see Table 6.4.

6.4.4 Predictions of the Shell Model

6.4.4.1 Abundances and Stability of the Closed Shell Nuclei

Direct mass measurements show marked breaks in the mass defect curve for nuclei with $Z = 20, 28, 50$ and $N = 20, 28, 50$. The number of stable species become markedly larger for $N = 20, 28, 50, 82$ than for nearby N values. Among the rare earths where chemical processes in nature can not much affect the original abundances, the isotopes with $N = 82$ are outstandingly abundant. The neutron binding energy is specially high for nuclei with $N = 50$ and 82. These facts show the extra stability of closed-shell nuclei. The strong tendency to asymmetric fission by thermal neutrons seems to be a dynamic effect of some complexity but again it depends on the marked stability of the closed nucleon shell, so that the most favored fragments are those with $N = 82$ and their compliments.

6.4.4.2 The Spins of Nuclear Ground States

All observed nuclei with even Z and even $N = A - Z$, are spherically symmetric in the ground state, with $J = 0$. This is the simplest line of evidence for the rule for pairing off the angular momenta for nuclei with Z odd, N even or with Z even, N odd which follows from the model nearly uniquely. With a handful of exceptions, assignment can be correctly made for about hundred and fifty such nuclei. As an example from Table 6.3 it follows that the lightest nucleus with spin 9/2 must occur when the neutron $1g_{9/2}$ state begins to be filled. This is at 41 neutrons. The lightest nucleus at $N = 41$ is ^{73}Ge, which indeed has a measured spin of 9/2. No lighter stable nucleus is known to have so high a spin. The scheme outlined so for enables us to deduce the angular momenta of nuclides consisting entirely of closed shells, and also of nuclides consisting entirely of closed shell plus or minus one particle. Because of the exclusion principle the former must have zero angular momentum, and in the latter case the total angular momentum is just that of the excess particle or of the "hole" that is the particle which would have to be added to complete the shell. Thus the angular momenta of $^{16}_{8}$O$_8$, $^{40}_{20}$Ca$_{20}$, $^{208}_{82}$Pb$_{126}$ should be and are zero, and those of $^{15}_{7}$N$_8$, $^{17}_{8}$O$_9$, $^{39}_{19}$K$_{20}$, $^{207}_{82}$Pb$_{125}$ and $^{209}_{83}$Bi$_{126}$ should be and are 1/2, 5/2, 3/2, 1/2 and 9/2 respectively.

Because of the pairing energy, an even number of like nucleons in the partly closed subshell form pairs so that their contribution to the ground-state angular momentum is 0^+. If the total number of neutrons (protons) in the subshell is odd, one will, of course, remain unpaired. These considerations lead to the following rules for the angular momenta and parities of nuclear ground states.

1. Even-even nuclei, that is, nuclei with even Z and even N, have total ground-state angular momentum $J = 0^+$. There is no known exception to the rule.
2. An odd nucleus, that is, a nucleus with odd Z or odd N, will have a total ground-state angular momentum equal to the half-integral angular momentum J and the parity $(-1)^l$ of the unpaired particle. These are no exceptions to this rule.
3. An odd-odd nucleus will have a total angular momentum which is the vector sum of the odd-neutron and odd-proton I-values

$$J = \mathbf{j_n} + \mathbf{j_p}$$

The quantum number J is therefore an integer between the limits

$$|j_n - j_p| \le J \le j_n + j_p$$

The parity will be the product of the proton and the neutron parity, that is, $\pi = (-1)^{l_n + l_p}$.

Observed angular momenta of nuclear ground states provide a more stringent test for the shell model than do the magic numbers. The level sequences are found to be in good agreement with the calculated schemes. However, some of the levels inside a major closed shell are quite close in energy so that the sequences are not always strictly followed. There is a tendency for pairs of particles to go into higher orbital angular momentum states rather than into s- or p-states when the competing states are close. This is due to the fact that the pairing energy increases with increasing l. The effect of the pairing energy is that a level will be depressed when it contains an even number of nucleons compared with its position when it contains an odd number. Further, the effect increases with increasing orbital angular momentum. Thus, the odd A nuclides with odd N above 58, should have angular momentum 7/2 or 11/2. Instead, they have angular momentum 1/2 which shows that the $1g_{1/2}$

and $1h_{11/2}$ levels are depressed below the $3s_{1/2}$ level when they are filled by even number of nucleons. The shell-model calculations are valid for spherical or near spherical nuclei. In regions where large quadruple moments are found, the ground-state angular momenta do not generally follow the predicted values. Among the light nuclei, 19F, ^{19}Ne and ^{23}Na have $J^{\pi} = 1/2^{+}$, $1/2^{+}$ and $3/2^{+}$ where as the shell model predicts $5/2^{+}$.

6.4.5 Magnetic Moments

Even-even nuclides having zero angular momentum also have zero magnetic moment. The magnetic dipole moment of a nucleus can be finite only if $J \geq 1/2$. One can calculate the magnetic moments of odd-A nuclides. It is assumed that the magnetic moment of such a nucleus is due entirely to the magnetic moment of the last nucleon. On the shell model each odd-even nucleus has a spherically symmetric set of closed neutron and proton shells, surrounded by paired neutrons and protons, again with $J = 0$ and therefore has no magnetic moment. Entire angular momentum J is assigned to the resultant spin and orbital motion of the one remaining odd nucleon and with that J, all the magnetic moment.

From this single-particle, the magnetic moment can be calculated, just as in the atomic Zeeman effect, by the formulae of the vector model.

The magnetic moment vector in the sum of two contributions, one from whatever current is produced by the orbital motion of the nucleon, the other from its intrinsic magnetic moment.

In the addition of spin and orbital angular momentum to a total angular momentum, the total magnetic moment vector no longer points in a direction parallel (or antiparallel) to the vector $\mathbf{j}$, because the g factors are different. However, since the total angular momentum is conserved in the absence of external torques, the vector

$$\boldsymbol{\mu} = \boldsymbol{\mu}_{l} + \boldsymbol{\mu}_{s}$$

will precess, together with $\mathbf{l}$ and $\mathbf{s}$ about the vector $\mathbf{j}$, so that the time average component of $\boldsymbol{\mu}$ in the direction of $\mathbf{J}$ will remain constant, see Fig. 6.4.

Odd Z (proton contribution only)

$$\boldsymbol{\mu} = \boldsymbol{\mu}_{orb} + \boldsymbol{\mu}_{int} = \frac{e\hbar}{2Mc}\left[l + \mu_{p}(2s)\right]$$

Odd N (neutron contribution only)

$$\mu = 0 + \mu_{n}(2s)\frac{e\hbar}{2Mc}$$

Fig. 6.4 Magnetic moments

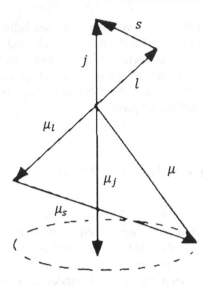

where the vector **l** and **s** are the orbital angular momentum and spin operators, and the coefficients μ_p and μ_n the intrinsic magnetic moments of the two nucleons in units of the nuclear magneton $e\hbar/2Mc$. The neutron orbital motion does not produce electric current and makes no contribution to the magnetic moment. The measured magnetic moments correspond to energies of orientation in a magnetic field along which the total angular momentum $\mathbf{j} = \mathbf{l} + \mathbf{s}$ is quantized while **l** and **s** precess about it. The tabulated magnetic moment μ refer to the maximum value of the projected component of the moment along the magnetic field direction, that is to that obtained when $\langle j_z \rangle = (M_j)_{\max} = j$.

The magnetic moment is just

$$\langle \mu_z \rangle_{\max} = \mu = \frac{(\boldsymbol{\mu} \cdot \mathbf{j})}{\langle j^2 \rangle} j$$

$$\mu = (g_l a_l + g_s a_s)\mathbf{j}$$

where $g_l = 1$ for proton and 0 for neutron, $g_s = 2.7g$ for proton and -1.91 for neutron. The coefficients $a_l j$ and $a_s j$ are the projections of **j** and **s** on **j**, that is

$$a_l = \frac{\mathbf{l} \cdot \mathbf{j}}{j^2} = \frac{j(j+1) + l(l+1) - s(s+1)}{2j(j+1)} \tag{6.40}$$

$$a_s = \frac{\mathbf{s} \cdot \mathbf{j}}{j^2} = \frac{j(j+1) + s(s+1) - l(l+1)}{2j(j+1)} \tag{6.41}$$

Now $s = 1/2$ and $l = j + 1/2$ or $j - 1/2$. If one of these is even, the other is odd and so l is a good quantum number, since a nucleus has a definite parity. Putting $l = j - 1/2$ and $s = 1/2$ in (6.40) and (6.41)

$$a_l = \frac{j(j+1) + (j - \frac{1}{2})(j + \frac{1}{2}) - \frac{3}{4}}{2j(j+1)}$$

$$= \frac{2j^2 + j - 1}{2j(j+1)} = \frac{(j - \frac{1}{2})}{j}$$

$$a_s = \frac{j(j+1) + \frac{3}{4} - (j - \frac{1}{2})(j + \frac{1}{2})}{2j(j+1)} = \frac{1}{2j}$$

For $l = j + \frac{1}{2}, s = \frac{1}{2}$

$$a_l = \frac{2j + 3}{2(j+1)}$$

$$a_s = -\frac{1}{2(j+1)}$$

$$\mu = \left(j - \frac{1}{2}\right)g_l + \frac{g_s}{2} \quad \text{for } l = j - \frac{1}{2} \tag{6.42}$$

$$\mu = \frac{j}{j+1}\left[\left(j + \frac{3}{2}\right)g_l - \frac{g_s}{2}\right] \quad \text{for } l = j + \frac{1}{2} \tag{6.43}$$

6.4.6 Schmidt Lines

The values of μ in (6.42) and (6.43) are known as the Schmidt values and are plotted against j in Fig. 6.5. Clearly, the agreement is not good although the experimental values follow the trend of the two theoretical lines, the "Schmidt lines" and almost without exception lie between them. In general, they are nearer to one line than to the other. It is then assumed that in the first approximation, the experimental values would be on the line to which they are nearer. It is possible to deduce the orbital angular momentum quantum number l of the last nucleon from the experimentally measured values of the total angular momentum j and the magnetic moment μ of the nucleus.

Notice that there is a very important difference between the shell model predictions of angular momentum and of magnetic moments. Angular momenta are quantized and predictions are therefore, right or wrong, in general they are right. Now if the basic simplifying assumption of the shell model that observable effects in odd-A nuclides are due to the last nucleon only, is not entirely correct, this will not affect the angular momentum values. It merely means that we are dealing with a mixture of states with the same angular momentum. The situation is quite different for magnetic moments. These are not quantized, and so different mixture of states yield different magnetic moments. The fact that the magnetic moments as a rule do not lie on the Schmidt lines is a proof that mixing of states occurs, that on the whole they do not lie far from the Schmidt line shows which single particle state of the shell model is the most important of the states in the mixture.

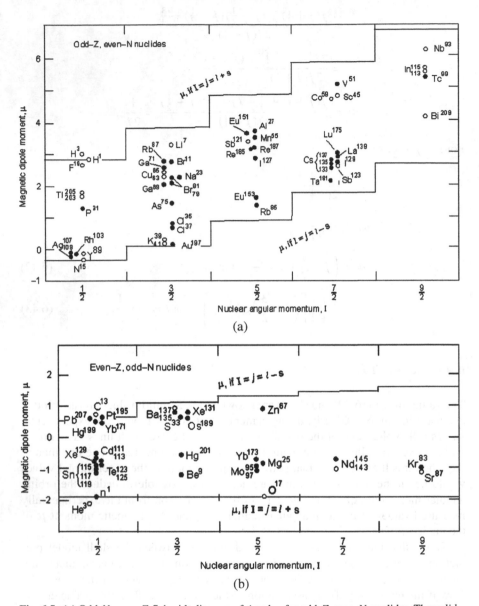

Fig. 6.5 (a) Odd N, even Z Schmidt diagram of j and μ for odd-Z even-N nuclides. The *solid–line* histogram corresponds to the Schmidt limits for each value of j, if j and μ were due entirely to the motion of one odd proton. *Open circles* represent nuclides with one proton in excess of, or with one proton less than a "closed shell" of 2, 8, 20, 28, 40, 50 or 82 protons. (b) Odd Z, even N Schmidt diagram of j and μ for even-Z odd-N nuclides. The histogram corresponds to the Schmidt limits for each value of j, if j and μ were due entirely to the motion of one odd neutron. *Open circles* represent nuclides which have one neutron more, or one neutron less, than the number required to form a "closed shell" of 2, 8, 20, 28, 40, 50, 82 or 126 neutrons

The formulae (6.42) and (6.43) define two curves for μ versus j with the values of $= l \pm 1/2$, for each class of odd-even nucleus. All measured values lie within the region bounded by the curves (with the exception of the pair H^3, He^3). The values lie mainly in two broad bands, roughly parallel to the two Schmidt curve, but are not very well defined. In a large number of cases the value lies nearer one of the curves than another, and nearly in all these cases, the l value so indicated is the shell-model value. The deviations from the Schmidt curves are under one nuclear magneton. These deviations are evidence of the approximate nature of the single-particle model but the qualitative correctness of the picture cannot be doubted.

6.4.7 Parity of Nuclei

With all the closed shells and the paired off nucleons having space symmetric wave functions, the parity of a nucleus on the shell model is just that of the orbital motion of the odd nucleon and is even or odd according to the character of l, that is parity is even if l is even and parity is odd if l is odd. Since l is determined by the magnetic moment in the ideal case, the measured magnetic moments provide one check on the parity. The other important determination of parity is provided by β-decay theory. The parity change in β-transitions, together with the spin change determines the life time for a given energy release. The predictions of the shell model have made possible a very orderly classification of β transitions.

6.4.8 Nuclear Isomerism

An excited nuclear state which lives long enough to have a directly measurable lifetime is called an isomeric state. Such a state decays by radioactivity which is different from the ground state but must be assigned the same values of Z and A. In a few cases several isomeric states are present. It is remarkable that of the sixty or seventy isomers of half life greater than one second known, all occur in "islands" of the periodic table, grouped just below the magic numbers, 50, 82 and 126. It is found that with two exceptions all isomers are found in four distinct groups, the so-called "Islands of isomerism" given by

(i) $19 \leq N$ or $Z \leq 27$
(ii) $39 \leq N$ or $Z \leq 49$
(iii) $63 \leq N$ or $Z \leq 81$
(iv) $91 \leq N$ or Z

Isomeric transitions always involve large changes of angular momentum ($\Delta I \geq 3$) and it is this large change which leads to the long life time.

Quadrupole Moment   The shell model also makes predictions about the electric quadrupole moment of an odd-A, odd-Z nuclide, on the assumption that this is the last proton in the nucleus. The quadrupole moment as measured is actually the greatest that is theoretically possible for a given j, that is it is the one for which the magnetic quantum number of the last proton $m_j = j$. For **l** and **s** parallel ($j = l + 1/2$) this corresponds to $m_l = l$ and $m_s = +1/2$. The proton wave function can be written as

$$\psi = \frac{u(r)}{r} N_l P_l^l(\cos\theta) e^{il\phi} \alpha$$

where α is the spin wave function and N_l is a normalization factor such that

$$\int \left| N_l P_l^l(\cos\theta) e^{il\phi} \right|^2 d\Omega = 1$$

Now Q, the quadrupole moment is given by

$$Q = \sum_{i=1}^{Z} (3Z_i^2 - r_i^2) |\psi(r_1 \cdots r_A)|^2 d\tau$$

where the integrations are over the $3A$ coordinates (x_i, y_i, z_i)

$$Q = \int_0^\infty r^2 [u(r)]^2 dr \int N_l^2 (3\cos^2\theta - 1) \left[P_l^l(\cos\theta) \right]^2 d\Omega$$

But $(2l + 1)\cos\theta p_l^l(\cos\theta) = p_{l+1}^l(\cos\theta)$

$$\int \left| P_n^m(\cos\theta) \right|^2 d\Omega = \frac{4\pi}{2n+1} \frac{(n+m)!}{(n-m)!}$$

$$\int \left| N_l P_l^l(\cos\theta) e^{il\phi} \right|^2 d\Omega = 1$$

$$N_l^2 \times 4\pi \frac{2l!}{(2l+1)} = 1 \quad \rightarrow \quad N_l^2 = \frac{2l+1}{4\pi 2l!}$$

Also

$$\int_0^\infty r^2 [u(r)]^2 dr = \langle r^2 \rangle$$

is the mean square radius of the orbit

$$Q = \langle r^2 \rangle \int (3\cos^2\theta - 1) \frac{(2l+1)}{4\pi 2l!} [P_l^l(\cos\theta)]^2 d\Omega$$

$$= \langle r^2 \rangle \int \frac{(3\cos^2\theta - 1)(2l+1)}{2 2l!} [P_l^l(\cos\theta)]^2 d\cos\theta$$

$$= \frac{\langle r^2 \rangle (2l+1)}{2 2l!} \left[3\int [\cos\theta P_l^l(\cos\theta)]^2 d\cos\theta - \int [P_l^l(\cos\theta)]^2 d\cos\theta \right]$$

$$= \frac{\langle r^2 \rangle (2l+1)}{2(2l!)} \left[3 \frac{(2l+1)!}{(2l+1)^2} \frac{2}{2(l+1)+1} - 2l! \frac{2}{2l+1} \right]$$

$$= \langle r^2 \rangle \left[\frac{3}{(2l+3)} - 1 \right] = -\frac{2l}{2l+3} \langle r^2 \rangle$$

$$Q = -\frac{2j-1}{2j+2} \langle r^2 \rangle \quad \left(\because \ j = l + \frac{1}{2} \right) \tag{6.44}$$

For l and s antiparallel the derivation is more complicated. However, the result turns out to be identical with (6.44)

$$Q = -\frac{2j-1}{2j+2} \langle r^2 \rangle = -\frac{2l-2}{2l+1} \langle r^2 \rangle \quad \left(j = l - \frac{1}{2} \right) \tag{6.45}$$

It was pointed out that mixing of angular momentum states leads to different magnetic moments. The same is true of quadrupole moments, in fact they are even more sensitive to the admixture of the single particle states. Admixtures ought to be least for nuclides with one proton outside a closed shell or a proton hole in a closed shell. It follows from (6.44) and (6.45) that a nuclide with a proton outside a closed shell has a negative quadrupole moment and one with a proton hole a positive one.

As regards magnitudes of quadrupole moments we first make an estimate of $\langle r^2 \rangle$. For a uniform charge distribution this is equal to $\frac{3}{5}R^2$, but for the least bound protons, which is more likely, $\langle r^2 \rangle$ is expected to be some what larger. Equation (6.44) shows that the absolute value of the quadrupole moment should be of the order of, but slightly less than R^2. This is found to be so for small A, but for $A > 100$ values as large as $10\ R^2$ occur. These deviations from the single particle model may well be due to mixing of states, but in any detailed calculations become quite prohibitive.

6.4.9 Criticism of the Shell Model

The shell model is an improvement upon the Fermi gas model in that a more realistic potential is used and the spin-orbit interaction is taken into account. All the magic numbers are correctly reproduced. The nuclei with closed shell are correctly predicted to have zero spin and positive parity in the ground state. Further, nuclei consisting of one nucleon outside a closed shell have the parity and spin of that nucleon. The same argument can be used to assign the J^π values for a shell which is deficit by one nucleon for its completion, that is the shell with a hole. Thus the shell model is very successful in predicting ground-state angular momenta but is not so successful in describing excited states and magnetic moments. It is therefore obvious that the shell model gives an over simplified picture of the actual situation inside the nucleus. The assumption of a spherical symmetric potential is incorrect in most of the cases, evidenced by appreciable quadrupole moments possessed by some of the nuclear states. The magnetic moments are also not predicted satisfactorily. Little can be said about the nuclei which are in the middle of a major shell.

6.5 The Liquid Drop Model

In this model the dynamics of a nucleus are compared with that of a liquid drop (N Bohr). The molecules of the liquid correspond to the nucleons in the nucleus. Certain features are analogous but others are not. The density of a liquid is almost independent of the size so that the radius R of a liquid drop is proportional to the cube root of the number A of the molecules, similar to the nuclear case. The energy necessary to evaporate the drop completely into well separated molecules is approximately proportional to the number A analogous to the binding energy of a nucleus. The surface tension of the liquid drop causes a correction to this relation since the molecules on the surface are somewhat loosely bound compared to those in the interior. This results in a correction term in the binding energy which is proportional to the surface area or $A^{2/3}$. Similar procedure is followed in the semi-empirical mass formula for nuclear binding energies. However the dynamics of liquid matter and nuclear matter are different in the quantum mechanical localization of the particles. The average kinetic energy of the molecules in the liquid is of the order of 0.1 eV. The corresponding de Broglie wavelength is of the order of 5×10^{-9} cm— a value which is much smaller than the distance between molecules. The average kinetic energy of the nucleons in nuclei is of the order of 10 MeV with a corresponding rationalized de Broglie wavelength $\lambda \sim 10^{-13}$ cm, which is just of the order of internucleon distance. Thus while in liquids the motion of its constituents can be described classically, and their positions can be well defined, compared to their mutual distance, in nuclei the motion must necessarily be described quantum mechanically since the uncertainty in the localization of the constituents is of the order of magnitude of their distance.

The underlying concepts of liquid drop model are diametrically opposite to those of Shell Model or independent-particle model. The interactions between nucleons are assumed to be strong rather than weak. Nuclear levels are pictured as quantized states of the entire nucleus and not as states of a single particle in an average field. The concept of Liquid-drop model originated in Bohr's assumption of compound nucleus in nuclear reactions. When an incident particle is captured by a nucleus its energy is quickly shared by all the nucleons. The mean free path of the captured particle is much smaller than the nuclear size. In order to explain such a behaviour, the interactions between nucleons have to be strong. Consequently, the particles cannot move independently with negligible collision cross sections with their neighbours. The Liquid-drop model attempts to explain the following nuclear properties

1. Substantially constant density of nuclei, with $R \propto A^{1/3}$
2. Systematic dependence of the neutron excess $(N - Z)$ on $A^{5/3}$ for stable nuclides
3. Approximate constancy of the binding energy per nucleon B/A as well as its trends with A
4. Mass difference in families of isobars and the energies of cascade β transitions
5. Systematic variation of α decay energies with N and Z
6. Fission by thermal neutrons of U^{235} and other odd $-N$ nuclides
7. Finite upper bound on Z and N of heavy nuclides produced in nuclear reactions and the nonexistence in nature of nuclides heavier than U^{238}

6.5.1 Semi-Empirical Mass Formula

If M is the total mass of an atom then

$$M(Z, A) = N M_n + Z M_p - a_v A + a_a \frac{(N - Z)^2}{A} + a_s A^{2/3}$$

$$+ a_c Z^2 A^{-1/3} + \delta(A, Z) \tag{6.46}$$

Total binding energy

$$B = (N M_n + Z M_p - M)c^2$$

$$= a_v A - a_a \frac{(N - Z)^2}{A} - a_s A^{2/3} - a_c Z^2 A^{-1/3} - \delta(A, Z) \tag{6.47}$$

where N and Z are the neutron and proton numbers respectively, M_n and M_p are the masses of neutron and proton respectively, A is the mass number, Z is the atomic number, $R = 1.3 \times 10^{-13} A^{1/3}$ metre is the nuclear radius, ε_0 is the permittivity, $\delta(A, Z)$, α, β and γ are constants to be determined experimentally.

The first term on the right side of (6.47) arises from nuclear binding energy which is proportional to A. The second term is due to the so-called asymmetry effect. It arises due to the fact that in general in a nuclide neutron and proton numbers are not equal. The third term arises due to the surface tension, and is proportional to the surface area which goes as $A^{2/3}$. The surface tension arises due to the fact that on the surface nucleons are less tightly bound than those in the interior. The fourth term is the Coulomb energy term which can be derived. The last term is the so-called odd-even effect term which results from the fact that the stability of nucleus depends on A and Z being odd or even.

Equation (6.46) is known as the semi-empirical mass formula due to Weisacker. In order to know the constants precisely, it is desirable to know these constants

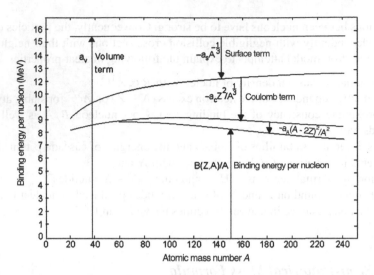

Fig. 6.6 The relative importance of the principal terms in the semi-empirical mass formula. It is notable that B/A is remarkably constant for $A > 50$ for the first three terms, and that the asymmetry term, although vitally important, is relatively small

from detailed theory of nuclear forces. The constant in the fourth term above can be derived, others are determined empirically from mass spectroscopy, that is one essentially uses precise masses of various nuclides in Eq. (6.47) and solves set of equations by the least square method. Inserting the accepted values of these constants Eq. (6.47) becomes

$$B \text{ (MeV)} = 15.56A - 17.23A^{2/3} - 0.7\frac{Z^2}{A^{1/3}}$$

$$- 23.285\frac{(\frac{A}{2} - Z)^2}{A} - \delta(A, Z) \qquad (6.48)$$

where $\delta(A, Z) = \pm 12A^{-1/2}$ (+ for A even, Z even, − for A even, Z odd, zero for A odd, Z any thing). The constants in (6.48) are not unique as they depend on the range of masses that have been used in their evaluation.

The binding energy per nucleon (B/A) is plotted against the mass number (A) in Fig. 6.6. The relative importance of various terms is (6.48) is indicated.

The δ-term in (6.48) or (6.50) arises due to pairing energy. We have noted (Chap. 4) that even-even nuclei containing protons as well as neutrons in pairs, are energetically favored as compared with odd-even, even-odd, and odd-odd nuclei. B/A is found by dividing the right side of (6.47) by A, Fig. 6.6. The decrease in B/A for small A is due to surface tension effect (Fig. 4.8) which is clearly important for small A since the nucleons lying on the surface is proportional to surface area i.e. R^2 or $A^{2/3}$ while total number of nucleons goes as volume i.e. R^3 or A and the fraction of nucleons lying on the surface is $A^{2/3}/A$ or $A^{-1/3}$. At higher A, Z will also increase.

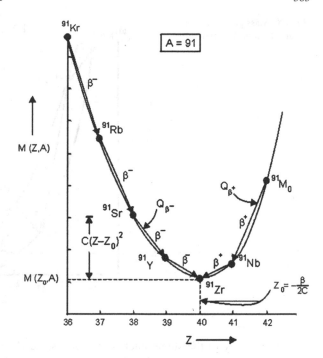

Fig. 6.7 Mass parabola for isobaric family with $A = 91$

Consequently, Coulomb's repulsive force being long range goes as Z^2 as every proton interacts with every other proton where as nuclear attractive force, being short range goes as A, as the nucleons interact only with their neighbours.

The first and the third terms are the most dominant terms. It is found that Eq. (6.48) for B agrees with the experimental values to better than 1 % for $A > 15$ and the difference in B values of nuclei of not two different A is often given correctly to 0.1 %. For a given A, the minimum of $M(A, Z)$ as a function of Z must correspond to the stable isobar. The condition $\delta M / \delta Z = 0$ yields

$$Z_0 = \frac{A}{2 + 0.015 A^{2/3}} \tag{6.49}$$

From formula (6.49) it is seen that the isobars for light nuclei (small A) those for which $Z \simeq A/2$. But for heavy nuclei there is a significant departure from $N = Z$ line (Chap. 4, Fig. 4.10) and the nuclei are stable for neutron excess.

6.5.2 Nuclear Instability Against β Emission

For odd A, the relationship between atomic mass M and nuclear charge Z is as shown in Fig. 6.7. The lowest isobar is the stable nuclide for the particular odd mass number A. Isobars of larger Z decay by β^+ or by electron capture. Isobar of

Fig. 6.8 Mass parabolas for
isobaric family, $A = 104$

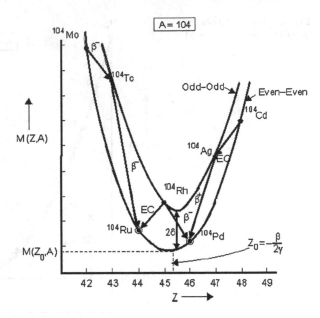

smaller Z decay to the stable nuclide by successive β^- decay. Note that for odd
A each isobar is either even Z odd N or odd Z, even N. Now, Eq. (6.46) can be
written as

$$M(Z, A) = aA + bZ + CZ^2 \pm \delta \tag{6.50}$$

which is an equation to a pair of parabolas. For odd A, $\delta = 0$. Therefore, only one
mass parabola exists. The minimum lies at some value of nuclear charge Z_0 which
determines the "most stable isobar" and which usually is a non-integer. The integer
Z which is nearest to Z_0 determines the stable isobar of odd A.

Even-A and odd-Z nuclei should be unstable and should not be found in nature.
Actually, there are a few, for example $^{40}_{19}K$, which is actually radioactive. The successive isobars no longer fall on a single parabola. The isobars of even Z, even N
fall on a lower parabola and, therefore, have more tightly bound nuclear structures
than the alternative odd Z, odd N isobars. The two parabolas differ by an amount
2δ and is associated with pairing energy, Fig. 6.8.

Figure 6.8 also shows that there can be two stable isobars for a particular even
value of A. The isobar on the lower parabola can decay only by β transition to
isobars on the upper parabola. Transitions can take place only by way of two successive transitions through intermediate odd Z, odd N isobar on the upper parabola.
When this is energetically impossible both even Z, even N isobar are stable. There
are known 54 stable pairs and 4 triads of even Z, even N. The only alternative
transition between a pair of even Z even N isobar would be by the simultaneous
emission of two β or double β decay, for which half life time is of the order of
10^{24} years.

6.5.3 Instability Against Neutron Decay

We can use Weisacker's formula to compute the binding energy of neutrons in a nucleus

$$BE(\text{neutron}) = \left[M(A-1, Z) + 1.008665 - M(A, Z) \right] c^2 \qquad (6.51)$$

This turns out to be positive for the stable elements showing that these do not emit neutrons spontaneously. However, in the case of the nuclei which are formed as the products of fission, usually they are highly excess in neutron number and can decay via neutron emission.

6.5.4 Instability Against Alpha Decay

$$BE(\alpha) = \left[M(A-4, Z-2) + 4.002603 - M(A, Z) \right] c^2 \qquad (6.52)$$

The BE for α particle becomes negative in the middle of the periodical table long before the natural α emitters are reached. The intervening elements are stable against α decay only because the α energies are so small that the lifetimes are prohibitively long (Geiger-Nuttal law). The periodical table ends in the region, $Z = 90 - 100$ because of the increasingly negative values of the BE for α emission and fission.

6.5.5 Fission and Fusion

The decrease of the binding energy at low mass numbers indicates that energy will be released if two nuclei of small numbers combine to form a single middle-class nucleus. The process is known as nuclear fusion, and is the reverse of fission. It occurs in the sun and other stars and is the underlying mechanism by which the sun generates the energy it radiates.

The decrease of the binding energy (Fig. 4.1) at high mass numbers indicates that they are more tightly bound when they are assembled into two middle-mass nuclei rather than into a single high-mass nucleus. Thus, energy can be released in the nuclear fission of a single massive nucleus into two smaller fragments.

It is easily shown that for $A \geq 85$ symmetric fission in exothermic. However, the nuclei would not undergo spontaneous fission since there is a potential barrier high enough to effectively prevent spontaneous fission (compare this with α-decay in which such a barrier accounts for long lives, Chap. 3).

Let the fission fragments be separated by distance r, the distance being measured between their centers. As r decreases, Coulomb's potential energy increases until the fragments touch each other at $r = a$, Fig. 6.9.

Fig. 6.9 Variation of
Coulomb's potential energy
with r, the distance between
the centres of two fragments

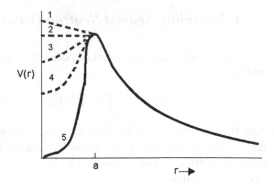

Fig. 6.10 Binary fission of
a fragment

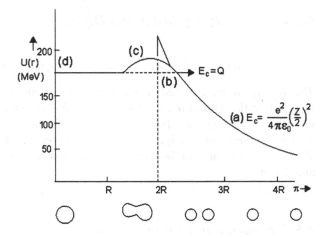

Inside 'a' (Fig. 6.9) several possibilities are open: for curve 1, spontaneous fission will occur and such a nucleus will not exist in nature. Curve 2 will correspond to the threshold for fission. All curves below 2 will call for excitation energy. But, for such nuclei the quantum mechanical penetration is still open. However, the mean life times are of the order of 10^{21} years or so. In transuranium elements, particularly among those recently discovered, several examples of relatively short half lives for spontaneous fission have been found.

Consider a binary fission in which each fragment has the mass number $A/2$ nuclear charge $Z/2$, and radius $R = r_0(A/2)^{1/3}$. When the separation r between the centres of two fragments is large compared with their radii (stage (a)), their mutual potential energy is simply given by the Coulomb energy $E_c = \frac{1}{4\pi\varepsilon_0}\frac{e^2}{r}(\frac{Z}{2})^2$. When r decreases until the two fragments are nearly touching, $r \sim 2R$, nuclear attractive forces begin to act. In that case the mutual potential energy is less than the Coulomb value, as indicated between positions (b) and (c) in Fig. 6.10.

If the fragments remained spherical, and attractive forces are most present then the Coulomb energy when the two spheres just touched, that is $r = 2R$, will be

$$E_c = \frac{e^2}{4\pi\varepsilon_0} \frac{(Z/2)^2}{2r_0(A/2)^{1/3}} = \frac{2^{1/3}}{8} \frac{e^2}{4\pi\varepsilon_0} \frac{Z^2}{A^{1/3}} \tag{6.53}$$

Introducing the Coulomb's constant $a_c = \frac{3}{20\pi\varepsilon_0}\frac{e^2}{r_0} \simeq 0.595$ MeV, Eq. (6.53) becomes

$$E_c = \frac{2^{1/3}}{8}\left(\frac{5}{3}a_c\right)\frac{Z^2}{A^{1/3}} = 0.262 a_c \frac{Z^2}{A^{1/3}} \simeq 210\ \text{MeV}$$

for spheres $(Z/2, A/2)$ in contact. This is shown as the extrapolated E_c curve at $r = 2R$ in Fig. 6.10.

When the two particles come closer together, $r < 2R$, the nuclear attractive forces become stronger and the two halves coalesce into the (Z, A) nucleus, where energy of symmetric fission, (d) in Fig. 6.10 is below the barrier height.

The nucleus (Z, A) will generally, be essentially stable against spontaneous fission if its dissociation energy Q is a few MeV below the barrier height.

6.5.5.1 Fission

Fission is the spontaneous disintegration of a heavy nucleus into two and sometimes three heavy fragments with the liberation of about 200 MeV energy. Natural fission is possible but the probability is small, being down by a factor of ten million compared with alpha emission. Fission may be readily realized by the absorption of negative pions or by bombardment of α particles or protons. This is in contrast with ordinary nuclear reactions in which only a small part of the heavy nucleus is chipped off and the energy liberated or absorbed is only a few MeV.

6.5.5.2 Characteristics

The process of fission is distinguished from all other nuclear reactions in regard to the amount of energy that is released, about 200 MeV, which is an order of magnitude greater than that in ordinary nuclear reactions. The masses of fission fragments are not equal to half of the original nucleus but exhibit considerable asymmetry. Fission is possible in heavy nuclei in which the binding energy per nuclei (B/A) decreases with increasing mass numbers beyond $A = 60$ (Fig. 4.1). Hence at the upper end of periodic table of the elements the division of the heavy nucleus with $\frac{B}{A} = 7.6$ MeV into approximately equal parts will release considerable energy due to increase in $B/A = 8.5$ for the fission products. We can calculate the approximate Q-values for the fission process. Consider the fission of a nucleus of mass numbers $A = 240$ into two fragments (binary fission) each of mass number $A = 120$

$$Q = 2 \times 120 \times 8.5 - 240 \times 7.6 = 216\ \text{MeV}$$

Alternatively, we can calculate the Q value from the actual masses. Consider a typical fission caused by the absorption of a neutron in the nucleus of U-235, in which two additional neutrons are produced

$$^{235}_{92}U + ^{1}_{0}n \rightarrow ^{141}_{56}Ba + ^{92}_{36}Kr + 3^{1}_{0}n$$

Initial masses: $235.0439 + 1.0087 = 236.0526$ amu
Final masses: $140.9139 + 91.8973 + 3.0261 = 235.8373$ amu

Difference in mass $= 0.2153$ amu

$$Q = 0.2153 \times 931.5 = 200.5 \text{ MeV}$$

In fission, the mass converted into energy is

$$\frac{200}{240 \times 940} = 0.001 \text{ or } 0.1 \text{ \%}$$

6.5.5.3 Mechanism

A heavy nucleus may be likened to a drop of liquid. In the ground state, the liquid drop is believed to be perfectly spherical, with a sharp surface of radius R. It is known that a liquid drop can be made to break up if mechanical vibrations of large amplitudes can be set up. For elements below uranium, decay by natural fission does not occur with appreciable probability. Some energy is therefore required to induce fission artificially. A minimum energy, called activation energy is needed for this purpose. When a neutron is absorbed, excitation energy is supplied to the nucleus. It remains only to decide whether the excitation energy supplied is less than, equal to or greater than the activation energy required for fission. Following the absorption of neutron, the liquid drop is distorted by a small amount resulting in the increased surfaces area. The surface tension forces tend to restore it to a spherical shape. The nucleus is stable against small amplitude oscillations because excitation energy associated with these oscillations is less than activation energy needed for fission. If enough excitation energy is provided then the mechanical vibrations of large amplitude can be set up. The energy associated with these vibrations is the analog of the activation energy. On the other hand any arbitrary distortion reduces the Coulomb energy because the centre of charge of the two 'halves' moves further apart. The surface effect and Coulomb effect thus work in opposite direction. The Coulomb effect lowers the activation energy needed for fission. Various stages in the deformation of the nucleus before and after fission are shown in Fig. 6.11. For a particular distortion, the nucleus may acquire a critical shape as in Fig. 6.11(e), which is the case of unstable equilibrium. The two halves are barely joined and the 'neck' is broken for a trifle increase of distortion. This is the point of no return

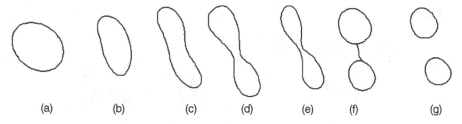

Fig. 6.11 Various stages in the deformation of the nucleus before and after fission

and fission is imminent. After the fission, the resulting fragments emit a couple of neutrons and reach a state of minimum energy. The theory is within the frame work of liquid drop model of the nucleus, originally due to Bohr and Wheeler.

^{235}U is fissionable by thermal neutrons ($\sim$0.025 eV) while ^{238}U is not. This is because when the neutron is absorbed in ^{235}U, an intermediate nucleus of ^{236}U (compound nucleus) is found. The excitation energy in ^{236}U compound nucleus is 6.8 MeV which is just equal to the required activation energy and fission takes place. However, in the case of absorption of a thermal neutron in the nucleus of ^{238}U nucleus, the excitation energy in the compound nucleus of ^{239}U is only 5.3 MeV. This is not enough to cause fission because activation energy is 7.1 MeV.

6.5.5.4 Fission Fragments

Fragments resulting from fission have too many excess neutrons compared to their stable isobars. The fragments tend to become stable through successive β^- decay followed by γ emission and alternatively by the evaporation of neutrons. These are called prompt neutrons as they are emitted in a time of the order of 10^{-14} s. On an average one neutron is produced per fission fragment. About 1 % of neutrons emitted by fission fragments are emitted at relatively long time after fission, that is 1 s to 1 minute later in the course of successive β decay. These are called delayed neutrons which play a decisive role in the control of nuclear reactor (Chap. 8).

About 85 % the energy released in the fission process appears as kinetic energy of fission fragments, the rest is associated with electrons photons and neutrinos produced from the chain radioactive decay. The neutrons produced in the fission process carry on average an energy of about 2 MeV If the fission occurs in a large absorber, then the energy of all the fission products save the neutrinos manifests itself as heat.

Example 6.1 Estimate the energy released in fission of $^{238}_{92}$U nucleus, given $a_c = 0.59$ MeV and $a_s = 14.0$ MeV (Osmania University 1962).

Solution

$$Q = M(Z, A) - 2M\left(\frac{Z}{2}, \frac{A}{2}\right)$$

$$= \left(a_s A^{2/3} + a_c \frac{Z^2}{A^{1/3}}\right) - 2\left[a_s \left(\frac{A}{2}\right)^{2/3} + a_c \frac{Z^2}{4(\frac{A}{2})^{1/3}}\right]$$

$$= a_s A^{2/3}\left(1 - 2^{4/3}\right) + a_c \frac{Z^2}{A^{1/3}}\left(1 - \frac{1}{2^{2/3}}\right)$$

Inserting $A = 238$, $Z = 92$, $a_c = 0.59$ and $a_s = 14$ we find $Q = 161$ MeV.

6.5.5.5 Stability Limits for Heavy Nuclei

A rough estimate of the mass and charge of a nucleus which is unstable against spontaneous fission can be obtained by finding (Z, A) such that Q for symmetric fission is as large as the Coulomb energy E_c for spherical fragments $(Z/2, A/2)$ in contact. A nucleus will be unstable if

$$Q \geq E_c \tag{6.54}$$

Now

$$Q = M(Z, A) - 2M\left(\frac{Z}{2}, \frac{A}{2}\right)$$

$$= a_s A^{2/3}\left(1 - 2^{1/3}\right) + a_c \frac{Z^2}{A^{1/3}}\left(1 - \frac{1}{2^{2/3}}\right)$$

$$= -0.260 a_s A^{2/3} + 0.370 a_c \frac{Z^2}{A^{1/3}} \quad \text{and}$$

$$E_c = 0.262 a_c \frac{Z^2}{A^{1/3}}$$

$$-0.260 a_s A^{2/3} + 0.370 a_c \frac{Z^2}{A^{1/3}} \geq 0.262 a_c \frac{Z^2}{A^{1/3}}$$

which upon simplification becomes

$$\frac{Z^2}{A} \geq 2.4 \frac{a_s}{a_c} = 2.4 \times \frac{13.0}{0.595} = 53$$

This is an upper limit because it ignores the possibility of finite barrier penetration. Note that the factor Z^2/A solely depends on the relative effective strengths of the forces associated with the Coulomb energy $(\propto Z^2/A^{1/3})$ and with the surface energy $(\propto A^{2/3})$.

Fig. 6.12 Axial symmetry of the surface of deformed nuclei in the lowest state of excitation

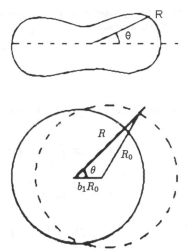

Fig. 6.13 Movement of an un-deformed sphere

A much better estimate of the critical value of Z^2/A can be obtained from the modes of oscillation of a drop of incompressible fluid, under the combined action of short-range surface tension forces and the long range Coulomb forces. In the ground state, the liquid drop is believed to be perfectly spherical, with a sharp surface of radius R_0. In the excited state the nuclei are deformed, but in the lowest excitation the surface will still have an axial symmetry (Fig. 6.12) so that it can be represented in terms of Legendre Polynomials

$$R(\theta) = R_0 \left[1 + \sum_{\iota=0}^{\infty} b_l P_l(\cos\theta) \right] \tag{6.55}$$

Now this formula can give both a deformation of the sphere and the translation of the sphere as a whole, the latter being of no interest to us. First we shall show that the coefficient of $P_1(\cos\theta)$ in (6.55) gives the distance through which the centre of mass of the nucleus has moved, and for pure deformation without translation we must set $b_1 = 0$.

Let the sphere move undeformed through a distance $b_1 R_0$ along the line $\theta = 0$ as in Fig. 6.13. Then

$$R_0^2 = R^2 + b_1^2 R_0^2 - 2 R R_0 b_1 \cos\theta$$

Put $\theta = 0$ then

$$R - R_0 = b_1 R_0$$

Thus, $b_1 = 0$ for the translation of centre of mass without deformation.

Further b_0 must be chosen in such a way that the total volume of the nucleus remains constant. The lowest excited states are given by $b_2 \neq 0$, $b_l = 0$ for $l > 2$.

We now express b_0 in terms of b_2 under the assumption that the original volume of the sphere, $(4/3)\pi R_0^3$ is unchanged. Volume of a spheroidal body with axial

Fig. 6.14 Spheroidal body
with axial symmetry

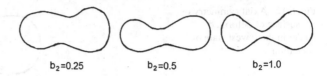

$b_2=0.25$ $b_2=0.5$ $b_2=1.0$

symmetry, as in Fig. 6.14, is given by

$$V = 2\pi \int_{-1}^{1} R^2 dR d\cos\theta = \frac{2\pi}{3} \int_{-1}^{+1} R^3 d\cos\theta$$

$$V = \frac{4\pi}{3} R_0^3 = \frac{2\pi}{3} R_0^3 \int_{-1}^{+1} \left(1 + b_0 + b_2 P_2(\cos\theta)\right)^3 d\cos\theta$$

$$= \frac{4\pi}{3} R_0^3 \left(1 + 3b_0 + \frac{3}{5}b_2^2\right)$$

where we have neglected the third power for b_2 and second and higher power for b_0 and $b_0 b_2^2$ and used the orthonormal properties of Legendre polynomials. We get

$$b_0 = -\frac{b_2^2}{5} \tag{6.56}$$

Inserting (6.56) in (6.55)

$$R = R_0 \left(1 - \frac{b_2^2}{5} + b_2 P_2(\cos\theta)\right) \tag{6.57}$$

Surface Area   The surface area of the axially symmetric object is given by

$$S = 2\pi \int R[R^2 + (dR/d\theta)^2]^{1/2} d\cos\theta \tag{6.58}$$

Now $\frac{dR}{d\theta} = R_0 b_2 \frac{dP_2}{d\theta}$

$$S = 2\pi R_0^2 \int_{-1}^{1} \left(1 - \frac{b_2^2}{5} + b_2 P_2(\cos\theta)\right)$$

$$\times \left[1 - \frac{b_2^2}{5} + b_2 P_2(\cos\theta) + b_2^2 \left(\frac{dP_2}{d\theta}\right)^2\right]^{1/2} d\cos\theta$$

Expanding the square root binomially and simplifying the integrand so that terms only up to b_2^2 are retained

$$S = 2\pi R_0^2 \int_{-1}^{1} \left[1 - \frac{2b_2^2}{5} + b_2^2 P_2^2(\cos\theta) + \frac{b_2^2}{2}\left(\frac{dP_2(\cos\theta)}{d\theta}\right)^2\right] d\cos\theta$$

Using $\frac{dP_2}{d\theta} = -\frac{3}{2}\sin 2\theta$, we finally obtain

$$S = 4\pi R_0^2 \left(1 + \frac{2}{5}b_2^2\right) \tag{6.59}$$

The corresponding surface tension energy is given by

$$E_s' = a_s A^{2/3}\left(1 + \frac{2}{5}b_2^2\right) \tag{6.60}$$

Coulomb Energy   We assume that the deformed surface is given by

$$R = R_0\left(1 + b_2 P_2(\cos\theta) - \frac{b_2^2}{5}\right) \tag{6.57}$$

and that the charge density is uniform

$$\rho = \frac{3Ze}{4\pi R_0^3} \tag{6.61}$$

The potential at an internal point is

$$V = V_1 + V_2 \tag{6.62}$$

where

$$V_1 = 2\pi\rho\left(R_0^2 - \frac{1}{3}r^2\right) \tag{6.63}$$

$$V_2 = 4\pi\rho R_0^2 \sum_{n=0}^{\infty} \frac{b_n}{2n+1}\left(\frac{R}{R_0}\right)^n \tag{6.64}$$

Coulomb energy for the first part is

$$E_c(1) = \frac{1}{2}\int \rho V_1 d\tau = \frac{1}{2}2\pi\rho^2 \int_0^{R_0}\left(R_0^2 - \frac{1}{3}r^2\right)4\pi r^2 dr$$

$$= \frac{3}{5}\frac{Z^2 e^2}{R_0} \tag{6.65}$$

where we have used (6.61).

Now for the second part

$$V_2 = 4\pi\rho R_0^2\left[b_0 + \frac{R^2}{5R_0^2}b_2 P_2(\cos\theta)\right]$$

$$= 4\pi\rho R_0^2\left[-\frac{b_2^2}{5} + \frac{R^2}{5R_0^2}b_2 P_2(\cos\theta)\right]$$

$$E_c(2) = \frac{1}{2}\rho^2 \times \frac{4\pi R_0^2}{5} \int \int \left(b_2 \frac{R^2}{R_0^2} P_2(\cos\theta) - b_2^2\right) 2\pi R^2 dR d\cos\theta$$

$$= \frac{4\pi^2 \rho^2 R_0^2}{5} \int_{-1}^{+1} \left(\frac{b_2 P_2(\cos\theta)}{5} \frac{R^5}{R_0^2} - \frac{b_2^2 R^3}{3}\right) d\cos\theta$$

Now

$$R^5 = R_0^5 \left(1 + b_2 P_2(\cos\theta) - \frac{b_2^2}{5}\right)^5 \simeq R_0^5 (1 + 5b_2 P_2(\cos\theta))$$

$$\int \frac{b_2 P_2(\cos\theta)}{5} \frac{R^5}{R_0^2} d(\cos\theta) = \int_{-1}^{+1} \frac{P_2(\cos\theta) b_2 R_0^5}{5R_0^2} (1 + 5b_2 P_2(\cos\theta)) d\cos\theta$$

$$= \frac{2}{5} b_2^2 R_0^3$$

Also

$$\int \frac{b_2^2}{3} R^3 d\cos\theta = \frac{b_2^2}{3} \int R_0^3 \left(1 + b_2 P_2(\cos\theta) - \frac{b_2^2}{5}\right)^3 d\cos\theta$$

$$= \frac{b_2^2 R_0^3}{3} \int (1 + 3b_2 P_2(\cos\theta))$$

$$= \frac{2}{3} b_2^2 R_0^3$$

$$\therefore \quad E_c(2) = \frac{4\pi^2 \rho^2 R_0^5 b_2^2}{5} \left(\frac{2}{5} - \frac{2}{3}\right) = -\frac{3}{25} b_2^2 \frac{Z^2 e^2}{R_0}$$

$$E'_{(c)} = E_c(1) + E_c(2) = \frac{3}{5} \frac{Z^2 e^2}{R_0} \left(1 - \frac{b_2^2}{5}\right)$$

$$= \frac{a_c Z^2}{A^{1/3}} \left(1 - \frac{b_2^2}{5}\right) \qquad (6.66)$$

Equation (6.60) shows that the surface tension energy has increased compared to that of the undeformed sphere

$$E_s = a_s A^{2/3} \qquad (6.67)$$

simply because the surface area has increased. The increase in surface energy is

$$\Delta E_s = E'_s - E_s = \frac{2}{5} b_2^2 a_s A^{2/3} \qquad (6.68)$$

Equation (6.66) shows that the Coulomb energy has decreased compared to the original sphere

$$E_c = a_c \frac{Z^2}{A^{1/3}} \qquad (6.69)$$

because the distance between the charges has increased. The decrease in Coulomb energy is given by

$$\Delta E_C = E_c' - E_c = -\frac{a_c Z^2 b_2^2}{5 A^{1/3}} \tag{6.70}$$

Condition for fission is then

$$|\Delta E_c| \geq |\Delta E_s| \tag{6.71}$$

using (6.68) and (6.70) in (6.71) and simplifying we find

$$\frac{Z^2}{A} \geq \frac{2a_s}{a_c} = \frac{2 \times 13.0}{0.595} \simeq 44$$

It must be stressed that when this condition is satisfied, fission is not only spontaneous but instantaneous. The fission of ^{238}U for which $Z^2/A = 36$ can not be of this type because of the very long life time for decay. It must in fact be due to a tunnel effect which can of course occur when the energy needed for the process is not available classically. Actually instantaneous fission does not occur until $Z \simeq 140$ and $A \simeq 390$. It is known that a liquid drop can be made to break up if mechanical vibrations of large amplitudes can be set up. The energy associated with these vibrations is the analog of the activation energy. For nuclei, excitation energy is supplied by the absorbed neutron. It remains only to decide whether the excitation energy is less than, equal to or greater than the activation energy required for fission. If the spherical nucleus is distorted by a small amount it will tend to return to its spherical shape. The nucleus is stable against small amplitude oscillations because excitation energy associated with these oscillations is less than the activation energy needed for fission. If the nucleus is stable against small shape distortions, it means that all such distortions tend to decrease E. If we increase the oscillation amplitude enough, fission occurs. If this happens it means that E must have passed through a maximum and then decreased, that is we must have surmounted the fission barrier. This maximum value of E will be smaller than for any other way. This smallest maximum is what we mean by the fission barrier. The corresponding shape of the nucleus is called the critical shape. It is a configuration of unstable equilibrium. If this distortion is reduced a trifle the nucleus moves back toward its original shape, if the distortion is increased a trifle, fission occurs.

6.5.6 Defects of Liquid Drop Model

Asymmetry in Fission Fragment Mass Distribution The fusion is invariably asymmetrical. There are comparatively few fissions which yield products of equal masses and the curve (Fig. 6.15) showing percentage yield plotted against mass

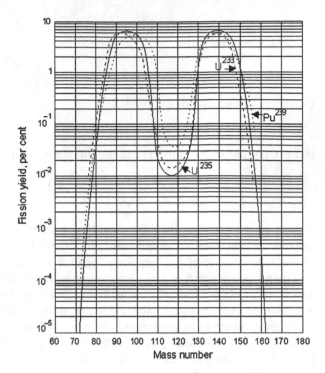

Fig. 6.15 Fission product yields for U^{235}, Pu^{239}, U^{233} (K. Way and N. Dismuke)

number has deep saddle depression in the centre with humps on each side. There is no adequate theory for this asymmetry. However, it is observed that the humps correspond to fragments with maximum stability. It is possible that more than two fragments result. But the probability for this is small. For fast neutrons the dip in the curve is less pronounced, indicating that symmetric fission is more likely.

Bohr's simple theory predicts symmetric distribution of mass. Further, Bohr's theory does not explain the following experimental facts.

(a) Mass asymmetry decreases with increasing energy.

(b) The asymmetry is well established as a part of the fission processes and exists whether the incident particles are neutrons, protons, alphas or it is the case of spontaneous fission. This is true for heavy elements near the uranium region. In the case of lighter elements, for example ^{226}Ra there are three humps unlike ^{235}U or ^{238}U fission mass spectrum.

(c) An angular anisotropy of the fission fragments with respect to the incident particles is observed in the photo fission of thorium and uranium, in that the fission fragments are emitted preferentially perpendicular to the γ-ray beam.

(d) Fission thresholds calculated on liquid drop model fail to agree with the experimental facts.

(e) The variation of fission cross-section with energy is not explained.

6.5.7 Criticism of Liquid Drop Model

Lord Rayleigh had derived the formula for the frequency of surface vibration of a deformed spherical drop:

$$\omega_l = \left[\frac{4\pi\alpha}{3M}l(l-1)(l+2)\right]^{1/2} \tag{6.72}$$

where M is the mass of the drop, the surface energy $E_s = \alpha s$, $S = 4\pi R^2$ for a spherical drop, $l =$ integer. The corresponding value for energy in nuclei would be

$$\hbar\omega_l \cong 14.7\left[\frac{l(l-1)(l+2)}{A}\right]^{1/2} \text{ MeV} \tag{6.73}$$

This value is somewhat too high to account for most low-lying nuclear states. The first excited states for nuclei with mass number A between 100 and 200 have an excitation energy of the order of 100 keV, where as (6.73) would predict several MeV. The frequencies w_l are reduced somewhat by the Coulomb effects. Whereas the surface tension increases if the drop in deformed, the Coulomb energy decreases upon deformation. This leads to somewhat smaller frequencies for heavier nuclei, but is still insufficient to represent the actual level distance. There are very many more closely spaced excited states than predicted. If the analogy with a liquid drop is valid at all, the surface vibrations must be considered as one very special type of nuclear motion. The actual excited states of nuclei correspond very probably to much more complicated types.

If the liquid is considered slightly compressible then compressional waves can also be set up in a drop with frequencies much higher than the frequencies of the surface waves.

The effect of the Coulomb field on the surface deformation becomes important for large Z. The condition of stability against surface deformation becomes

$$\frac{3}{5}\frac{e^2/r_0}{a_s}\frac{Z^2}{A} < 2 \quad \text{or} \quad \frac{Z^2}{A} < 42.2 \tag{6.74}$$

Any nucleus violating this condition should get deformed and finally undergo fission. Note that the heaviest nuclei are very close to this limit ($Z^2/A = 35.5$ for ^{238}U). The condition (6.74) is the main reason for the non-existence of nuclei heavier than those observed.

Nuclei for which Z^2/A is near its limit, a small perturbation from the outside is necessary to induce an instability, that is a breakup of the nucleus. Thus in this model nuclei near the limit of Z^2/A are easily induced to undergo fission by the additional supply of small energy.

To summarize, the liquid drop model of the nucleus is not very successful in describing the actual excited states. It gives too large level distances. It follows that the dynamical motions in the nucleus which give rise to the excited states are much more complicated than those envisaged in this model. The liquid drop model is more successful when used to determine the stability of ground states against deformation. The limit for stability against fission is well reproduced and the underlying idea is well supported by the fact that nuclei near this limit show the phenomenon of induced fission.

6.6 The Collective Model or Unified Model

In this model the concepts of the liquid drop and the shell model are combined. Consequently, the collective model is able to explain much large experimental data. Both collective model and unified model embody the collective effects. However, there are certain differences. The unified model is a hybrid of the liquid-drop model and distorted-shell model in which nucleons move approximately independently in nonspherical potential rather than being strongly coupled as in the case of the liquid-drop model. In the collective model, the nucleus is considered to consist of a core and extra core particles with the core being treated as a liquid drop. The main assumption of unified model that differs from that of the independent-particle model is that a number of nearly loose particles move in a slowly varying potential that originate from nuclear deformation. This deformation in the shape of the nucleus leads to excitation modes which are classified as vibrational and rotational.

6.6.1 Rotational States

The observed excitations of even-even and odd-A nuclei far away from the closed shells indicate level-spacings characteristic of the vibrational and rotational spectra. Figure 6.16 shows an example of rotational energy levels for ^{238}U. The rotational levels show remarkable regularity

  (i) All the levels have the same parity
 (ii) Successive levels have angular momentum increased by $2\hbar$
(iii) Spacing between adjacent levels increases with increasing spin. The energy eigen values are given by

$$E_J = \frac{\hbar^2}{2I} J(J+1), \quad J = 0, 2, 4, \ldots$$

and I is the rotational inertia associated with the nuclear deformation. Since the parity is given by $(-1)^J$, for reasons of symmetry the required parity is even and the permissible values are $J = 0, 2, 4, \ldots$. $I = 0$ for spherical nuclei, for example the even-even spherical nucleus ^{208}Pb, so that it is not expected to show rotational spectrum. This is confirmed by experiments.

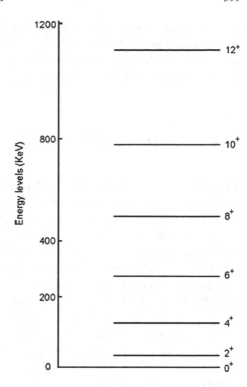

Fig. 6.16 Example of rotational energy levels for ^{238}U

6.6.2 Vibrational States

In the collective vibrational motion of nucleons, if the frequency of the vibrational mode of the core is ω then the corresponding energy is quantized in the units of $\hbar\omega$. These are analogous to the energy quanta in vibrations of solids, the energy quanta being phonons. The vibrational levels are evenly spaced, their energies are in the multiples of $\hbar\omega$.

6.6.3 Electric Quadrupole Moments

The shell model cannot explain quadrupole moments observed in many nuclei. For example, the measured reduced quadrupole moment of the middle of the shell nuclei, $^{175}_{71}$Lu is $+0.25$, where as the single-particle estimate give a value of -0.014. The reduced quadrupole moment is given by dividing the quadrupole moment by the nuclear charge and the square of the nuclear radius. The single-particle estimate of quadrupole moment is not only wrong in magnitude but also wrong in sign. On the other hand for a doubly magic plus one proton nucleus like $^{209}_{83}$Bi, $Q_{reduced}(\text{expt}) = -0.014$ in good agreement with the shell-model calculated value of -0.012. It is therefore concluded that the shell-model is not able to account for the large observed quadrupole moments for the middle of the shell nuclei. Rainwater suggested that for the nuclei like ^{175}Lu, the core is not spherical, as assumed by the shell-model, but is permanently deformed by the nucleus in the outer shell. It is this deformation that produces large quadrupole moments.

6.6.4 Shortcomings of the Shell Model

Inspite of the overwhelming success which the shell model had enjoyed in regard to the ground states and low-lying excited states of atomic nuclei, there remained a number of problems which were completely inexplicable in terms of the shell model of a spherical nucleus. These were concerned with (1) the magnitude of the nuclear quadrupole moments, (2) the ground states of odd nuclei in the range $150 \leqq A \leqq 190$ and at $A \geq 220$, (3) magnetic moments of certain nuclei, (4) excited states of even-even nuclei, (5) probabilities of radiative transitions and nuclear Coulomb excitations. Rainwater [13] made the first attempt to explain the quadrupole moments by pointing out the connection between particle motion and nuclear surface deformation. Hill and Wheeler pointed out that the surface oscillations of a heavy nucleus are mainly responsible for the fission process which determine the ground state and low-lying properties of their ground state.

Bohr and Mottelson had drawn attention to the striking properties of the spectra of heavy even-even nuclei. These nuclei exhibit the spin parity sequence 0^+, 2^+, 4^+, ... The ratio of the energies of the states 2^+ and 4^+ is 10/3, and the quadrupole transition probabilities are very high.

A similar spectrum is revealed by the quantum-mechanical rotator (symmetric top), for which

$$E = \frac{\hbar^2 J(J+1)}{2I} \tag{6.75}$$

where I is the moment of inertia of the rotator, and J assumes only even values if the body has plane symmetry and intrinsic angular momentum zero.

In quantum mechanics, the moment of inertia for spherical systems is equal to zero. Such a body cannot be set into rotation since all directions are equivalent. For small deviations from a spherical form, the moment of inertia is very small and the rotational energy would have to be very large. However due to interaction with other degrees of freedom, no rotational levels exist. Formula (6.75) shows that the moment of inertia should be large which is possible if the nuclear shape differs considerably from a sphere.

Bohr and Mottelson introduced a model of the deformed nucleus based on an interaction between the single-particle and the so-called collective degrees of freedom which are the rotational and vibrational degrees of freedom of the nucleus as a whole.

6.6.5 General Theory of Deformed Nuclei

A nucleus has number of oscillatory degrees of freedom. The frequency of the volume oscillations can be estimated using the Fermi gas model. Assuming equal number of neutrons and protons, the Fermi energy is given by the formula

$$E_f = \frac{p_{max}^2}{2m} = \left(\frac{9\pi}{8}\right)^{2/3} \frac{\hbar^2}{2mr_0^2}$$

$$\text{where } r_0 = \frac{R}{A^{1/3}} \tag{6.76}$$

For $r_0 = 1.2$ fm, $E_f = 33$ MeV. Total kinetic energy is given by

$$E_{kin} = \frac{3}{5} A E_F \tag{6.77}$$

For a nucleus in the ground state, the Fermi gas will be in a steady state and so its total energy is at its minimum. Now

$$E_{tot} = \frac{3}{5} A E_F + A V \tag{6.78}$$

where V is the nucleon potential energy. Then the first derivative $\partial E_{tot}/\partial R = 0$ while the second derivative $\partial^2 E_{tot}/\partial R^2 = K$, the elastic force for volume deformation. Hence

$$K = \frac{18}{5} A \left(\frac{9\pi}{8}\right)^{2/3} \frac{\hbar^2}{2mr_0^2} \frac{1}{R^2} + A \frac{\partial^2 V}{\partial R^2} \tag{6.79}$$

As a first approximation, the term $\frac{\partial^2 V}{\partial R^2}$ may be ignored. The energy of the corresponding oscillation will be

$$E = \left(N + \frac{1}{2}\right)\hbar\omega = \left(N + \frac{1}{2}\right)\left[\sqrt{\frac{18\hbar^2}{5mR^2}}\right]^{1/2} E_f = \left(N + \frac{1}{2}\right)\frac{1.75E_F}{A^{1/3}} \tag{6.80}$$

Thus the minimum excitation energy is $\frac{60}{A^{1/2}}$ MeV, or for a heavy nucleus ($A = 200$), $\hbar w = 10$ MeV. Note that a similar estimate of the volume oscillation energy is obtained within the frame work of the liquid drop model.

Another type of oscillation consists of oscillation of neutrons against protons. The excitation energy of such dipole oscillations is of the order of 15 MeV. Such oscillations are responsible for the occurrence of giant resonances; Sect. 6.6.9.

The collective or unified model was developed by Bohr and Mottelson in early 50's. We have noted that certain properties of nuclei are explained by the liquid drop model and certain others by independent particle shell model and that the underlying assumptions for these model are diametrically opposite. The truth might lie in between.

Fig. 6.17 Spectrum of
γ-rays emitted from ^{232}Th
nuclei excited by collisions
with ^{208}Pb. The regularly
spaced peaks indicate a
simple feature of nuclear
structure [14]

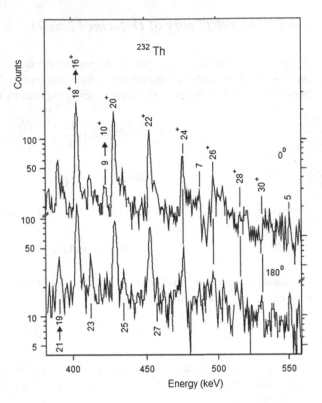

The excited states of a nucleus can be determined by raising it to higher energy
and measuring the γ-ray energy as it cascades down to the ground state. These
measurements serve the purpose of constructing the energy level diagram. The ex-
amination of the energy level diagram shows that as we move away from the closed
shells, the spectra of excited states become more and more complicated, and then
suddenly become simple again. As an example the lower states of ^{232}Th exhibit a
simple sequence of energies and angular momenta, Fig. 6.17. This is an unexpected
result from the stand point of the simple shell model Fig. 6.17 shows the spectrum
of γ-rays emitted from ^{232}Th nuclei excited by collisions with ^{208}Pb. The regularly
spaced peaks indicate the simplicity of nuclear structure. The energy level diagram
for ^{232}Th constructed from the γ-ray spectrum. It consists of the ground state band
and the octupole band based on an excited 1-state [14].

The explanation of the said regularity in the observed spectra cannot be given
in terms of single shell model. In the collective model a number of nearly loose
particles move in a slowly varying potential that arises from nuclear deformation.
This deformation leads to modes of excitation which are classified as rotational and
vibrational. For closed-shell nuclei, the pairing forces are dominant and the nucleus
retains its spherical shape. However deformation occurs when the additional nucleons start to fill the unfilled shell outside the closed shell. The nucleus gradually

reaches ellipsoidal shape, and the collective motion appears in the form of vibrational and rotational modes of excitation.

The failure to explain the quadrupole moment of non-spherical nuclei by the independent particle shell model emphasizes the inadequacy of the model. For example, the measured quadrupole moment of ^{177}Lu is 5.5b—a value which is 25 times larger than we could obtain from the simple shell model by considering the orbits of particles outside the closed shells and assuming an inert core. This can of course be understood using the shell model, but this requires the involvement of so many nucleons as to render the exercise utterly unnatural. Since the spectra of energy states is simple, an equally simple mechanism is called for. The excitations involving the motion of nucleons must be described in a correlated or collective way.

Another phenomenon which points to the collective model is inelastic scattering whose cross-section is much larger than that can be explained by assuming that the excitation is caused by raising a single nucleon to a higher state. It is mainly the collective states that are excited by inelastic scattering. In that case the projectile interacts with the target nucleus as a whole and not just with one of its constituent nucleons. This is a collective excitation and is described as a coherent sum of many particle-hole excitations. Calculations are made using either the rotational or the vibrational models for the wave functions of the nuclear states and a collective model for the interaction. Since the interaction is strong, the coupled-channels theory is used. This takes into account the coupling between various reaction channels. A good agreement is found with the measured elastic differential cross-sections of a few MeV protons with medium weight nuclei. Thus the data favour the assumption that many particles are excited together, corresponding to the collective motion. Further, the strengths of the γ-ray transition probabilities in these nuclei are more than hundred times greater than those given by single-particle transition.

Two types of collective motion are considered (1) rotational states, (2) vibrational states. Rotational motion is quite complex in that it is not a rigid-body rotation but a rotation of the shape of the deformed surface enclosing free particles. The vibrational states of nuclei are formed by flexings of nuclear surface. Both nuclear rotational motion and vibrational motion involve orderly displacements of many nucleons and both types are therefore classified as nuclear collective motion.

6.6.6 Rotational Model

The rotational type of excitation is responsible for the low-lying excited states of nuclei with large quadrupole deformations. The frequency of such rotation is low, so that to a good approximation the internal motion of nucleons and the rotational motion can be treated separately. Countless energy levels have been identified as rotational levels. They are recognized by the use of the of the formula for even-even nucleus

$$E = \frac{\hbar^2 J(J+1)}{2I} - BJ^2(J+1)^2, \quad \text{with } J = 0, 2, 4, 6 \qquad (6.81)$$

Fig. 6.18 Energy level diagram of $^{180}_{72}$Hf. With experimental and calculated excitation energies [1]

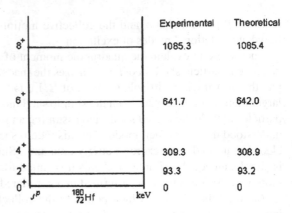

	Experimental	Theoretical
8^+	1085.3	1085.4
6^+	641.7	642.0
4^+	309.3	308.9
2^+	93.3	93.2
0^+	0	0

J^p $^{180}_{72}$Hf   keV

The equation has two adjustable parameters I and B.

The moment of inertia I is greater than that for a rigid body and increases with increasing angular momentum of rotation because of the centrifugal force which is taken care of by the second term in (6.81).

The ground state for the even-even nucleus has $J^\pi = 0^+$. The appropriate statistics is Bose-Einstein because there is an even number of nucleons in each half of the nucleus, and therefore the resulting angular momentum in each half is integral. With a 180° rotation, one half is exchanged with the other and the wave function ψ_R should be symmetric. We therefore expect a rotational energy-level diagram with angular momentum-parity values, $J^\pi = 0^+, 2^+, 4^+, \ldots$ for a deformed even-even nucleus.

A symmetrical deformed body is described by the equation

$$R(\theta) = R_0\big(1 + \beta_2 P_2(\cos\theta)\big) \tag{6.82}$$

where R_0 is the equilibrium radius of the undeformed spherical body, $P_2(\cos\theta)$ is the Legendre polynomial of second order, and β_2 is the deformation parameter.

It is found that the ratio I/I_{rigid} increases with increasing deformation parameter β_2. For nuclei with small deformation parameter β_2, no rotational spectrum is found at all.

As an example, Fig. 6.18 shows the energy-level diagram of $^{180}_{72}$Hf with experimental and calculated excitation energies.

The excitation energies of the low lying states of an even-even rotational nucleus closely follow the formula

$$E = \frac{\hbar^2 J(J+1)}{2I}, \quad J = 0, 2, 4, \ldots \tag{6.75}$$

with spacing $2J\hbar^2/I$. The sequences of states conforming to (6.75) are said to form rotational bands. Note that they have unequal spacing.

The expression (6.75) predicts that the ratio of the energies of the first two excited states $E_4/E_2 = 10/3$, so this provides a convenient signature of rotational motion.

Fig. 6.19 Energies of lowest
2^+ states of even-Z, even-N
nuclei. The *lines* connect
sequences of isotopes

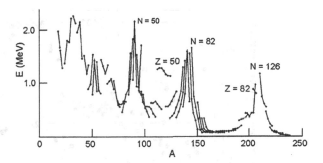

Four different properties of even even nuclei that reveal collective behaviour are as
follows:

1. Except in regions near closed shells, the energy of the first 2^+ excited state
   (Fig. 6.19) seems to decrease rather smoothly as function of A.
2. Figure 6.19 shows that except near the closed shells the ratio of the first two
   excited states $E(4^+)/E(2^+)$, Fig. 6.20 is roughly 2.0 for nuclei below $A = 150$
   and very much constant at 3.3 for nuclei between the shell closure at 82, 126
   and 184. They are the rare earth and actinide nuclei, respectively. We will see in
   Sect. 6.6.7, the value of the ratio E_4/E_2 for vibrational nuclei is 2. This is evident
   in Fig. 6.20. There is, however, much more scatter around the theoretical value
   compared to the rotational nuclei. This is understood in view of the fact that the
   vibrational nuclei are soft and wobbly and are easily affected by small perturbation not included in the simple model. On the other hand, rotational nuclei are
   hard and rigid and are not so much affected by small perturbations.
3. The magnetic moments of the 2^+ states (Fig. 6.21) are fairly constant in the range
   0.7–1.0.
4. The electric quadrupole moments, Fig. 6.22, are small for $A < 150$ and much
   larger for $A > 150$.

Thus, there are two types of collective structures, one set of nuclei with $A < 150$
are characterized by vibration while the other set of nuclei with $150 < A < 190$
characteristic of rotations.

We can gain some insight into the structure of deformed nuclei by considering
the moment of inertia in two extreme cases. First consider the rigid rotation. The
classical moment of inertia for a uniform ellipsoid of mass M with shape given by
the lowest power in β is

$$I_{rig} = \frac{2}{5}AMR_0^2(1+0.31\beta) \tag{6.83}$$

which reduces to the familiar value for a sphere when $\beta = 0$. The comparison is
only classical. A symmetrical spherical quantum mechanical system can not rotate
at all. Now the expression for the intrinsic quadrupole moment

$$Q_0 = \frac{3Z}{\sqrt{5\pi}}R_0^2\beta(1+0.16\beta) \tag{6.84}$$

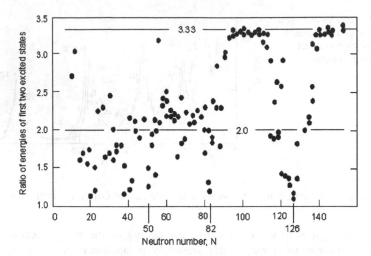

Fig. 6.20 Ratio of the energies of the first two excited states of many even-even nuclei as a function of neutron number N. There are clusters of points about the value 10/3 characteristics of rotational nuclei, and to a much less extent around the value 2 characteristic of vibrational nuclei [3]

Fig. 6.21 Magnetic moments of lowest 2^+ states of even-Z, even-N nuclei. Shell model nuclei showing noncollective behavior are indicated

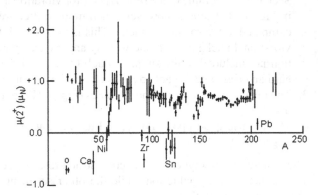

Fig. 6.22 Electric quadrupole moments of lowest 2^+ states of even-Z, even-N nuclei. The *lines* connect sequences of isotopes

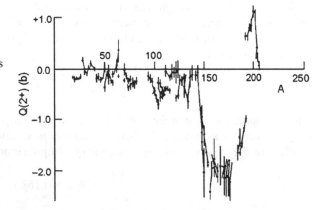

Thus I_{rig} and Q_0 are related through the parameter β. Note that Q_0 depends on the departure from sphericity and vanishes for $\beta = 0$. The experimental value of Q_0 can be used to determine β which in turn enables I_{rig} to be measured. As an example, for ^{178}Hf, $Q_0 = 8.1 \times 10^{-24}$ cm^2 and the relation $R_0 = 1.2A^{1/3}$ gives $\beta \sim 0.3$. It follows that $\frac{\hbar^2}{2I_{rig}} = 3$ keV. This is to be compared with the experimental value of 15.5 keV. Thus the rigid body approximation overestimates the moment of inertia by a large factor. In the other extreme the nucleus may be considered as a frictionless fluid in the deformed envelope. If the envelope rotates, only the outer regions of the fluid rotate, the central part remaining stationary. In that case the moment of inertia is given by

$$I_{fluid} = \frac{9}{8\pi} A M R_0^2 \beta^2 \qquad (6.85)$$

which gives $\frac{\hbar^2}{2I_{fluid}} = 60$ keV for ^{178}Hf.

Thus the experimental values of moment of inertia lie between the two extremes that is $I_{rig} > I > I_{fluid}$. The experimental values of the moment of inertia determined from the rotational energy states vary from 0.2 to 0.5 of I_{rig} for nuclei in the rare earth region. That the rotational behaviour is intermediate between a rigid object, in which the particles are tightly bonded together and a fluid in which the particles are weakly bonded, is ascribed to the short range strong forces between a nucleon ad its immediate neighbors only and the absence of long range force that characterizes the rigid body, secondly the lack of rigidity of nucleus from the increase in moment of inertia at higher angular momentum or rotational frequency. This effect known as 'centrifugal stretching' is seen frequently in heavy-ion reactions.

Many nuclei, particularly in the rare earth and actinide regions can have excited states characterized by more complicated deformation than the ellipsoidal

$$R(\theta) = R_0\left(1 + \beta_2 p_2(\theta) + \beta_3 p_3(\theta) + \beta_4 p_4(\theta) + \cdots\right) \qquad (6.86)$$

where β_2, β_3, β_4 are the quadrupole, octupole and hexadecapole deformation parameters.

Let us examine the concept of the shape of a rotating nucleus. The average kinetic energy of a nucleon in a nucleus is of the order of 20 MeV, corresponding to a speed of $\sim 0.2c$. The angular velocity of a rotating state is $\omega = \sqrt{2E/I}$ where E is the energy of the state. For the first rotational state, $\omega \simeq 10^{20}$ rad/s and a nucleon near the surface would rotate with a tangential speed of $v \simeq 0.002c$. The rotational motion is therefore far slower than the internal motion. The correct picture of a rotating deformed nucleus is therefore a stable equilibrium shape determined by nucleons in rapid internal motion in the nuclear potential, with the particles rotating so slowly that their rotation has little effect on the nuclear structure or on the nucleon orbits.

The rotational energies are of the order of a few keV's ($\hbar^2/2I \simeq 10$–20 keV) while the vibrational states and particle states occur at energies of about 1 MeV.

As the angular velocity of rotation of a nucleus is increased there comes a stage when the centrifugal forces overcome the attractive forces that hold the nucleus together and the nucleus is ruptured into two pieces as in fission. The value of the angular momentum l_{crit} at which this occurs can be estimated using the liquid drop model.

Both the vibrational and rotational collective motions give rise to the magnetic moment of the nucleus. The movement of the protons may be considered as an electric current, and a single proton moving with angular momentum quantum number l would give a magnetic moment $\mu = l\mu_N$. We assume that the neutrons do not contribute to the magnetic moment from the collective motion and that the protons and neutrons are all coupled pairwise so that the spin magnetic moments also do not contribute. We then expect the protons to contribute a fraction Z/A to the total nuclear angular momentum. The collective model therefore predicts for the magnetic moment of a vibrational or rotational state of angular momentum J

$$\mu(J) = J\frac{Z}{A}\mu_N \tag{6.87}$$

For light nuclei, $Z/A \simeq 0.5$ and $\mu(Z) \simeq +1\mu_N$, while for heaver nuclei, $Z/A \simeq 0.4$ and $\mu(Z) = +0.8\mu_N$. Figure 6.21 shows, barring the closed shell nuclei (for which the collective model is not valid any way) the magnetic moments of the 2^+ states are in good agreement with this prediction.

6.6.7 Vibrational Model

The second mode of collective excitation is that of vibration. The nucleus is considered as a dynamic system which can perform small oscillations about the equilibrium shape. The oscillations can be described in terms of normal modes which for small amplitudes can be assumed to be independent. Since the energies involved in vibrational excitations are of the order of several hundred keV to MeV, the coupling between the vibrational and intrinsic motions need no longer be weak as was the case with rotational motion. We should not, therefore, expect as good an agreement for the vibrational spectra with the experiment compared to the rotational spectra. Unlike the rotational spectra, the vibrational spectra can be exhibited by spherically symmetric nuclei as well. As we move away from the closed shell nuclei, the nucleons outside the core make them easily deformable so that they may be set into vibrational mode. The quanta of energy involved in these vibrations are called phonons by analogy with the quanta of atomic vibrations in crystals.

The vibrational states are known to occur not only among spherical even-even nuclei but also among deformed nuclei.

Fig. 6.23 A vibrating
nucleus with a spherical
equilibrium shape with
instantaneous coordinate $R(t)$
locate at a point on the
surface in the direction (θ, ϕ)

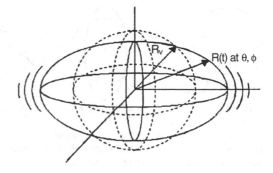

6.6.8 Collective Oscillations

The nucleus is considered as an incompressible liquid drop with a sharp surface. If
R_0 is the radius of the nucleus which is spherical, the equation for its surface can be
written as

$$R(t) = R_0\left[1 + \sum_{\lambda,\mu} \alpha_{\lambda\mu}(t)Y_{\lambda\mu}(\theta,\phi)\right]$$ (6.88)

where $Y_{\lambda\mu}$ are the spherical harmonics. $R(t)$ is the instantaneous coordinate of a
point on the nuclear surface at (θ, ϕ) as shown is Fig. 6.23.

Each spherical harmonic component will have an amplitude $\alpha_{\lambda\mu}(t)$. They are
also the deformation parameters which determine the nuclear shape. The subscript
μ takes the values $-\lambda$ to $+\lambda$, so that there are $2\lambda + 1$ modes of deformation of
order λ. The $\alpha_{\lambda\mu}$ are not completely arbitrary, reflection symmetry requires that
$\alpha_{\lambda\mu} = \alpha_{\lambda-\mu}$. Further, under the assumption that the nuclear fluid is incompressible,
the deformation of order $\lambda = 1$ is equivalent to a translation of the whole system that
is the net displacement of the centre of mass is zero and is irrelevant to a vibration,
Fig. 6.24. The lowest mode of surface deformation corresponds to quadrupole mode
($\lambda = 2$). For $\lambda = 2$, the five values of $\mu = -2$ to $+2$ correspond to five independent
modes which represent ellipsoidal shapes. The mode with $\mu = 0$ (for all λ values)
has symmetry with respect to arbitrary rotation about the z-axis and therefore represents an axially symmetric nuclear shape.

The surface oscillations are caused by the variation of deformation parameters
$\alpha_{\lambda\mu}$, which although quantum mechanical, may be treated classically and as time
dependent. The surface oscillation gives rise to the collection transport of nuclear
matter within the nucleus. The kinetic energy T of nuclear mass transport in the
nucleus is

$$T = \frac{1}{2}\sum B_\lambda |\dot{\alpha}_{\lambda\mu}|^2$$ (6.89)

where B_λ corresponds to the moment of inertia of the nucleus with respect to
changes in deformation and is calculated under the assumption of irrotational flow

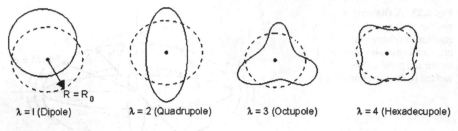

$\lambda = 1$ (Dipole) $\lambda = 2$ (Quadrupole) $\lambda = 3$ (Octupole) $\lambda = 4$ (Hexadecupole)

Fig. 6.24 The lowest four vibrations of a nucleus. *In each diagram* a slice through the midplane is shown. The *dashed lines* show the spherical equilibrium *shape* and the *solid lines* show an instantaneous view of the vibrating surface

$$B_\lambda = \frac{\rho R_0^5}{\lambda} \tag{6.90}$$

where C_λ is the density of nuclear matter.

The potential energy for collective motion is

$$V = \frac{1}{2} \sum_{\lambda,\mu} C_\lambda |\alpha_{\lambda\mu}|^2 \tag{6.91}$$

where C_λ are the deformability coefficients which represent the resistance of the nucleus against deformation. The total Hamiltonian H is given by

$$H = E_0 + \sum_{\lambda,\mu} H_{\lambda\mu} \tag{6.92}$$

where

$$H_{\lambda\mu} = \frac{1}{2} B_\lambda |\dot{\alpha}_{\lambda\mu}|^2 + \frac{1}{2} C_\lambda |\alpha_{\lambda,\mu}|^2 \tag{6.93}$$

and E_0 is the energy of the nucleus for a spherically symmetric shape. The C_λ has been determined by Bohr and Wheeler and is given by

$$C_\lambda = \frac{\lambda - 1}{4\pi} \left[(\lambda + 2)E_s - \frac{10}{2\lambda + 1} E_c \right] \tag{6.94}$$

where E_s and E_c are the surface and Coulomb energies, respectively for a spherical shape. The classical frequency of oscillation ω_λ is given by

$$\omega_\lambda = \sqrt{\frac{C_\lambda}{B_\lambda}} \tag{6.95}$$

The Hamiltonian (6.92) can be quantized whose energy eigen values (energy levels) are the harmonic-oscillator energies:

Fig. 6.25 Multi-phonon quadrupole spectrum. n_2 is the phonon number. J^P are the possible angular momenta and parity [8]

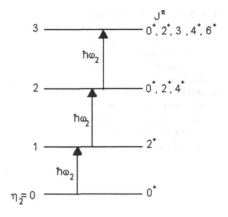

$$E = E_0 + \sum_{\lambda,\mu} \left(n_{\lambda\mu} + \frac{1}{2} \right) \hbar\omega_\lambda \tag{6.96}$$

where $n_{\lambda\mu}$ is the number of oscillators or phonons in the $\lambda\mu$-mode of oscillation. From (6.94), (6.95) and (6.90), the excitation energy for the $\lambda\mu$ mode, neglecting Coulomb energy is

$$\hbar\omega_\lambda \simeq \frac{13\lambda^{3/2}}{A^{1/2}} \text{ MeV} \tag{6.97}$$

For low lying states it is sufficient to consider only small values of λ as the frequency of the emitted radiation ω_3 (for $\lambda = 3$) is $\simeq 2\omega_2$ and $\omega_4 = 3\omega_2$.

For surface oscillations $\lambda \geq 2$, the angular momentum of a phonon in the state $\lambda\mu$ is λ, its z-component being μ, and the parity is $(-1)^\lambda$. Consequently the spin and parity of the ground and first excited state should be respectively 0^+ and 2^+, which have been confirmed by experiments for even-even nuclei. The spin and parity of the second excited state can be obtained from the consideration that the energy of one $\lambda = 3$ phonon is approximately equal to the energy of two $\lambda = 2$ phonons. Hence the spin and parity of the second excited state are 3^- or any of the values 0^+, 2^+, and 4^+ obtained from the combination of two angular momenta of two units. Experiments give the values 2^+, some times 4^+ and less frequently 3^-.

Equation (6.92) shows that the energy of the γ-rays emitted should decrease with increasing mass number, which is supported by experimental observation. Compared to the single-particle model, the collection oscillations causing the electric quadrupole transition between the first excited state and the ground state should be stronger since a large number of nucleons are involved.

Experiments show that the E_2 transition is invariably atleast one order of magnitude large than the predictions of the shell model.

A typical multiphonon quadrupole spectrum is shown in Fig. 6.25, indicating the excitation of one, two, three ... phonons. Residual interactions not included in the

Fig. 6.26 β and γ vibrations of a distorted nucleus. The *symmetry axis OZ'* is shown and the *diagrams at the right* are sections in the equatorial plane. The *full line* is the equilibrium contour, the *dotted line* is one extreme excursion. The *arrow* shows the direction of rotation in a rotational band [2]

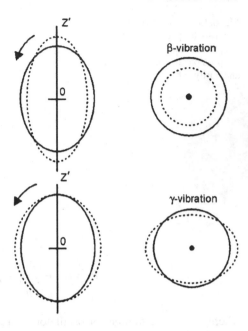

model remove the degeneracy of the phonon multiplets and enable the individual states to be identified.

Deformed nuclei can also vibrate and in this case the vibrations are of two types, called β and γ, depending on the projection μ of the phonon angular momentum along the symmetry axis. β-vibrations characterised by $\mu = 0$, preserve the axis of symmetry, while the γ-vibrations with $\mu = 2$, do not, Fig. 6.26. Rotational bands can be associated with each of these vibrational states, similar to rotation-vibrational levels in molecular spectroscopy. These are classified as β and γ bands.

6.6.9 Giant Resonances

β and γ vibrations were first detected in the measurement of cross-section of (γ, p) and (γ, n) reactions, which revealed pronounced resonances, with a few MeV width, Fig. 6.27. These are known as giant resonances and are found to occur in all but the lightest nuclei. Their characteristics are mainly determined by the bulk properties of the nuclei. They are interpreted as collective vibrations of protons relative to neutrons. The incident γ-rays are assumed to provide an oscillating electric field that exerts a force on the charged protons but not on the uncharged neutrons, causing the protons to be pulled away from the neutrons. When this force is switched off the protons and neutrons come together again. If the γ-ray frequency coincides with their natural oscillation then resonance oc-

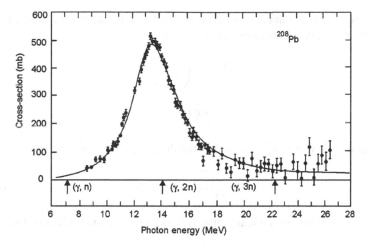

Fig. 6.27 Cross-section for the (γ, n) reaction on ^{208}Pb showing the giant diapole resonance at 14 MeV [15]

curs, leading to a peak in the cross-section of any reaction of γ-rays with the nucleus.

6.6.10 Nilsson Model

In this approach the single-particle and collective aspects of nuclear structure are unified by deforming the single particle potential. Nilsson's model is similar to that of the shell model except that the potential is deformed. The energies of single particle states are calculated by using an anharmonic oscillator potential. The levels in a spheroidal well have been calculated by a number of authors. The calculations by Nilsson are the best known.

In his calculations [11] used a Hamiltonian of the form

$$H = H_{kin} + \frac{m}{2}\left[\omega_1^2\left(x^2 + y^2\right) + \omega_2^2 z^2\right] + C\mathbf{l}\cdot\mathbf{S} + Dl^2 \qquad (6.98)$$

Such a Hamiltonian is characterized by axial symmetry, and differs from an oscillator potential with spin-orbit interaction by the inclusion of an anisotropy and by the term proportional to l^2. The potential is axially symmetric. The z-axis is taken along the symmetry axis. The frequency ω in the harmonic oscillator potential is taken as ω_1 along x and y axes, and ω_2 along z-axis.

By virtue of the term Dl^2 (D is negative), the potential effectively decreases for large orbital angular momenta, particularly at large distance, which leads to a lowering of the corresponding states.

The frequencies ω_1 and ω_2 are connected with the frequency ω_0 by means of the deformation parameter, according to the formulae

$$\omega_1^2 = \omega_0^2\left(1 + \frac{2}{3}\delta\right)$$

$$\omega_2^2 = \omega_0^2\left(1 - \frac{4}{3}\delta\right)$$

(6.99)

The quantity δ is very simply related to the deformation parameter β_2

$$\delta \cong \frac{3}{2}\sqrt{\frac{5}{4\pi}}\beta_2 \cong 0.95\beta_2$$

(6.100)

The incompressibility of nuclear matter (constancy of volume) gives us the relation

$$\omega_1^2\omega_2 = \omega_0^3$$

(6.101)

In (6.98) the constant C determines the strength of the spin-orbit coupling. The term Dl^2 lowers the energy of the large angular momentum states. Both C and D are negative.

The deformed potential is governed by four parameters ω_0, C, D and δ. Of these δ only depends on the nuclear shape. Now

$$\delta = \frac{\Delta R}{R_0}; \qquad R_0 \sim \frac{a+b}{2} \quad \text{and} \quad \Delta R = b - a$$

where a and b are the semi-axis of a homogeneously charged ellipsoid with charge Ze (b along z-axis), then the quadrupole moment

$$Q = \frac{2}{5}Z(b^2 - a^2) = \frac{4}{5}ZR_0^2\delta$$

(6.102)

Measurement of Q yields δ and hence β_2. The other three parameters (ω_0, C, D) are practically shape-independent. They are determined from levels of spherical nuclei for which $\beta_2 = 0$. Once ω_0, C and D are determined the energy eigen values can be explored as a function of the deformation parameter β_2. For $\beta_2 = 0$ (spherical nuclear shape) the energy levels agree to these shown in Table 6.3, and they can be labeled by quantum numbers Λ, l and j. For a non-spherical potential, the angular momentum l is no longer a "good" quantum number, that is we cannot identify states by their spectroscopic notation (s, p, d, f, etc.) as was done for the spherical shell model. To put it differently, we have to deal with mixtures of different l values but belonging to the same parity, that is even or odd.

In the cases of spherical potential, the energy levels of each single particle state have a degeneracy of $(2j + 1)$, that is relative to any arbitrary choice, all $2j + 1$ possible orientations of $\mathbf{j}$ are equivalent. For a deformed shape this is no longer valid. The energy levels depend on the spatial orientation of the orbit. To be more precise, the energy depends on the component of $\mathbf{j}$ along the symmetry axis of the

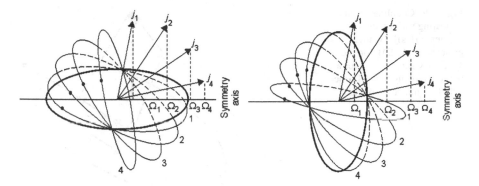

Fig. 6.28 Single-particle orbits with $j = 7/2$ and their possible projections along the symmetry axis, for prolate (*left*) and oblate (*right*) deformations. The possible projections are $\Omega_1 = 1/2$, $\Omega_2 = 3/2$, $\Omega_3 = 5/2$ and $\Omega_4 = 7/2$. (For clarity, only the positive projections are shown.) Note that in the prolate case, orbit 1 lies closest (on the average) to the core and will interact most strongly with the core; in the oblate case, it is orbit 4 that has the strongest interaction with the core (Krane)

core. For example an $f_{7/2}$ nucleon can have eight possible components of $\mathbf{j}$, ranging from $-7/2$ to $+7/2$. This component of $\mathbf{j}$ along the symmetry axis is generally denoted by Ω. On account of the reflection symmetry the components $+\Omega$ and $-\Omega$ will have the same energy, giving the levels a degeneracy of 2. In general number of Ω values will be $\frac{1}{2}(2j + 1)$. Thus, in our example the $f_{7/2}$ state splits up into four states which are labeled as $\Omega = \frac{1}{2}, \frac{3}{2}, \frac{5}{2}, \frac{7}{2}$, all with negative parity.

Figure 6.28 indicates different possible "orbits" of the odd particle for prolate and oblate deformations. For prolate deformations, the orbit with the smallest possible Ω (here $1/2$) interacts most strongly with the core and is thus more tightly bound and lowest in energy. On the other hand, for oblate deformations the orbit with maximum Ω (equal to j, here $7/2$) has the strongest interaction with the core and is in the lowest energy. For finite deformation, the only constant of motion of the Hamiltonian (6.98) are the parity and Ω, the projection of $\mathbf{j} = \mathbf{l} + \mathbf{s}$ along the symmetry axis z. However, for an infinite deformation the terms $\mathbf{l} \cdot \mathbf{s}$ and l^2 become negligible in comparison to the deformation term and the Hamiltonian has cylindrical symmetry. Thus the wave functions are asymptotically characterized by the quantum numbers Λ, n_z, Δ, where n_z is the number of nodal planes in the z-direction and Δ is the component of l along z. The coupling scheme indicating the nucleon angular momenta and their projection along the symmetry axis for nucleon moving in the deformed axially symmetric potential is shown in Fig. 6.29.

Since the eigen values of the energy vary continuously with β_2 and so do the wave functions, each wave function may be characterized by the quantum numbers $\Omega^{\pi} (\Lambda, n_z, \Delta)$ where the numbers in parentheses, which become good quantum numbers only for infinite β_2, are called asymptotic quantum numbers. The lower energy levels in the axially deformed potential are shown in Fig. 6.30 as a function of the deformation parameter. Note that just as the spin-orbit potential breaks the degeneracy of the major shells, so the deformation breaks the degeneracies of the

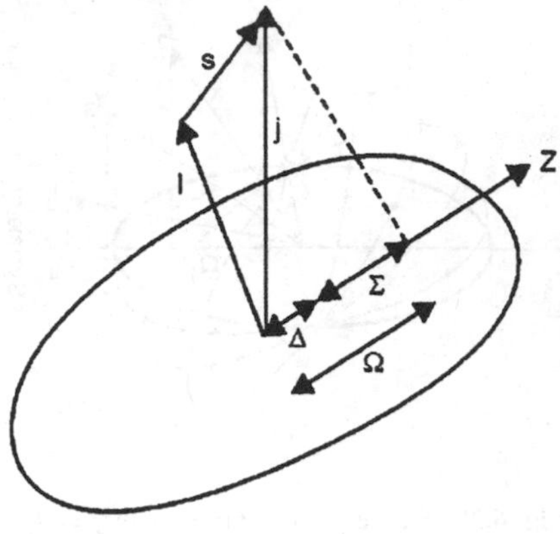

Fig. 6.29 Coupling scheme indicating various angular momenta $(\mathbf{l}, \mathbf{s})\mathbf{j}$ and their projection $(\Delta, \Sigma)\Omega$ on the symmetry axis (z-axis) for a particle moving in a deformed axially symmetric potential

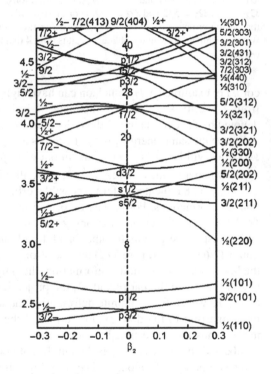

Fig. 6.30 Energy eigenvalues $\hbar\omega_0$ in units of β_2 plotted against the deformation parameter XX for the Nilsson model in the oscillator shells $N = 1, 2$ and 3. These energy levels are to be associated with odd-neutron or -proton nuclei in the $1pm2s$–$1d$ and $2p$–$1f$ shells [7]

magnetic quantum numbers. The nucleon states labeled with the quantum numbers $(\Lambda, n_z, \Delta$ are also shown in Fig. 6.30. The positive deformations correspond to probate ellipsoids and negative ones for oblate ellipsoids.

For small δ the scheme is close to that of a spherical nucleus. Each state with given j and π is resolved into a number of substates characterized by Ω and by the parity. Already for very small deformations, terms which arise from various values of j for a spherical nucleus, cross. Furthermore, an interaction occurs between levels of different origin but with the same Ω and π.

From the knowledge of nuclear deformation it is possible to successfully predict the spins and parities of the ground states for odd-A nuclei, using Nilsson diagrams. It is also possible to predict ground-state moments. Low-lying excited states can also be successfully predicted.

Example 6.2 Deduce that with $\varepsilon = 0.72$ MeV and $\gamma = 23$ MeV the ratio $Z_{min}/A \cong$ 0.5 and a for typical light nuclei and for heavy nuclei respectively.

Solution

$$\frac{Z_{min}}{A} = \frac{0.5}{1 + 0.25A^{2/3}\varepsilon/\gamma} = \frac{1}{2 + (0.5 \times 0.72/23)A^{2/3}} = \frac{1}{2 + 0.0156A^{2/3}}$$

For light nuclei the second term in the denominator is small in comparison with 2.0, so that $Z_{min}/A \rightarrow 0.5$. For heavy nuclei, typically $A \sim 100$ and $Z_{min}/A \rightarrow 0.43$.

Example 6.3 For large A it is found that $B/A = 9.402 - 7.7 \times 10^{-3}A$.

Given that the binding energy of alpha particles is 28.3 MeV. Determine A above which alpha emission becomes imminent.

Solution The condition to be satisfied is

$$B(A - 4) + B(\alpha) > B(A)$$

Using

$$B(A) = \left(9.402 - 7.7 \times 10^{-3}A\right)A$$

$$B(A - 4) = \left[\left(9.402 - 7.7 \times 10^{-3}\right)(A - 4)\right](A - 4)$$

$$B(\alpha) = 28.3$$

and solving, we find $A > 151$.

Example 6.4 For the nucleus ^{16}O the neutron and proton separation energies are 15.7 and 12.2 MeV respectively. Estimate r_0 assuming that the particles are removed from its surface and that the difference in separation energies is due to the Coulomb potential energy of the proton.

Solution

$$\frac{e^2}{4\pi\varepsilon_0} = 1.44 \text{ MeV fm}$$

$$S_n - S_p = 15.7 - 12.2 = \frac{3}{5}\frac{e^2}{4\pi\varepsilon_0 R}\left[Z^2 - (Z - 1)^2\right] = \frac{3}{5}\left(\frac{e^2}{4\pi\varepsilon_0}\right)\frac{(2Z - 1)}{R}$$

$$3.5R = 0.6 \times 1.44(2 \times 8 - 1)$$

$$R = 3.7 \text{ fm}$$

$$r_0 = \frac{R}{A^{1/3}} = \frac{3.7}{16^{1/3}} = 1.47 \text{ fm}$$

Example 6.5 The shell model energy levels are in the following way $[1s_{1/2}][1p_{3/2},$ $1p_{1/2}][1d_{5/2}, 2s_{1/2}, 1d_{3/2}][1f_{7/2}][2p_{3/2}, 1f_{5/2}, 2p_{1/2}, 1g_{9/2}][1g_{7/2}, 2d_{5/2}, 2d_{3/2},$ $3s_{1/2}, 1h_{11/2}] \cdots$ Assuming that the shell are filled in the order written, what spins and parities should be expected for the ground state of the following nuclei? ${}_{3}^{7}\text{Li}$, ${}_{8}^{16}\text{O}$, ${}_{8}^{17}\text{O}$, ${}_{19}^{39}\text{K}$, ${}_{21}^{45}\text{Sc}$.

Solution

${}_{3}^{7}\text{Li}$: Spin comes from the third proton in $p_{3/2}$ state. Hence $J^{\pi} = \frac{3^{-}}{2} (\because l = 1)$

${}_{8}^{16}\text{O}$: This a doubly magic nucleus, $J^{\pi} = 0^{+}$

${}_{8}^{17}\text{O}$: Spin comes from the 9th neutron is $d_{5/2}$ state. Hence $J^{\pi} = \frac{5^{+}}{2} (\because l = 2)$

${}_{19}^{39}\text{K}$: Spin comes from the proton in shell minus hole in $d_{3/2}$ state hence $J^{\pi} = \frac{3^{+}}{2}$

${}_{21}^{45}\text{Sc}$: Spin comes from the 21st proton in the $f_{7/5}$ state. Hence $J^{\pi} = \frac{7^{-}}{2}$

Example 6.6 Find the gap between the $1p_{1/2}$ and $1d_{5/2}$ neutron shells for nuclei with mass number $A \approx 16$ from the total binding energy of the ${}^{15}\text{O}$ (111.9556 MeV), ${}^{16}\text{O}$ (127.6193 MeV) and ${}^{17}\text{O}$ (131.7627 MeV) atoms.

Solution The ${}^{15}\text{O}$ nucleus may be considered as the ${}^{16}\text{O}$ nucleus with a deficit neutron in the $1P_{1/2}$ shell. The energy of this level is then $B({}^{15}\text{O}) - B({}^{16}\text{O})$. An ${}^{17}\text{O}$ nucleus may be considered as an ${}^{16}\text{O}$ nucleus with an additional neutron in the $1f_{5/2}$ shell, the energy of this level being $B({}^{16}\text{O}) - B({}^{17}\text{O})$. The gap between the shells is thus

$$E(1f_{5/2}) - E(1P_{1/2}) = B({}^{16}\text{O}) - B({}^{17}\text{O}) - [B({}^{15}\text{O}) - B({}^{16}\text{O})]$$

$$= 2B({}^{16}\text{O}) - B({}^{17}\text{O}) - B({}^{15}\text{O})$$

$$= 2 \times 127.6193 - 131.7627 - 111.9556$$

$$= 11.52 \text{ MeV}$$

Example 6.7 Compute the expected shell-model quadrupole moment of ${}^{209}\text{Bi}(9^{-}/2)$.

Solution The single-particle quadrupole moment of an odd proton in a shell-model state j

$$Q = -\frac{2j - 1}{2(j + 1)} \langle r^{2} \rangle$$

For a uniformly charged sphere, $\langle r^{2} \rangle = \frac{3}{5} R^{2} = \frac{3}{5} r_{0}^{2} A^{2/3}$. Using $j = 9/2$, $r_{0} = 1.2 \text{ fm}$, $A = 209$, we find $Q = -0.25b$. This value may be compared with the observed value of $-0.37b$.

Example 6.8 The masses of $^{15}_7N$ and $^{15}_8O$ are $15.000108u$ and $15.003070u$ respectively. The masses of neutron and proton are $1.008665u$ and $1.007825u$ respectively. Find the Coulomb's coefficient a_c in the semi-empirical mass formula.

Solution For mirror nuclei

$$M_{Z+1} - M_Z = M_p - Mn + a_c A^{2/3}$$

$$a_c = \frac{(15.00307 - 15.000108 + 1.008665 - 1.007825)931.5}{15^{2/3}}$$

$$= 0.58 \text{ MeV}$$

Example 6.9 The chlorine isotope of mass number 33 decays by positron emission as follows: $^{33}_{17}Cl \rightarrow ^{33}_{16}S + \beta^+ + \mu$ and the maximum positron energy is 4.3 MeV. Calculate r_0 from these data.

Solution

$$\Delta E_c = (M_n - M_p + M_e)c^2 + E_{\max}$$

$$= 1.29 + 4.3 = 5.59 \text{ MeV}$$

$$\Delta E_c = \frac{3}{5} \frac{e^2(2Z - 1)}{4\pi\varepsilon_0 R} = \frac{0.6 \times 1.44 \times 33}{r_0(33)^{1/3}} = 5.59$$

$$r_0 = 1.59 \text{ fm}$$

Example 6.10 Determine the most stable isobar with mass number $A = 64$.

Solution

$$Z_0 = \frac{A}{2 + 0.015A^{2/3}}$$

$$= \frac{64}{2 + 0.015 \times 64^{2/3}} = 28.57$$

$$Z_0 = 29$$

Example 6.11 Using the Fermi gas model show that the Fermi pressure is given by $p = \frac{2}{5}\rho_N E_F$, where ρ_N is the nucleon density.

Solution At constant entropy S, the pressure is given by the thermodynamic relation

$$p = -\left(\frac{\partial U}{\partial V}\right)_S$$

where V is the volume and U is the internal energy of the system.

$$U = \frac{3}{5}AE_F$$

$$p = -\frac{3}{5}A\frac{\partial E_F}{\partial V}$$

From the gas model

$$N = \frac{V(p_F(n))^3}{3\pi^2\hbar^3}, \qquad Z = \frac{V(p_F(p))^3}{3\pi^2\hbar^3}$$

Neutrons and protons, respectively, $p_F(n)$ and $p_F(p)$ are the Fermi momenta for neutrons and protons.

For $N = Z = A/2$

$$A = \frac{2Vp_F^3}{3\pi^2\hbar^3} = 2\frac{V(2ME_F)^{3/2}}{3\pi^2\hbar^3}$$

$$\frac{\partial E_F}{\partial F} = -\frac{2}{3}\frac{E_F}{V}$$

$$p = \frac{2}{5}\frac{A}{V}E_F = \frac{2}{5}\rho_N E_F$$

6.7 Questions

6.1 Give important applications of the Fermi gas model.

6.2 What is the evidence to support the reality of Fermi momentum?

6.3 Derive an expression for the total kinetic energy of all the proton in terms of the maximum Fermi energy (E_F).

6.4 Write down the semi-empirical mass formula and identify various terms.

6.5 Why does fission occur with thermal neutrons in ^{235}U but not in ^{238}U?

6.6 How is the asymmetric fission explained?

6.7 Show that the Coulomb energy of a uniformly charged sphere with total charge Q and radius R is $\frac{3}{5}[\frac{Q^2}{4\pi\varepsilon_0 R}]$.

6.8 The spins and parities of the ground and four excited states of $^{207}_{82}Pb$ are: (π), $1/2(-), 5/2(-), 3/2(-), 13/2(+), 7/2(-)$. Comment on the shell model description of these states.

6.9 Review briefly the evidence for a shell model of the nucleus.

6.10 Explain how a shell model uses potentials such as a square well and a harmonic oscillator to try and predict the magic numbers.

6.11 Explain how the spins, parities and magnetic moments of a nucleus in both the ground state and excited states may be predicted by the shell model and give three nuclei as examples.

6.12 Briefly outline the merits and defects of the shell model.

6.13 What are magic numbers? Explain.

6.14 What are Schmidt lines? How are they explained?

6.15 What is the reason for the $J = l + 1/2$ state to lie deeper than the $J = l - 1/2$ state in a nucleus?

6.16 The Q-value for fission reaction is positive. What prevents ^{238}U from undergoing spontaneous fission?

6.17 What are the predictions of the collective model of the nucleus, and how far have they been verified?

6.18 Many of the nuclei which are long lived isomer states have N or Z in the ranges $39 \cdots 49$ and $69 \cdots 81$. Explain.

6.19 Write down the shell model state of the odd nucleon in (a) $^{25}_{12}$Mg, (b) $^{63}_{29}$Cu.

6.20 Draw the energy level diagrams, showing the filling of the levels by neutron and protons in (a) $^{7}_{3}$Li, (b) $^{41}_{20}$Ca.

6.21 How is the spacing in the rotation and vibration energy level differ?

6.22 What is a giant dipole resonance?

6.8 Problems

6.1 The empirical mass formula (neglecting a term representing the odd-even effect) is

$$M(A, Z) = Z(m_p + m_e) + (A - Z)m_n - \alpha A + \beta A^{2/3}$$
$$+ \gamma (A - 2Z)^2 / A + \varepsilon Z^2 A^{-1/3}$$

where α, β, γ and ϵ are constants. Show that the most stable isobar is characterized by

$$Z_{min} = 0.5A\left(1 + 0.25A^{2/3}\varepsilon/\gamma\right)^{-1}$$

for the value of Z which corresponds to the most stable nucleus for a set of isobars of mass number A.

6.2 $^{23}_{12}$Mg undergoes positron decay to the mirror nucleus $^{23}_{11}$Na. If E_{max} of positron is 3.5 MeV show that $r_0 \cong 1.6$ fm.
[Ans. 1.63 fm]

6.3 Use the shell model to determine the spin and parity of the ground states of the nuclei

(a) ^{3}He
(b) ^{21}Ne
(c) ^{27}Al

[Ans. (a) $\frac{1}{2}^+$, (b) $\frac{3}{2}^+$, (c) $\frac{5}{2}^+$]

6.4 Show that the electrostatic energy of a uniformly charged sphere of radius R is $(\frac{3}{5})(Q^2/R)$ where Q is the total charge of the sphere.

6.5 Consider a proton as a uniform solid sphere of radius $R = 1$ fm.

(a) what angular velocity is needed to give it an angular momentum of h.
(b) what rotational kinetic energy does it correspond to?

[Ans. (a) 1.57×10^{23} rad s^{-1}, (b) 8.2×10^{-10} J]

6.6 Show that for a homogeneous ellipsoid of semi axes a, a, b the quadrupole moment is given by $Q = (\frac{2}{3})(b^2 - a^2)$.

6.7 Show that for a rotational ellipsoid of small eccentricity and uniform charge density, the quadrupole moment is given by $Q = (\frac{4}{5})ZR\Delta R$.

6.8 With the mass formula calculate the energy that is released in the binary fission of uranium.
[Ans. ~170 MeV]

6.9 For nuclei with atomic mass number A greater than 100, the average binding energy per nucleon is given by the approximate expression.

$$\frac{B}{A} \simeq 8.97 - 0.0068A \text{ MeV}$$

Given that the binding energy of the alpha as particle is 28.3 MeV, estimate the minimum atomic mass number for which alpha decay is energetically possible.
[Ans. 142]

6.10 The masses of the mirror nuclei $^{27}_{13}$Al and $^{27}_{14}$Si are 26.981539 and 26.986704 respectively, the neutron and proton masses are $1.008665u$ and $1.007825u$ respectively. Determine the Coulomb's coefficient in the semi-empirical mass formula. [Ans. 0.62 MeV]

6.11 $^{27}_{14}$Si and $^{27}_{13}$Al are mirror nuclei. The former is a positron emitter with $E_{max} = 3.48$ MeV. Determine r_0. [Ans. 1.63 fm]

References

1. A. Bohr, *Rotational States of Atomic Nuclei* (1954)
2. Burcham (1979)
3. Burge (1977)
4. Chrien, Dover (1986)
5. Chrien, Dover (1989)
6. R.E. Chrien, C.V. Dover, Annu. Rev. Nucl. Part. Sci. **39**, 113 (1989)
7. Davidson (1968)
8. Heyd (1994)
9. A.A. Kamal, *Particle Physics* (Springer, Berlin, 2014)
10. J.F. Mainz, in *Particles and Nuclei*, ed. by B. Povh, K. Rith, C. Scholz, F. Zetsche (Springer, Berlin, 1995)
11. Nilsson (1955)
12. B. Povh, Prog. Part. Nucl. Phys. **5**, 245 (1981)
13. Rainwater (1950)
14. Simon et al. (1982)
15. Van der Woude (1987)

Chapter 7
Nuclear Reactions

7.1 Types of Reactions

In the collision of two particles different processes may take place. A typical nuclear reaction may be written as

$$a + X \rightarrow b + Y + Q \tag{7.1}$$

where X represents the target nucleus 'a' the projectile, while Y is the residual nucleus (unobserved) and b is the particle observed. For brevity this reaction may be written, $X(a, b)y$. Isotopes are indicated by the use of their mass number as a superscript on the left of the chemical symbol. Special symbols are used to designate elementary particles, and some of the light nuclei; for example, e for electron, p for proton, n for neutron, d or ^2H for deuteron, t or ^3H for triton, $\propto$ or ^4He for alpha particle, γ for photon or gamma ray, π for pion, μ for muon etc. Sometimes, b or y may be produced in an excited state. This is indicated by the use of an asterisk, Y^* etc.

The symbol Q in (7.1) is the energy released in a reaction; if both b and y are left in the ground state, this is denoted by Q_0. If $Q \neq 0$, it means that a part of kinetic energy has gone into excitation energy and/or new type of nuclei. If E_f and E_i are the total kinetic energy in the final and initial state, then

$$Q = E_f - E_i \tag{7.2}$$

If Q is positive, the reaction is said to be exoergic (or exothermic as in chemical reactions) and a negative value of Q signifies that the reaction is endoergic or endothermic. In this case a definite minimum kinetic energy, called the threshold energy is required for the projectile to initiate the reaction. The threshold energy needed is equal to $-Q$ in the centre of mass system.

If $Q = 0$, then it represents elastic scattering in which case total kinetic energy is conserved. Given enough energy for the bombarding particle a collision may result in more than two particles in the final state. At sufficiently high energy a collision

A. Kamal, *Nuclear Physics*, Graduate Texts in Physics,
DOI 10.1007/978-3-642-38655-8_7, © Springer-Verlag Berlin Heidelberg 2014

may result in an appreciable number of reaction products, called a spallation reaction, although the break-up of a nucleus into all the constituents is an unlikely
event.

There are several major types of reactions.

(i) *Elastic Scattering* Here $b = a$ and $y = x$. The internal states are unchanged so
that $Q = 0$ and the kinetic energy of the particles in the CMS is unchanged before
and after the scattering. In general

$$a + x \rightarrow a + x$$

for example

$$n + {}^7\text{Li} \rightarrow n + {}^7\text{Li} \quad \text{or} \quad {}^7\text{Li}(n, n){}^7\text{Li}$$

(ii) *Inelastic Scattering* Here $b = a$, but 'X' is raised to an excited state, $Y = X^*$,
so that $Q = -E_x$, where E_x is the excitation of the state. Since "a" is emitted with
reduced energy, it is usually written as a'

$$a + X \rightarrow a' + X^* - E_x$$

For example

$$^{10}\text{B} + \alpha \rightarrow {}^{10}\text{B}^* + \alpha'$$

or

$$^{10}\text{B}(\alpha, \alpha'){}^{10}\text{B}^* \tag{7.3}$$

If 'a' is itself a complex nucleus, it may get excited instead of the target, or both
may be excited. An example of the latter is

$$^{12}\text{C} + {}^{16}\text{O} \rightarrow {}^{12}\text{C}^* + {}^{16}\text{O}^* \tag{7.4}$$

(iii) *Nuclear Reaction* Here $b \neq a$ and $y \neq x$ so that there is a rearrangement of the
constituent nucleons between the colliding pair, known as transmutation. A number
of possibilities are open; $x + a \rightarrow Y_1 + b_1 + Q_1$ or $\rightarrow Y_2 + b_2 + Q_2$ etc. Examples
are

$$\alpha + {}^{10}\text{B} \rightarrow {}^{13}\text{C} + p \tag{7.5}$$

$$^{26}\text{Mg} + {}^{14}\text{N} \rightarrow {}^{27}\text{Mg} + {}^{13}\text{N} \tag{7.6}$$

$$^7\text{Li} + p \rightarrow {}^7\text{Be} + n \tag{7.7}$$

(iv) *Capture Reactions* This is a special case of class (iii); the pair $x + a$ coalesce,
forming a compound system in an excited state which decays via one or more
γ-rays,

$$x + a \rightarrow C^* \rightarrow C + \gamma + Q \tag{7.8}$$

For example

$$^{197}\text{Au}(p, \gamma){}^{198}\text{Hg} \tag{7.9}$$

(v) *Fission* A neutron absorbed by a heavy nucleus like ^{235}U causes the nucleus to split it into two and sometimes three large fragments with the emission of a few neutrons. An example is

$$n + {}^{235}U \to {}^{141}Ba + {}^{92}Kr + 3n + 189 \text{ MeV} \tag{7.10}$$

The energy released is much greater than in other types of reactions. Fission can also be caused by other projectiles like p, α or pions.

(vi) *Other Reactions* If sufficient energy is available then in the final state there can be more than two particles. In general, $x + a \to y + b + c + Q$. For example

$$\alpha + {}^{40}Ca \to p + \alpha' + {}^{39}K \quad \text{or} \quad {}^{40}Ca(\alpha, \alpha' p){}^{39}K \tag{7.11}$$

The bombarding particle and the target particle $a + x$ constitute the entrance channel, while the products such as $b + y$ form the exit channel. Open channels are those which are energetically available.

7.2 Energy and Mass Balance

Q-value has been defined as the energy that is released. Hence it is the change in the sum of the kinetic energies of the colliding particles and the reaction products

$$E_f - E_i = Q \tag{7.12}$$

The Q-value can also be expressed in terms of the rest masses of the particles and using the relativistic relation $E = mc^2$. Consider the reaction $X(a, b)Y$

$$m_X + m_a = m_Y + m_b + \frac{Q}{c^2} \tag{7.13}$$

where m_i is the mass of the ith particle.

Alternatively, Q is equal to the change in the binding energies B_i of the particles

$$B_y + B_b = B_x + B_a + Q \tag{7.14}$$

For an elementary particle, such as a nucleon we regard $B_i = 0$ in this equation. Further (7.13) says that the sum of masses in the initial state is heavier then that in the final state. On the other hand, (7.14) says that the particles in the initial state are less tightly bound than those in the final state in case of exoergic reactions and more tightly bound in the entrance channel that those in the exit channel for endoergic reactions. Obviously Q-value deduced from (7.13) or (7.14) must be identical. As an example consider the reaction

$$d + d \to {}^3He + n + Q \tag{7.15}$$

By (7.13)

$$Q = \left[2m_d - (m_{^3\text{He}} + m_n)\right] \times 931.5 \text{ MeV}$$
$$= \left[2 \times 2.014102 - (3.016030 + 1.008665)\right] \times 931.5$$
$$= 3.27 \text{ MeV}$$

By (7.14)

$$Q = (B_{^3\text{He}} + B_n) - 2B_d$$
$$= (7.72 + 0) - 2 \times 2.225$$
$$= 3.27 \text{ MeV}$$

It is noteworthy that in nuclear reactions the energies evolved or absorbed are of the order of a few MeV while those in chemical reactions they are of the order of a few eV. This is closely connected with the fact that nuclear reactions involve rearrangement of nucleons whose binding energy is of the order of a few MeV, while chemical reactions involve rearrangement of atoms in molecules whose dissociation energy is of the order of a few eV.

7.3 Conservation Laws for Nuclear Reactions

The conservation laws may be stated and explained with reference to a specific example

$$^{10}_{5}\text{B} + ^4_2\text{He} \rightarrow ^1_1\text{H} + ^3_6\text{C} \tag{7.16}$$

(i) *Charge* Total charge is conserved in every type of reaction. In $^{10}\text{B}(\alpha, p)^{12}\text{C}$; there are seven protons initially, and also in the products of the reaction. In all such reactions, we may for brevity write

$$\Sigma \frac{Q_i}{e} = \text{const} \tag{7.17}$$

where e is the electronic charge.

(ii) *Mass Number* The total number of nucleons entering and leaving the reaction is constant. In the above example there are fourteen nucleons initially and in the products of the reaction. In general we may write

$$\Sigma A = \text{const} \tag{7.18}$$

(iii) *Statistics* Both sides of a reaction such as (7.16) involve the same total number of fermions, hence the statistics is either Fermi-Dirac through out (for odd ΣA) or Bose-Einstein throughout (for even ΣA). In our example Bose-Einstein statistics will be applicable

(iv) *Angular Momentum* The total nuclear angular momentum is always a constant of motion. In our example ^{10}B has nuclear spin $I = 3$, while α particle has spin zero. If the initial capture takes place by a s-wave, that is $l_i = 0$, then the total angular momentum is also 3. Now, in the final state both ^1H and ^{13}C have nuclear spin 1/2, which add up vectorially to 0 or 1. The mutual angular momentum of ^1H and ^{13}C is restricted to $l_F = 2$, 3 or 4.

(v) *Parity* As nuclear reactions are actuated by strong forces, total parity is conserved. The parity of a system is given by the product of the parity of its components. In our example if low energy alphas are used, then only s-waves ($l = 0$) are involved in the collision process. Hence the contribution from orbital angular momentum will be $(-1)^l = (-1)^0 = +1$. Shell model indicates that the parity of the ground level of ^{10}B and ^4He is even while that of ^{13}C is odd. Hence the total parity of the system in the initial state is even. In the final state proton has even parity while ^{13}C in the ground state has odd parity. In order to conserve the over all parity, it is necessary that the parity from the product particles be odd, that is $l_F = 1, 3, 5, \ldots$. Combining the results from the conservation of both angular momentum, and parity, we conclude that $l_F = 3$ alone is allowed provided the alpha particle is captured in the $l = 0$ state.

(vi) *Linear Momentum* In all nuclear reactions, the total linear momentum before and after the reaction is constant.

(vii) *Energy* In any nuclear reaction the sum of kinetic energy and the rest mass energy (mc^2) is constant.

(viii) *Isospin* Nuclear reactions will proceed if the total isospin is conserved. In the entrance channel ^4He has $T = 0$, so also ^{10}B so that initially total isospin is $I = 0$. In the final state ^1H has $T = 1/2$ while ^{13}C has $T = 1/2$ (the other member being ^{13}N) so that $I = 0$ or 1. Conservation of isospin requires that the nuclear reaction can proceed only through $I = 0$ channel.

7.3.1 Quantities that Are not Conserved

In nuclear reactions, quantities like magnetic dipole moments and electric quadrupole moments of the reacting nuclei which depend upon the internal distribution of mass, charge and current within the nuclei are not conserved.

7.4 Cross-Sections

In order to measure the probability quantitatively that a given nuclear reaction will take place we introduce the concept of cross-section. Consider a reaction of the type $x(a, b)\gamma$. If I_0 is the flux of particles 'a' incident per unit area on a target consisting N nuclei of type x, then the number of particles b emitted per unit time (I) will

Fig. 7.1 Scattering of the
incident beam after hitting the
target

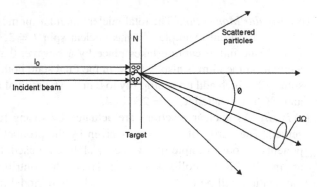

be proportional to both I_0 and N. The constant of proportionality σ is called the
cross-section which has the dimensions of area. Thus

$$I = I_0 N \sigma \quad \text{or} \tag{7.19}$$

$$\sigma = \frac{I}{I_0 N} \tag{7.20}$$

In words

Cross-section

$$= \frac{\text{number of particles } b \text{ emitted/s}}{(\text{number of particles 'a' incident/unit area/s})(\text{number of a target nuclei within the beam})}$$

In nuclear Physics the unit of cross-section is a Barn (b) $1 \text{ b} = 10^{-24} \text{ cm}^2 = 10^{-28} \text{ m}^2$; The sub-multiples are: millibarn, $1 \text{ mb} = 10^{-3} \text{ b}$, microbarn, $1 \text{ μb} = 10^{-6} \text{ b}$, nanobarn, $1 \text{ nb} = 10^{-9} \text{ b}$.

The number of particles b emitted per unit time within an element of solid angle $d\Omega$ in the direction with polar angles (θ, ϕ) with respect to the incident beam will be proportional to $d\Omega$ as well as I_0 and N, Fig. 7.1. The constant of proportionality is known as the differential cross-section $d\sigma(\theta, \phi)/d\Omega$, also written as $(d\sigma/d\Omega)$ or $\sigma(\theta, \phi)$ so that

$$\frac{d\sigma}{d\Omega} = \frac{I}{I_0 N d\Omega} \tag{7.21}$$

the unit of $d\sigma/d\Omega$ is barn/steradian. If the particles are unpolarised then the scattered particles will not depend on the azimuth angle ϕ, and the scattering will be symmetrical about the beam axis. In that case the differential cross-section will depend only on the polar angle θ, and will be written as $d\sigma(\theta)/d\Omega$ or $\sigma(\theta)$.

The total elastic cross-sections σ and $d\sigma/d\Omega$ are related by

$$\sigma = \int_0^{4\pi} \left(\frac{d\sigma}{d\Omega}\right) d\Omega \tag{7.22}$$

Fig. 7.2 Effect of collision
between two spheres and
point particle and disc

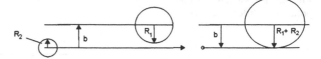

But $d\Omega = \sin\theta d\theta d\phi$; so that

$$\sigma = \int_0^\pi \sin\theta d\theta \int_0^{2\pi} d\phi \left(\frac{d\sigma}{d\Omega}\right)$$

In the absence of spin polarization, $d\sigma/d\Omega$ is independent of ϕ, we get

$$\sigma = 2\pi \int_0^\pi \left(\frac{d\sigma}{d\Omega}\right) \sin\theta d\theta \tag{7.23}$$

For the same entrance channel a number of exit channels will be open corresponding to different reaction products at a given energy. As the exit channels are independent, there will not be any quantum interference and the cross-section of different reaction channels may be added. The sum of all these non-elastic channels cross-section is called the reaction or absorption cross-sections and is denoted by σ_r. When the elastic cross-section is also added we speak of the total cross-section

$$\sigma_{total} = \sigma_r + \sigma_{el} \tag{7.24}$$

Strictly speaking the finite dimensions of the projectile must also be taken into account in the calculation of cross-section. Let a sphere 1 of radius R_1 be at rest and sphere 2 proceed toward it with impact parameter b, see Fig. 7.2. The two spheres will collide only if the impact parameter $b \leq R_1 + R_2$.

The effect is the same as for the collision of a point particle with a disc of radius $R_1 + R_2$, the disc being perpendicular to the axis joining the centres of the spheres. The area of the disc which is the projected area of the two spheres touching each other, is equal to $\pi(R_1 + R_2)^2$. This is the cross-section for the collision. This then means that if the radius of target nucleus is to be determined the radius of the bombarding particle must be taken into account.

7.5 Exoergic and Endoergic Reactions

We may illustrate these reactions in the following examples:

$$D + D \rightarrow {}_2^3\text{He} + n + Q \tag{7.25}$$

$$P + 3\text{Li}^7 \rightarrow {}_4^7\text{Be} + n + Q \tag{7.26}$$

The energy Q that is released is given by subtracting the sum of masses on the RHS from the sum on the LHS and multiplying by c^2. In (7.25), $Q = +3.29$ MeV,

Fig. 7.3 Emission of reaction
products at different angles

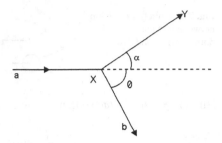

and is positive; the reaction is called exoergic reaction, where as in (7.26), $Q = -1.64$ MeV, being negative, is called endoergic reaction. The excess of energy manifests itself as kinetic energy of the product particles.

The Coulomb's repulsion between the incident and the target particle means that the incident particle must have certain minimum kinetic energy in order to get close to the nucleus and thence induce a nuclear reaction. Certain conservation laws must be valid for a reaction to proceed, e.g. the conservation of charge, linear momentum, nucleon number, angular momentum, energy etc.

In general, we may write the reaction of the incident particle 'a' of kinetic energy E_a with the target nucleus x, resulting in the products, b and y

$$a + x \rightarrow b + y \tag{7.27}$$

Conservation of energy gives us

$$(m_x + m_a)c^2 + E_a = (m_y + m_b)c^2 + E_b + E_y \tag{7.28}$$

Now

$$Q = (m_x + m_a - m_y - m_b)c^2 = E_b + E_y - E_a \tag{7.29}$$

Note that Q is independent of E_a and can be positive or negative; or zero for elastic scattering.

Let the reaction products b and y be emitted at angles θ and α, respectively, with respect to the original direction (see Fig. 7.3).

Conservation of energy and momentum gives us

$$Q = E_y + E_b - E_a \tag{7.30}$$

$$p_a = p_b \cos \theta + p_y \cos \alpha \tag{7.31}$$

$$0 = -p_b \sin \theta + p_y \sin \alpha \tag{7.32}$$

where p is the momentum. Writing, $p^2 = 2mE$, and eliminating α and E_y, we obtain

$$Q = E_b \left(1 + \frac{m_b}{m_y} \right) - E_a \left(1 - \frac{m_a}{m_y} \right) - \frac{2}{m_y} \sqrt{m_a m_b E_a E_b} \cos \theta \tag{7.33}$$

If the nucleus y is produced in the ground state, then the Q-value deduced from (7.33) and (7.29) would be identical. On the other hand, if the nucleus y is produced in an excited state, then the Q-value calculated from (7.33) would be lower corresponding to the energy of the excited state. The experimental Q-values, therefore, allow the energy levels to be determined. We shall now study the variation of E_b for a fixed value of Q. Writing

$$C = \frac{\sqrt{E_a m_a m_b}}{m_b + m_y} \cos \theta \tag{7.34}$$

$$D = \frac{E_a(m_y - m_a) + Q m_y}{m_b + m_y} \tag{7.35}$$

Equation (7.33) may then be written as

$$E_b - 2C\sqrt{E_b} - D = 0 \tag{7.36}$$

The solution of the above equation is

$$\sqrt{E_b} = C \pm \sqrt{C^2 + D} \tag{7.37}$$

Now, E_b must always be real and positive. The factors which can make the emission of b at an angle θ impossible are, (i) negative Q-value, (ii) $m_a > m_y$, (iii) large θ so that $\cos \theta$ may be negative.

7.5.1 Exoergic Reactions (Q-Value is Positive)

Example

$$n + {}^{10}\mathrm{B} \rightarrow \alpha + {}^{7}\mathrm{Li} \quad (Q = +2.8 \text{ MeV}) \tag{7.38}$$

When the bombarding energy (energy of the incident particle) is very small, i.e. $E_a \rightarrow 0$ then $C \rightarrow 0$, and (7.37) shows that

$$E_b = \frac{Q m_y}{m_b + m_y} \tag{7.39}$$

The particle b is, therefore, emitted with the same energy at all angles. When the bombarding energy, E_a is finite two cases may be distinguished

(A) $m_a < m_y$

In example (7.38), we choose $a = n$, $b = \alpha$, $y = {}^{7}\mathrm{Li}$. This means that D is always positive, and only one of the solutions of (7.37) viz, $\sqrt{E_b} = C + \sqrt{C^2 + D}$ is acceptable. Further, E_b will depend on θ, the angle of emission; being maximum for $\theta = 0$ and minimum for $\theta = 180°$, Fig. 7.4.

Fig. 7.4 Energy of the produced particle as a function of the incident energy, for the exoergic reaction $n + {}^{10}\text{B} \rightarrow \alpha + {}^{7}\text{Li} + 2.8$ MeV

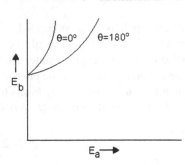

(B) $m_a > m_y$

Consider the reaction

$$\alpha + {}^{10}\text{B} \rightarrow {}^{13}\text{C} + p \tag{7.40}$$

Here, we choose $m_b = 13$, $m_y = 1$, and $m_a = 4$. The energy of the product particle ${}^{13}\text{C}$ is being studied. It follows that $D = (Q - 3E_\alpha)/14$, which goes negative for α energies greater than $Q/3$, i.e. >1.33 MeV, since $Q = 4$ MeV. For α energy <1.33 MeV, only one solution for E_b is acceptable. E_b is positive for all angles $0 < \theta < 180°$. For α energy >1.33 MeV, D is negative and there are two solutions for E_b in the interval $0 < \theta < 90°$. But, for $\theta > 90°$, only one solution is possible.

7.5.2 Endoergic Reactions (Q-Value is Negative)

There exists a threshold below which the reaction cannot proceed. The threshold energy is the minimum energy required in the lab system so that in the center of momentum system, the product particles are at rest.

Calculations in the CM System   Let the particle 'a' of mass m_a be moving with velocity v_a and kinetic energy E_a in the Lab system, and hit the target particle x, initially at rest (Fig. 7.5). In the CM System, both the particles will be observed to be approaching each other with equal and opposite momentum (by definition of the CM System). Let the CM system itself be moving with velocity v_c along the incident direction. Let v_a^* and v_x^* be the velocity of 'a' and 'x' respectively, in the CMS. Clearly

$$v_x^* = v_c \tag{7.41}$$

since to an observer fixed to the CM System, the particle x will be seen to approach with velocity v_c. Also

$$v_a^* = v_a - v_c \tag{7.42}$$

By definition

Fig. 7.5 Mass, energy and velocity of the particles in lab system and the CM system

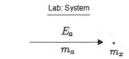

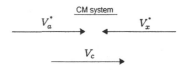

$$m_a v_a^* = m_x v_x^* \tag{7.43}$$

$$\therefore \quad m_a(v_a - v_c) = m_x v_c \tag{7.44}$$

whence, we find

$$v_c = v_x^* = \frac{v_a m_a}{m_a + m_x} \tag{7.45}$$

Also

$$v_a^* = \frac{v_a m_x}{m_a + m_x} \tag{7.46}$$

The sum of the energy of the particles in the CMS is then,

$$E^* = \frac{1}{2}m_a v_a^{*2} + \frac{1}{2}m_x v_x^{*2} = \frac{1}{2}m_a v_a^2 \frac{(m_x)^2}{(m_a + m_x)^2} + \frac{1}{2}m_x v_a^2 \frac{(m_a)^2}{(m_a + m_x)^2}$$

$$= \frac{1}{2}m_a v_a^2 \cdot \frac{m_x}{m_a + m_x} = \frac{E_a m_x}{m_a + m_x} \tag{7.47}$$

where, we have used (7.45) and (7.46), and $E_a = (1/2)m_a v_a^2$. If the reaction is to barely proceed then we must write, $E^* = -Q$

$$E_{a(threshold)} = |-Q|\left(1 + \frac{m_a}{m_x}\right) \tag{7.48}$$

At E_a corresponding to the threshold energy, the particle b appears at $0°$, with energy

$$E_b = \frac{E_a m_a m_b}{(m_b + m_y)^2} = \frac{E_a m_a m_b}{(m_a + m_x)^2} \tag{7.49}$$

As E_a is raised above the threshold, b appears at angles $>0°$. At angles $\geq 90°$, the particle b first appears with $E_b = 0$ when the terms C and D vanish separately. Thus, at $\theta = 90°$

$$E_{a(90°)} = \frac{-Q m_y}{m_y - m_a} \tag{7.50}$$

Fig. 7.6 Energy of the
produced particle as a
function of the incident
energy for the endoergic
reaction, $p + {}^7\text{Li} \to {}^7\text{Be} + n$
($Q = -1.64$ MeV)

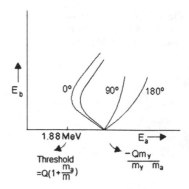

The particle b can appear in the forward direction with double solution for E_b,
provided $E_a \leq -Q m_y/(m_y - m_a)$, and $E_a \geq E_{a(0°)}$, i.e. $C^2 + D \geq 0$. In other
words, for the backward hemisphere ($\theta > 90°$) there is a single solution.

In endoergic reactions, E_b becomes single valued for all θ when $E_a > -Q m_y/(m_y - m_a)$. It may be remarked that the heavier fragment can never be
projected in the backward direction. Also, E_b can be zero only at 90°. The energy
E_b as a function of E_a is graphically represented in Fig. 7.6, for the typical endoergic reaction

$$p + {}^7\text{Li} \to {}^7\text{Be} + n - 1.64 \text{ MeV}$$

7.6 Behaviour of Cross-Sections near Threshold

For a given pair of nuclei a large variety of nuclear reactions are possible. However,
the general trends for the variation of cross-sections with energy can be investigated
for different classes.

Consider the reaction of the type $X(a, b)Y$. Let n_a be the number of particles
of type 'a' per unit volume and v_{ax} the velocity of 'a' relative to X. The product
$n_a v_{ax}$ represents the flux of 'a', that is the number of particles of type 'a' crossing
unit area per second. We can write

$$v_{ax} n_a \sigma_{X \to Y} = W \tag{7.51}$$

where W is the number of transitions per second. Now the golden rule gives

$$W = \frac{2\pi}{\hbar} \langle |H_{if}|^2 \rangle \frac{dN}{dE} \tag{7.52}$$

where $\langle |H_{if}| \rangle$ is the matrix element averaged over individual states and dN/dE is
the density of final states. The statistical factors due to spin are

$$g_i = (2S_a + 1)(2S_x + 1) \quad \text{and} \quad g_f = (2S_b + 1)(2S_y + 1) \tag{7.53}$$

for the initial state i and the final state f. Experimentally the initial system is normalized to have a weight 1. In the final state, since different spin orientations are not distinguished, $\langle |H_{if}|^2 \rangle$ must be multiplied by g_f. The matrix element H_{if} is

$$H_{if} = \int \psi_f^* V \psi_i d\tau \tag{7.54}$$

As the wave functions ψ_f^* and ψ_i are normalized, we put $\Omega = 1$. The density of final states is

$$\frac{dN}{dE} = \frac{\text{phase space}}{\hbar^3} = \frac{(\text{physical space}) \times (\text{momentum space})}{\hbar^3}$$

$$= \Omega 4\pi p_b^2 \frac{dp_b}{dE} = \frac{4\pi p_b^2 dp_b}{v_{by} dp_b} = \frac{4\pi p_b^2}{v_{by}} \tag{7.55}$$

Combining (7.51) and (7.55) and calling $v_a = v_{ax}$ and $v_b = v_{by}$ for simplicity, and noting $n_a = 1/\Omega = 1$, we find

$$\sigma_{x \to y} = \frac{1}{\pi \hbar^4} \langle |H_{if}|^2 \rangle \frac{p_b^2}{v_a v_b} (2I_y + 1)(2I_b + 1) \tag{7.56}$$

Equation (7.56) permits us to investigate the variation of cross sections for different class of reactions. Here we avoid the regions of resonances, that is nuclear levels which play a dominant role

1. *Elastic scattering* (Both bombarding particle and scattering particle are uncharged) as for neutron scattering, $v_a = v_b$, therefore $p_b^2/v_a v_b = (M_{neutron})^2 = $ const. At low energy H_{if} is approximately constant so that $\sigma \simeq$ constant at low energy, Fig. 7.7(a).
2. *Exoergic reaction* The bombarding particle is uncharged, for example neutron at low energy. The Q-value is positive and is of the order of few MeV and neutron energy is a few eV so that $v_b \simeq$ const, and $p_b^2/v_a v_b \propto (1/v_a)$. Now $|H_{if}|^2 \propto e^{-2(G_n + G_b)}$, where the exponential is the barrier factor. But $G_n = 0$ for a neutral particle and $G_b = (\pi Z_b Z_y e^2/\hbar v_b) \simeq$ const. It is therefore concluded that $\sigma \propto 1/v_n$, the famous $1/v$ law, Fig. 7.7(b). The examples to be considered are (n, γ), (n, p), (n, α), (n, f) reactions where f is for fission.
3. *Exoergic reaction, charged incoming particle* Examples are (p, n), (α, n), (p, γ), (α, γ) reactions. At incident energies $\ll Q$, the factor $\frac{p_b^2}{v_a v_b} \propto \frac{1}{v_a}$ and the barrier factor e^{-G_a} alone is of consequence so that $\sigma \propto (1/v_a)e^{-2G_a}$, Fig. 7.7(c).
4. *Inelastic scattering* (n, n') This is a particular case of an endoergic reaction. Q is negative and $-Q$ is the excitation energy of the nucleus. At incident neutron energies slightly above the threshold, $v_n \simeq$ const; since the fractional change in incident energy is small. But $v_{n'} \propto (\Delta E)^{(1/2)}$ where $\Delta E = $ excess of energy above the threshold. Consequently, the factor $\frac{p_{n'}^2}{v_n v_{n'}} \propto v_n'$ or $\propto (\Delta E)^{(1/2)}$. Therefore, near the threshold $\sigma \propto (\text{energy excess})^{1/2}$, Fig. 7.7(d).

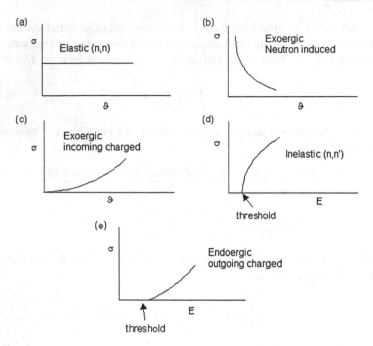

Fig. 7.7 Schematic behavior of various cross-sections near the threshold

5. *Endoergic reaction; charged outgoing particles* Examples are (n, p), (n, α) reactions. The dependence is the same as in case 4, except the barrier factor e_b^{-G} operates and $\sigma \propto$ (energy excess)$^{1/2} \times e^{-2G_b}$, or $\propto (\Delta E)^{1/2} e^{c1} / E_b^{1/2}$, Fig. 7.7(e).

6. *Endoergic reaction; charged outgoing particles* Examples are (n, p), (n, α) reactions. The dependence is the same as in case 4, except the barrier factor e^{-G_b} operates and $\sigma \propto$ (energy excess)$^{1/2} \times e^{-2G_b}$ or $\propto (\Delta E)^{\frac{1}{2}} e^{-c'/E_b 1/2}$, $(c' = \text{const})$ Fig. 7.7(e).

7.7 Inverse Reaction

If the equation describing the reaction process

$$a + x \rightarrow b + y + Q$$

is invariant under time reversal (changing the sign of the time variable) then it also describes the process

$$y + b \rightarrow x + a - Q$$

At a given total energy in the CMS, the forward reaction cross-section $\sigma (a \rightarrow b)$ and the backward reaction cross-section $\sigma (b \rightarrow a)$ are not identical but are simply

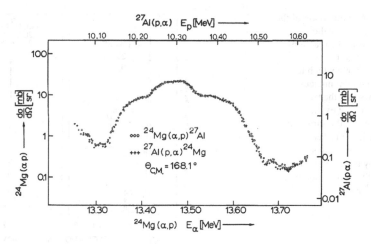

Fig. 7.8 An experimental test of the reciprocity theorem, Eq. (7.58). The vertical scales for the cross-sections of the reaction and its inverse have been adjusted to compensate for the statistical weights due to spin and momentum which appear in Eq. (7.58) [31]

related by the density of final states, that is by the phase space available in the respective exit channels. The number of states available for momenta between p and $p + dp$ is proportional to p^2. Hence $\sigma(a \rightarrow b)$ is proportional to p_b^2 where p_b is the relative momentum of b with respect to y, and $\sigma(b \rightarrow a)$ is proportional to p_a^2 if p_a is the relative momentum of "a" with respect to x. We then have

$$\frac{\sigma(b \rightarrow a)}{\sigma(a \rightarrow b)} = \frac{p_a^2}{p_b^2} \tag{7.57}$$

This is known as the reciprocity theorem or the principle of detailed balance. It is valid for differential as well as total cross-sections.

For particles with spins we must also take into account the corresponding statistical weights. Assuming that the participating particles are unpolarised, there will be $(2I + 1)$ states of orientations available for a particle of spin I. For particles with spin I_a, I_x, I_b and I_y (7.57) becomes

$$\frac{\sigma(b \rightarrow a)}{\sigma(a \rightarrow b)} = \frac{(2I_x + 1)(2I_a + 1)p_a^2}{(2I_y + 1)(2I_b + 1)p_b^2} \tag{7.58}$$

The reciprocity theorem has been tested in numerous experiments. The agreement is excellent and demonstrates the validity of time-reversal invariance of the underlying equations. In case one or more particles are polarised more complicated relations hold for these relations. The measured differential cross-sections for the reactions $^{24}Mg(\alpha, p)^{27}Al$ and $^{27}Al(p, \alpha)^{24}Mg$ at the same energy in CMS and at the same angle are compared and are shown to verify the predictions of time-invariance to a high degree of accuracy, Fig. 7.8. The principle of detailed balance has been ap-

Fig. 7.9 A comparison of the energy spectra for the products from the reactions $^{12}C(^{12}C, \alpha)$ and $^{16}O(^7Li, t)^{20}Ne$ which lead to the same final nucleus. This shows that the two reactions do not excite the states of Ne in the same way; both reactions are very 'selective'. The peaks are labeled by the excitation energy of the corresponding states in ^{20}Ne [8]

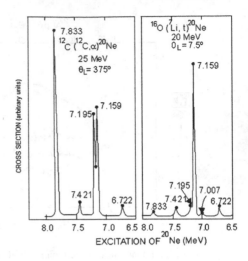

plied to the photo-disintegration of deuteron and the capture of neutron by proton. Another important application is the determination of pion spins [24], Chap. 3.

7.8 Qualitative Features of Nuclear Reactions

Nuclear reactions display a bewildering variety. Nevertheless some general characteristics can be studied and certain types can be classified. Nuclear reactions often result in the production of a variety of nuclei which are left in various excited states. It is possible to obtain the same product nucleus with the use of different pairs of incident and target particle. As an example consider the production of ^{20}Ne nucleus in the reactions $^{12}C(^{12}C, \alpha)^{20}Ne$ and $^{16}O(^7Li, t)^{20}Ne$. Referring to Fig. 7.9 it is found that not all the states are excited with equal probability. Further the excited states in ^{20}Ne are populated differently in the above mentioned reactions. Thus, there is certain amount of selectivity in various channels which varies with the bombarding energy. It often permits us to obtain useful information about the mechanism of the reactions and the nuclear structure. The type of information obtained from reaction measurements also depends upon the nature of the projectile and the bombarding energy. Thus, the collision of a proton of several hundred MeV with a nucleus will be like a nucleon-nucleon collision in which pions or strange particles may be produced and a target nucleon may be knocked out. The collision of a heavy ion such as ^{40}Ca on ^{84}Kr of the same energy will be different as the energy will be deposited over a large volume, setting up a large scale collective motion of the compound system. Further, the heavy ion may input larger amount of angular momentum. For example, a proton of 400 MeV incident upon an ^{107}Ag nucleus will not strike the nucleus if its angular momentum is greater than about 30 h, while a ^{84}Kr with 400 MeV can interact with the same target nucleus when their relative angular momentum is up to 470 h.

7.9 Reaction Mechanisms

Three different reaction mechanisms are compound nucleus reactions, direct reactions and pre-equilibrium reactions. In the compound nucleus reactions the incident particle is captured by the target nucleus and its energy (kinetic + binding energy $\simeq$ 8 MeV) is shared among the nucleons of the compound nucleus until it attains a state of statistical equilibrium. After a time of the order of 10^{-14}–10^{-15} s at low incident energies and 10–100 times greater at high energies, a nucleon or a group of nucleons near the surface may, by a statistical fluctuation, receive enough energy to escape, in the manner of evaporation of a molecule from a heated drop of liquid. This statistical process favours the emission of low energy particles which form the Maxwellian distribution. If the excitation energy of the compound nucleus is high enough, several particles may be emitted in succession until the energy of the nucleus has dropped below the threshold for particle emission. Then the nucleus emits γ rays until the ground state is reached. The nucleus may decay in a variety of other ways such as fission into two large fragments if the compound nucleus is very heavy or through the production of radioisotopes. The type of information which the study of compound nucleus yields includes the properties of energy levels of the compound nucleus which are excited, the mechanism of nuclear de-excitation, the density of high energy states, the role of angular momentum and nuclear deformation in affecting the evaporation process. The measurement of γ ray energies and intensities and their angular correlations find important applications for the structure studies of low energy levels.

The direct reactions take place in the time the incident particle takes to traverse the target nucleus which is typically of the order of 10^{-22} s. Here the incident particle may interact with a nucleon or a group of nucleons or the entire nucleus and emission takes place immediately. The simplest direct process is the elastic scattering in which the target nucleus is left in the ground state. In non-elastic processes the states of the residual nuclei which are excited bear a simple structural relationship to the ground state of the target nucleus. Inelastic scattering predominantly excites collective states, one nucleon transfer excites single-particle states and multi nucleon transfer excites cluster states. Measurements of cross-sections of these states, the angular distribution of the emitted particles, and their state of polarization permit the study of these states. Much of our knowledge of nuclear structure has originated from the study of direct reactions.

It is possible that after the interaction the particle may not be emitted immediately as in the direct reactions nor after a long time in the statistical way as from the compound state. The particle may be emitted before reaching the statistical equilibrium, such processes are termed as pre-compound or pre-equilibrium reactions and constitute the third category.

The interaction of two heavy ions requires yet one other mechanism.

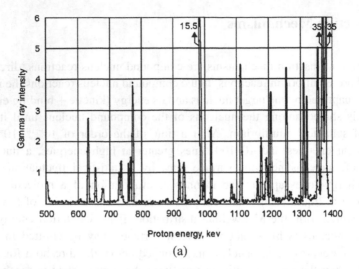

Fig. 7.10 (a) Some resonance peaks that show up when ^{27}Al is bombarded with protons. The ordinate measures the gamma radiation from the target; the abscissa is the proton energy in MeV [11]

7.10 Nuclear Reactions via Compound Nucleus Formation

In early 1930's a large number of nuclear reactions were studied. Detailed studies showed sharp peaks in cross sections at selective bombarding energies, Fig. 7.10(a).

This led Bohr [6] and independently Breit and Wigner to postulate that these reactions proceed through two stages, first the formation of an intermediate state, called compound nucleus state, second its break up in a relatively long time into the observed products. According to Bohr:

(i) The same compound nucleus can be formed in a variety of ways. For example, the compound nucleus in a particular excited state in the compound state designated as ^{64}Zn* can be produced by

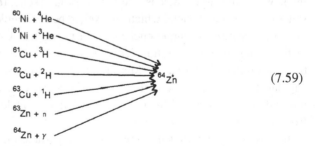

$$(7.59)$$

The asterisk (∗) shows that the compound nucleus is produced in an excited state.

(ii) A particle which hits the nucleus is captured and the energy released consists of its binding energy plus the kinetic energy. In its strong interaction with one or more nucleons, the available excitation energy is dissipated and shared by

Fig. 7.10 (Continued)

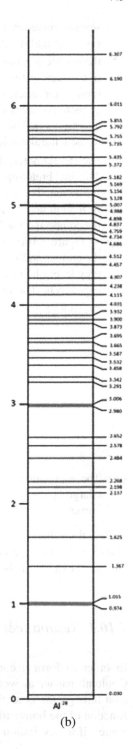

(b)

several nucleons in the collision processes. The excitation energy (E_{exc}) is given by adding the binding energy (B) and the available kinetic energy in the c-system. Thus, $E_{exc} = B + \frac{E_a m_x}{(m_a + m_x)}$, where E_a is the bombarding energy in the Lab system. The new nucleus is thus formed in the intermediate stage (compound nucleus). Sooner or later lot of energy may be deposited on one or more particles of the compound state nucleus, resulting in the emission of a particle (s). The compound state nucleus is long lived ($\sim 10^{-16}$ s) compared to the natural nuclear time which may be taken as the time taken to cross the nuclear diameter, that is $t \sim \frac{10^{-12} \text{ cm}}{10^9 \text{ cm/s}} = 10^{-21}$ s.

(iii) The final break-up of the compound state nucleus is independent of the mode of formation. The time involved in break-up is so long that 'memory' is lost. The formation and break-up can be regarded as independent events. For example, the compound nucleus ^{64}Zn* can decay into a variety of ways as in (7.60).

Figure 7.10(a) shows the resonance peaks formed in the reaction ($p, ^{27}$Al) the corresponding energy level diagram is indicated in Figs. 7.7–7.10, the energy level diagram is constructed from the measured Q values for various energy levels. The ground energy level E_0 is taken as zero. The energy level E_n is given by $E_n = Q_n - Q_0$, where n is the energy level

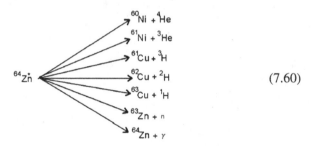

$$(7.60)$$

(iv) The exit channel with neutron emission is much more favoured than the charged particles like proton and α as the latter have to overcome the Coulomb barrier.

Reaction Channels-Various ways in which a compound state is formed are called entrance channels, and the various ways in which a compound nucleus breaks up are called exit channels.

7.10.1 Resonances in the Formation of the Compound Nucleus

In order to form a compound nucleus the incident particle must penetrate the Coulomb barrier as well as the nuclear surface. Now the probability of transmission through the potential step at the "surface" of the nucleus is not a monotonic function of the bombarding energy but shows very large values for certain selected values. If the excitation energy of the compound nucleus is just equal to one of the

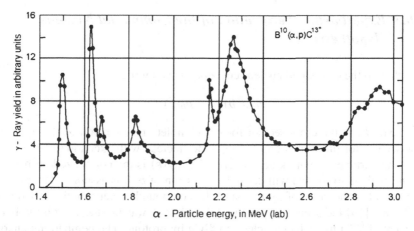

Fig. 7.11 Relative yield of the 3.9 MeV γ-rays from C^{13*}, showing resonances in the formation of the compound nucleus in the reaction $B^{10}(\alpha, p)C^{13*}$. A thin target ($\sim$10 keV stopping power for α-rays) of isotopically enriched ^{10}B was used. The γ-rays were observed with a scintillation spectrometer, at 90° from the α-particle beam (Talbott and heydenburry)

excited levels then a resonance formation of the compound nucleus is expected. Figure 7.11 shows the resonances in the formation of the compound nucleus $^{14}N^*$ in the reaction $^{10}B(\alpha, \rho)^{13}C^*$. Relative yield of 3.9 MeV γ-rays from $^{13}C^*$ is measured as a function of α-energy. Peaks of γ-ray intensity in Fig. 7.11 correspond to resonance penetration of the $^{10}B + \alpha$ barrier and formation of $^{14}N^*$ in a succession of excited levels.

7.10.2 Width of Resonance Levels

From Fig. 7.11 we can also estimate the width T of each of the virtual levels. Because of its finite lifetime, the level can not be said to have a perfectly sharply defined energy and the uncertainties in energy and time are related by the Heisenberg uncertainty principle $\Delta E \Delta t \geq \hbar$.

Any level which has a number of possible competing modes of decay will have a corresponding number of "partial width" Γ_j, each corresponding to the probability of decay by a particular mode.

The total width of the level, which corresponds to the total probability of decay, is the sum of all the partial widths

$$\Gamma = \Gamma_1 + \Gamma_2 + \Gamma_3 + \cdots \tag{7.61}$$

The total width is defined as the full width of the resonance peak measured at one-half the maximum height of the peak, Fig. 7.11.

7.10.3   Experimental Verification of the Compound Nucleus Hypothesis

According to the compound nucleus concept, we can write

$$\sigma(a, b) = \sigma_c p_b(\varepsilon) \tag{7.62}$$

where $\sigma(a, b)$ is the cross section for the complete reaction $X(a, b)Y$; σ_c is the cross section for the formation of compound nucleus with excitation energy ε by absorbing particle 'a' with kinetic energy E_a; $p_b(\varepsilon)$ is the normalized probability that the nucleus so formed will decay by emission of b. It is assumed that $p_b(\varepsilon)$ is independent of the mode of formation of the compound nucleus. In the experiment of Ghoshal [19] the same compound nucleus $^{64}Zn^*$ was produced with the bombardment of $^{60}_{28}Ni$ by alpha particles and $^{63}_{29}Cu$ by protons. The bombarding energy for each of the reactions was adjusted to yield the same excitation energy for the compound nucleus.

The reactions observed were

(1) $^{60}Ni(\alpha, n)^{63}Zn$
(2) $^{60}Ni(\alpha, 2n)^{62}Zn$
(3) $^{60}Ni(\alpha, pn)^{62}Cu$
(4) $^{63}Cu(p, n)^{63}Zn$
(5) $^{63}Cu(p, 2n)^{62}Zn$
(6) $^{62}Cu(p, pn)^{62}Cu$

Since the excitations produced through the two processes are the same the decay rate through the channel b is the same, as it depends only upon the excitation produced in the compound nucleus, and not upon the mode of formation. If the compound nucleus assumption is true then it is expected from Fig. 7.12

$$\sigma(p, n) : \sigma(p, 2n) : \sigma(p, pn) = \sigma(\alpha, n) : \sigma(\alpha, 2n) : \sigma(\alpha, pn) \tag{7.63}$$

These expectations were borne out by this and other experiments. In John's experiment on reactions with heavy elements, excitation functions of (α, Xn) reactions in ^{206}Pb were compared with those of (p, Xn) in ^{209}Bi where $X = 2, 3, 4$. For the same excitation in the compound nucleus ^{210}Po, the expected ratios

$$\sigma(p, 2n) : \sigma(p, 3n) : \sigma(p, 4n) = \sigma(\alpha, 2n) : \sigma(\alpha, 3n) : \sigma(\alpha, 4n) \tag{7.64}$$

were experimentally confirmed.

7.10.4   Energy Level Density

The density of nuclear levels depends strongly on the excitation energy and on the mass number. In the light nuclei, near the ground state the levels are about 1 MeV

Fig. 7.12 Experimental cross section for the reactions of Eq. (7.59). The scales of alpha energy are shifted by 7 MeV to take account of the fact that an alpha particle must have more energy than a proton (by this amount) if they are to produce compound nuclei with the same excitation energy [19]

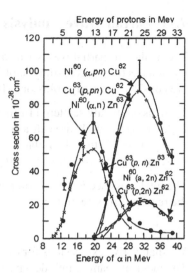

apart, while in heavy nuclei about 50 keV apart, except in magic number nuclei. When the excitation energy is in the vicinity or the neutron binding energy (8 MeV) slow neutron resonances in medium-heavy elements are a few electron volts apart. Using Fermi's gas model (Chap. 6) one can obtain the level density $\rho(E)$ for the total system energy E by an approximate formula

$$\rho(E) = \rho(0)e^{2(aE)1/2}$$

where $\rho(0)$ and 'a' are empirical constants.

Decay of Compound Nucleus The compound nucleus formed by the absorption of neutron is de-exited by evaporating one or more neutrons or other particles. If the excitation energy is sufficient then the evaporation of one particle may leave enough energy to enable the second particle to leave and so on. We may then have (p, Xn), for example with $X = 1, 2, 3, \ldots$ upto 6 or 7. The probability of decay through a larger number of particles increases with greater excitation. When little excitation is left particle emission ceases and only gamma emission is possible. The energy distribution of the evaporated neutrons is described by the Maxwellian distribution at the temperature T of the residual nucleus. Number or neutrons emitted between E_n and $E_n + dE_n$, being

$$n dE_n \simeq \text{const} \cdot E_n e^{-E_n/kt} dE_n$$

In the case of charged particles the Coulomb barrier inhibits the emission of low energy particles. The energy distribution is modified by multiplying the Maxwell distribution by the Coulomb barrier penetration factor.

7.11 Partial Wave Analysis of Nuclear Reactions

The method of partial waves was used for $n-p$ and $p-p$ scattering (Chap. 5). This method has been successfully applied by Weisskopf and others to nuclear reactions. We consider two-body collisions. Two possibilities are:

1. 'a' is elastically scattered from X.
2. 'a' reacts with X in such a way that it is removed from the incident beam by any process other than elastic scattering. Process, 1 is described by an elastic scattering cross-section σ_{el} and process 2 by the reaction cross-section σ_r.

The total cross-section σ_t is composed of the elastic scattering cross-section σ_s and the reaction cross-section σ_r

$$\sigma_t = \sigma_s + \sigma_r \tag{7.65}$$

We shall derive expressions for σ_s and σ_r for the lth partial wave for neutron beam. The incident beam of unit density and of flux v along the z-axis in the partial-wave expansion can be written as

$$e^{irz} = \sum_{\iota=0}^{\infty} i^l (2l+1) j_l(kr) P_l(\cos\theta)$$

$$\simeq \sum_{r\to\infty} \frac{i^{l+1}}{2kr} (2l+1) \left\{ \exp\left[-i\left(kr - \frac{l\pi}{2}\right)\right] - \exp\left[i\left(kr - \frac{l\pi}{2}\right)\right] \right\} P_l(\cos\theta)$$

$$\tag{7.66}$$

Equation (7.66) is the superposition of both the incoming and outgoing spherical waves. In scattering or nuclear reaction the amplitude of the outgoing spherical wave part of the plane wave is modified.

The wave function $\psi(r)$ describing the outgoing wave after interaction is written as

$$\psi(r) \underset{r\to\infty}{\simeq} \sum_l \frac{i^{l+1}}{2kr} (2l+1) \left\{ \exp\left[-i\left(kr - \frac{l\pi}{2}\right)\right] - \eta_l \exp\left[i\left(kr - \frac{l\pi}{2}\right)\right] \right\} P_l(\cos\theta)$$

$$\tag{7.67}$$

where

$$\eta_l = e^{2i\delta_l} \tag{7.68}$$

is a complex amplitude for the lth partial wave. If $|\eta_l| = 1$, there is no change in the number of particles in the lth wave and only elastic scattering will occur. If, however $|\eta_l| < 1$, then both elastic scattering and nuclear reaction will take place, a condition that is valid for all values of l. In case the incident particles are charged then the Coulomb function must be incorporated in the exponential factors in (7.67). When the target is polarized and the non-central forces are present, the factor $2l + 1$ must be replaced by a weighted sum over the magnetic quantum number m.

The elastically scattered wave function ψ_{el}, which is the difference of (7.66) and (7.67) is given by

$$\psi_s = \sum_l \frac{i^{l+1}}{2kr}(2l+1)(1-\eta_l)\exp\left[i\left(kr-\frac{l\pi}{2}\right)\right]P_l(\cos\theta) \qquad (7.69)$$

Now the quantum mechanical expression for current density is

$$j = \frac{\hbar}{2mi}\left(\frac{\partial\psi_s}{\partial r}\psi_s^* - \frac{\partial\psi_s^*}{\partial r}\psi_s\right) \qquad (7.70)$$

The ingoing flux corresponding to ψ_s through a sphere of radius r_0 is given by

$$\phi = \frac{\hbar r_0^2}{2im}\int\left(\frac{\partial\psi}{\psi r}\psi_s^* - \frac{\partial\psi_s^*}{\partial r}\psi_s\right)d\Omega\bigg|_{r=r_0} \qquad (7.71)$$

$$\sigma_s = \frac{\phi}{v} = \frac{\pi\hbar k}{2k^2mv}\int_{-1}^{+1}\left|\Sigma(2l+1)(1-\eta_l)P_l(\cos\theta)\right|^2 d\cos\theta$$

$$\sigma_s = \sum_{l=0}^{\infty}\sigma_s^l \qquad (7.72)$$

where

$$\sigma_s^l = \frac{\pi}{k^2}(2l+1)|1-\eta_l|^2 \qquad (7.73)$$

The total flux entering a large sphere of radius r_0 may be computed from (7.71), by using $\psi(r)$ and not ψ_s

$$\phi_r = -\frac{\hbar r_0^2}{2im}\int\left(\frac{\partial\psi}{\partial r}\psi^* - \frac{\partial\psi^*}{\partial r}\psi\right)d\Omega$$

$$= -\frac{\hbar\pi}{2imk^2}\int\sum_l(2l+1)\left\{\left[-ik\exp\left(-i\left(kr-\frac{l\pi}{2}\right)\right)\right.\right.$$

$$\left.- ik\eta_l\exp\left(i\left(kr-\frac{l\pi}{2}\right)\right)\right]\left[\exp i\left(kr-\frac{l\pi}{2}\right) - \eta_l^*\exp\left(-\left(kr-\frac{l\pi}{2}\right)\right)\right]$$

$$\left. - \text{complex conjugate}\right\}d\cos\theta$$

$$= \frac{\hbar\pi}{mk}\sum_l(2l+1)\left(1-|\eta_l|^2\right) \qquad (7.74)$$

$$\sigma_r = \frac{\phi_r}{v} = \sum_{l=0}^{\infty}\sigma_r^l \qquad (7.75)$$

Fig. 7.13 The *cross-hatched area* shows permissible value for the scattering and absorption cross sections multiplied by β [5]

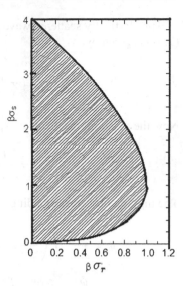

where

$$\sigma_r^l = \frac{\pi}{k^2}(2l+1)\left(1 - |\eta_l|^2\right) \tag{7.76}$$

The total cross-section is the sum of the scattering and reaction cross-sections

$$\sigma_t^l = \sigma_s^l + \sigma_r^l = \pi \lambda^2 (2l+1)|1 - \eta_l|^2 = \pi \lambda^2 (2l+1)\left(2 - \eta_l - \eta_l^*\right) \tag{7.77}$$

Equations (7.72) and (7.75) suggest that there is a relationship between σ_r and σ_s. This becomes quite apparent if we consider the partial wave cross-sections σ_r^l and σ_s^l. These quantities divided by $\pi(2l+1)\lambda^2 = (1/\beta)$, are plotted in Fig. 7.13.

All possible values of these quantities lie in the cross-hatched area. Values outside this area are not realized.

Thus $\sigma_r^l/\pi \lambda^2 (2l+1)$ cannot exceed unity and $\sigma_s^l/\pi \lambda^2 (2l+1)$ cannot exceed 4. Note that (7.73) and (7.76) cannot be directly compared with experiment because δ_l's and η_l's are undetermined. When there is only scattering and no reaction, $|\eta_l|^2 = 1$. In terms of phase-shifts, since $\eta_l = e^{2\delta_l}$, δ_l is real: In the presence of any reaction, atleast one of the δ_l's is complex so that η_l has magnitude less than 1.

The maximum value of σ_s^l is obtained for $\delta_l = 180°$ for which $\eta_l = -1$; $\sigma_s^l(\max) = 4\pi \lambda^2(2l+1)$ in which case $\sigma_r^l = 0$.

The maximum value of σ_r^l is obtained for $\eta_l = 0$ in which case $\sigma_s^l \neq 0$. Blatt and Weisskopf explain this fact by suggesting that the outgoing part of the wave is weakened. Such a weakening can be caused by the coherent elastic scattering of a part of the incoming wave with a phase shift of 180°. Thus there can be scattering without absorption but the converse is not true.

Consider a limiting case of a nucleus with a sharp edge of radius $R \gg \lambda$ so that a semiclassical concept of trajectories is valid. We assume that all particles that strike the nucleus react. If the impact parameter is b_l then the angular momentum

$$L = l\hbar = pb_l = \hbar\frac{b_l}{\lambda} \tag{7.78}$$

Hence all the incident particles with $b_l < R$ or $\leq (R/\lambda)$, react and all those with $l > R/\lambda$ do not, the last condition gives $|\eta_l|^2 = 1$ for $l > (R/\lambda)$. For $l \leq (R/\lambda)$, the reaction is maximum, which is equivalent to the condition $\eta_l = 0$. Then, since $(R/\lambda \geq 1)$ we find from (7.75) and (7.76)

$$\sigma_r = \pi\lambda^2 \sum_{l=0}^{R/\lambda}(2l+1) = \pi\lambda^2 \left[\frac{2(R/\lambda)(1+\frac{R}{\lambda})}{2} + 1 + \frac{R}{\lambda}\right]$$

$$= \pi(R+\lambda)^2 \simeq \pi R^2 \tag{7.79}$$

where we have summed over natural numbers. Similarly, the scattering cross-section is given by

$$\sigma_s \simeq \pi R^2 \tag{7.80}$$

Adding Eqs. (7.79) and (7.80), the total cross-sections is

$$\sigma_t \simeq 2\pi R^2 \tag{7.81}$$

Equation (7.81) implies that σ_t is approximately twice the geometrical cross-section of the nucleus for fast neutrons reacting with totally absorbing nucleus. The nucleus absorbs the partial wave completely and acts as a black disc. However even in this case there is scattering. Equations (7.79) and (7.80) show that the elastic cross-section is then equal to the reaction cross-section. Classically one might expect that a reaction cross-section equal to πR^2 should not be accompanied by elastic scattering. In the high energy limit $(kR \gg 1)$. It is then possible to make wave packets that are small in comparison with the size of the scattering region, and these can follow the classical trajectories without spreading appreciably. However, the apparently anamolous result (7.81) is explained by the fact that the asymptotic form of the wave function is so set up in Eq. (5.112) $\psi_{r\to\infty} = A[e^{ikz} + (1/r)f(\theta)e^{ikr}]$ that in the classical limit the scattering is counted twice; once in the true scattering which turns out to be spherically symmetrical, and again in the shadow of the scattering sphere strongly in the forward direction, since this is produced by interference between the incident plane wave e^{ikz} and the scattered wave $f(\theta)(e^{ikr}/r)$. So long R/λ is finite and diffraction around the sphere in the forward direction does take place. In order to suppress completely the incoming wave behind the obstacle, a source must be placed on it with the same amplitude but opposite phase uniformly spread over the obstacle. This source produces a beam, in a cone of angular aperture $\simeq (\lambda/R)$ of intensity approximately equal to that intercepted by the obstacle. This is known as shadow-scattering and is detectable provided R is a few times λ.

At very high energies the wavelength of the projectile becomes small compared with R, and collisions occur with single nucleons in the nucleus. Between these two limits the nuclei are partially transparent for the incident particles. They are said to be 'grey' obstacles.

7.12 Slow Neutron Resonances and the Breit-Wigner Theory

The resonances shown in Fig. 7.11 correspond to various energy levels (states) of the compound nucleus characterized by definite spin, parity and width. The life time of a given level is related to the width of the level through the uncertainty relation.

$$\tau \Gamma \simeq \hbar \tag{7.82}$$

The lifetime of compound states τ is of the order of 10^{-15} s. For lighter nuclei, the widths of the resonances are greater and the corresponding lifetimes shorter, but in any case they would be much larger than the characteristic time on nuclear scale (10^{-22} s). Bohr assumed that when the compound nucleus decays it has lost the 'memory' and does not 'remember' how it was formed. This independence has been experimentally verified.

Consider only the s-wave ($l = 0$) neutron of low energy. We shall derive Breit-Wigner formulae for elastic scattering and reactions in the neighbourhood of a single isolated resonance level. These formulae describe the variation of cross-sections for slow neutrons as a function of energy. To begin with we ignore the spins of the particles. In Chap. 5 we observed for zero-energy neutrons the connection of the phase shift 6 with the scattering length a_k, and the effective range r_0, the corresponding equations are

$$\lim_{k \to 0} \frac{\delta}{k} = -a_k \tag{7.83}$$

$$k \cot \delta = -\frac{1}{a_k} + \frac{r_0 k^2}{2} \tag{7.84}$$

where $k = 2\pi/\lambda$. We define $a(k)$ through the relation

$$k \cot \delta = -\frac{1}{a(k)} \tag{7.85}$$

for all values of k. If there is no reaction, $a(k)$ is real. For $\sigma_r \neq 0$, $a(k)$ is in general complex. By putting $a(k) = a$

$$\eta = e^{2i\delta} = (\cos \delta + i \sin \delta)^2$$

$$= \left(\frac{1 - ika}{\sqrt{1 + k^2 a^2}} \right)^2$$

$$= \frac{1 - ika}{1 + ika} \tag{7.86}$$

The scattering cross-section σ_{sc} in Eq. (7.73) and the reaction cross-section σ_r in Eq. (7.76) are expressed in terms of the scattering length 'a' by substituting the value of η from (7.86). Remembering that $l = 0$

$$\sigma_s = \frac{\pi}{k^2}|1 - \eta|^2 = \frac{4\pi}{|\frac{1}{a} + ik|^2} \tag{7.87}$$

$$\sigma_r = \frac{\pi}{k^2}(1 - |\eta|^2) = \frac{4\pi}{k}\frac{Im(\frac{1}{a})}{|\frac{1}{a} + ik|^2} \tag{7.88}$$

For real a, $\sigma_r = 0$. At resonance, the phase-shift is $\pi/2$, so that

$$\frac{1}{a(E_0)} = 0 \tag{7.89}$$

where E_0 is the energy at which the resonance occurs. Expanding $1/a(E)$ about E_0 by Taylor's series

$$\frac{1}{a(E)} = 0 + (E - E_0)\left[\frac{d}{dE}\left(\frac{1}{a}\right)\right]_{E_0} + \cdots \tag{7.90}$$

Defining

$$\left[\frac{d}{dE}\left(\frac{1}{a}\right)\right]_{E_0} = \frac{2k}{\Gamma_s} \tag{7.91}$$

Retaining only the linear term in the expansion and substituting in (7.87) for σ_s, we obtain

$$\sigma_s = \frac{4\pi}{|(E - E_0)\frac{2k}{\Gamma_s} + ik|^2} = \frac{\pi\lambda^2\Gamma_s^2}{|(E - E_0) + \frac{i\Gamma_s}{2}|^2}$$

$$= \frac{\pi\lambda^2\Gamma_s^2}{(E - E_0)^2 + (\frac{\Gamma_s}{2})^2} \tag{7.92}$$

where Γ_s is defined as the width of the level from which the scattering takes place. It is the width at half maximum, Fig. 7.14.

Equation (7.92) is known as the Breit-Wigner single-level formula for scattering when the absorption is absent.

We can follow the shape of the resonance curve by invoking for the phase-shift which is more physical than the scattering length. At the resonance energy E_0, the cross-section is a maximum and this occurs when $\sigma_l = \pi/2$. We expand the phase-shift around $\pi/2$ to obtain

$$\delta_l = \frac{\pi}{2} - (E_0 - E)\frac{d\delta_l}{dE} \tag{7.93}$$

Fig. 7.14 This *curve* gives
the schematic representation
of the width Γ_s of a level of
energy E_0 with a spread in
energy of $\pm \Gamma_s/2$

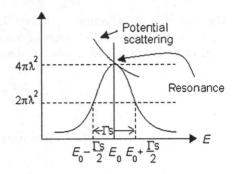

where $d\delta_l/dE$ determines the sharpness of the resonance. If it is small the resonance is broad, if it is large the resonance is sharp. Assume

$$\frac{d\delta_l}{dE} = \frac{2}{\Gamma_s} \tag{7.94}$$

where the factor 2 ensures that Γ refers to full width at half maximum. From (7.93) and (7.94) we find

$$\tan \delta_l \simeq \delta_l = \frac{\Gamma_s/2}{(E_0 - E)} \tag{7.95}$$

$$\sin^2 \delta_l = \frac{\Gamma_s^2}{\Gamma_s^2 + 4(E_0 - E)^2} \tag{7.96}$$

Using (7.96) in the formula for scattering cross-section

$$\sigma_s^l = \pi \lambda^2 (2l + 1) \sin^2 \delta_l$$

for s-waves we get Eq. (7.92). Note that at $E = E_R$, the cross-sections for resonance scattering and potential scattering (without the compound nucleus formations are equal, each equal to $4\pi \lambda^2$ Fig. 7.14).

We shall now derive the formulae for σ_{sc} the scattering cross-section in the presence of absorption and the cross-section for absorption (σ_r). In this case $a(E)$ is a complex function of the variable E.

Let $1/a(E_0) = 0$ at $E_0 = \varepsilon_0 - i\Gamma_R/2$, where ε_0 and Γ_R are real. Expanding $1/a(E)$ about E_0,

$$\frac{1}{a(E)} = 0 + (E - E_0) \left[\frac{d}{dE} \left(\frac{1}{a} \right) \right]_{E_0} + \cdots \tag{7.97}$$

Let

$$\text{Re} \left[\frac{d}{dE} \left(\frac{1}{a} \right) \right]_{E_0} = \frac{2k}{\Gamma_s} \tag{7.98}$$

$$\text{Im} \left[\frac{d}{dE} \left(\frac{1}{a} \right) \right]_{E_0} = k\alpha \tag{7.99}$$

Combining (7.97), (7.98) and (7.99) and substituting for E_0,

$$\frac{1}{a(E)} = \left(E - \varepsilon_0 + \frac{i\Gamma_R}{2}\right)\left(\frac{2k}{\Gamma_s} + ik\alpha\right)$$

$$= \frac{2k}{\Gamma_s}(E - E_R) + ik\left[\frac{\Gamma_R}{\Gamma_s} + \alpha(E - E_0)\right] \tag{7.100}$$

where

$$E_R = \varepsilon_0 + \frac{\alpha\Gamma_R\Gamma_s}{4} \tag{7.101}$$

The scattering cross-section is given by

$$\sigma_s = \frac{4\pi}{[\frac{1}{a} + ik]^2} = \frac{\pi\lambda^2\Gamma_s^2}{(E - E_R)^2 + (\frac{\Gamma_s + \Gamma_R + \alpha\Gamma_s(E - \varepsilon_0)}{2})^2}$$

For sharp resonance, α is small, so that near resonance, $\alpha\Gamma_s(E - E_0)$ is very small. Writing for the total width

$$\Gamma = \Gamma_s + \Gamma_R \tag{7.102}$$

$$\sigma_s = \frac{\pi\lambda^2\Gamma_s^2}{(E - E_R)^2 + (\frac{\Gamma}{2})^2} \tag{7.103}$$

Substituting (7.100) in (7.88),

$$\sigma_r = \frac{\pi\lambda^2[\Gamma_R\Gamma_s + \Gamma_s^2\alpha(E - E_0)]}{(E - E_R)^2 + (\frac{\Gamma_s + \Gamma_R + \alpha\Gamma_s(E - \varepsilon_0)}{2})^2} \tag{7.104}$$

Again neglecting $\alpha(E - E_0)$ and using (7.102), (7.104) becomes

$$\sigma_r = \frac{\pi\lambda^2\Gamma_R\Gamma_s}{(E - E_R)^2 + (\frac{\Gamma}{2})^2} \tag{7.105}$$

Equations (7.103) and (7.105) are the Breit-Wigner formulae for a single isolated level for neutrons with $l = 0$. Here $E_R = E_0$, is the resonance energy at which the cross section is maximum. So far we have not taken the spins of the particles into account. If I_a is the spin of the projectile and I_x that of the target nucleus and I_c that of the compound nucleus then both formulae (7.103) and (7.105) must be multiplied by the statistical weight

$$g = \frac{(2I_c + 1)}{(2I_a + 1)(2I_x + 1)} \tag{7.106}$$

which is the probability that the two randomly directed spins I_a and I_x couple to give I_c, this will be so for unpolarised projectile and target particles. In the preceding formulae Γ is the level width defined by (7.82), $\Gamma_s = \Gamma_a$ and $\Gamma_R = \Gamma_b$ are the "partial level widths" defined from

$$\Gamma_a \tau_a = \hbar \quad \text{and} \quad \Gamma_b \tau_b = \hbar \tag{7.107}$$

where τ_a and τ_b are the mean lifetimes that the compound nucleus would have if (1) the elastic scattering of "a" or (2) the emission of "b" were the only possible modes of decay. Clearly

$$\Gamma = \Gamma_a + \Gamma_b + \Sigma_i \Gamma_i \tag{7.108}$$

where the summation is over all possible modes of decay except the two mentioned. The concept of "width" is not always applicable so that the definition of the Γ's in terms of a mean life time or its inverse, the decay probability is to be preferred. The energy $E = E_a$ is the relative energy of the system $a + x$, and $E_0 = E_R$ is a constant, the resonance energy. When $E_a = E_R$

$$\sigma(a, b) = \sigma_0 = \frac{4\pi \Gamma_a \Gamma_b}{k^2 \Gamma^2} \tag{7.109}$$

7.12.1 Resonance Absorption and the $1/v$ Law

For reactions involving neutrons of energy <1 keV, (7.105) leads to the well known $1/v$ variation for the cross-section for energies away from the resonance. Suppose the particle b is a gamma ray (the usual case when a slow neutron is absorbed). In order to find out the variation of cross-sections for both the types of channels with neutron energy it is necessary to know the behaviour of Γ_b and Γ_a. Now

$$\lambdabar^2 \propto \frac{1}{v_n^2} \tag{7.110}$$

When a slow neutron is absorbed an excitation energy equal to the binding energy (~ 8 MeV) $+ E_n$ the kinetic energy of neutron (few eV–few keV) is imparted to the compound nucleus. Clearly the excitation energy E_{ex} is insensitive to small variation of E_n since the former will be of the order of few MeV and $E_n \sim$ few eV–few keV. Hence the energy available for γ-ray transition is practically constant.

It follows that $\Gamma_r = \Gamma_b$ which represents the probability for γ emission also remains constant.

$$\Gamma_\gamma \simeq \text{const} \tag{7.111}$$

For particle emission, however, the situation is completely different. Since the excitation energy is shared among many particles in the compound nucleus the possibility of emission of a particle depends on the concentration of sufficient energy on this particle to allow it to escape. Upon leaving, the emitted particle must return the binding energy to the nucleus. Let τ_a' be the average time between such rearrangements of the nuclear constituents as would permit the emission of particle a. The frequency of emission of the particle "a" is given by the product of the frequency for the favourable configuration and the probability P_a that the particle a is given the required amount of energy to penetrate through the nuclear surface

$$\Gamma_n = \frac{\hbar}{\tau_n} = \frac{\hbar P_a}{\tau_a'}$$

But

$$P_a = \frac{4kK}{(k+K)^2} \tag{7.112}$$

where K and k are the inside and outside wave numbers, respectively. Equation (7.112) is the formula for the penetration of a rectangular barrier.

For small E_n, $k \ll K$. Therefore $P_a \simeq \frac{4k}{K} = \frac{4\lambda_{in}}{\lambda_{out}}$, giving

$$P_a \propto \frac{1}{\lambda_{out}}, \quad \text{so that}$$

$$\Gamma_n \propto v_n \tag{7.113}$$

Further

$$\Gamma_\gamma \gg \Gamma_n \quad \text{and} \quad \Gamma = \Gamma_\gamma + \Gamma_n \simeq \text{const} \tag{7.114}$$

since $\Gamma_r \simeq$ const. When $E_n \ll E_R$ (resonance energy) the denominator in (7.103) and (7.105) are nearly constant. Using (7.110), (7.111), (7.113) and (7.114) in (7.105), we get (away from resonance energy)

$$\sigma_{(n,r)} \propto \frac{1}{v} \tag{7.115}$$

As $E_n \to E_R$, $\sigma(n, \gamma)$ varies much more rapidly and $\sigma(n, \gamma) \to \sigma_0 = \sigma_{max}$ when $E_n = E_R$. Subsequently, when E_n exceeds E_R, $\sigma(n, \gamma)$, will decrease at first sharply, then more slowly with increasing E_n. Note that the absorption curve is asymmetric because of $\frac{1}{v}$ or $(1/\sqrt{E})$ factor in the expression.

When $\Gamma \gg E_n - E_R$, then again $\sigma(n, r) \propto (1/v_n)$. Thus the $(1/v)$ law is valid either when

(1) $E_n \ll E_R$
(2) $\Gamma \gg E_n - E_R$

In (2) no maximum occurs in the $\sigma(n, \gamma)$–E_n plot, $\sigma(n, r)$ decreasing with increasing neutron energy E_n.

In the case of emission of charged particles the Gamow factor is introduced in the final formula. The energy levels in the compound nucleus are relatively far apart near the ground state and become closer at higher excitation energy. At energies $\sim$15–20 MeV, energy levels are practically continuous. For nuclei in the medium range of $A = 100$–150 the spacing near the ground state is $\sim$0.1 MeV. However, when the energy is in the region of 8 MeV above ground state, the spacing is 1–10 eV. For

Fig. 7.15 Total cross-section
for neutron interactions with
^{238}U, showing many very
narrow resonances with
intrinsic widths of order
10^{-2} eV corresponding to
excited states of ^{239}U [10]

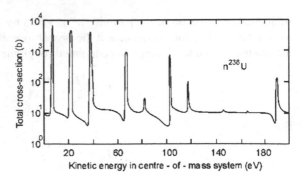

Fig. 7.16 The total
cross-section for the
scattering of neutrons by
^{238}U showing resonant
structure [16]

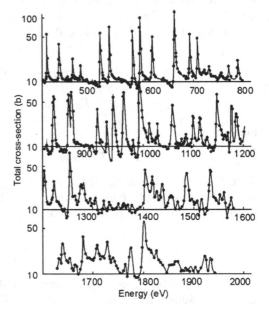

light nuclei, the spacing is ~1 MeV near the ground state and ~10 keV when the
internal energy is ~8 MeV above the ground state.

For elements of moderate and high mass numbers, resonances occur for neutron
energy ~1 eV–10 eV. For example ^{238}U exhibits resonance capture of neutrons in
the eV and keV range as in Figs. 7.15 and 7.16. If the neutron kinetic energy is
such that the excited level coincides with one of the levels of the compound nucleus
then a high absorption takes place. Note that at high neutron energies, 1 MeV or
more, the compound nucleus will acquire 9 MeV or more above the ground level,
Fig. 7.17.

In this region the levels widths are frequently comparable with the level spacing.
Hence resonance effects will not show up.

For nuclei of low mass number the spacing of the energy levels in the 8 MeV
region is larger than for nuclides of moderate or high atomic weights. Hence resonance absorption occurs with E_n ~ 10 keV or so. But the absorption is not marked

Fig. 7.17 Hypothetical levels for a compound nucleus $E_R = 8 + E_n$

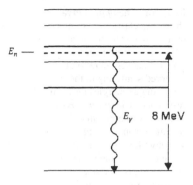

as there is a general tendency for the absorption to decrease with the increase of neutron energy.

7.12.2 Elastic Scattering

Two types of elastic scattering must be distinguished (1) potential scattering, (2) resonance scattering. The potential scattering also known as shape-elastic scattering is caused by the interaction of the neutron wave with the potential at the nuclear surface. Effectively the incident neutron does not enter the target nucleus and the compound nucleus is not formed. It results from the diffraction of those neutrons which pass close by, but not into the nucleus. In the case of elastic resonance scattering the neutron is captured by the target nucleus when its energy is close to one of the quantum states and it is re-emitted. Re-emission of the captured neutron is expected from the fraction (Γ_n/Γ) of the compound nucleus. Formula (7.103) for $\sigma(n,n)$ is applicable only if this resonance elastic scattering is the only kind of elastic scattering (Fig. 7.14). However, because of the presence of potential scattering the two types are superimposed coherently. For $l = 0$ and the neutrons assumed as projectiles

$$\sigma_{el} = \frac{\pi}{k^2} \left| \frac{i\Gamma_n}{(E_n - E_R) + \frac{i\Gamma}{2}} + \left[\exp(2ikR) - 1 \right] \right|^2 \qquad (7.116)$$

The quantity between the vertical bars is a complex number. The first term is called the resonance scattering term, the second is called the potential scattering term. We can write (7.116) as

$$\sigma_{el} = \frac{\pi}{k^2} |A_{res} + A_{pot}|^2 \qquad (7.116a)$$

with

$$A_{res} = \frac{i\Gamma_n}{(E_n - E_R) + \frac{1}{2}i\Gamma} \quad \text{and} \quad A_{pot} = \exp(2ikR) - 1$$

Fig. 7.18 The *upper curve* shows the elastic scattering of protons from aluminum as a function of proton energy. The ordinate is the ratio of the observed scattering to the scattering at energies far from resonance. The *lower curve* shows how the gamma radiation from the target varies with proton energy; it is a measure of the resonance absorption. *Both curves* are drawn to give a best fit, without regard to theory [2]

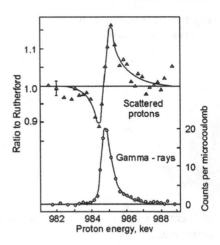

where R is the nuclear radius. The potential energy term varier smoothly with energy. The resonance term rise to large values when $E_n = E_R$, otherwise it is small.

For $E_n < E_R$ the two terms interfere destructively yielding a low value of σ_{el}.
For $E_n > E_R$ they interfere constructively (Fig. 7.18).

Except near resonance, the resonance scattering is usually much less than the potential scattering. Away from the resonance energy $(E_n \ll E_R)$, in (7.103), the denominator is constant as before, $\lambda^2 \propto (1/v^2)$ and $\Gamma_n^2 \propto v^2$. Hence $\sigma(n, n) \simeq \text{const}$ (Fig. 7.18). The cross-sections for resonance scattering and potential scattering are of the same order of magnitude, except near the resonance. It is generally accepted that the total elastic scattering cross-section is independent of E_n. This is specially true for neutrons with energies <0.1 MeV when scattered by nuclei of fairly low mass number.

7.13 Optical Model

It is of interest to consider cross-sections averaged over energy ranges greater that the average resonance spacing. At such energies the characteristic features of a particular compound nucleus are suppressed and features common to all nuclei are revealed. It turns out that these energy averaged cross-sections vary smoothly with projectile energy in passing from one nucleus to the neighbouring one. As the incident particle energy is varied broad resonances are encountered with life times of 10^{-21}–10^{-22} s. This smooth variation of the cross-sections is displayed. In Figs. 7.19(a) and 7.19(b) for neutrons interaction with element heavier than Mn, as a function of the neutron energy and the target nucleon mass [1] and [15]. These cross-sections appear to be related to the bulk properties of the nucleus, the details of its structure being unimportant. This remarkable feature of nuclear scattering cross-section can be explained by the optical model which employees a complex potential based on the nuclear dimension and nuclear shape.

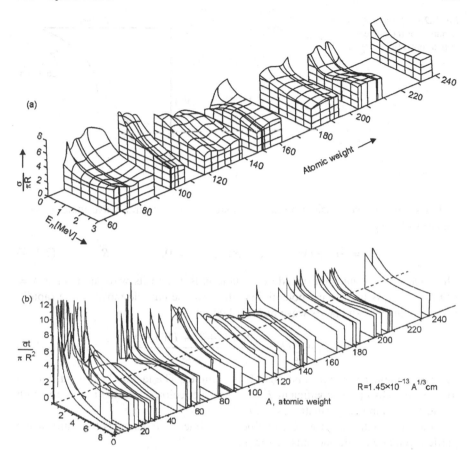

Fig. 7.19 (a) Measured cross-section averaged over resonances for E and A [1]. (b) Observed "gross structure" of total neutron cross section (compiled by Feshbach et al. [15])

It is assumed that all the individual nucleon-nucleon interactions between the projectile and the target nucleus can be replaced by a one-body interaction that can be represented by a potential $V(r)$, where r is the separation of the projectile and the nucleus. This assumption is similar to that underlies the shell model (Chap. 6). $V(r)$ is expected to be uniform in the interior and fall off exponentially toward the surface following the nuclear matter distribution, Fig. 7.20. Outside the nucleus it is zero because of short range of nuclear forces. A real potential alone is not adequate to describe the experimental data since besides scattering absorption will also be present by which the incident particles can be removed by various non-elastic channels via compound nucleus reactions, direct or pre-equilibrium reactions.

Now, it is well known that the scattering and absorption of light can be successfully explained mathematically by using a complex potential. It is therefore tempting to use a complex potential for the nuclear case.

Fig. 7.20 The radial
variation of the nuclear
optical potential

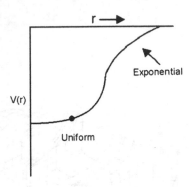

For simplicity we consider a one-dimensional model with a wave incident on a
square well potential

$$V = -U - iW, \qquad z \le R; \qquad V = 0, \qquad z > R \qquad (7.117)$$

The energy of the nucleon outside the nucleus is E and its wave number is $k = (2mE)^{1/2}/\hbar$. As the nuclear potential is attractive the particle moves more rapidly inside the nucleus so that its wave number becomes

$$k + k_1 + \frac{1}{2} iK = \frac{[2m(E + U + iW)]^{(1/2)}}{\hbar} \qquad (7.118)$$

The wave number is complex because V is complex. The factor $(1/2)$ is introduced
to make the absorption coefficient K equal to the reciprocal of the mean free path
of the nucleon in nuclear matter as in (7.125).

In analogy with optics we can define the refractive index n as the ratio of the
particle velocity inside and outside the potential well

$$n = \left(\frac{E + U + iW}{E} \right)^{(1/2)} = 1 + \frac{k_1}{k} + \frac{iK}{2k} \qquad (7.119)$$

At high energies $kR \gg 1$, $k_1 \ll k$, and $K \ll k$ so that $n \simeq 1$ and to a high degree of
accuracy

$$n^2 - 1 \simeq 2(n - 1) \qquad (7.120)$$

Combining (7.119) and (7.120), and equating real and imaginary terms, we easily
find

$$U = \hbar v k_1 \quad \text{and} \qquad (7.121)$$

$$W = \frac{\hbar}{2} v K \qquad (7.122)$$

where $v = \sqrt{(2E/m)}$ is the velocity of the incident nucleon. The nucleon wave
function inside the nucleus is given by

$$\psi = e^{inkz} = e^{i(k+k_1)z} e^{-(1/2)Kz} \qquad (7.123)$$

Fig. 7.21 Schematic representation of the path of a high energy neutron within a nucleus

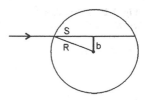

Thus the nucleon wave is attenuated exponentially according to

$$|\psi|^2 = e^{-Kz} \tag{7.124}$$

and it implies that the imaginary potential removes particles from the elastic channel. In (7.124), K is clearly the reciprocal of the mean free path in the nuclear matter in (7.122), the product $vK = v/\lambda$ represents the number of collisions per unit time (collision frequency) of the incident nucleon with the nucleons of the nucleus. If ρ is the nuclear density, that is number of nucleons per unit volume and σ the average nucleon-nucleon cross-section in nuclear matter then

$$K = \frac{1}{\lambda} = \rho\sigma = \frac{3\sigma}{4\pi r_0^3} \tag{7.125}$$

because $\rho = A/V$ and nuclear radius $R = r_0 A^{1/3}$. The average cross-section $\overline{\sigma}$ must be corrected for the fact that the interaction occurs with nucleons endowed with Fermi moment varying in magnitude and direction and that the collision cross-sections are energy dependent. Further, the cross-sections are reduced due to Pauli's principle which forbids nucleon-nucleon interactions leading to nucleon to a state already occupied. According to Goldberger the reduction factor is obtained from

$$\sigma_{np} = \frac{2}{3}\sigma_{np}(\text{free}) \tag{7.126}$$

We shall now derive expressions for the reaction and elastic cross-sections for a high energy neutron interacting with a nucleus of radius R, see Fig. 7.21.

The portion of the wave which strikes a distance b from a line through the centre of the sphere emerges after travelling a distance $2S$, with $S^2 = R^2 - b^2$, on emerging has the amplitude $a = e^{(-K+2ik_1)S}$ so that

$$\sigma_r = 2\pi \int_0^R \left(1 - |a|^2\right)bdb = 2\pi \int_0^R \left(1 - e^{-2Ks}\right)sds$$

$$= \pi R^2 \left\{1 - \left[\frac{1 - (1 + 2KR)}{2K^2R^2}e^{-2KR}\right]\right\} \tag{7.127}$$

According to (7.123), the wave transmitted through a distance z in the nucleus is

$$\psi_t = e^{(-\frac{K}{2}+ik_1)z}e^{ikz} = ae^{ikz} \tag{7.128}$$

Fig. 7.22 Curves for
$k_1/K = 1.5$

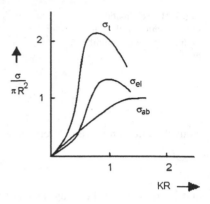

The scattered wave is

$$\psi_s = 2\pi \int_0^R |1-a|^2 b\, db = 2\pi \int_0^R |1 - e^{(-K+2ik_1)S}|^2 S\, dS \tag{7.129}$$

The expressions (7.127) and (7.129) reproduce the experimental data quite accurately at high energies. Note that $\sigma_r/\pi R^2$ is a function of KR alone while $\sigma_s/\pi R^2$ depends on k_1/K as well. A direct confirmation of the theory, quite independent of the parameters is obtained by plotting $\sigma_{total}(\text{exptl})/A^{2/3}$ against $A^{1/3}$ where one finds that the experimental points from Li to U lie on a smooth curve which reaches a maximum and comes down as predicted by the theory. By contrast, the curves would be a horizontal line if the nucleus were taken as a totally black sphere, that is a body with complete absorption.

In Fig. 7.22 the curves are drawn for $k_1/K = 1.5$. For a given sss and K we can find R, the radius required for each nucleus to give the observed total scattering cross-sections. A value of $K = 2.2 \times 10^{12}$ cm^{-1} corresponds to a mean free path in nuclear matter, $\lambda = 4.5$ fm. The radii calculated from the measured cross-sections (Fig. 7.22) yield $r_0 = 1.37$ fm. The associated value of 3.3×10^{12} cm^{-1} corresponds to $U = 30.8$ MeV.

The concept that the imaginary part of the complex potential in the potential model has the effect of removing particle flux from the elastic channel is equally valid for the three-dimensional problem. To this end we set up the Schrodinger equation for scattering by the complex potential (7.117)

$$\nabla^2 \psi + \frac{2m}{\hbar^2}(E + U + iW)\psi = 0 \tag{7.130}$$

Multiply (7.130) by ψ^* and subtract the complex conjugate of this equation multiplied by ψ to obtain

$$\psi^*\nabla^2\psi - \psi\nabla^2\psi^* = -\frac{4imW}{\hbar^2}\psi\psi^* \tag{7.131}$$

The quantum mechanical expression for the current density is

$$j = \frac{\hbar}{2im} \left(\psi^* \nabla \psi - \psi \nabla \psi^* \right) \tag{7.132}$$

so that (7.131) becomes

$$\text{div } j = -\frac{2}{\hbar} W \psi \psi^* \tag{7.133}$$

Since $\psi \psi^*$ is the probability density and W is given by (7.122), this equation is equivalent to the classical continuity equation

$$\frac{\partial \rho}{\partial t} + \text{div } j = -\frac{v}{\lambda} \rho \tag{7.134}$$

where v is the particle velocity. When steady state is reached, the term $\partial \rho / \partial t$ in (7.134) vanishes. Provided $W > 0$, the imaginary part of the complex potential has the effect of absorbing flux from the incident channel.

7.14 Direct Reactions

It was pointed out that direct reactions are those which proceed without the formation of the compound nucleus. They include a variety of nuclear reactions such as inelastic scattering, stripping and its inverse, pick-up reaction, Knock-out reaction and heavy ion reaction. The time during which the incident and the target nucleus interact is very much shorter (of the order of 10^{-21} s) than the life of a compound nucleus (10^{-15}–10^{-16} s). Because of this difference, the reactions products have characteristics which are entirely different from those observed in compound nucleus reactions.

These two processes represent extreme views of the mechanism of nuclear reactions. It is difficult to state at which energy one or the other mechanism will operate. As a rule at low energies, compound nucleus formation is more likely while at high energies direct reaction will dominate.

While compound nucleus reaction is a two-step process, direct reaction is a one-step process. The possibility of the direct reaction was first recognized by Oppenheimer and Phillips while analyzing the low-energy (d, p) reactions. It was experimentally observed that (d, p) reactions were more frequent than (d, n) reactions. This is opposite to what is expected if the reaction proceeds through the compound nucleus formation. In the collision of deuteron, the neutron is captured by the target nucleus, while the proton is repelled due to Coulomb forces, leading to a preponderance of (d, n) reactions over (d, p) reactions.

Butler's theory [13] has demonstrated that the forward peak in the angular distribution is given by the square of the spherical Bessel function of order l, where

l is the angular momentum of the state is which the neutron (for (d, p) reaction) and the proton (for (d, n) reaction) is captured. (See Fig. 7.27.) The spherical Bessel functions are given in [24], Appendix C. The uncaptured nucleon proceeds in the forward direction giving a forward peak. Such reactions are called "stripping" reactions. Later it was found that a large number of reactions such as $^{31}P(\alpha, p)^{34}S$, $^{13}C(^3He, \alpha)^{12}C^{23}$, $Na(d, p)^{24}Na$, $^7Li(p, d)^6Li$ etc. show characteristics of "stripping" or "pick-up" reactions recognized by the forward peaking. The reactions (p, d) and (n, d) are known as the inverse of the (d, p) and (d, n) reactions. In the process of pick-up, the incident proton on approaching close to the target nucleus strongly interacts with an outer neutron and forms a deuteron which is emitted. A similar explanation is given for the (n, d) reaction. The forward peaking in (α, p) reaction is explained as the stripping of a triton from the alpha particle. On the other hand, the (p, α) reaction is the pick-up of a triton by the proton from the target nucleus to form an alpha-particle. Reactions of the type (α, p) and (n, p) are known as knock-on reactions in which the incident particle strikes a nucleon or a cluster of nucleons and ejects it. The stripping or pick-up reactions mentioned above are special cases of a general class of direct reactions known as transfer reactions or rearrangement collisions.

Reactions which correspond to inelastic scattering such as (p, p'), (α, α') and (d, d') may also fall under direct reaction category and analyzed similarly.

Direct reactions have provided valuable information for nuclear structure. The Born approximation which is based on plane-wave approximation does not give satisfactory results to explain the direct approximation. Since the incident wave is distorted due to nuclear reaction, one resorts to the distorted-wave Born approximation (DWBA) which is outside the scope of this book. We shall now describe various types of direct reactions in the order of complexity.

7.14.1 Inelastic Scattering

In the inelastic scattering process the incident particle interacts with the target nucleus and imparts some of its energy, raising it to an excited state. Measurement of the energy loss of the scattered particle indicates the energy of the excited level and the differential cross-section and polarization of the scattered particle throw light on the nuclear structure.

At low energy the target nucleus can be excited by pure Coulomb field between the projectile and the target, a process known as *Coulomb excitation* and is important only for high Z particles. This aspect will be discussed in Sect. 7.17.2 which is devoted to heavy ion reactions. At higher energies the excitation is caused by the nuclear interaction in the presence of Coulomb field.

The simplest model for the inelastic scattering is the shell model which assumes that the interaction raises a single nucleon to a higher state. However, the calculated cross-section is far below the experimental values.

The observation that the collective states are strongly excited by inelastic scattering led to the assumption that the incident particle interacts with the target nucleus

Fig. 7.23 Elastic and
inelastic differential
cross-sections for the
interaction of 12 MeV
protons with a medium
weight nucleus, compared
with calculations made using
the coupled channel theory.
The spin of the Ni state
excited in the inelastic
process is 2^+ [12]

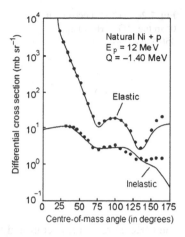

as a whole. This then is the collective excitation which is described as the coherent sum of many particle hole excitation. In heavy ions such as uranium highly developed rotational bands have been found with only even values $I = 0, 2, 4, \ldots$ The symmetry of the problem demands that I can take only even values with even parity. The nucleus is excited by quadrupole transitions in successive jumps produced by the electromagnetic field of the same projectile. Calculations are therefore made using the collective model for the rotational or vibrational states.

Reasonably good results are obtained by using the weak coupling expression

$$\frac{d\sigma_{if}}{d\Omega} = \frac{k_f m_i m_f}{4\pi^2 \hbar^4 k_i} |\langle f|V|i\rangle|^2 \tag{7.135}$$

for the differential cross-section for the transition from the initial state i to the final state f. Here $\langle f|V|i\rangle$ is the matrix element. However, the interaction being strong, the so-called coupled channel theory is employed. This theory is much more accurate and takes into account the coupling between various reaction channels. The theory gives both the differential cross-section and polarization. As an example the theoretical predictions are compared with the experimental data on differential cross-sections for both the elastic and inelastic scattering of 12 MeV protons from Ni nuclei, in Fig. 7.23 from the work of Buck [12]. Further the spin and parity of the excited state of Ni(2^+) is extracted from the theory.

7.14.2 Charge-Exchange Reactions

In a charge-exchange reaction both energy and charge are exchanged between the projectile and the target nucleus. The most important charge exchange reactions are the (p, n) and the $(^3\text{He}, t)$ reactions. Both provide useful information on nuclear structure. The cross-sections for the former are one order of magnitude greater than the latter. However, the (p, n) reactions occur throughout the nuclear volume while

Fig. 7.24 Triton spectrum at
0° in the ^{71}Ga(^{3}He, t) at
450 MeV [18]

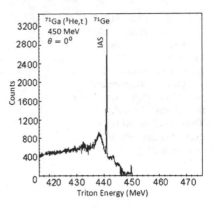

the (^{3}He, t) reaction is confined to the nucleus surface since both ^{3}He and t easily
dissolve into their constituents in their passage through the nuclear interior.

In such a charge exchange process the target nucleus is converted into an isobar. That the isospin T is a good quantum number is evidenced by the parity of
the isobaric analogue state (IAS) excited in the high resolution measurements of
(^{3}He, t) reactions. As an example Fig. 7.24 shows the IAS excited in the reaction
$^{71}_{31}$Ga(^{3}He, t)$^{71}_{32}$Ge obtained by Fugiwara et al. [17].

If the reaction goes to the isobaric state of the target nucleus only the isospin
vectors are flipped and the isobar analogue will have the same isospin as the target
nucleus. Such reactions are quite similar to elastic scattering and are referred to as
quasi-elastic scattering.

7.14.3 Nucleon Transfer Reactions

Here one or more nucleons are transferred from the projectile to the target nucleus
(stripping reactions) or from the target nucleus to the projectile (pick-up reactions).
Deuteron stripping reaction is the simplest example of transfer reaction in which
one neutron or proton from deuteron is transferred to the target nucleus. Direct reactions in which a single neutron or proton is transferred from the projectile to the
target nucleus are known as single transfer reactions. Examples of this in heavy ion
collisions are

$$^{32}\text{S}\left(^{14}\text{N}, \, ^{13}\text{N}\right)^{33}\text{S} \quad \text{and} \quad ^{25}\text{Mg}\left(^{14}\text{N}, \, ^{13}\text{N}\right)^{26}\text{Mg}$$

Double transfer means that two particles, for example two neutrons or two protons
have been transferred from one nucleus to the other. Multiple transfer is then the
transfer of many particles, all in one direction.

The neutron-transfer reaction has several fairly well established features. Below
the Coulomb barrier, the excitation function decreases more slowly with decreasing
energy than the excitation function for the compound nucleus formation. This is to

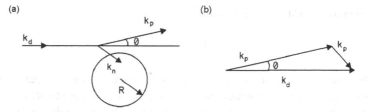

Fig. 7.25 Schematic representation of a (d, p) reaction

be expected for a direct reaction which takes place on the nuclear surface and which proceeds without the formation of compound nucleus. The total cross-sections for neutron-transfer reactions tend to level off at values of tens of mb. Further, cross-sections are strongly dependent on Q values.

The angular distribution of neutron transfer of the type $(^{14}N^{13}N)$ show a peak in the forward direction. Consider the transfer reaction to be quasi-elastic scattering modified by the passage of a neutron from one nucleus to the other. Neutrons are not transferred for very distant collisions (small angle scattering) and for very close collisions for which the compound nucleus reactions strongly compete with the transfer processes. The position of the maximum depends on the energy of the incident ion. The peak is shifted to wider angles at higher incident energy.

We shall focus on the (d, p) reaction as historically it was the first to be studied in detail and was found to be valuable in nuclear structure studies. In this reaction the neutron from the projectile is transferred to an unfilled single particle state of the residual nucleus. The same analysis is valid for (d, n) reaction and other one nucleon transfer reactions. The (d, n) reaction is not favorable to study as the energy resolution of the resultant neutrons is so low that it is not possible to resolve the neutrons corresponding to protons captured in nearby states.

The angular distribution of protons from (d, p) reactions was observed to be peaked either in the forward or backward direction. this observation led Serber to suggest the direct reaction model. When the capture of neutron by the target nucleus takes place at the surface in a peripheral collision, the proton is detached and continues its flight in the forward direction at a small angle. This explains the forward peak. On the other hand if a deuteron at low energy makes a head-on collision on a heavy nucleus, the neutron may be captured and the detached proton would be repelled in the backward direction due to Coulomb forces, leading to the backward peak.

In the stripping reaction of deuteron (d, p) let the energies of the incident deuteron and the outgoing proton be large enough so that these particles are unaffected by the Coulomb field. Let the incident deuteron have linear momentum $k_d \hbar$, the outgoing proton $k_p \hbar$ and the captured neutron $k_n \hbar$ directed toward target nucleus. The momentum of the deuteron is then given by vector sum of momenta of proton and neutron, Fig. 7.25.

The momentum triangle, Fig. 7.25(b) gives

$$k_n^2 = k_p^2 + k_d^2 - 2k_p k_d \cos\theta \qquad (7.136)$$

The momentum transfer to the neutron is

$$|k_d - k_p| = k_n = \frac{L_n}{R} \qquad (7.137)$$

Now, the absolute value of the angular momentum carried by the neutron into the nucleus is $\hbar k_n b$, where the impact parameter b ranges from zero to R, the radius of the nucleus. But the neutron is captured from an orbital of angular momentum $l\hbar$, so that

$$k_n R \geq l_n \qquad (7.138)$$

Squaring (7.138) and combining with (7.136)

$$2k_p k_d \cos\theta \leq k_p^2 + k_d^2 - \frac{l_n^2}{R^2} \qquad (7.139)$$

For $n = 0$, $l_n = 0$, and (7.139) will be satisfied, at $\theta = 0°$. For a given transition k_p^2 and k_d^2 are constants determined by energy conservation, and (7.139) shows that the preferred angle of scattering θ will increase with increasing l_n. Actually the semi-classical argument is oversimplified. In any case (7.137) will be satisfied for a larger angle when k_n has increased. Sophisticated calculations are made using DWBA. The shape of the angular distribution permits the determination of l. The analysis is simple if the target nuclei have $J = 0$. If the neutron is captured with orbital angular momentum L, then the total angular momentum of the final state consisting of target + neutron, is $J = L + (1/2)$ so that $J = L \pm 1/2$. Thus the determination of L gives two possible values for J. The ambiguity is removed from the measurement of polarization of the outgoing proton which has opposite sign for the two possibilities. Furthermore, the product of the parities of the initial state and final state is determined by $(-1)^l$. This means that both spin and parity of the final state is determined with the knowledge of initial state and the value of l. Figure 7.26 shows the fitting of theoretical curves with the experimental data. There are many other types stripping and pick-up reactions in which one or more nucleons are transferred. They have been used to determine nuclear structure of numerous nuclei.

One-nucleon transfer reactions give valuable information on the single-particle structure of nuclei, two neutrons transfer reactions such as (p, t) and (t, p) reactions provide information on pairing energy, while α-transfer reaction like $(^6Li, d)$ reaction throw light on the α-cluster structure.

The contributions of different angular momentum (l) values shown in Fig. 7.26 are reminiscent of Butler's semi classical theory, using approximations which are equivalent to Born's approximation. The energetics of the stripping reactions are indistinguishable from those of compound nucleus reactions. However, the angular distribution does not have fore-and aft symmetry about $\theta = 90°$, but shows a pronounced maximum in the forward direction. Note that for l_n or $l_p = 0$, this maximum lies at $\theta = 0$ and progressively advances to wider angles for larger values of l_n or l_p. Besides, there are also secondary maxima for each value of l_n or l_p value

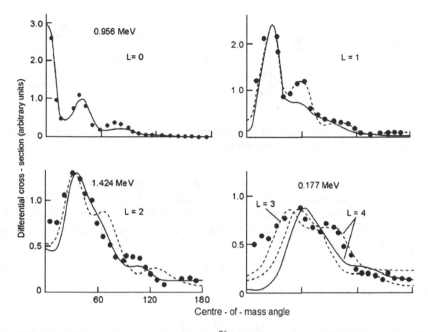

Fig. 7.26 The differential cross-section for the ^{76}Se(d, p) reaction at 7.8 MeV to various final states showing how the peak angle increases with the orbital angular momentum transfer L. The *curves* were obtained from distorted wave calculations with two sets of optical potentials [26]

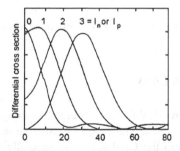

Fig. 7.27 Angular distribution of the uncaptured particle in the stripping reaction (d, p) and (d, n). The captured particle transfers orbital angular momentum I_n or I_p directly into a level in the final nucleus. In general, the differential cross section is largest for I_n or $I_p = 0$ and decreases as the angular momentum transfer increases. The illustrative angular distributions shown referto any stripping reaction for which the incident deuteron energy is 14.9 MeV and the uncaptured particle has 19.4 MeV, both in center-of-mass coordinates [13]

(Fig. 7.27). The contributions to $\sigma(\theta)$ is directly proportional to the square of the spherical Bessel function of the corresponding order. The spherical Bessel functions $j_0(x)$, $j_1(x)$ and $j_2(x)$ are given in [24], Appendix C.

At high energy the stripping mechanism can be described semi-classically by Serber's model [29] in which one of the nucleons from deuteron is absorbed by the

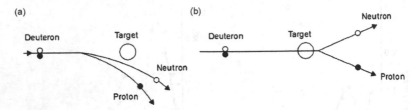

Fig. 7.28 Schematic representation of (**a**) Coulomb break-up at low energies on a heavy target nucleus; (**b**) nuclear break-up at higher energies

target nucleus and the other continues its flight almost unperturbed with the initial velocity of the deuteron, and therefore with half of deuteron energy. The energy and angular distribution governed by the addition of the internal momentum of the deuteron to half the momentum of its centre of mass. The energy spread of the stripped nucleon is given by

$$\Delta E = 1.5 (B_d E_d)^{1/2} \tag{7.140}$$

the angular width by

$$\Delta \theta = 1.6 \left(\frac{B_d}{E_d} \right)^{1/2} \tag{7.141}$$

and the energy of the stripped nucleon by

$$E_N = \frac{1}{2} E_d \tag{7.142}$$

where B_d is the binding energy of the deuteron and E_d is its kinetic energy.

7.14.4 Break-up Reactions

In the Coulomb field at few MeV incident energy a composite particle may undergo a break-up into its constituents. The simplest example is a (d, pn) reaction. The break-up may take place in the Coulomb field specially when the target nucleus has high Z, without the neutron being captured. This is known as Coulomb break-up. As there are three particles in the final state, energy can be shared in a number of ways, the angular distribution of proton and the neutron is not a line spectrum but a continuous one, with a broad peak centred at about one-half of the incident deuteron energy. As the deuteron traverses a Coulomb orbit, the neutron and proton break up, the neutron and proton are emitted on the same side of the incident direction, the proton being repelled and the neutron continuing in a straight line, Fig. 7.28(a).

At higher energies (of the order of 100 MeV or more) the break-up may occur in the nuclear field. In this process the proton and neutron may be emitted on the opposite side and the target nucleus may be left in the ground state or raised to an excited state. These processes are known as elastic and inelastic break-up respectively.

Fig. 7.29 Schematic
representation of the $(p, 2p)$
knock-out reaction

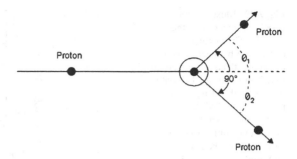

7.14.5 *Knock-out Reactions*

These consist of reactions $X(a, bc)Y$ in which there are three nuclei in the final
state. The simplest example of a knock-out is the $(p, 2p)$ reaction. At high energies
this can occur as a simple billiard-ball collision with a target nucleon, Fig. 7.29 with
the two outgoing proton emerging at an angular separation of $\theta = \theta_1 + \theta_2 = 90°$ in
the Lab system.

The condition for the occurence of the knock-out reaction is that the wavelength
of the incident particle is small compared with the internucleon separation in the
nucleus and that the projectile energy is much higher than the binding energy of
the struck nucleon. The first condition ensures that the collision occurs with a single nucleon and not with the entire nucleus; the second one implies that the ejected
proton results from a direct process and not via a compound nucleus. This then
means that a few hundred MeV energy is required for the projectile. The knock-
out process is recognized by detecting the two protons in coincidence. This is accomplished by fixing one proton counter at a definite angle say $\theta = 45°$, on one
side of the beam axis, and measuring coincidence rates with the second proton
counter placed on the other side of the beam, in the same plane, at variable angle. There will be a peak in the coincidence rates at the angle $\theta_2 = 45°$ for the
knock-out protons. This is singled out from the background protons resulting from
other processes, such as pre-equilibrium reactions, which have continuous distribution.

If the energies of the two outgoing protons are measured then the energy balance indicates the binding energy that the struck proton had in the target nucleus. One finds peaks in the summed energy spectrum corresponding to the binding energies of the various shell model single-particle orbitals. This is one of the
most direct pieces of evidences for the existence of these orbits and the results
are analogous to those obtained by Franck-Hertz experiments on the ionization of
atoms.

The statement that the emitted protons subtend a right angle needs some qualifi-
cation. The struck proton will not be completely at rest but will be moving about in
the target nucleus. This will permit the emergence of the two protons at angle of separation different from 90°, and also allow non-coplanar events. This leads to a peak
centred around the favoured angle but broadened by the initial motion as shown in

Fig. 7.30 Illustrating a
knock-out reaction of the type
$(a, a'b)$. The angular
correlation between protons
emitted with equal deflections
to left and right (that is, with
$\theta_a = \theta_b$ in the ^{6}Li$(p, 2p)$
reaction at 450 MeV
bombarding energy for
knock-out from the $1s_{1/2}$ and
$1p_{3/2}$ orbits. The maximum
and minimum, respectively,
are shifted slightly from
$\theta_a = \theta_b = 45°$ because of
refraction of the incident and
outgoing protons [23]

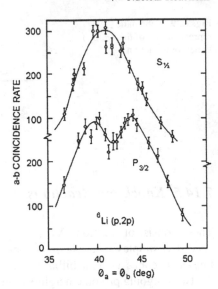

Fig. 7.30. Only if the struck proton is in a s-state then it can be found at rest. If
it has a finite amount of angular momentum, it will always be in motion, although
this motion will be slow compared to the projectile's motion provided the incident
energy is high. In such cases, there will be actually a minimum at the favoured angle $\theta = \theta_1 + \theta_2$, corresponding to a free particle, with a peak on either side. This
is illustrated in Fig. 7.30 for the knock-out process ^{6}Li$(p, 2p)$ from the p-state, in
the work of Jacob and Maris [23]. From such measurements on angular correlation
one can get information on the angular momentum of the state from which the target proton was ejected as well as its energy. Furthermore from the observed shape
of the peak, information can be obtained about the momentum distribution of the
nucleon in the target nucleus before it was struck and emitted. A great advantage
of the knock-out reactions is that the deep single-particle states can be studied from
the missing energy while nucleon transfer reactions permit access only to such states
which lie near the Fermi surface.

7.15 Comparison of Compound Nucleus Reactions and Direct Reactions

In Table 7.1 are compared the distinguishing features of the two extreme types of
reactions.

7.16 Pre-equilibrium Reactions

So far we have discussed two extreme views of nuclear reactions. The direct reaction takes place with a single nucleon or a cluster of nucleons and the reaction

Table 7.1 Comparison of compound nucleus reactions and direct reactions

Feature	Compound nucleus reactions	Direct reactions
Times involved	10^{-14}–10^{-16} s	$\approx 10^{-20}$–10^{-21} s
Dominance of reaction	Low energy	High energy
Nature of reaction	Surface phenomenon	Nuclear interior
Cross-section	$\approx$b	$\approx$mb
Angular distribution	Isotropic in the centre of mass system	Peaked in the forward hemisphere
Location of peaks	Energy dependent	Orbital angular momentum dependent
Energy distribution	Boltzmann distribution	Peaked toward higher energy side
Deuteron stripping at low energy	(d, n) reactions more frequent	(d, p) reactions more frequent
Variation of excitation function with energy	Rapid and irregular	Smooth

is completed in a very short time ($\sim 10^{-20}$–10^{-21} s). In the compound nucleus reaction the projectile is captured and initiates numerous collisions and the energy is distributed over the entire nucleus so that a thermal equilibrium is reached in a relatively long time ($\sim 10^{-14}$–10^{-16} s). The hot liquid drop cools off by emitting one or more particles. We can expect some events to occur after the first stage of direct reaction is over but in relatively few collisions the compound nucleus evaporates long before the thermal equilibrium is reached. Such reactions are called pre-equilibrium reactions or pre-compound reactions. Their time scale is intermediate between the very fast direct reactions and the relatively slow compound nucleus reactions. Direct evidence for pre-equilibrium reactions is provided by the energy spectra of particles at relatively high energy particles. They show conspicuous deviation from the Maxwellian distribution expected from compound nucleus reactions and by the presence of peak from the excitation of low lying energy level due to direct reactions. As an example Fig. 7.31 shows the energy spectra of inelastic protons emitted from the bombardment of ^{55}Fe by protons of 29, 39 and 62 MeV. The spectra can be divided into three distinct regions, the one on the left side corresponds to low energy particles evaporated from a hot nucleus with the characteristic Maxwellian distribution; this is followed by a somewhat structure less continuum in the middle, which is weakly dependent on ejectile energy; finally on the right side toward the higher energies, sharp peaks corresponding to protons associated with single particle excited states.

Analysis of the angular distribution of the observed particles reveals that the Maxwellian peak is symmetric about 90°, thereby confirming its compound state origin. On the higher side the sharp peaks reveal a forward-peaked structure confirming their origin in the direct reactions. Further, these peaks shift toward higher energy with the increase of bombarding energy, showing that higher states are excited. The structure less continuum in the middle can neither be attributed to the

Fig. 7.31 The energy spectra of inelastic protons emitted after bombardment of ^{55}Fe by protons of 29 (*upper left*), 39 (*upper right*) and 62 MeV (*lower*). The highest-energy protons emitted (right hand end of spectra) correspond to the excitation (mainly by direct reactions) of discrete, low-lying, excited states in ^{56}Fe; the peaks to the left correspond to evaporation from the compound nucleus ^{57}Co [3]

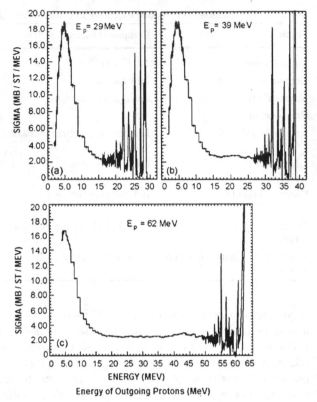

compound state reactions nor to the direct reactions. On the lower energy side of this continuum the angular distribution is symmetric about 90° but the cross-sections are much greater than those predicted by the compound state theory. On the higher energy side the angular distributions are peaked forward but lack both diffraction structure and the nuclear structure dependence which characterize particles originating from direct reactions.

These observations are explained by assuming that the nucleon cascade generated in the high energy collisions causes the excitation of particle-hole states (excitons) through nucleon-nucleon interactions. At various stages particles may be emitted from these particle-hole states in the intermediate nuclei much before they have attained thermal equilibrium. Such particles are said to result from pre-equilibrium reactions.

Usually the individual final states cannot be resolved as they occur at considerable energy. The total cross-section for the pre-equilibrium reactions comes from the continuum of final states. As there is no interference in the reactions occuring at various stages, the total pre-equilibrium reaction cross-section is evaluated by simply adding the contribution from various stages of the nucleon-nucleon interaction cascade.

The first successful theory to explain these processes was the excitation model, introduced by Griffin [22]. It is outside the purview of this book. Such reactions may be studied by Monte Carlo methods.

7.17 Heavy-Ion Reactions

A heavy ion is defined as a nucleus with mass number $A > 4$. Reactions between two heavy nuclei offer a richer variety of phenomena than those between a light ion and a heavy target nucleus.

7.17.1 Characteristics of Heavy Ion Reactions

(1) The large mass implies a greater linear momentum compared to a light ion of the same energy. This also means a greater angular momentum about the target nucleus ($l\hbar <\sim pR$). For example, consider a Zn nucleus ($A_1 = 30$) at 500 MeV making a grazing collision with a Sn nucleus ($A_2 = 50$). Then the reduced mass $\mu = 30 \times 50/(30+50) = 18.75$. The relative velocity of Zn nucleus will be $v = c\sqrt{2E/M_1c^2} = c\sqrt{2 \times 500/(63.93 \times 931.5)} = 0.13c$. The inter nuclear distance $r = R_1 + R_2 = 1.3(A_1^{1/3} + A_2^{1/3}) = 1.3(30^{1/3} + 50^{1/3}) = 6.79$ fm. $J = \mu v r = l\hbar$

$$\therefore \quad l = \frac{18.75 \times 931.5 \times 0.13 \times 6.79}{197} \simeq 78$$

Thus an angular momentum of $78\hbar$ is involved in the grazing collision of Zn nucleus about the centre of the Sn target nucleus. This affords the excitation of nuclear states with very high spin. States as large as $60\hbar$ or more have been excited in (HI, xn) reactions. Here HI means heavy ion projectile which is absorbed by a target nucleus resulting in the evaporation of x neutrons.

A large amount of kinetic energy and angular momentum is distributed over a large number of nucleus leading to the formation of a compound nucleus. It is interest to know whether there is some critical limit to the angular momentum that can be sustained by the compound nucleus beyond which the compound nucleus becomes unstable against fission by virtue of centrifugal forces. Indeed, experiments have demonstrated that for angular momenta larger that the critical value the compound nucleus breaks into two large fragments.

(2) For reactions intermediate between extreme peripheral (grazing collisions corresponding to direct reactions) and complete fusion reaction (compound nucleus reactions) resulting from head-on collision, the cross-sections are by far larger for heavy ions than for light ions. There are events in which two heavy nuclei stick together for a time longer than that associated with direct reactions but not

long enough to fuse into a compound nucleus. Large loss of kinetic energy occurs in such collisions, the energy being converted into heat (excitation energy)
the system, however, retains the 'memory' as evidenced by the forward peak
in the angular distribution. Such events have been referred to as deep inelastic
or strongly damped collisions. After complete fusion, fission into two heavy
fragments is more probable than evaporation of neutrons.

(3) The large mass of a heavy ion means that its de Broglie wave length will be
shorter than for light ion of the same energy. In a number of experiments the
wavelength will be so short that classical concepts such as trajectory become
valid, this has lead to the revival of semi-classical theories in heavy ion physics,
atleast for qualitative understanding.

(4) As the charge in a heavy ion may be quite large, high projectile energies are
required to overcome Coulomb barrier between the incident ion and the target
nucleus. Thus the Coulomb barrier to be surmounted before two lead nuclei
come into contact is about 600 MeV this requires a bombarding energy in the
Lab system of about 1200 MeV. The subject of heavy ion interactions has now
become fascinating and accelerators have been constructed to produce heavy
ions of energy ranging from few MeV per nucleon to as much as 200 GeV per
nucleon, and to accelerate ions as heavy as uranium. In the collisions of such
ultra relativistic ions with targets of heavy nuclei elementary particles and their
resonant states will be produced copiously.

(5) The velocity of sound in nuclear matter is estimated to be about $0.2c$ corresponding to ~ 20 MeV nucleon. This leads to the division of low energy into
two parts, subsonic and supersonic. In the subsonic region nuclei are expected to
behave as incompressible while in the supersonic region compressibility comes
into the picture and shock waves are assumed to be generated in nuclei.

(6) It is suggested that ultra heavy nuclei with $A \sim 310$ may be stable although they
have not yet been found in nature. One other interest in heavy ion physics is to
explore the possibility of creating super heavy elements by fusing two heavy
ions.

7.17.2 Types of Interactions

It was pointed out that in the semi-classical picture it was legitimate to use the ion
trajectories. It E_{cm} is the centre of mass energy of the two interacting ions then the
minimum distance of approach is given by

$$r_{min} = \frac{b}{\sqrt{1 - \frac{V(r_{min})}{E_{cm}}}} \tag{7.143}$$

where b is the impact parameter and $V(r_{cm})$ is the nuclear potential acting between
the two ions. The three regions in which different reactions dominate may be distinguished in terms of the minimum distance of approach in the decreasing order:

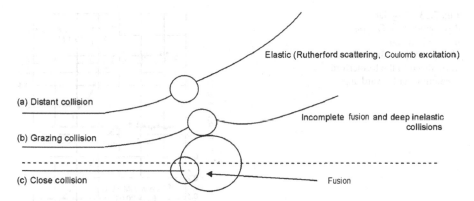

Fig. 7.32 Schematic representation of heavy ion trajectories showing distant, grazing and close collisions

(a) The Coulomb region with $r_{min} > R_N$, where R_N is the distance for which the nuclear interactions are ineffective.
(b) The deep inelastic and the incomplete fusion region with $r_{min} = R_1 + R_2$.
(c) The fusion region with $0 \leq r_{min} \leq R_1 + R_2$.

The ion orbits corresponding to the above regions are schematically shown in Fig. 7.32.

7.17.3 Distant Collisions

Elastic and Inelastic Scattering   At energies below the Coulomb barrier two ions can suffer only Coulomb interaction. For distinguishable particles (projectile and the target) the angular distribution is accurately described by the Rutherford scattering. However, for the indistinguishable (identical) ions the quantum mechanical interference produces a conspicuous departure from the classical predictions, the scattering being modulated by damped oscillations. The problem of scattering of identical particles was elaborately discussed for Fermions and Bosons in Chap. 5.

Figure 7.33 shows the angular distributions of ^{16}O ions on magnesium and aluminium, showing the characteristic exponential decrease following a local increase above the Rutherford predictions. The local increase is attributed to interference between Coulomb and nuclear amplitudes. From such data it is possible to extract the interaction radii.

Figure 7.34 shows typical angular distributions for oxygen elastically scattered from oxygen. In both the diagrams the energies are below the Coulomb barrier. The data are in excellent agreement with the Mott scattering predictions along with the marked modulation.

Fig. 7.33 Angular distributions for O^{16} on magnesium and aluminium [9]. The data are normalized to the Rutherford predictions at forward angles

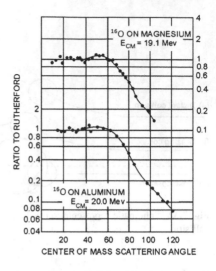

The appropriate formula for Mott scattering is

$$\frac{d\sigma}{d\omega}\bigg|_{Mott} = \left(\frac{Ze}{4E}\right)^2 \bigg|\cos^4\frac{\theta}{2} + \sec^4\frac{\theta}{2}$$

$$+ (-1)^{2s}\frac{2}{2s+1}\cos\eta\ln\tan^2\frac{\theta}{2}\left(\csc^2\frac{\theta}{2}\sec^2\frac{\theta}{2}\right)\bigg| \qquad (7.144)$$

where S in the nuclear spin and $\eta = \frac{(Ze)^2}{\hbar v}$, is the Sommerfield parameter. The first two terms are clearly the Rutherford prediction for identical particles while the third one is due to quantum mechanical interference. The third term actually modulates the classical predictions through the $\ln\tan^2(\theta/2)$ term as well as η which depends on both Z and v. Compared to the scattering of $\propto$ particles on helium, the heavy

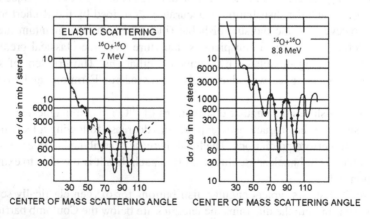

Fig. 7.34 Angular distributions for ^{16}O on oxygen. The *dashed* and *solid curves* are the Rutherford and Mott predictions, respectively

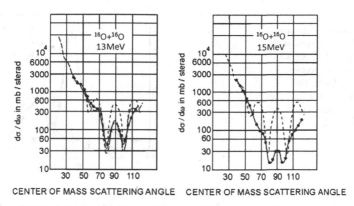

Fig. 7.35   Angular distributions for ^{16}O on oxygen. The legend is that of Fig. 7.34 with the addition of the *dot dashed* Blair prediction for $l_{max} = 6$

ion scattering tends to *amplify* the modulation markedly. Note that for $\theta = 90°$ the energy modulation disappears. Secondly, the cross-section falls smoothly inversely with the square of energy as for Rutherford scattering.

Figure 7.35 shows angular distributions measured at energies above the Coulomb barrier, showing the characteristic decrease in cross-section below the Mott prediction, but without marked change of shape [9]. The departure from the Mott scattering is obviously caused by nuclear absorption. Blair's sharp cut-off model describes the elastic scattering reasonably well even in its very simple form [4]. In this model all partial waves in the incident beam with an impact parameter corresponding to a classical distance of closest approach (7.143) less than the nuclear interaction radius are totally absorbed, while those with higher impact parameters are subjected to pure Coulomb scattering. This corresponds to the subtraction of the low partial wave contributions to the Coulomb scattered as shown in (7.145). For $S = 0$ bosons such as ^{16}O or ^{12}C the expression takes a simple form

$$
\begin{aligned}
\left.\frac{d\sigma}{d\omega}\right|_{Blair} = \left(\frac{ze^2}{4E}\right)^2 \Bigg| & csc^2\frac{\theta}{2} \exp\left(2i\delta_0 - i\eta \ln \sin^2\frac{\theta}{2}\right) \\
& + \sec^2\frac{\theta}{2} \exp\left(2i\delta_0 - i\eta \ln \cos^2\frac{\theta}{2}\right) \\
& - \frac{2i}{\eta}\sum_{l=0}^{l_{max}}(2l+1)\exp(2i\delta_l)P_l(\theta) \Bigg|^2
\end{aligned}
\tag{7.145}
$$

where $\exp(2i\delta_l) = \frac{\Gamma(l+1+i\eta)}{\Gamma(l+1-i\eta)}$, is Coulomb phase of order l. Here only even values are permitted. Since the Coulomb phases are completely defined in the Mott scattering formalism (7.144), it is possible to compare the experimental data with the predictions for absorption of different limiting values. Blair model predictions for ^{16}O on oxygen in Fig. 7.35, are shown to fit the experimental data with $l_{max} = 6$.

Fig. 7.36 Differential
cross-section for production
of projectile-like fragments in
the interaction of ^{136}Xe with
^{209}Bi at incident energies of
940 and 1130 MeV. The
arrows indicate the two
incident ion experimental
grazing angles [27]

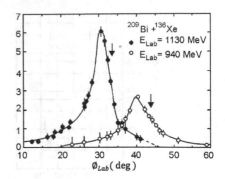

At large impact parameters inelastic scattering and transfer reactions also contribute to the reaction cross-sections.

For large impact parameters and corresponding small scattering angles the two interacting nuclei do not come very close to each other so that the probability for nucleon transfer is small. The Coulomb excitation is simply the excitation of the nucleus by Coulomb interaction. When the Sommerfield parameter n is large this process may be comparable to the excitation by nuclear interaction, specially for the low multipoles (small L values). These two contributions are coherent and produce interference effects. Because of the long range character of Coulomb interaction ($\sim (1/r^{L+1})$ for the Lth *multipole*) the Coulomb excitation becomes important for $L = 2$ and $L = 3$. Below the Coulomb barrier the Coulomb excitation dominates and provides an important tool for measuring transition probabilities. Because the Coulomb interaction is large for heavy ions, Coulomb excitation to very high spin states, particularly in nuclei displaying rotational spectra through repeated $E2$ transitions up the band becomes important.

7.17.4 Deep Inelastic Collisions

When the impact parameter is reduced to the point that $r_{\min} = R_1 + R_2$, then (a) grazing collision occurs as in the trajectory (b) of Fig. 7.32, the two ions make sufficiently strong contact resulting in large loss of energy in the form of excitation energy; hence the name 'deep inelastic scattering or collision' or damped reactions. In this process a few nucleons may be transferred, yet they do not stay long enough to be considered as a compound nucleus. The excited residues that are produced resemble the original ions, one projectile like and the other target like. This is evidenced by the angular distribution of the residues which peaks very close to an angle close to the grazing angle which is the angle at which the projectile would have suffered Rutherford scattering in the grazing collision. Figure 7.36 shows the angular distribution of residues in the collision of ^{136}Xe with ^{209}Bi at lab energy of 940 and 1130 MeV [28]. The extreme peripheral collisions are the direct reactions discussed in Sect. 7.14.

The products of deep inelastic collisions are found to be primarily binary, their masses and charges differing by only a few units from those of the projectile and target, meaning that only a few nucleons have been exchanged although considerable energy transfer has occured.

The angular distributions of the deep-inelastic products are strongly correlated with the energy loss. These distributions are approximately Gaussian with a width which increases with decreasing total kinetic energy. This suggests a strong correlation between the kinetic energy loss and the number of nucleons exchanged. Peripheral collisions involve less energy loss and a quick separation time leading to the ejectiles projected in the forward direction concentrated in a small range of angles. Greater energy loss would imply a more intimate contact (as for trajectory (c) in Fig. 7.32) in which the two ions rotate about one another for a longer time. During this time they are assumed to retain their dinuclear configuration, without complete fusion. When they break-up they would have exchanged a large number of nucleons. Such events are characterized by a broader angular distribution. The excited ejectiles decay by particle or γ-ray emission. Measurements of neutron spectra from these ejectiles show that the two fragments have the same temperature, meaning that when they separated after the statistical equilibrium was reached.

When the bombarding energy is below or close to the Coulomb barrier the reaction products exhibit angular distribution of "bell shape" centred around the deflection angle for the grazing Rutherford orbit. As the energy increases above the Coulomb barrier, the angular distribution reveals the structure. With further increase of energy, a diffraction appears typical of direct reactions, Fig. 7.37.

Another important aspect of direct reactions is the transfer reactions involving the transfer of an α-particle-like cluster of two neutrons and two protons to explore the validity of α-particle models for the lighter nuclei and the clustering of α-particles in heavy nuclei. To this end reactions such as (^{16}O, ^{12}C), (^{20}Ne, ^{16}O), (^{6}Li, d) and their inverse have been used.

7.18 Fusion

Fusion in heavy ion reactions produce nuclei with high excited states and high spins. Further, these reactions can produce super heavy nuclei and proton rich nuclei far from the stability curve. We summarize the characteristics of fusion reactions.

1. At low incident energies light projectile are found to have fusion cross-section, a large fraction of reactions cross-section.
2. For larger ionic charges the fusion cross-section σ_F falls off abruptly.
3. The plot of σ_F with $1/E_{cm}$ shows a linear increase up to a maximum after which it decreases linearly.
4. The variation of σ_F with the bombarding energy is generally smooth, except the oscillatory behaviour when the interacting ions are light.

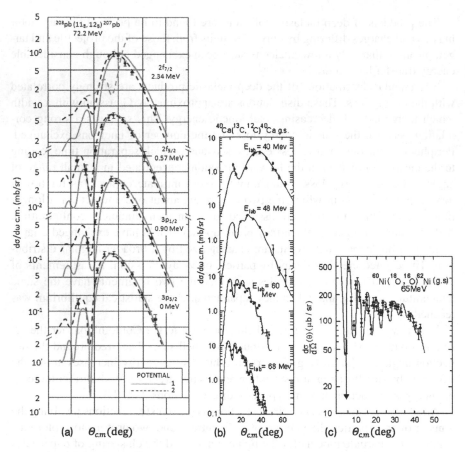

Fig. 7.37 Typical angular distributions for peripheral or direct reactions between heavy ions in which one or two nucleons is transferred and the residual nuclei are left with little or no excitation energy. (**a**) 'Bell-shaped' curves typical of reactions initiated by bombarding energies close to the Coulomb barrier Toth et al. [30]. (**b**) Transition from a bell-shape to a diffraction pattern as the bombarding energy is increased (from Bond et al. [7]). (**c**) An example of an oscillating angular distribution when two neutrons are transferred (Levine et al. [25]). In each case, the *curves* represent theoretical calculations using the distorted-wave Born approximation. The notation g.s. means a transition to the ground state of the final nucleus

5. The same composite system may be formed from different combinations of ions, but the energy dependence of σ_F may be quite different.

Quite a few of the above features can be explained at least qualitatively by using the critical model which is based on the optical potential. The real part of the effective one body potential describes the refraction of the incident ion and the imaginary part the absorption of all inelastic processes. For the present discussion we shall focus only on the real part which consists of three terms, the Coulomb potential, the nuclear potential and the centrifugal potential. To a good approximation the Coulomb potential may be taken as that between two uniformly charged spheres.

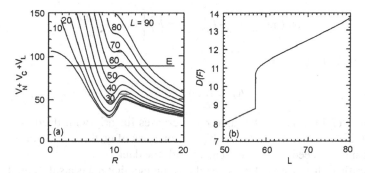

Fig. 7.38 (a) Sum of the nuclear, Coulomb, and centrifugal potentials for ^{18}O + ^{120}Sn as a function of radial distance for various values of orbital angular momentum L. The nuclear potential has the Saxon-Woods form with $V = 40$ MeV, $\gamma_0 = 1.31$, $a = 0.45$. The *horizontal line* at $E_{cm} = 87$ MeV corresponds to an incident energy of 100 MeV. The turning points for various values of I are marked by dots. (b) The distance of closest approach for various partial waves, showing the discontinuity at $L = 57$ [20]

The nuclear potential may be taken that of Saxon-Woods form

$$V_n(r) = \frac{V_0}{1 + \exp(\frac{r-R}{a})} \tag{7.146}$$

where $R = r_0(A_1^{1/3} + A_2^{1/3})$. The centrifugal term is

$$V_L(r) = \frac{h^2}{2\mu} l \frac{(l+1)}{r^2} \tag{7.147}$$

The relative contributions of these three potentials is determined by the energy, masses and charges of the interacting ions. As an example, Fig. 7.38(a) shows the total potential $V = V_N + V_C + V_l$, as a function of R, for typical values of l, in the collision of ^{18}O + ^{120}Sn [20]. Note that for smaller angular momentum a pocket is formed in the potential which vanishes for larger values of l. There is a range of partial waves for which the l values are such that the ions are permitted sufficiently close to be trapped in the pocket leading to fusion. Otherwise the two ions are reflected back and do not fuse. Figure 7.38(a) shows that fusion may take place up to a critical angular momentum l_{crit}. The figure also suggests that the pocket is located at a distance R_F which is approximately constant. The occurence of l_{crit} causes a discontinuity in the curve for the distance of closest approach $D(F)$ as a function of l as in Fig. 7.38(b). In this example it occurs for $l = 57$.

The maximum value or the fusion reaction occurs at an energy corresponding to the relative momentum

$$k(R_F) = k\sqrt{1 - \frac{V(R_F)}{E_{cm}}} = \frac{L_{crit}}{R_F} \tag{7.148}$$

$$\sigma_F(\text{max}) \simeq \frac{\pi}{R^2} L_{crit}^2 = \pi R_F^2 \left[1 - \frac{V(R_f)}{E_{cm}} \right] \tag{7.149}$$

with the experimental value

$$R_F \simeq r_0 \left(A_1^{1/3} + A_2^{1/3} \right) \tag{7.150}$$

Expression (7.149) shows that upto σ_F increases linearly with decreasing $1/E_{cm}$ and then it decreases as $1/E_{cm}$ at higher energies after the maximum is reached since l cannot exceed l_{crit}, in agreement with the data.

The critical-distance model also explains the fact that σ_F is a substantial fraction of σ_r, the reaction cross-section for light projectiles. Further, the abrupt decrease in the ratio σ_F/σ_r for the higher Z ions is attributed to the increase of repulsive Coulomb potential which results to a decrease of l_{crit}, thereby reducing the probability of pocket formation.

When the composite system is imparted a large amount of excitation energy as well as a large angular momentum, it is set into rapid rotation. Consequently, the original spherical shape of the nucleus in the ground state changes into a highly deformed shape.

Now the rotational energy of a body the moment of inertia I rotating with angular velocity ω is

$$E = \frac{1}{2} I \omega^2 \tag{7.151}$$

Its angular momentum is $J = I\omega$, so

$$E = \frac{J^2}{2I} \tag{7.152}$$

Consider a deformable body rotating in space with a fixed angular momentum. The rotation deforms the body and it will take up a shape that minimizes the energy, and by (7.152) it tends to maximize the moment of inertia. At small deformations corresponding to small angular velocities the favoured shape is oblate, rotating about its axis of symmetry. This explains why earth is an oblate spheroid flattened at the poles. At high angular velocities, the favoured shape is a prolate spheroid with the axis of rotation perpendicular to the axis of symmetry. If the composite nucleus is imparted sufficiently high angular speed then at a certain stage the centrifugal force overcomes the attractive nuclear force and the system becomes unstable against fission into two fragments. The excitation energy can also be dissipated by particle emission. In the initial stages the particle emission occurs rapidly, the orderly motion of the interacting ions being transformed into a chaotic motion due to the cascade of nucleon-nucleon interactions while thermalization of the system goes on. The pre-equilibrium emission takes place in a time less than 10^{-21} s, this is followed by evaporation from the equilibrated nucleus in a time of the order of 10^{-16} s. The particles thus emitted are recorded in coincidence with the residue nuclei. Fission requires much longer time for deformation and so does not compete with particle

Fig. 7.39 Angular distribution of fission fragments in the center-of-mass system for fission of ^{197}Au induced by ^{12}C ions of 123-MeV energy. *Solid line*, experimental curve; *broken line*, $1/\sin\theta$ [21]

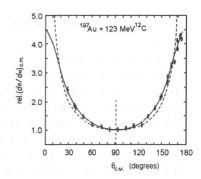

Fig. 7.40 Schematic picture of the contributions of different partial waves to the reaction cross-section for a collision between two heavy ions. The relative proportions of the various non-elastic events varies with energy and the masses of the ions

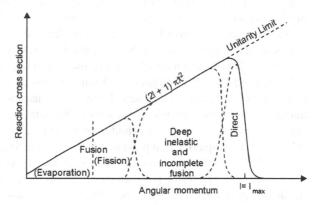

emission during the nuclear thermalization but only during evaporation stage. On the other hand if initially the compound nucleus receives an angular momentum greater than about $65\hbar$ then fission into two fragments is the most favoured process for the decay because the fission barrier would be quite small. In the evaporation regime if the excitation energy is below the neutron binding energy, then neutron emission is inhibited, and the de-excitation occurs via γ-emission, α particle and even heavy ions emission at the end of the evaporation chain.

The angular distribution of fission fragments as well as the evaporated particles is characterized by $1/\sin\theta$ form peaking in the forward and backward direction along the collision axis and symmetric about $90°$ in the CMS as in Fig. 7.39. The form is predicted from simple classical considerations.

The angular momentum structure of the reaction or absorption cross-section for the heavy ion collisions may be schematically represented by a diagram such as in Fig. 7.40, which shows how various contributions vary with the angular momentum in the entrance channel. The diagram is only of a qualitative nature as the relative contributions depend upon the energy and nature of the two ions. However, the figure does indicate qualitatively that in the event of genuine fusion, systems possessing larger angular momenta are more likely to decay by fission into two large fragments, while those with smaller spins tend to decay via evaporation of low energy particles specially neutrons and gamma rays.

7.18.1 Quark-Gluon Plasma

The study of heavy ion reactions at relativistic energies is another fascinating area. At ultra relativistic energies (≥ 100 GeV per nucleon). The approaching ions are Lorentz contracted in the form of flat discs. Upon collision the quark-gluon plasma (separation of quarks from gluons) may be formed in the hot and dense region, and when the ions separate the region between the ions expands and cools off. Thus the quark-gluon plasma may last for a time of the order of collision time. Its existence is based on the study of particles from a highly excited state. Experiments made at CERN [14] by bombarding heavy nuclei with 200 GeV/nucleon sulphur ions suggested that the conditions necessary for the formation of quark-gluon plasma might have already been realized-a conclusion which has been controversial.

Such experiments have been repeated in the past 15 years both at CERN and Brookhaven laboratory. At CERN lead ions were used and at Brookhaven lab gold ions. In the latter heavy ion Collider was used and gold ions bombarded each other at beam energy of 130 GeV/nucleon. Temperatures as high as 10^5 times the central temperature of the sun were reached and the nuclear density 20 times the normal nuclear density was established. All sorts of elementary particle were recorded perpendicular to the beams. All this had to be done in the time of the order of 10^{-23} s during which the quark gluon plasma would last. It is found that the production rate of particles is less than that predicted by the standard theory. It is appears that under these conditions the quark-gluon plasma, the fifth state of matter, could exist. But the discovery is still uncertain. This problem is of paramount importance for the Big Bang theory of the early universe ([24], Chap. 8).

Example 7.1 Calculate the energy of protons detected at $90°$ when 4.0 MeV deuterons are incident on ^{27}Al to produce ^{28}Al with an energy difference $Q = 5.5$ MeV.

Solution   Consider the reaction

$$a + X \rightarrow b + Y + Q$$

$$Q = E_b\left(1 + \frac{m_b}{m_Y}\right) - E_a\left(1 - \frac{m_a}{m_Y}\right) - \frac{2}{m_Y}\sqrt{m_a m_b E_a E_b}\cos\theta$$

Given reaction is

$$d + {}^{27}\text{Al} \rightarrow p + {}^{28}\text{Al}$$

put

$$m_a = m_d = 2, \qquad m_b = m_p = 1, \qquad m_Y = m_{\text{Al}} = 28$$

$$E_a = 4.0 \text{ MeV}, \qquad \theta = 90°, \qquad Q = +5.5 \text{ MeV}$$

Solving the equation we find $E_p = 9.03$ MeV.

Example 7.2 Determine the range of neutrino energies in the solar fusion reaction, $p + p \rightarrow d + e^+ + \nu$. Assume the initial protons have negligible kinetic energy and that the binding energy of the deuteron is 2.2 MeV.

Solution The minimum energy of ν is zero when d and e^+ go off in opposite direction and ν carries negligible energy.

The maximum energy of ν corresponds to a situation in which both e^+ and d move in a direction opposite to ν

$$Q = 2.2 - 0.51 = 1.69 \text{ MeV}$$

(i) $T_{ed} + T_\nu = Q = 1.69$ (energy conservation)

(ii) $P_{ed}^2 = P_\nu^2$ (momentum conservation)

(iii) $2Mc_{ed}^2 T_{ed} = T_\nu^2$

Eliminating T_{ed} between (i) and (iii) and using $M_{ed}c^2 = 1875$ MeV, we find $\Gamma_\nu = 1.69$ MeV. Thus the energy range of ν will be zero to 1.69 MeV.

Example 7.3 A nucleus has a neutron resonance at 65 eV and no other resonances nearby. For this resonance, $\Gamma_n = 4.2$ eV, $\Gamma_\gamma = 1.3$ eV and $\Gamma_\alpha = 2.7$ eV and all other partial widths are negligible. Find the cross-sections for (n, γ) and (n, α) reactions at 65 eV.

Solution The low energy cross-section for the compound nuclear reaction

$$a + X \rightarrow c^* \rightarrow Y + b$$

is given by

$$\sigma_{ab} = \frac{\lambda^2 \Gamma_a \Gamma_b}{4\pi((E - E_0)^2 + \frac{\Gamma^2}{4})}$$

$$\Gamma = \Gamma_n + \Gamma_\gamma + \Gamma_\alpha = 4.2 + 1.3 + 2.7 = 8.2 \text{ eV}$$

$$\lambda = \frac{h}{p} = \frac{2\pi\hbar c}{cp} = \frac{2\pi\hbar c}{\sqrt{2Mc^2 T}}$$

$$= \frac{2\pi \times 197}{\sqrt{2 \times 939 \times 65 \times 10^{-6}}} fm = 3.544 \times 10^{-12} \text{ m}$$

Put $a = n$, $b = \gamma$ or α, $E = 65$, $E_0 = 65$

$$\sigma_{n\gamma} = 616 B; \qquad \sigma_{n\alpha} = 1280 B$$

Example 7.4 Find the energy of the helium nucleus in the fusion reaction $d + p \rightarrow {}^3\text{He} + \gamma + 5.3$ MeV, where initially proton ans deuteron are essentially at rest.

Solution

$$(i) \quad T_{He} + E_\gamma = 5.3 \quad \text{(energy conservation)}$$

$$(ii) \quad P_{He} = P_\gamma \quad \text{(momentum conservation)}$$

$$(iii) \quad c^2 p_{He}^2 = 2T_{He} M_{He} c^2 = E_\gamma^2$$

Eliminating E_γ between (i) and (iii) and solving the quadratic equation, with $M_{He} c^2 = 2809$ MeV, we find $T_{He} = 5$ keV.

Example 7.5 Calculate the threshold energy of the projectiles in a $^3H(p, n)^3He$ reaction, given, $Q = -0.74$ MeV.

Solution   For the endoergic reaction $X(a, b)Y$

$$E_a(\text{thres}) = |-Q| \left(1 + \frac{m_a}{m_X} \right)$$

$$= 0.74 \times \left(1 + \frac{1}{3} \right) = 0.987 \text{ MeV}$$

Example 7.6   Consider the reactions

$$d + {}^{16}O \rightarrow n + {}^{17}F - 1.631 \text{ MeV}$$

$$\rightarrow p + {}^{17}O + 1.918 \text{ MeV}$$

Which is the unstable member of the pair $({}^{17}O, {}^{17}F)$, and calculate the maximum energy of the β-particles it emits? (n-^1H mass difference is 0.782 MeV.)

Solution

$$d + {}^{16}O \rightarrow n + {}^{17}F - 1.631 \text{ MeV}$$

$$d + {}^{16}O \rightarrow p + {}^{17}O + 1.918$$

$$n + {}^{17}F - 1.631 = p + {}^{17}O + 1.918 \quad \text{or}$$

$${}^{17}O - + {}^{17}F = (n - p) - 3.549$$

$$= (n - {}^1H + e) - 3.549$$

$$= (0.782 + 0.511) - 3.549$$

$$= -2.256 \text{ MeV}$$

$\therefore$ ^{17}F is heavier. $+^{17}F$ decays to $+^{17}O$ by β^+ emission:

$$^{17}_9F \rightarrow +^{17}_8O + \beta^+ + \nu$$

$$Q = E_{\max}(\beta) + 0.511 \times 2$$

Fig. 7.41 Energy level
diagram

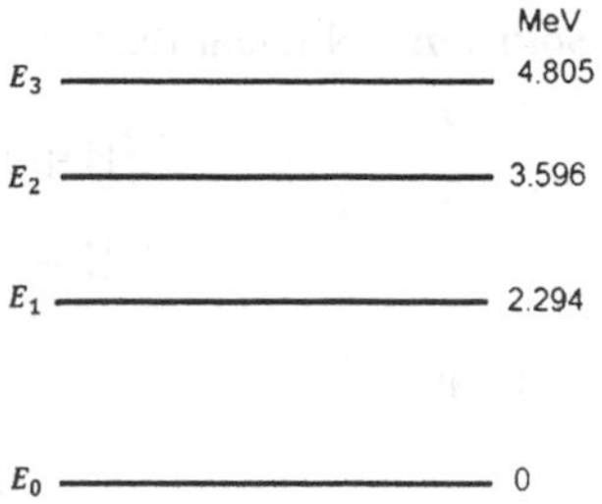

$$E_{max}(\beta) = 2.25 - 1.02 = 1.23 \text{ MeV}$$

Example 7.7 An aluminum target is bombarded by α particles of energy 7.68 MeV, and the resultant proton groups 90° were found to posses energies 8.63, 6.41, 5.15 and 3.98 MeV. Draw an energy level diagram of the residual nucleus, using the above information.

Solution

$$^4\text{He} + {}^{27}\text{Al} = p + {}^{30}\text{Si} + Q$$

$$a + X \rightarrow b + Y$$

$$Q = E_b\left(1 + \frac{m_b}{m_Y}\right) - E_a\left(1 - \frac{m_a}{m_Y}\right) - \frac{2}{m_Y}\sqrt{m_a m_b E_a E_b}\cos\theta$$

$$Q = E_p\left(1 + \frac{1}{30}\right) - 7.68\left(1 - \frac{4}{30}\right)$$

For

$$E_p = 8.63, \qquad Q_0 = 2.262$$
$$E_p = 6.41, \qquad Q_1 = -0.032$$
$$E_p = 5.15, \qquad Q_2 = -1.334$$
$$E_p = 3.98, \qquad Q_3 = -2.543$$
$$Q_0 - Q_0 = 0.000$$
$$Q_0 - Q_1 = 2.294$$
$$Q_0 - Q_2 = 3.596$$
$$Q_0 - Q_3 = 4.805$$

The energy level diagram of the residual nucleus is as in Fig. 7.41.

Example 7.8 If the Q-value for the ${}^3\text{H}(p, n){}^3\text{He}$ reaction is—0.7637 MeV and tritium β^- decays to ${}^3\text{He}$ with end point energy E_{max}. Given the difference in mass between the neutron and the hydrogen atom as 0.78 MeV, calculate E_{max}.

Solution   Nuclear reaction

$$^3\text{H} + p \rightarrow {}^3\text{He} + n - 0.7637 \text{ MeV} \quad \text{or} \tag{1}$$

$$^3\text{H} - {}^3\text{He} = n - p - 0.7637 \text{ MeV} \tag{2}$$

β-decay:

$$^3\text{H} \rightarrow {}^3\text{He} + \beta^- + \bar{\nu} + E_{\text{max}} \tag{3}$$

On nuclear scale

$$^3\text{H} \rightarrow {}^3\text{He} + m_e c^2 + 0 + E_{\text{max}} \tag{4}$$

Combining (2) and (4),

$$E_{\text{max}} = n - \left(p + e^-\right) - 0.7637$$

$$= n - {}^1\text{H} - 0.7637$$

$$= 0.781 - 0.7637 = 0.0173 \text{ MeV}$$

$$= 17.3 \text{ keV}$$

Example 7.9   The reaction $^3\text{H}(d, n)^4\text{He}$ has a Q-value of 17.6 MeV. What is the range of neutron energies that may be obtained from this reaction for an incident deuteron beam of 300 keV?

Solution

$$a + x \rightarrow b + y + Q$$

$$Q = E_b \left(1 + \frac{m_b}{m_y}\right) - E_a \left(1 - \frac{m_a}{m_y}\right) - \frac{2}{m_y}\sqrt{m_a m_b E_a E_b} \cos\theta \tag{1}$$

Identify

$$a = d; \qquad x = {}^3\text{H}; \qquad b = n; \qquad y = {}^4\text{He}$$

$$Q = 17.6 \text{ MeV}; \qquad E_a = 0.3 \text{ MeV}$$

• For $E_n(\text{max})$ put $\theta = 0$ in (1) and solve for E_n

$$E_n(\text{max}) = 15.41 \text{ MeV}$$

• For $E_n(\text{min})$ put $\theta = 180°$ in (1) and solve for E_n

$$E_n(\text{min}) = 13.08 \text{ MeV}$$

Thus, the range of neutron energies obtained is 13.08 to 15.41 MeV.

Example 7.10 The cross-section of the $^{207}Pb(n, \gamma)^{208}Pb$ reaction for neutrons of a certain energy is 10^{-24} cm^2. By how much will the intensity of a beam of neutrons of this energy be reduced on passing through 20 cm of ^{207}Pb (density 11 g/cm^3)?

Solution

$$\Sigma = \sigma \frac{Na_v \rho}{A} = \frac{10^{-24} \times 6 \times 10^{23} \times 11}{207} = 0.032 \text{ cm}^{-1}$$

$$\frac{I}{I_0} = e^{-\Sigma x} = e^{-0.032 \times 20} = 0.527$$

The intensity will be reduced to 0.527.

Example 7.11 A target of hydrogen is bombarded with 897.7 MeV ^{181}Ta ions to form ^{182}W in an excited state. Calculate the energy of the excited state (ignore the Coulomb barrier and assume the target nuclei at rest). If ^{182}W in the same excited state were produced by bombarding a ^{181}Ta target with energetic protons, what energy would be needed? The atomic masses in amu are:

$$^1_1H = 1.007825; \qquad ^{181}_{74}Ta = 180.948007; \qquad ^{182}_{74}W = 181.948301$$

Solution If E_{Ta} is the kinetic energy of Ta ions, energy available in the CMS will be

$$W^* = E_{Ta} \frac{m_p}{m_p + m_{Ta}} \tag{1}$$

If the projectile and the target are interchanged, the same excitation energy W^* is available with E_p given by

$$W^* = E_p \frac{m_{Ta}}{m_p + m_{Ta}} \tag{2}$$

Using the masses in (1) $W^* = 4.972$ MeV Combining (1) and (2)

$$E_p = 5.0 \text{ MeV}$$

Example 7.12 Consider a general type of reaction $a + x \rightarrow b + y$. Then the principle of detailed balance gives the result

$$(2S_a + 1)(2S_x + 1)P_a^2 \sigma_{ab} = (2S_b + 1)(2S_y + 1)P_b^2 \sigma_{ba}$$

Apply this principle to the reaction $d + {}^{14}N \rightarrow \alpha + {}^{12}C$, to determine the spin of ^{14}N, given the ratio $\sigma_{dN}/\sigma_{\alpha C}$ at the same CMS energy.
    Atomic masses of ^{14}N, ^2H and ^4He are 14.003074, 2.014102, 4.002603 amu, respectively. 1 amu = 931.44 MeV.

Solution

$$Q = \left[(m_d + m_N) - (m_\alpha + m_c)\right] \times 931.44$$

$$= \left[(2.014102 + 14.003074) - (4.002603 + 12.0)\right] \times 931.44$$

$$= 13.57 \text{ MeV}$$

Forward reaction.

Energy available in the CMS

$$Q + T_F^* = Q + T_d \frac{m_N}{m_N + m_d} = 13.57 + \frac{20 \times 14}{14 + 2} = 31.07 \text{ MeV}$$

The energy of 31.07 MeV is shared between α and ^{12}C. From energy and momentum conservation we find $P_\alpha^* = 416.8$ MeV/c.

The inverse reaction is endoergic, an energy of $31.07 + 13.57 = 44.64$ MeV must be provided. In that case α in the Lab system must have kinetic energy

$$T_\alpha = 44.64 \times \frac{12 + 4}{12} = 59.5 \text{ MeV}$$

In the CMS, the energy of 44.64 MeV is shared between ^{2}H and ^{14}N. The momentum of deuteron would be 383.23 MeV/c

$$\frac{\sigma_{dN}}{\sigma_{\alpha C}} = \frac{(2S_\alpha + 1)(2S_C + 1)}{(2S_d + 1)(2S_N + 1)} \frac{P_\alpha^{*2}}{P_d^{*2}}$$

$$= \frac{1 \times 1}{3 \times (2S_N + 1)} \times \left(\frac{416.8}{383.2}\right)^2 = \frac{1.18}{2S_N + 1}$$

since $S_\alpha = S_C = 0$ and $S_d = 1$. From the experimental ratio of the cross-sections the spin of ^{14}N can be determined. We find $S_N = 1.01$ or 1.

Example 7.13   Neutrons incident on a heavy nucleus with spin $J_N = 0$ show a resonance at an incident energy $E_R = 200$ eV in the total cross-section with a peak magnitude of 1500 b, the observed width of the peak being $\Gamma = 25$ eV. Find the elastic partial width of the resonance.

Solution   Since $J_N = 0$, statistical factor will be absent. At resonance

$$\sigma_{total} = \frac{\lambda^2}{\pi} \frac{\Gamma_n}{\Gamma} = \frac{(0.286)^2}{E\pi} \frac{\Gamma_n}{\Gamma} \times 10^{-16}$$

$$= \frac{(0.286)^2}{200\pi} \frac{\Gamma_n}{25} \times 10^{-16} = 5.21 \times 10^{-22} \Gamma_n$$

$$\therefore \quad \Gamma_n = \frac{\sigma_{total}}{5.21 \times 10^{-22}} = \frac{1500 \times 10^{-24}}{5.21 \times 10^{-22}} = 2.88 \text{ eV}$$

Example 7.14 1.0 g of ^{23}Na of density 0.97 is placed in a reactor at a region where the thermal flux is 10^{11}/cm^2/s. Set up the equation for the production of ^{24}Na and determine the saturation activity that can be produced. The half-life of ^{24}Na is 15 h, and the activation cross-section of ^{23}Na is 536 mb.

Solution If Q is the number of atoms of ^{23}Na at any time t, the rate of production of ^{24}Na is

$$\frac{dQ}{dt} = \phi \Sigma_a - \lambda Q$$

For saturation activity, $\frac{dQ}{dt} = 0$. Then

$$\lambda Q_s = \phi \Sigma_a = \phi \sigma_a \frac{N_{av} \rho}{A}$$

$$= 10^{11} \times 536 \times 10^{-27} \times \frac{6 \times 10^{23}}{23} \times 0.97$$

$$= 1.36 \times 10^9 \text{ s}^{-1}$$

Example 7.15 Suppose 100 mg of gold ($^{197}_{79}$Au) film is exposed to a thermal neutron flux of 10^{10} neutrons/cm^2/s in a reactor. Calculate the activity and the number of atoms of ^{198}Au in the sample at equilibrium. (Thermal neutron activation cross-section for ^{197}Au is 98 b and half-life for ^{198}Au is 2.7 h.)

Solution At equilibrium number of ^{198}Au atoms in m grams

$$Q_s = \frac{\phi \Sigma_a}{\lambda} = \frac{\phi \sigma_a N_{av} m}{0.693 A} T_{\frac{1}{2}}$$

$$= \frac{10^{10} \times 98 \times 10^{-24} \times 6.02 \times 10^{23} \times 0.1 \times 2.7 \times 3600}{0.693 \times 197}$$

$$= 4.2 \times 10^{10}$$

Activity $= Q_s \lambda = \frac{Q_s \times 0.693}{T_{1/2}} = \frac{4.2 \times 10^{10} \times 0.693}{2.7 \times 3600} = 3 \times 10^6$/s.

Example 7.16 For neutrons with kinetic energy 100 MeV incident on nuclei with mass number $A \simeq 120$, the real and imaginary parts of the complex potential are approximately -24 and -8.0 MeV respectively. On the basis of these data, estimate.

(i) The de Broglie wavelength of the neutron inside the nucleus.
(ii) The probability that the neutron is not absorbed in passing diametrically through the nucleus.

Solution

(i)

$$cp = \sqrt{2m(E + U)}$$

$$\lambda = \frac{h}{p} = \frac{2\pi\,\hbar c}{\sqrt{2mc^2(E+U)}} = \frac{2\pi \times 197\ \text{MeV-fm}}{\sqrt{2 \times 939 \times (100+24)}\ \text{MeV}}$$

$$= 2.56\ \text{fm} = 2.56 \times 10^{-15}\ \text{m}$$

(ii)

$$W = \frac{1}{2}\hbar v K = \frac{hc}{2}\left(\frac{v}{c}\right)K$$

$$\frac{v}{c} = \sqrt{\frac{2E}{mc^2}} = \sqrt{\frac{2 \times 100}{939}} = 0.4615$$

$$8.0 = \frac{197}{2} \times 0.4615\,K \quad \rightarrow \quad K = 0.176\ \text{fm}^{-1}$$

$$2R = 2r_0 A^{1/3} = 2 \times 1.3 \times (120)^{1/3} = 12.82\ \text{fm}$$

Probability that the neutron will not be absorbed in passing diametrically through the nucleus

$$= e^{-K2R} = e^{-0.176 \times 12.82} = 0.1$$

Example 7.17 In the reaction, $^{48}\text{Ca} + {}^{16}\text{O} \rightarrow {}^{49}\text{Sc} + {}^{15}\text{N}$, the Q-value is -7.83 MeV. What is the minimum kinetic energy of bombarding ^{16}O ions to initiate the reaction. At this energy, estimate the orbital angular momentum in units of the ions for a grazing collision. Take $R = 1.1A^{1/3}$ fm.

Solution

$$T_{thr} = |Q|\left(1 + \frac{m_O}{m_{Ca}}\right) = 7.83\left(1 + \frac{16}{48}\right) = 10.44\ \text{MeV}$$

$$v = c\sqrt{\frac{2T}{Mc^2}} = 3 \times 10^8 \sqrt{\frac{2 \times 10.44}{931 \times 16}} = 1.123 \times 10^7$$

$$b = R_1 + R_2 = 1.1\left(16^{1/3} + 48^{1/3}\right) \times 10^{-15} = 6.153 \times 10^{-15}\ \text{m}$$

$$J = M_0 vb = n\hbar$$

$$n = \frac{M_0 vb}{\hbar} = \frac{16 \times 1.66 \times 10^{-27} \times 1.123 \times 10^7 \times 6.153 \times 10^{-15}}{1.05 \times 10^{-34}} \simeq 18$$

Example 7.18 In a scattering experiment, an aluminium foil of thickness 10 μm is placed in a beam of intensity 3×10^{12} particles per second. The differential scattering cross-section is known to be of the form

$$\frac{d\sigma}{d\Omega} = A + B\cos^2\theta$$

where A, B are constants, θ is the scattering angle and Ω is the solid angle.

When a detector of area 0.1×0.1 m^2 is placed at a distance of 5 m from the foil, it is found that the mean counting rate is 20.0 s^{-1} when θ is $30°$ and 15.75 s^{-1} when θ is $60°$. Find the values of A and B. The mass number of aluminium is 27 and its density is 2.7 g cm^{-3}.

Solution

$$\frac{d\sigma}{d\Omega} = \frac{I}{I_0 N d\Omega} = A + B \cos^2 \theta$$

$$I_0 = \frac{3 \times 10^{12}}{\text{m}^2\text{-s}} = \frac{3 \times 10^8}{\text{cm}^2\text{-s}}$$

$$d\Omega = \frac{0.1 \times 0.1}{5^2 \text{ m}^2} \text{ m}^2 = 4 \times 10^{-4}$$

$$N = \frac{N_{Av}\rho t}{A} = \frac{6 \times 10^{23} \times 2.7 \times 10 \times 10^{-4}}{27} = 6 \times 10^{19}$$

$$A + B \cos^2 30° = \frac{20}{3 \times 10^8 \times 6 \times 10^{19} \times 4 \times 10^{-4}} = 2.778 \times 10^{-22}$$

$$A + B \cos^2 60° = \frac{15.75}{3 \times 10^8 \times 6 \times 10^{19} \times 4 \times 10^{-4}} = 2.188 \times 10^{-22}$$

Solving the above equations

$$A = 89.3 \text{ b/sr}, \qquad B = 118 \text{ b/sr}$$

Example 7.19 A beam of 460 MeV deuterons impinges on a target of bismuth. Given the binding energy of the deuteron is 2.2 MeV, compute the mean energy, spread in energy and the spread in the angle of the cone in which the neutrons are emitted.

Solution

$$\overline{E}_n = \frac{1}{2}E_d = 0.5 \times 460 = 230 \text{ MeV}$$

$$\Delta E_n = 1.5(B_d E_d)^{\frac{1}{2}} = 1.5(2.2 \times 460)^{\frac{1}{2}} = 47.7 \text{ MeV}$$

$$\Delta\theta = 1.6\left(\frac{B_d}{E_d}\right)^{1/2} = 1.6\left(\frac{2.2}{460}\right)^2 = 0.11 \text{ rad}$$

7.19 Questions

7.1 State the quantities which are conserved in nuclear reactions and those which are not.

7.2 What is the importance of measuring the Q-value of various reactions?

7.3 Draw rough graphs to indicate the behaviour of neutron cross-sections of various types as a function of bombarding energy near the threshold.

7.4 State the reciprocity theorem.

7.5 How can the spin of a particle be determined from the inverse reaction cross-sections?

7.6 How is it known that the emission of a particle from the compound nucleus is independent of the formation of the compound nucleus?

7.7 Under what conditions is the $(1/v)$ law for neutron absorption valid?

7.8 Draw rough graphs to indicate interference effects in elastic scattering of neutrons on nuclei.

7.9 What is the purpose of introducing a complex potential in the optical model?

7.10 State the salient features of direct reactions and compound state reactions.

7.11 State various types of direct reactions.

7.12 How is the kock-out reaction $(p, 2p)$ distinguished from other reactions.

7.13 What are the various types, of phenomena associated with the heavy ion reactions?

7.14 In what way the heavy ion reactions differ from the light ion reactions?

7.15 State the distinguishing features of fusion reactions with heavy ions.

7.16 Describe Blair's model for the scattering of ions with complex nuclei.

7.17 What are pre-equilibrium reactions?

7.18 What is deep inelastic scattering?

7.20 Problems

7.1 ^{13}N is a positron emitter with an end point energy of 1.2 MeV. Determine the threshold of the reaction $^{13}C + P \rightarrow {}^{13}N + n$, if the neutron-hydrogen atom mass difference is 0.78 MeV.
[Ans. 3.23 MeV]

7.2 Consider the reaction

$$^7_3\text{Li} + p \rightarrow ^7_4\text{Be} + n - 1.62 \text{ MeV}$$

Calculate the total energy released in the decay of K capture

$$e^- + ^7_4\text{Be} \rightarrow ^7_3\text{Li} + \nu$$

Calculate the energy carried by ν and ^7Li, respectively. ($M_p c^2 = 938.23$ MeV, $M_n c^2 = 939.52$ MeV, $M_e c^2 = 0.51$ MeV.)
[Ans. $Q = 0.84$ MeV, $E_\nu = 0.84$ MeV, $E_{\text{Li}} = 125$ eV]

7.3 If a target nucleus has mass number 20 and a level at 1.41 MeV excitation, what is the minimum proton energy required to observe scattering from this level?
[Ans. 1.48 MeV]

7.4 The nuclear reaction which results from the incidence of sufficiently energetic α-particles on nitrogen nuclei is, $^4_2\text{He} + ^{14}_7\text{N} \rightarrow ^{17}_8\text{O} + ^1_1\text{H}$. What is the decay product X?

What is the threshold energy required to initiate the above reaction?
(Atomic masses in amu: $^1\text{H} = 1.0081$; $^4\text{He} = 4.0039$; $^{14}\text{N} = 14.0075$; $^{17}_8\text{O} = 17.0045$.)
[Ans. 1.437 MeV]

7.5 Consider the fusion reactions

$$d + d \rightarrow ^3\text{He} + n + 3.27 \text{ MeV}$$
$$d + d \rightarrow ^3\text{H} + p + 4.03 \text{ MeV}$$

(a) Calculate the difference between the binding energy of triton and helium-3 nuclei.
(b) Show that this is approximately the magnitude of coulomb energy due to the two protons of the ^3He nucleus at a distance of 1.87 fm.

[Ans. 0.76 MeV]

7.6 Thermal neutrons are absorbed by $^{10}_5\text{B}$ to form $^{11}_5\text{B}$ which decays by α-emission to Li. Assuming that α-emission takes place from $^{11}_5\text{B}$ at rest, find

(a) the Q-value of the decay of α-particle
(b) the energy of α-particle

(Atomic masses, $^{10}_5\text{B} = 10.01611$ amu, $^1_0 n = 1.008987$ amu, $^7_3\text{Li} = 7.01822$ amu, $^4_2\text{He} = 4.003879$ amu and 1 amu $= 931$ MeV.)
[Ans. 2.79 MeV, 1.78 MeV]

7.7 The end-point of positrons spectrum from the decay of ^{27}Si is found be 3.5 MeV. Find the threshold proton energy for the reaction ^{27}Al$(p, n)^{27}$Si, given that the neutron-proton mass difference is 0.8 MeV.
[Ans. 5.0 MeV]

7.8 Calculate the threshold proton energy for the reaction

$$p + \alpha \rightarrow {}^3\text{He} + d - 18.4 \text{ MeV}$$

[Ans. 23 MeV]

7.9 Calculate the threshold energy for the appearance of the neutrons (a) in the forward direction (b) at 90° for reaction ^{3}H$(p, n)^3$He which has $Q = -0.764$ MeV.
[Ans. (a) 1.019 MeV, (b) 1.146 MeV]

7.10 Consider the reaction ^{115}Sn$(n, r)^{116}$Sn, with thermal neutrons. Calculate the γ-ray energy. (Atomic masses: ^{115}Sn $= 114.903346$ amu, ^{116}Sn $= 115.901745$ amu.)
[Ans. 9.56 MeV]

7.11 The end-point energy of positron spectrum from the decay of $^{11}_{6}$C is found to be 0.98 MeV. Calculate the difference between the rest mass energy of $^{11}_{6}$C and $^{11}_{5}$B.
[Ans. 2.0 MeV]

7.12 In the reaction ^{2}H$(^2$H, ^{3}He$)n$, $Q = 3.26$ MeV. Calculate the mass of neutrons in amu. (Atomic masses: ^{2}H $= 2.014102$ amu and ^{3}He $= 3.016030$ amu.)
[Ans. 1.008675 amu]

7.13 Find the Q-value of the reaction

$$^{30}\text{Si} + d \rightarrow {}^{31}\text{Si} + p + Q \tag{1}$$

given

$$^{30}\text{Si} + d \rightarrow {}^{31}\text{P} + n + 5.10 \text{ MeV} \tag{2}$$

$$^{31}\text{Si} \rightarrow {}^{31}\text{P} + \beta^- + 1.51 \text{ MeV} \tag{3}$$

$$n \rightarrow p + \beta^- + 0.78 \text{ MeV} \tag{4}$$

[Ans. 4.37 MeV]

7.14 The nucleus ^{12}C has an excited state at 4.43 MeV. You wish to investigate whether this state can be produced in inelastic scattering of protons through 90° by a carbon target. It you have access to a beam of protons of kinetic energy 20 MeV, what is the kinetic energy of the scattered protons for which you must look?
[Ans. 12.83 MeV]

7.15 A beam of 2 MeV neutrons is used to give the reaction $^{14}_{7}N + ^{1}_{0}n \rightarrow ^{11}_{5}B + ^{4}_{2}He$. Determine

(a) The threshold energy of this reaction
(b) The maximum energy of the α-particles

Given the atomic masses in amu: $^{14}_{7}N = 14.003074$; $^{1}_{0}n = 1.008665$; $^{4}_{2}He = 4.002603$; $^{11}_{5}B = 11.009305$; amu $= 931.6$ MeV.
[Ans. 169 keV, 1.68 MeV]

7.16 Calculate the energy of proton ejected in the forward direction when 51 MeV gamma rays undergo elastic scattering with hydrogen.
[Ans. 0.5 MeV]

7.17 Calculate the thickness of Indium foil which will absorb 2 % of neutrons incident at the resonance energy for indium (1.44 eV) where $\sigma = 28000$ b. At wt of Indium $= 114.7$ amu, density of Indium $= 7.3$ g/cm^3.
[Ans. 93 μm]

7.18 Show that the de Broglie wavelength for neutron is given by the formula $\lambda = \frac{0.286}{\sqrt{E}}$ Å, where the kinetic energy E is in electron volts.

7.19 Calculate the thickness of ^{113}Cd required to reduce the beam of neutrons to 0.1 % of the original intensity. Density of cadmium is 8.67 g/cm^3 and $\sigma_a = 20800$ b.
[Ans. 71.8 μm]

7.20 Protons of energy 5 MeV scattering from $^{10}_{5}B$ at an angle of 45° show a peak in the energy spectrum of the scattered protons at an energy of 3.0 MeV

(a) To what excitation energy of $^{10}_{5}B$ does this correspond?
(b) What is the expected energy of the scattered protons if the scattering is elastic?

[Ans. (a) 1.45 MeV, (b) 4.2 MeV]

References

1. H.H. Barschall, Phys. Rev. **86**, 431 (1952)
2. Bender et al.
3. F.E. Bertrand, R.W. Peele, Phys. Rev. C **8**, 1045 (1973)
4. J.S. Blair, Phys. Rev. **95**, 1218 (1954)
5. Blatt, Weisskoph
6. Bohr (1936)
7. P.D. Bond et al., Phys. Lett. B **47**, 231 (1973)
8. Bromley (1974)
9. D.A. Bromley, J.A. Euehner, E. Almquist, in *Proceedings of the Second Conference in Reactions Between Complex Nuclei* (1960)
10. Brookhaven National Laboratory

11. Brostrom, Huus, Tangent
12. B. Buck, Phys. Rev. **130**, 712 (1963)
13. S.T. Butler, Proc. R. Soc. Lond. A **208**, 559 (1951)
14. Chiavassa et al. (1994)
15. H. Feshbach, C.E. Porter, V.F. Weisskopf, Phys. Rev. **96**, 448 (1954)
16. Fink et al. (1960)
17. Fugiwara et al. (World Scientific, 1995)
18. Fujiwara et al. (1995)
19. S.N. Ghoshal, Phys. Rev. **80**, 939 (1950)
20. Glendenning (1974)
21. G.E. Gordon, A.E. Larsh, T. Sikkeland, from E. Hyde, UCRL 9065
22. Griffin, (1966)
23. G. Jacob, A.J. Maris, Rev. Mod. Phys. **45**, 6 (1966)
24. A.A. Kamal, *Particle Physics* (Springer, Berlin, 2014)
25. M.J. Levine et al., Phys. Rev. C **10**, 1602 (1974)
26. B.E.F. Macefield, R. Middleton, D. Pullen, J. Nucl. Phys. **44**, 309 (1963)
27. Schroder, Huizenga (1984)
28. W.V. Schrröder, J.R. Huizenga, in *Treatise on Heany-Ion science* ed. D.A. Bromley, 2, p. 115 (1984)
29. Serber (1947)
30. K.S. Toth et al., Phys. Rev. C **14**, 1471 (1976)
31. von Witsch et al. (1968)

Chapter 8
Nuclear Power

8.1 Nuclear Fission Reactor

It is estimated that in near future the available conventional power, i.e. the power harnessed from coal, oil, hydro-electricity etc would become totally inadequate to match the ever increasing demands. This grim situation, therefore, calls for the exploitation of new types of power, e.g. the nuclear power obtained from the nuclear reactors, both fission and fusion.

The basic idea of a fission reactor is that if a fissile material like U^{235} undergoes fission, then 2 or 3 neutrons are produced in each fission process, and if by some trick we could minimize the leakage of neutrons from the assembly called pile or reactor, and reduce the losses of neutrons in non-fissionable processes (radiative, capture, i.e. absorption of neutrons leading to production of γ-rays) then it is possible that the neutrons thus produced will cause further fissions in their encounters with other nuclei of the fuel element, the number of neutrons in one generation being at least equal to the number in the previous generation; and if this process called the Chain Reaction, is sustained for sufficiently long time, then we have a convenient way of obtaining power from the reactor. This is so because a very large number of fissions take place per second, and in each fission, the fission fragments carry bulk of the kinetic energy (85 %), which means that heat is produced at a constant rate. In principle, the reactor will continue to work until the fuel is exhausted or the assembly rendered 'poisonous' owing to the constant accumulation of fission products which merely absorb the neutrons via non-fissionable processes.

The direct physical consequence of the operation of a reactor is that heat is produced. The reactor is then nothing more than a furnace but capable of yielding much greater power. This is because in nuclear reactions the energy released is of the order of million times greater than in chemical reactions. The remaining problem is that of extraction of heat with high efficiency, and to shield it suitably so as to minimize the health hazards, and above all to keep it under control so that the neutron production and hence the nuclear power does not exceed the danger level.

The first reactor was built by Enrico Fermi at the Chicago university in 1942 which produced a small power of 0.5 W, and later raised to 200 W. Today we have reactors capable of yielding power up to 500 Mega Watts. The smaller ones give 1 to 5 Mega Watts and are mostly used for research work and to produce isotopes. The power reactors are currently in use for the production of electricity and for driving submarines and ships. In the early days, the 'pile' was used for reactor or assembly. Reactors may be divided into two classes (1) thermal reactors in which fissions take place mainly at thermal energy ($\sim$0.025 eV), (2) fast reactors in which fissions take place at a few MeV neutron energy.

Reactors may be divided into homogeneous and heterogeneous types. In the former, the moderator is homogeneously mixed, while in the latter the fuel in the form of lumps is embedded heterogeneously in the moderator. More details are given in Sect. 8.19. First, we shall consider the thermal reactors.

8.2 The Thermal Reactor

8.2.1 Moderation of Neutrons

The neutrons produced from fission are relatively fast (1 to 2 MeV). Now, if the fuel element used is U^{235} then, σ (fission) for neutrons of energy 1 to 2 MeV is quite small ($\sim$1.5 b) where as for neutrons of small energy (say 0.025 eV), it is quite large ($\sim$600 b). It is therefore, necessary to moderate the neutrons (slow down). This can be accomplished if the neutrons in their traversal through the medium suffer a number of elastic collisions. But, if the scattering takes place with the uranium nuclei, then the average energy loss per collision is so small that an unduly large number of collisions are required (about 2000) to slow them down to the desired energies, and by this time the neutrons would have wandered off outside the reactor and thus lost for setting up the chain reaction. This then means that some other lighter element must be incorporated in the reactor in order to slow down the neutrons after reasonably small number of collisions. An element or a mixture used for this purpose is called a *moderator*.

In what follows, we shall study the simple kinematics of elastic scattering. Let the mass of neutron be unity and that of the target nucleus A. The target nucleus is assumed to be initially at rest. This is justified since the small energy of the scatterer is negligible in comparison with neutron energy until neutrons have reached thermal energies (i.e. of the order of 0.025 eV corresponding to kT, where k is the Boltzmann constant and T is the absolute temperature of the medium), at which point the gain and loss of energy in the scattering collisions with the target nuclei would be equally probable. This state of affairs is called thermal equilibrium.

Let v_0 be the velocity of neutron and E_0 its kinetic energy before collision in the Lab System (LS). In the Center of Mass System (CMS) i.e. a system in which the total linear momentum is equal to zero, the neutron will be seen to approach with velocity $\frac{Av_0}{(A+1)}$ along the incident direction, while the target nucleus would be moving with velocity $\frac{v_0}{(A+1)}$ in the opposite direction. The CMS itself moves with velocity $v_c = v_0/(A + 1)$ along the incident direction.

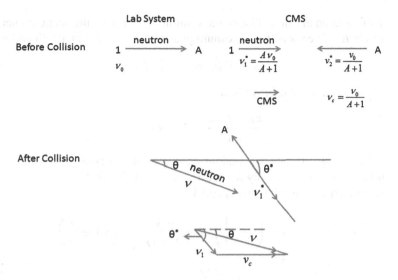

Fig. 8.1 Velocity of neutron before and after collision in LS and CMS

After the collision, let the neutron be scattered at an angle θ in the LS and be moving with velocity v and kinetic energy E. Let the corresponding scattering angle in the CMS be θ^* (Fig. 8.1). Since the scattering is assumed to be elastic, kinetic energy is conserved. It follows that after the collision, neutron will be moving with the same velocity (v_1^*) in the CMS as before the collision. From the diagram, it is seen that the velocities v_1^* and v_c combine vectorially to yield v. This gives us

$$v^2 = v_1^{*2} + v_c^2 + 2v_1^* v_c \cos \theta^*$$

Substituting for v_1^* and v_c, we have

$$v^2 = \frac{(Av_0)^2}{(A+1)^2} + \frac{(v_0)^2}{(A+1)^2} + \frac{2\cos\theta^* Av_0^2}{(A+1)^2}$$

$$\therefore \quad \frac{E}{E_0} = \frac{v^2}{v_0^2} = \frac{A^2 + 2A\cos\theta^* + 1}{(A+1)^2} \tag{8.1}$$

For a glancing collision

$$\theta^* = 0 \quad \text{and} \quad E_{\max} = E_0 \tag{8.2}$$

For a head-on collision

$$\theta^* = \pi \quad \text{and} \quad E_{\min} = \alpha E_0 \tag{8.3}$$

where

$$\alpha = \frac{(A-1)^2}{(A+1)^2} \tag{8.4}$$

It follows that in an elastic collision, the minimum energy of the scattered neutron is equal to $\propto E_0$. Therefore, the maximum possible energy loss in a collision is

$$E_0 - E_{min} = (1 - \alpha)E_0 \tag{8.5}$$

The maximum fractional energy loss is

$$\frac{E_0 - E_{min}}{E_0} = 1 - \alpha = \delta \tag{8.6}$$

Thus, in the scattering with the nuclei of Carbon ($A = 12$), $\delta = 0.28$, i.e. 28 %, while the corresponding quantity for hydrogen ($A = 1$) is 100 %.

For large A, it is convenient to expand δ as

$$\delta = 1 - \alpha = 1 - \left(\frac{A-1}{A+1}\right)^2 = \frac{4A}{(A+1)^2} = \frac{4}{A}\left(1 + \frac{1}{A}\right)^{-2}$$

$$\simeq \frac{4}{A}\left(1 - \frac{2}{A} + \cdots\right) = 4/A \tag{8.6a}$$

We note that for heavy target nuclei (large A), the maximum fractional loss of energy is small. Thus, for example, if $A = 200$, $\delta = 4$ %, whilst for $A = 100$, $\delta = 2$ %. We conclude that the smaller A, the larger is δ, the maximum fractional energy loss and vice versa.

We shall now show that if the neutron scattering (at a fixed incident energy E_0) is isotropic in the CMS the energy distribution in the LS is rectangular (uniform) i.e independent of the energy E of the scattered neutrons.

Let $n(E)$ neutrons scattered between the angles θ^* and $\theta^* + d\theta^*$ in the CMS appear with energy E and $E + dE$ in the LS. Then we can write

$$n(E)dE = -\frac{2\pi \sin\theta^* d\theta^*}{4\pi} = -\frac{1}{2}\sin\theta^* d\theta^* \quad \text{or} \tag{8.7}$$

$$n(E)dE = \frac{1}{2}d\cos\theta^* \tag{8.8}$$

The negative sign on the RHS of (8.7) is introduced because an increase in θ^* implies a decrease in E. Differentiating (8.1)

$$\frac{dE}{E_0} = \frac{2Ad\cos\theta^*}{(A+1)^2} \tag{8.9}$$

Eliminating $d\cos\theta^*$ between (8.8) and (8.9), we get

$$n(E)dE = \frac{(A+1)^2}{4A}\frac{dE}{E_0} = \frac{dE}{(1-\infty)E_0} \tag{8.10}$$

The last equation shows that the energy spectrum of the scattered neutrons is independent of the energy E, the spectrum extending uniformly from the minimum value $E_{min} = \alpha E_0$, to the maximum value E_0 (Fig. 8.2).

Fig. 8.2 Energy spectrum
extending uniformly for
different values of E

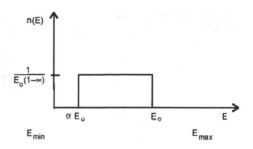

It is readily seen that $\int_{E_{min}}^{E_{max}} n(E|)dE = 1$, a result which is expected since the
distribution was assumed to be normalized.

The height of the distribution is easily seen to be equal to $\frac{1}{E_0(1-\alpha)}$ since the area
under the spectrum must be equal to 1.

8.2.2 The Average Energy Decrement

It is of importance to know the average energy loss per collision. It turns out that
the quantity $\langle E_0/E \rangle$ is not suitable since the fluctuations about this quantity would
be very large. We therefore choose energy on the logarithmic scale and introduce
another variable ξ defined by

$$\xi = \left\langle \ln \frac{E_0}{E} \right\rangle \tag{8.11}$$

called the average logarithmic energy decrement

$$\xi = \int_{\alpha E_0}^{E_0} \ln \frac{E_0}{E} dE \Big/ \int_{\alpha E_0}^{E_0} dE = -\frac{1}{1-\alpha} \int_{\alpha E_0}^{E_0} \ln \frac{E}{E_0} d\frac{(E)}{E_0} = 1 + \frac{\alpha \ln \alpha}{1-\alpha} \tag{8.12}$$

Substituting (8.4) for α in (8.12)

$$\xi = 1 + \frac{(A-1)^2}{2A} \ln \frac{A-1}{A+1} \tag{8.13}$$

If $A \gg 1$, then:

$$\ln \frac{A-1}{A+1} = \ln \left(1 - \frac{1}{A}\right)\left(1 + \frac{1}{A}\right)^{-1} \simeq \ln \left(1 - \frac{2}{A}\right) \simeq -2/A$$

and

$$\xi \simeq \frac{2A-1}{A^2} \tag{8.14}$$

We note that ξ depends only on A, the target mass and is independent of the initial
energy of the neutron. Thus on an average neutron always loses the same fraction of

Fig. 8.3 Logarithmic scale
of energy loss

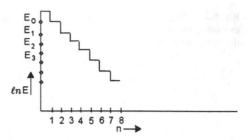

Table 8.1 Various elements
with number of collisions and
their corresponding values
of ξ

Element	A	ξ	n
H	1	1	18
D	2	0.73	25
Li	7	0.27	67
C	12	0.165	114
O	16	0.143	127
U	238	0.0084	2168

energy it had before the collision. This is true as long as the scattering is isotropic
in the CMS. When the neutron energy is comparable with the thermal energy of
the moderator, neutron is likely to gain as much as lose energy in a collision, i.e.
remain in thermal equilibrium, until it is captured. For substances other than the
monoatomic gases, the formula breaks down even before the region of thermal energies owing to the binding energy of the atoms. In fact at energies of the order of
0.3 eV, the collisions can not be considered as elastic.

Since in each collision, the average energy loss on logarithmic scale is constant
(Fig. 8.3), it follows that the number of collisions needed to reduce the initial energy
E_0 to final energy E_n is given by

$$n = \frac{1}{\xi} \ln \frac{E_0}{E_n} \tag{8.15}$$

Thus, the number of collisions required to thermalize neutrons of energy $E_0 =$
2 MeV, in graphite ($A = 12$), $n = \frac{1}{0.16} \ln[\frac{2 \times 10^6}{0.025}] = 114$, where we have assumed
$E_n = 0.025$ eV and $\xi = 0.165$ obtained by putting $A = 12$ in (8.13).

Table 8.1 shows n the number of collisions required on an average to thermalize
neutrons in various targets ($E_0 = 2$ MeV, $E_n = 0.025$ eV). Also, are indicated the
corresponding values of ξ.

The greater is the value of ξ, the smaller is the number of collisions required to
thermalize neutrons. But, this is not enough. The cross-section for scattering must
be large so that scattering may take place with appreciable probability. At the same
time, the absorption cross-section should be small so that too many neutrons are not

lost. The quantity

$$\phi = \frac{\xi \Sigma_s}{\Sigma_a} \tag{8.16}$$

called moderating ratio is the best practical measure of the effectiveness of the moderator. Here, Σ_s and Σ_a are the macroscopic scattering and absorption cross-sections respectively. In case the moderator is a mixture or a compound then the moderating ratio is defined as

$$\phi = \frac{\xi_1 \Sigma_{s1} + \xi_2 \Sigma_{s2} + \cdots}{\Sigma_{a1} + \Sigma_{a2} + \cdots} \tag{8.17}$$

8.2.3 Forward Scattering

It is of interest to know the departure from isotropic scattering in the LS since this has a bearing on the mean distance the neutrons drift away from the source as compared with isotropic scattering. The quantity, $\overline{\cos}\,\theta$ is a direct measure of the departure from spherical symmetry in the LS.

From Fig. 8.1, we note that

$$v \cos \theta = \frac{v_0}{A+1} + \frac{A v_0 \cos \theta^*}{A+1} = \frac{v_0}{A+1}(1 + A \cos \theta^*) \tag{8.18}$$

$$v \sin \theta = \frac{A v_0 \sin \theta^*}{A+1} \tag{8.19}$$

Dividing (8.19) by (8.18), we get

$$\tan \theta = \frac{A \sin \theta^*}{A \cos \theta^* + 1}$$

whence we find

$$\cos \theta = \frac{1 + A \cos \theta^*}{\sqrt{1 + 2A \cos \theta^* + A^2}}$$

Assuming that scattering is isotropic in the CMS, we find:

$$\overline{\cos}\,\theta = \int_0^\pi \cos\theta \frac{1}{2} \sin\theta^* d\theta^* = \frac{1}{2}\int_{-1}^{+1} \frac{(1 + A \cos\theta^*)d(\cos\theta^*)}{(1 + 2A\cos\theta^* + A^2)^{1/2}}$$

A straight forward integration yields

$$\overline{\cos}\,\theta = 2/3A \tag{8.20}$$

If A is large, then $2/3A \rightarrow 0$ and the scattering is very nearly isotropic in the LS. This is merely a consequence of the fact that the CMS velocity would be very small

Fig. 8.4 Moderation of
neutrons by elastic collisions
with moderator nuclei

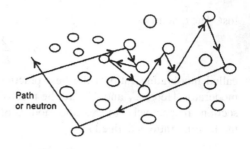

Path
or neutron

so that the spherical symmetry in the CMS is preserved in the Lab system. Any
positive value of $\overline{\cos}\,\theta$ would imply excess forward scattering in the LS. Even for
carbon nucleus, $\overline{\cos}\,\theta = 0.056$, and the scattering is roughly isotropic. For hydrogen,
$\overline{\cos}\,\theta = 2/3$, and the scattering is strongly in the forward direction.

8.3 Thermal Neutrons

It was pointed out that neutrons produced from fission are fast, their energy being
a few MeV. However, as a result of elastic collisions with the nuclei of the moderator which is incorporated in the thermal reactor, their energy is soon reduced.
The slowing down process continues until the average kinetic energy of neutrons is
of the order of the thermal energy of the moderator nuclei (Fig. 8.4). The neutrons
are said to be in thermal equilibrium with their immediate surroundings. When this
state of affairs is reached, neutrons are said to be thermalized, and their energy distribution will be approximately Maxwellian, corresponding to the temperature of
the medium.

The velocity distribution will be given by the Maxwell-Boltzmann expression

$$dn = n(v)dv = \frac{4\pi n_0 v^2 e^{-\frac{1}{2}mv^2/kT}\,dv}{(2\pi kT/m)^{3/2}} \qquad (8.21)$$

where, $dn = n(v)dv =$ number of neutrons with speed between v and $v + dv$.

$n_0 =$ total number of neutrons
$m =$ neutron mass
$T =$ Absolute temperature
$k =$ Boltzmann constant

$$= 1.38 \times 10^{-23}\,\mathrm{J\,K^{-1}}, \quad \text{or} \quad 8.55 \times 10^{-5}\,\mathrm{eV\,K^{-1}}$$

The velocity distribution is shown in Fig. 8.5. By maximizing (8.21), we find $v_{\max}$
the most probable value of v

$$v_{\max} = (2kT/m)^{1/2} \qquad (8.22)$$

The energy E_p that corresponds to this velocity is given by

$$E_p = \frac{1}{2}mv_{max}^2 = kT \tag{8.23}$$

The average speed $\bar{v}$ is given by

$$\bar{v} = \frac{\int_0^\infty vn(v)dv}{\int_0^\infty n(v)dv} = (8kT/\pi m)^{\frac{1}{2}} \tag{8.24}$$

It follows that

$$\bar{v} = \frac{2v_{max}}{\sqrt{\pi}} \tag{8.25}$$

Another quantity of interest for the speed distribution is the root-mean-square speed (v_{rms}). It is obtained from

$$v_{rms}^2 = \frac{\int_0^\infty v^2 n(v)dv}{\int_0^\infty n(v)dv} = \frac{3kT}{m} \tag{8.26}$$

whence

$$v_{rms} = \left(\frac{3kT}{m}\right)^{1/2} \tag{8.27}$$

Thus

$$v_{max} : \bar{v} : v_{rms} :: 1 : \frac{2}{\sqrt{\pi}} : \sqrt{3/2} \tag{8.28}$$

for the Maxwellian distribution.

The energy distribution, i.e. the number to be found between E and $E + dE$, is given by

$$dn = n(E)dE = \frac{2\pi n_0 E^{\frac{1}{2}} e^{-E/kT}}{(\pi kT)^{3/2}} dE \tag{8.29}$$

The energy distribution is shown in Fig. 8.6. The number of neutrons that are found within a small velocity or energy interval, given by (8.21) and (8.29) respectively, are shown in Figs. 8.5 and 8.6.

The average energy is given by

$$\bar{E} = \frac{\int_0^\infty E dn}{\int_0^\infty dn} = 3kT/2 \tag{8.30}$$

The most probable value of E is obtained by maximizing expression (8.29), and occurs at

$$E_{max} = \frac{1}{2}kT \tag{8.31}$$

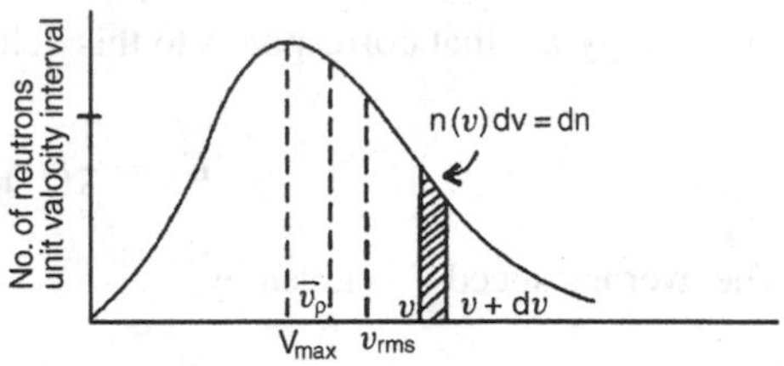

Fig. 8.5 Velocity distribution of neutrons according to Maxwell-Boltzmann distribution

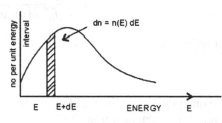

Fig. 8.6 Maxwell-Boltzmann energy distribution

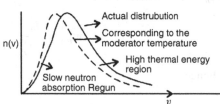

Fig. 8.7 The neutron energy distribution

The most probable value of velocity occurs at a value of $E = kT$, which is actually twice that given by (8.31). The energy distribution of neutrons is closely given by the Maxwellian distribution at intermediate energies (Fig. 8.7) but deviates markedly at lower energies because of strong absorption, and also at higher energies because of continual production of those neutrons by the slowing down of fission neutrons. At room temperature, say 30 °C, the average energy $\overline{E} = 3kT/2 = 3 \times 8.55 \times 10^{-5} \times 300/2 = 0.04$ eV. Neutrons in the neighborhood of this energy are called thermal neutrons.

If the neutron absorption obeys $1/v$ law, then the absorption cross section may be expressed by $\sigma_a = c_0/v$, where $c_0 = \text{const}$ and the average cross section $\overline{\sigma}_a$ is given by

$$\overline{\sigma}_a = \frac{\int_0^\infty \sigma_a(E)\phi(v)dv}{\int_0^\infty \phi(v)dv} = \frac{\int_0^\infty \sigma_a(E)n(v)vdv}{\int_0^\infty n(v)vdv} \tag{8.32}$$

where, ϕ is the neutron flux defined by $\phi = n(v)v$.

We, therefore, obtain

$$\overline{\sigma}_a = \frac{c_0 \int_0^\infty n(v)dv}{\int_0^\infty vn(v)dv} = \frac{c_0}{\overline{v}} \tag{8.33}$$

where, we have used (8.24). This shows that the effective cross-section for neutrons is equal to the cross-section with speeds equal to the average speed for thermal neutron distribution. The constant c_0 is conventionally taken with reference to the most probable velocity v_{max} for the thermal neutrons so that,

$$c_0 = v_{max}\sigma_{max} \tag{8.34}$$

where, σ_{max} is the cross-section for neutrons of speed v_{max}. It follows that

$$\bar{\sigma} = (v_{max}/\bar{v})\sigma_{max} = \frac{\sqrt{\pi}}{2}\sigma_{max} \tag{8.35}$$

The effective cross-section is smaller than the cross-section for the most probable velocity by the factor $\sqrt{\pi}/2 = 0.886$ since the average speed for Maxwellian distribution is greater than the most probable speed by the same factor.

The thermal neutron cross-sections are normally quoted for the speed 2200 m/s which corresponds to the most probable speed of a Maxwellian neutron distribution at 293.6 K and corresponds to a neutron energy of 0.0253 eV.

The effective neutron cross-section to be used in reactor calculations for a neutron distribution at a temperature T is given by

$$\sigma = \frac{\sqrt{\pi}}{2}(293.6/T)^{\frac{1}{2}}\sigma_{max}(2200)g \tag{8.36}$$

where, we used (8.35) and (8.32); g is a correction factor, close to unity, and is called 'Not $1/v$ factor'; it corrects for the fact that the $1/v$ absorption law may not be strictly valid.

8.4 Scattering Mean Free Path (M.F.P.)

Let $p(x)$ be the probability that the neutron does not scatter after having travelled a distance x. Then, if λ_s is the mean free path for scattering, the probability of scattering in a distance dx is given by dx/λ_s. Therefore the probability that the neutron does not scatter between x and $x + dx$ is given by $(1 - dx/\lambda_s)$. Now the probability that the neutron does not scatter in $(0, x + dx)$ is given by the product of the probability that it does not scatter in $(0, x)$ and the probability that it does not scatter in $(x, x + dx)$ and is equal to the product of these probabilities since the probabilities are independent, i.e.

$$p(x + dx) = p(x)p(dx) = p(x)(1 - dx/\lambda_s)$$

Expanding the LHS by Taylor's theorem up to two terms

$$p(x) + \frac{dp(x)}{dx}dx = p(x) - p(x)dx/\lambda_s \quad \text{or}$$

$$\frac{dp(x)}{p(x)} = -dx/\lambda_s$$

Direct integration gives

$$\ln p(x) = -\frac{x}{\lambda_s} + \ln C$$

where, $\ln C$ is the constant of integration

$$\therefore \quad p(x) = Ce^{-x/\lambda_s}$$

Now, at $x = 0$, $p(0) = 1$ since neutron certainly would not scatter. This gives us, $c = 1$; we then get

$$p(x) = e^{-x/\lambda_s} \tag{8.37}$$

for the probability that the neutron does not scatter in distance x. The probability that the neutron scatters in a distance x is then

$$p_s(x) = \left(1 - e^{-x/\lambda_s}\right) \tag{8.38}$$

$\therefore$ $dp_s(x) = \frac{1}{\lambda_s} e^{-\frac{x}{\lambda_s}} dx$ is the probability for it to scatter in the interval $(x, x + dx)$. Note that

$$p(x + dx) = e^{-(x+dx)/\lambda_s} = e^{-x/\lambda_s} e^{-dx/\lambda_s} = e^{-x/\lambda_s}\left(1 - \frac{dx}{\lambda_s}\right)$$

$$= p(x)p(dx)$$

(neglecting higher power of dx), as required. It is readily seen that the mean distance of scattering is equal to λ_s

$$\bar{x} = \int_0^\infty x \, dp_s(x) = \int_0^\infty \frac{x}{\lambda_s} e^{-x/\lambda_s} dx = \lambda_s \tag{8.39}$$

8.4.1 Transport Mean Free Path

Because of the predominant forward scattering in the Lab System, the average distance travelled by the neutron from the source in a given number of collisions would be more than that for isotropic scattering. The increased effective mean free path for non-isotropic scattering is called the Transport mean free path λ_{tr}, and may be defined as the mean distance travelled between collisions by a neutron projected on to the direction of its original motion (Fig. 8.8)

$$\lambda_{tr} = \lambda_s + \lambda_s \overline{\cos\theta} + \lambda_s \overline{\cos\theta}\,\overline{\cos\theta} + \cdots$$

$$= \lambda_s(1 + \overline{\cos\theta} + (\overline{\cos\theta})^2 + (\overline{\cos\theta})^3 + \cdots$$

$$= \frac{\lambda_s}{1 - \overline{\cos\theta}} \quad \text{or}$$

Fig. 8.8 Transport mean free
path

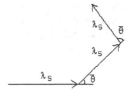

$$\lambda_{tr} = \frac{\lambda_s}{1 - 2/3A} \qquad (8.40)$$

where we have used (8.20).

8.4.2 The Mean Square Distance of Scattering

The mean distance of scattering is given by,

$$\langle x^2 \rangle = \int_0^\infty \frac{x^2}{\lambda_s} e^{-x/\lambda_s} dx = 2\lambda_s^2 \qquad (8.41)$$

For isotropic scattering, after n collisions, the net mean distance square is

$$\langle R^2 \rangle = n\langle x^2 \rangle = 2n\lambda_s^2 \qquad (8.42)$$

However, in the Lab System, the forward direction is preferred and Eq. (8.42) is
modified as follows:

$$\langle R^2 \rangle = \frac{2}{1 - 2/3A} \left(\frac{1}{\xi} \ln E_0/E_n \right) \lambda_s^2 \qquad (8.43)$$

8.5 Slowing-Down Density

Slowing-down density, $q(E)$ is defined as the rate at which neutrons per unit volume
of a moderater slow down past a particular energy E. Since the energy loss in elastic scattering is discontinuous, and because of absorption resonances for neutrons
of discrete energies, the variation of q with energy is not expected to be simple.
However, the problem is rendered simpler if the energy loss per collision is small
as would be the case for sufficiently heavy moderator so that slowing down process
may be regarded as virtually a continuous process. In what follows we shall assume
the validity of this model. The assumption of continuous slow-down would be approximately correct for graphite as a moderator but would be grossly incorrect for
hydrogen.

We assume that fission neutrons are produced throughout the assembly and that a
steady state has reached, which means that the rate at which neutrons enter an energy

Fig. 8.9 Slowing-down
density is the rate at which
the neutron density crosses
the energy E

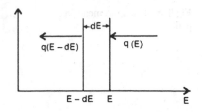

interval between E and $E + dE$ is equal to the rate at which they leave it (Fig. 8.9).
Supposing that the rate of production of neutrons with an initial high energy E_0 is
constant and equal to Q, then in the absence of losses of neutrons through absorption
or through leakage (escape from the assembly), before they have reached thermal
energies, the slowing-down density $q(E)$ will remain constant at all energies and
will be equal to Q. The leakage can be reduced to negligible proportion by use of
large amount of moderator material, so that neutron absorption will remain the only
loss of neutrons.

We shall first assume that $\Sigma_a = 0$, i.e the absorption is absent. We can then write,

$$q(E) = Q = \text{const} \tag{8.44}$$

Now, the number of neutrons that lie in the energy interval E and $E + \Delta E$ per cm^3
is given by (8.29) and is equal to $n(E)\Delta E$. If Δt is the time that is required in
their transit through the energy interval ΔE, the number of neutrons that are present
within the energy interval E and $E + \Delta E$ is given by the product of the rate of flow
$q(E)$ and time interval Δt, i.e.

$$-q(E)\Delta t = n(E)\Delta E \tag{8.45}$$

In the above equation, negative sign is introduced because with increasing t, E decreases.

The neutron density that crosses the energy E per second, $q(E)$ is also given
by the product of the neutron density that crosses E per collision with an energy
loss ΔE per collision, and the number of collisions this neutron group suffers per
second. But neutron density crossing E per collision $= n(E)\Delta E$; and number of
collisions per second $= v/\lambda_s = v\Sigma_s$

$$\therefore \quad q(E) = \left[n(E)\Delta E\right]\left[v\Sigma_s\right] \tag{8.46}$$

If ΔE is chosen sufficiently small, we can set

$$\frac{\Delta E}{E} = \Delta(\ln E) = \xi$$

whence we obtain

$$\Delta E = \xi E \tag{8.47}$$

Substituting ΔE from (8.47) in (8.46), we find

$$q(E) = n(E)E\xi v\Sigma_s = \phi(E)E\xi\Sigma_s \tag{8.48}$$

where, $\phi(E) = n(E)v$ is the neutron flux per unit energy. Since in the absence of neutron absorption, $q(E) = Q$, we can write

$$\phi(E) = \frac{Q}{E\xi\,\Sigma_s} \tag{8.49}$$

so that slowing down neutron flux per unit energy interval is proportional to the energy E. If (8.49) is recast as follows:

$$\phi(E)\Sigma_s\,\Delta E = \frac{Q\Delta E}{E\xi} \tag{8.50}$$

then the left side of (8.50) represents the number of neutrons within the energy interval ΔE (per cm^3) that undergo scattering collisions with the moderater nuclei and are scattered out of it per s. The quantity $\phi(E)\Sigma_s$ is called the collision density.

If a steady state prevails then the number of neutrons scattered out of the energy interval ΔE per s per cm^3 must be equal to the number of neutrons scattered into this energy interval per s: per cm^3; i.e.

Scattering loss per cm^3 per s = Neutron influx gain per cm^3 per s

As the left side of (8.50) has already been identified with scattering loss, it follows that the right side must represent the rate of neutron influx.

8.5.1 Slowing-Down Time

Combination of (8.45) and (8.48) yields T, the slowing-down time, which is the time required for the neutrons to slow down from an initial energy E_0 to final energy E_f. We, therefore, have

$$\Delta t = -\frac{\Delta E}{E\xi v\Sigma_s} = -\sqrt{m/2}\frac{\Delta E}{\xi\,\Sigma_s E^{3/2}} \tag{8.51a}$$

If we now assume that ξ and Σ_s remain substantially constant over the slowing down range of energies, we can integrate (8.51a) to obtain the slowing down time T

$$T = \int_0^T \Delta t = -\sqrt{\frac{1}{2}m}\frac{1}{\xi\,\Sigma_s}\int_{E_0}^{E_f} E^{-3/2}\Delta E = \frac{\sqrt{2m}}{\xi\,\Sigma_s}\left[E_f^{-\frac{1}{2}} - E_0^{-\frac{1}{2}}\right] \tag{8.51b}$$

As an example, we may calculate the slowing-down of 2 MeV neutrons to thermal energies (0.025 eV) in graphite. We accept $\xi = 0.158$, and $\sigma_s = 4.8$ b

$$\therefore \quad \Sigma_s = \frac{6.03 \times 10^{23} \times 4.8 \times 10^{-24} \times 1.62}{12} = 0.385 \text{ cm}^{-1}$$

Fig. 8.10 Neutron absorption cross-section as a function of energy for ^{238}U; the peaks corresponding to resonances are clearly indicated

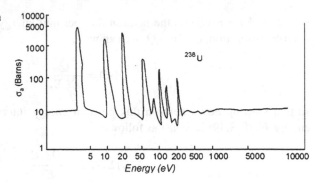

Also, since $E_0 \gg E_f$, the second term in the parenthesis in (8.51b) can be neglected

$$T = \frac{(2mc^2/0.025)^{1/2}}{0.158 \times 0.385 \times c} = \frac{(2 \times 940 \times 10^6/0.025)^{1/2}}{0.385 \times 0.158 \times 3 \times 10^{10}} = 1.5 \times 10^{-4} \text{ s}$$

8.6 Resonance Escape Probability

Previously we have assumed that the moderator has $\Sigma_a = 0$. But, when this assumption is no longer correct, i.e. absorption is present, then q would not be constant (equal to Q) but would depend on neutron energy. The change in q, as the neutrons pass from an energy E to an energy $E - \Delta E$ caused by neutron absorption is equal to the rate of neutron absorption per cm^3. We can, therefore, write

$$\text{Absorption loss} = \Delta q = n(E)\Delta E v \Sigma_a = \phi(E)\Delta E \Sigma_a \qquad (8.52)$$

In the presence of absorption, the slowing-down density would only be a fraction of its value Q, in the absence of absorption. We may express this fact by writing

$$q(E) = Qp(E) \qquad (8.53)$$

where $p(E) < 1$ is the measure of fraction of neutrons that have escaped absorption in the process of their slowing down from an initial energy E_0 to an energy E.

In the process of slowing down, the neutrons are largely absorbed in the material, when their energies are in the neighborhood of various resonances. The resonances in the epithermal energy region are shown in Fig. 8.10 for ^{238}U. The factor $p(E)$ is a measure of the extent to which the neutrons escape resonance capture, and is called the Resonance escape probability.

With the inclusion of neutron absorption, the steady-state condition within energy interval ΔE is modified as:

$$\text{Scattering loss} + \text{Absorption loss} = \text{Influx}$$

Substituting (8.52) for absorption loss and left side of (8.49) for scattering loss, the state condition becomes

$$\phi(E)\Sigma_s\Delta E + \phi(E)\Sigma_a\Delta E = \frac{q(E)\Delta E}{E\xi} \tag{8.54}$$

Combining (8.52) and (8.54) we get:

$$\frac{\phi(E)\Sigma_a\Delta E}{\phi(E)\Sigma_a\Delta E + \phi(E)\Sigma_s\Delta E} = \frac{\Delta q E\xi}{q(E)\Delta E} \quad \text{or} \tag{8.55}$$

$$\frac{\Sigma_a}{(\Sigma_s + \Sigma_a)} \cdot \frac{\Delta E}{E\xi} = \frac{\Delta q}{q(E)} \tag{8.56}$$

Physically, the left side of Eq. (8.55) implies the ratio of the number of neutron collisions terminating in an absorption and the total number of collisions leading to absorption and scattering. The factor $\frac{\Delta E/E}{\xi} = \frac{\Delta(\ln E)}{\xi}$ is the average number of collisions corresponding to an increase in neutron lethargy by an amount $du = -d(\ln E)$, so that the product of the two factors on left side of (8.56) represents the probability of neutron absorption as the neutron energy changes by an amount corresponding to a change in lethargy of $\Delta u = \Delta(\ln E)$.

Integration of (8.56) between the limits E and E_0 gives

$$p_a = \int_E^{E_0} \frac{\Sigma_a}{\Sigma_s + \Sigma_a} \frac{\Delta(\ln E)}{\xi} = \ln\left[\frac{q(E_0)}{q(E)}\right] \tag{8.57}$$

The integral merely represents the summation over partial probabilities of neutron absorption, i.e. p_a represents the total probability of absorption when neutrons are moderated from energy E_0 to E.

We can re-write (8.57) as

$$q(E) = q(E_0)\exp(-p_a) \tag{8.58}$$

Since the initial slowing-down density $q(E_0)$ is equal to the rate of neutron production Q, we can substitute this in (8.58) and write

$$q(E) = Q\exp(-p_a) \tag{8.59}$$

Comparing (8.59) and (8.53), we find

$$p(E) = \exp(-p_a) \tag{8.60}$$

In all cases of practical interest, the exponent is small, so that we can write approximately

$$p(E) = 1 - p_a \tag{8.61}$$

As p_a is the probability for neutron absorption, $p(E)$ which is equal to $1 - p_a$, must represent the probability of neutron escaping absorption during the moderation from E_0 to E, an interpretation which is consistent with the definition of $p(E)$ as resonance escape probability (8.53).

8.6.1 The Effective Resonance Integral

The exponent in (8.60) as given by the integral (8.57) is called Resonance Integral, which with the assumption of constancy of Σ_s and ξ under the range of integration, can be conveniently written as

$$p_a = \int_E^{E_0} \frac{\Sigma_a}{\Sigma_a + \Sigma_s} \frac{d(\ln E)}{\xi} = \frac{1}{\xi \Sigma_s} \int_E^{E_0} \frac{\Sigma_a}{1 + \Sigma_a/\Sigma_s} \frac{dE}{E} \quad \text{or} \quad (8.62)$$

$$p_a = p = \frac{N_0}{\xi \Sigma_s} \int_E^{E_0} \frac{\sigma_a}{1 + \frac{N_0 \sigma_a}{\Sigma_s}} \frac{dE}{E} \quad (8.63)$$

where N_0 is the number of atoms per unit volume.

The integral in this expression is known as the effective resonance integral

$$\int_E^{E_0} \frac{\sigma_a}{1 + \frac{N_0 \sigma_a}{\Sigma_s}} \frac{dE}{E} = \int_E^{E_0} (\sigma_a)_{eff} \frac{dE}{E} \quad (8.64)$$

and the integrand

$$\frac{\sigma_a}{1 + \frac{N_0 \sigma_a}{\Sigma_s}} = (\sigma_a)_{eff} \quad (8.65)$$

as the effective absorption cross section. The quantity $(\sigma_a)_{eff}$ is a smoothed-out effective value of the absorption cross section in a region where the variation of absorption with energy is extremely complicated because of the presence of pronounced resonances. The limits of integration in the effective resonance integral are the energies that bound the resonance absorption region.

Since the resonant absorptions of various fuels that are of interest in a reactor assembly vary in irregular manner, the value of resonance integral must be found empirically. It depends mainly on the ratio of the fissionable to moderator material as well as on other factors determined by the type of the reactor employed. The number of nuclei per cm^3, N_0, which appears in the preceding formulae refer to the number of resonance absorber nuclei present. In particular, if all the resonance absorption takes place in uranium alone, then N_0 is simply the number of uranium nuclei per cm^3. If different kinds of nuclei are present in the moderator as would be the case when the moderator is a mixture or a compound then $\xi \Sigma_s$ in the denominator of (8.63) must be replaced by the sum terms $\xi_i \Sigma_{s,i}$ for the ith type of scatterer.

Further, in so far as the scattering is concerned, the effect of the uranium nuclei can be neglected because of the small value of ξ which is 0.0084 for ^{238}U and because these nuclei are only a small fraction of the total number of scatterers present, we can write to a very good approximation, $\Sigma_s = \Sigma_{s,m}$, where m stands for moderator.

Fig. 8.11 Diffusion of
neutrons

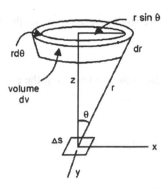

8.7 Diffusion of Neutrons

Consider a large volume of moderator in which neutrons are diffusing as a result of
successive elastic collisions with the nuclei of the moderator. We shall calculate the
net flow of neutrons in a given direction, and also the leakage of neutrons from a
unit volume. Consider a small area ΔS located in the XY-plane within this volume
of the moderator. We shall determine the flow of neutrons along the Z-direction.
We assume that the scattering of neutrons is isotropic, and that the absorption cross-
section is negligible. We further assume that the scattering is independent of energy
and position.

We introduce the polar co-ordinates, r and θ. Imagine an element of volume dV
located centrally about the z-axis (Fig. 8.11) in which the neutrons get scattered. We
note that

$$dV = (2\pi r \sin\theta)(rd\theta)(dr)$$

The number of scattering collisions taking place per second in the volume dV is
equal to $\phi \Sigma_s dV$, where ϕ is the neutron flux (total neutron path traversed per s: per
cm^3).

The probability that a neutron scattered in the volume dV should proceed along
r so as to pass through ΔS, is given by $\frac{\Delta s\cos\theta}{4\pi r^2}$, since the area ΔS is inclined at an
angle θ with respect to r. But we must allow for the fact that a neutron scattered in
the volume dV and heading towards the correct direction (i.e. along r) should escape
another scattering collision, the probability for which is given by $\exp(-r/\lambda_s)$, or
$\exp(-\Sigma_s r)$.

Thus the number of neutrons dn scattered in volume dV that reach the surface
ΔS is given by

$$dn = (\phi\Sigma_s)\frac{(2\pi r^2 \sin\theta d\theta dr)}{4\pi r^2}\Delta S \cos\theta \exp(-\Sigma_s r)$$

If J_- is the neutron current in the downward direction through ΔS expressed as
number of neutrons/cm²/s, then

$$J_- = n/\Delta s = \int_\omega^{r=\infty}\int_0^{\theta=\pi/2}\frac{\phi\Sigma_s}{2}\sin\theta\cos\theta\exp(-\Sigma_s r)d\theta dr \qquad (8.66)$$

We shall first assume that the flux ϕ is constant, then we find

$$J_- = \phi/4$$

Because of symmetry, the current in the positive Z-direction would also be equal to $\phi/4$; i.e.

$$J_+ = \phi/4$$

So that the net current

$$J = J_+ - J_- = 0$$

Next, we suppose that the flux ϕ is not constant in a direction normal to the surface, but varies only slowly then using Maclaurian expansion, we can write

$$\phi = \phi_0 + z\left(\frac{\partial\phi}{\partial z}\right) \tag{8.67}$$

The subscript 0 means that the quantity is to be evaluated at the origin.

Since $z = r\cos\theta$, we have

$$\phi = \phi_0 + r\cos\theta\left(\frac{\partial\phi}{\partial z}\right)_0 \tag{8.68}$$

Inserting (8.68) in (8.66), a straight forward integration yields

$$J_- = \frac{\phi_0}{4} + \frac{1}{6\Sigma_s}\left(\frac{\partial\phi}{\partial z}\right)_0 \tag{8.69}$$

where, J_- is the neutron current in the downward direction through Δs. For J_+, the integration must be performed between $\frac{\pi}{2}$ and π, giving us

$$J_+ = \frac{\phi_0}{4} - \frac{1}{6\Sigma_s}\left(\frac{\partial\phi}{\partial z}\right)_0 \tag{8.70}$$

The net current in the vertical direction is then given by

$$J = J_+ - J_- = -\frac{\lambda_s}{3}\left(\frac{\partial\phi}{\partial z}\right)_0 \tag{8.71}$$

The net flow of neutrons through a unit area located in the XY-plane depends on the gradient of the flux in the z-direction. Variation of the flux in the x and y directions would not affect the flow of neutrons in the vertical direction, and may be ignored. The assumption under which (8.71) is valid is that the gradient of the flux is small over distances of a few scattering M.F.P. It will be invalidated in regions close to the source or sink or near a boundary separating two media with different scattering cross-sections. Actually, (8.71) is more valid than this treatment would allow, for, when the second order term is included in the Maclaurian expansion then it gets

Fig. 8.12 Leakage of neutrons in orthogonal directions

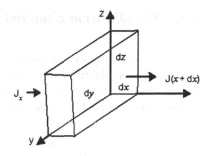

cancelled out in the calculation of J. The anisotropic scattering may be allowed for by replacing λ_s by λ_{tr} in (8.71)

$$J = -\frac{\lambda_{tr}}{3}\left(\frac{\partial \phi}{\partial z}\right)_0 \tag{8.72}$$

8.7.1 Leakage of Neutrons

Consider a small volume element of sides dx, dy, dz within the moderator (Fig. 8.12). Let J_x be the net neutron current in the x-direction, then

$$dJ_x = \left(\frac{\partial J}{\partial x}\right)dx = -\frac{\lambda_{tr}}{3}\left(\frac{\partial^2 \phi}{\partial^2 x}\right)dx$$

where, dJ_x represents the excess of neutron current passing through the volume element in the x-direction. If L_x is the leakage in the x-direction defined by the excess number of neutrons leaving the volume element per second in the x-direction, then

$$L_x = dJ_x dy dz = -\frac{\lambda_{tr}}{3}\left(\frac{\partial^2 \phi}{\partial^2 x}\right)dx dy dz \tag{8.73}$$

Similar terms will be obtained in the y- and z-direction.

The total leakage L out of unit volume expressed as the number of neutrons leaving per cm^3 per second is

$$L = \frac{L_x + L_y + L_z}{dx dy dz} = -\frac{\lambda_{tr}}{3}\nabla^2 \phi \tag{8.74}$$

where ∇^2 is the Laplacian.

8.7.2 The Diffusion Equation for Thermal Neutrons

We shall now set up the diffusion equation which governs the distribution of neutrons throughout the reactor. We shall assume that all the neutrons to be considered having the same energy. Let n neutrons be present at any time t in a unit volume.

The change of neutron density with time may be written as

$$\frac{\partial n}{\partial t} = \text{Production} - \text{Leakage} - \text{Absorption}$$

Let S be the rate of production of neutrons/cm^3/s; the absorption rate is given by $\phi \Sigma_a$ per cm^3/s; then using (8.74) for the leakage we may write the balance equation as

$$\frac{\partial n}{\partial t} = S + \frac{\lambda_{tr}}{3} \nabla^2 \phi - \phi \Sigma_a \tag{8.75}$$

This is the general form of diffusion equation and is applicable only to monoergic neutrons and at distances of about three M.F.P. from strong sources or absorbers and from boundaries of dissimilar material.

If the reactor is working at a steady state then the neutron flux, and hence the power would be constant. In this case

$$\frac{\partial n}{\partial t} = 0, \quad \text{and the balance equation becomes}$$

$$\text{Production} = \text{Leakage} + \text{Absorption}$$

As the neutrons are produced only at the point located at the source, neutron production will be zero at all other points. Putting $S = 0$ and $\frac{\partial n}{\partial t} = 0$ in (8.75) the steady state equation for these regions reduces to

$$\nabla^2 \phi - \frac{3}{\lambda_{tr} \lambda_a} \phi = 0 \quad \text{or} \tag{8.75a}$$

$$\nabla^2 \phi - \frac{1}{L^2} \phi = 0 \tag{8.75b}$$

with

$$L^2 = \frac{\lambda_{tr} \lambda_a}{3}$$

The value of the Laplacian in (8.75) would depend on the geometry of the system. In Table 8.2, we give the explicit form of the Laplacian for typical geometry.

Note that the Laplacian for Cylindrical and Spherical geometry is much simpler than the usual expressions because of the absence of the terms involving azimuth angle and the polar angle. This is so because we expect the flux not to depend on these angles.

Table 8.2 Geometrical shapes and their Laplacian

Geometry	∇^2
1. Parallelpiped	$\frac{\partial^2}{\partial x^2} + \frac{\partial^2}{\partial y^2} + \frac{\partial^2}{\partial z^2}$
2. Infinite slab	$\frac{\partial^2}{\partial x^2}$
3. Cylindrical	$\frac{\partial^2}{\partial r^2} + \frac{1}{r}\frac{\partial}{\partial r} + \frac{\partial^2}{\partial z^2}$
4. Spherical	$\frac{\partial^2}{\partial^2} + \frac{2}{r}\frac{\partial}{\partial r}$

Fig. 8.13 Extrapolation distance

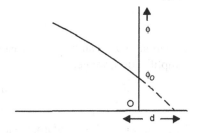

8.7.3 Extrapolation Distance

At the outside surface of the reactor, there cannot be any scattering of neutrons back into the reactor, so that at the boundary, the neutron current in the negative direction is zero:

$$J_- = \frac{\phi_0}{4} + \frac{\lambda_{tr}}{6}\left(\frac{\partial\phi}{\partial z}\right)_0 = 0 \qquad (8.76)$$

where ϕ_0 is the flux at the outside surface of the reactor. If we extrapolate the neutron flux beyond the outside surface using a straight line with the same slope as at the boundary (Fig. 8.13), then

$$\frac{\partial\phi}{\partial z} = \frac{-3}{2}\frac{\phi_0}{\lambda_{tr}} = -\phi_0/d \quad \text{or}$$

$$d = \frac{2}{3}\lambda_{tr} \qquad (8.77)$$

where d is the extrapolation distance at which the flux falls to zero.

The value of d as given by (8.77) is only approximate since diffusion theory is not valid near the boundary. A more refined treatment using the Transport theory gives

$$d = 0.71\lambda_{tr} \qquad (8.78)$$

Example 8.1 Calculate the steady state neutron flux distribution about a point source emitting Q neutrons/s isotropically in an infinite homogenous diffusion medium. Assume that inside a region of interest, neutrons are not produced.

Solution   A steady state implies that $\frac{\partial n}{\partial t} = 0$. Further the source term $S = 0$. Using the Laplacian for spherical geometry in (8.75), we have

$$\frac{\lambda_{tr}}{3}\left(\frac{\partial^2 \phi}{\partial r^2} + \frac{2}{r}\frac{\partial \phi}{\partial r}\right) - \phi \Sigma_a = 0 \quad \text{or}$$

$$\frac{d^2 \phi}{dr^2} + \frac{2}{r}\frac{d\phi}{dr} - k^2 \phi = 0 \tag{8.79}$$

where

$$k^2 = 3\Sigma_a/\lambda_{tr}$$

Equation (8.79) can be solved easily by the change of variable $v = \phi r$. After some simplification, we get

$$\frac{d^2 v}{dr^2} - k^2 v = 0 \tag{8.80}$$

The solution of this standard equation is known to be

$$v = A_1 e^{kr} + A_2 e^{-kr}$$

where A_1 and A_2 are constants of integration

$$\therefore \quad \phi = \frac{A_1 e^{kr}}{r} + \frac{A_2 e^{-kr}}{r} \tag{8.81}$$

Since k is positive, $e^{kr} \to \infty$ as $r \to \infty$. But flux must be finite everywhere including at large distances. Therefore, we must put $A_1 = 0$. Thus

$$\phi = \frac{A_2 e^{-kr}}{r} \tag{8.82}$$

We also have:

$$\frac{d\phi}{dr} = \frac{-A_2}{r^2}(kr + 1)e^{-kr} \tag{8.83}$$

We shall now evaluate the constant A_2. Consider a small sphere of radius r surrounding the point source. The net current through this sphere is

$$J = \frac{-\lambda_{tr}}{3}\frac{\partial \phi}{\partial r} = \frac{\lambda_{tr}}{3r^2} A_2(kr + 1)e^{-kr}$$

where we have used (8.83).

The net number of neutrons leaving the sphere per second is,

$$4\pi r^2 J = \frac{4}{3}\pi \lambda_{tr} A_2(kr + 1)e^{-kr}$$

But as $r \to 0$, the total number of neutrons leaving the sphere per s must be equal to the source strength q

$$\therefore \quad q = \frac{4}{3}\pi \lambda_{tr} A_2 \quad \text{or}$$

$$A_2 = \frac{3q}{4\pi \lambda_{tr}}$$

The complete solution is

$$\phi = \frac{3q e^{-kr}}{4\pi r \lambda_{tr}} \tag{8.84}$$

Example 8.2 A point source of thermal neutrons is placed at the centre of a large sphere of Beryllium. Estimate what its radius must be if less than 1.5 % of the neutrons are to escape through the surface.

At: wt: of beryllium = 9
Density of beryllium = 1.85 g/cc
Avagadro number = 6×10^{23} atoms/g atom
Thermal neutron scattering cross section on Beryllium = 5.6 b
Thermal neutron capture cross-section on Beryllium = 10 mb (at velocity $v = 2200$ m/s)

Solution Using the results of Example 8.1

$$\frac{n}{q} = (1 + kr)e^{-kr}$$

$$\therefore \quad \frac{1.5}{100} = (1 + 0.0157r)e^{-0.0157r}$$

Solving the above equation by the method of successive approximation, we find $r = 393$ cm.

Example 8.3 Calculate the steady state neutron flux distribution about a plane source emitting Q neutrons/s/cm^2 in an infinite homogeneous diffusion medium. Assume that neutrons are not produced in any region of interest.

Solution Since we are interested only in the x-direction, we use the Laplacian $\frac{d^2}{dx^2}$ in (8.75), and remembering that $S = 0$, and $\frac{\partial n}{\partial t} = 0$, we have

$$\frac{\lambda_{tr}}{3} \frac{d^2\phi}{dx^2} - \phi \Sigma_a = 0 \quad \text{or} \tag{8.85}$$

$$\frac{d^2\phi}{dx^2} - k^2\phi = 0 \tag{8.86}$$

Fig. 8.14 Diffusion from
plane (infinite) source

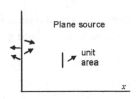

with

$$k^2 = \frac{3\Sigma_a}{\lambda_{tr}} = \frac{3}{\lambda_a \lambda_{tr}} \qquad (8.87)$$

The solution of (8.86) is

$$\phi = A_1 e^{kx} + A_2 e^{-kx} \qquad (8.88)$$

where A_1 and A_2 are constants of integration. The requirement that the flux should not diverge at any point including at infinity implies that $A_1 = 0$

$$\therefore \quad \phi = A_2 e^{-kx} \qquad (8.89)$$

We shall now calculate A_2. Consider a unit area located at a small distance x from the plane source, Fig. 8.14. On an average, half of the neutrons will travel along positive x-direction. As $x \to 0$, the net current flowing in the positive x-direction would be equal to $\frac{1}{2}Q$; the diffusion of neutrons through unit area would have a cancelling effect since from symmetry equal number of neutrons would diffuse in opposite direction at the surface ($x = 0$). Now, the current

$$J = \frac{-\lambda_{tr}}{3}\frac{\partial\phi}{\partial x} = A_2 k \frac{\lambda_{tr}}{3} e^{-kx}$$

But, as

$$x \to 0, \qquad J = Q/2 = A_2 k \frac{\lambda_{tr}}{3}$$

whence we find

$$A_2 = \frac{3Q}{2k\lambda_{tr}}$$

The complete solution is

$$\therefore \quad \phi = \frac{3Q e^{-kx}}{2k\lambda_{tr}} \qquad (8.90)$$

Fig. 8.15 Thermal neutron traveling from point of production to point of absorption

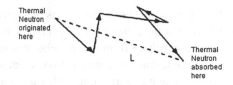

8.7.4 Diffusion Length

Equation (8.90) shows that the flux falls off exponentially with distance from the source. The distance required for the flux to fall by a factor e is

$$L = 1/k = \sqrt{\frac{\lambda_{tr}\lambda_a}{3}} \qquad (8.91)$$

Apart from the factor $1/3$, the diffusion length L is seen to be given by the geometric mean of λ_{tr} and λ_a. The diffusion length is a measure of the average distance a thermal neutron travels from the point of production to the point of absorption (Fig. 8.15). The average crow-flight distance traveled by a neutron away from an infinite plane source in a moderator before being absorbed is calculated using (8.90)

$$\bar{x} = \frac{\int_0^\infty x e^{-x/L}dx}{\int_0^\infty e^{-x/L}dx} = L \qquad (8.92)$$

as expected.

The mean square distance travelled $\langle x^2 \rangle$ from the infinite plane source is given by

$$\langle x^2 \rangle = \frac{\int_0^\infty x^2 e^{-x/L}dx}{\int_0^\infty e^{-x/L}dx} = 2L^2 \qquad (8.93)$$

8.7.5 Relationship Between $\langle r^2 \rangle$ and L^2 for a Point Source

Consider a spherical shell surrounding the source of radius r from the point source and thickness dr. The volume of the shell is $4\pi r^2 dr$. The neutron capture rate in this shell is then equal to $4\pi r^2 dr \phi \Sigma_a$. The mean square distance that a neutron travels from the source before getting captured is

$$\langle r^2 \rangle = \frac{1}{Q} \int_0^\infty r^2 4\pi r^2 dr \phi \Sigma_a \qquad (8.94)$$

Substituting the value of ϕ from (8.84) in (8.94) we get

$$\langle r^2 \rangle = \frac{3\Sigma_a}{\lambda_{tr}} \int_0^\infty r^3 e^{-kr}dr = 6/k^2 = 6L^2 \qquad (8.95)$$

It may be pointed out that $\langle r^2 \rangle$ is the mean square crow flight distance the neutron travels, but the actual path length is much greater because of the zig-zag path of the neutron resulting from various elastic scatterings with the nuclei of the moderator. Since the leakage of neutrons from a reactor depends on the average crow flight distance the neutrons travel, the diffusion length is one of the factors that is intimately connected with the criticality of a reactor.

Example 8.4 The spatial distributions of thermal neutrons from a plane neutron source kept at a face of a semi-infinite medium of graphite was determined and found to fit $e^{-0.03x}$ law where x is the distance along the normal to the plane of the source. If the only impurity in the graphite is boron, calculate the number of atoms of boron per cm^3 in the graphite if the mean free path for scattering and absorption in graphite are 2.7 and 2700 cm, respectively. The absorption cross-section of boron is 755 b.

Solution   We ignore the scattering of neutrons in boron. Let there be N atoms of boron per cm^3

$$\lambda_a(\text{graphite}) = 2700 \text{ cm}$$

$$\therefore \quad \Sigma_a(\text{graphite}) = \frac{1}{2700} = 3.7 \times 10^{-4} \text{ cm}^{-1}$$

$$\Sigma_a(\text{boron}) = 755 \times 10^{-24} N$$

$$\Sigma_a = \Sigma_a(\text{graphite}) + \Sigma_a(\text{boron}) = 3.7 \times 10^{-4} + 755 \times 10^{-24} N$$

$$\lambda_{tr} = \frac{\lambda_s}{1 - \frac{2}{3A}} = \frac{2.7}{1 - \frac{2}{3 \times 12}} = 2.86 \text{ cm}$$

Given

$$L = \frac{1}{0.03} = 33.33 \text{ cm}$$

but

$$L^2 = \frac{\lambda_{tr} \lambda_a}{3} = \frac{\lambda_{tr}}{3 \Sigma_a}$$

$$\therefore \quad (33.33)^2 = \frac{2.86}{3(3.7 \times 10^{-4} + 755 \times 10^{-24} N)}$$

Solving for N, we get

$$N = 6.47 \times 10^{17} \text{ boron atoms/cm}^3$$

Table 8.3 Values of L and λ_{tr} for different moderators

Moderator	L (cm)	λ_{tr} (cm)
Water	2.88	0.426
Heavy water	100	2.4
Graphite	50	2.71

8.7.6 Experimental Measurement of Diffusion Length

Calculated values of L and λ_{tr} for thermal neutrons can be in considerable error for various reasons. First, thermal neutrons have wide energy spectrum and hence variable properties. Secondly, the simple scattering laws break down owing to complications arising due to chemical binding and crystalline effects at low energies. Further, the scattering cross-section is not strictly constant in the thermal region. For these reasons, the reliable values of L and λ_{tr} are those which are determined experimentally. One way of measuring L is to measure the flux by the foil activation method at various distances from a plane source of thermal neutrons placed inside a large body of moderator. The plot of $\ln \phi$ against x can be fitted by a straight line the slope of which yields $1/L$. Table 8.3 gives the values of L and λ_{tr} for the moderators water, heavy water and graphite.

8.7.7 The Albedo

The loss of neutrons which escape from the reactor can be minimized by surrounding it with a reflector which has the ability to scatter a number of neutrons back into the reactor. The efficiency of the reflector may be measured in terms of the reflection coefficient or Albedo which is the ratio of the number of neutrons reflected back to the number entering the reflector

$$\text{Albedo} = \frac{\frac{\phi_0}{4} + \frac{\lambda_{tr}}{6}\left(\frac{\partial \phi}{\partial x}\right)}{\frac{\phi_0}{4} - \frac{\lambda_{tr}}{6}\left(\frac{\partial \phi}{\partial x}\right)}$$

If the reflector is an infinite slab, we have the one-dimensional problem where the neutron current entering the reflector acts as a plane source of neutrons (Fig. 8.16). But

$$\phi = A_2 e^{-kx}; \qquad \frac{d\phi}{dx} = -kA_2 e^{-kx}$$

$$\therefore \quad \text{Albedo} = \frac{3 - 2k\lambda_{tr}}{3 + 2k\lambda_{tr}} = \frac{3 - \frac{2\lambda_{tr}}{L}}{3 + \frac{2\lambda_{tr}}{L}} \tag{8.96}$$

Thus, the smaller the value of λ_{tr} and the greater the L, the Albedo approaches unity. The Albedo depends on the size and the shape of the reflector. The Albedo

Fig. 8.16 Reflector as an
infinite slab

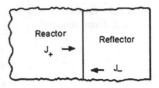

Fig. 8.17 A rectangular
column composed of the
moderator containing the
neutron source

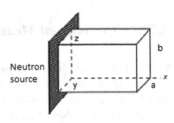

will be smaller for a reflector of finite thickness than the one of infinite thickness
which in practice is reached for thickness equal to about $2L$. Further, for a reflector
about a spherical reactor, the Albedo will be smaller for a given thickness than in
the case of an infinite slab because *a* neutron sees a smaller reactor surface in the
case of a sphere so that the probability of its getting scattered back into the reactor
is smaller.

8.7.8 Determination of Diffusion Length from the 'Exponential' Pile

If the plane source is finite then the previous treatment needs modification. Consider
a rectangular column (pile) composed of the moderator with one of its faces containing the neutron source. The neutron source may be either an artificial neutron
source or thermal neutrons from a nuclear reactor. Let the sides of the base of this
column have lengths a and b (Fig. 8.17).

The diffusion equation is

$$\nabla^2\phi - k^2\phi = 0$$

with

$$k^2 = \frac{3}{\lambda_{tr}\lambda_a}$$

We shall choose the x-axis along the length of the column; y- and z-axes are taken
along the sides of the base of the column. Since the flux is no longer independent of
the y and z co-ordinates, we must use the full Laplacian

$$\frac{\partial^2\phi}{\partial x^2} + \frac{\partial^2\phi}{\partial y^2} + \frac{\partial^2\phi}{\partial z^2} - k^2\phi = 0 \tag{8.97}$$

We solve this equation by the method of separation of variables. Let

$$\phi = X(x)Y(y)Z(z) \tag{8.98}$$

Substituting (8.98) in (8.97), and dividing through out by XYZ

$$\frac{1}{X}\frac{d^2X}{dx^2} + \frac{1}{Y}\frac{d^2Y}{dy^2} + \frac{1}{Z}\frac{d^2Z}{dz^2} - k^2 = 0 \quad \text{or}$$

$$\frac{1}{X}\frac{d^2X}{dx^2} = k^2 - \frac{1}{Y}\frac{d^2Y}{dy^2} - \frac{1}{Z}\frac{d^2Z}{dz^2} \tag{8.99}$$

Since left side of (8.99) is a function of x only, while the right side is a function of y and z only, the only way in which the above equation can be satisfied is when each side equals the same constant, say A^2

$$\therefore \quad \frac{1}{X}\frac{d^2X}{dx^2} = A^2 \tag{8.100}$$

The solution for this equation is known to be

$$X = c_1 e^{-Ax} + d_1 e^{+AX}$$

where c_1 and d_1 are constants. But the requirement that the flux must not diverge as $x \to \infty$ gives us $d_1 = 0$

$$\therefore \quad X = c_1 e^{-Ax} \tag{8.101}$$

Also

$$k^2 - \frac{1}{Y}\frac{d^2Y}{dy^2} - \frac{1}{Z}\frac{d^2Z}{dz^2} = A^2 \quad \text{or}$$

$$\frac{1}{Y}\frac{d^2Y}{dy^2} = k^2 - A^2 - \frac{1}{Z}\frac{d^2Z}{dz^2} \tag{8.102}$$

Left side of (8.102) is a function of y only and right side is a function of z only. Therefore, each side must be equal to constant, say $-B^2$

$$\therefore \quad \frac{1}{Y}\frac{d^2Y}{dy^2} = -B^2$$

gives us $Y = c_2 \sin By + d_2 \cos By$, where c_2 and d_2 are constants. An approximate boundary condition is that, $\phi = 0$ at $y = 0, y = a, z = 0, z = b$. The first of the boundary conditions gives us, $d_2 = 0$. The second condition gives us

$$Ba = n\pi \tag{8.103}$$

Note that a positive constant $+B^2$ cannot give the required boundary conditions.

Taking the first mode ($n = 1$)

$$B = \pi/a \tag{8.104}$$

Also

$$\frac{1}{Z}\frac{d^2 Z}{dz^2} = k^2 - A^2 + B^2 = -D^2 \tag{8.105}$$

giving us

$$Z = c_3 \sin Dz + d_3 \cos Dz$$

Application of the third boundary condition gives us $d_3 = 0$, and the fourth one gives us

$$D = \pi/b \tag{8.106}$$

∴. The flux

$$\phi = ce^{-Ax} \sin\frac{\pi y}{a} \sin\frac{\pi z}{b} \tag{8.107}$$

where

$$c = c_1 c_2 c_3$$

Further, from (8.105) we find

$$A^2 = k^2 + B^2 + D^2 \tag{8.108}$$

We, therefore find that the flux along the x-axis varies as e^{-x/L_1} where $L_1 = 1/A$. From (8.108), we note that

$$\frac{1}{L_1^2} = \frac{1}{L^2} + \frac{\pi^2}{a^2} + \frac{\pi^2}{b^2} \tag{8.109}$$

The quantity L_1 is called the effective diffusion length. We can, therefore, first determine L_1 in the manner previously described and then obtain L, the true diffusion length using (8.109).

Note that if the sides a and b are much larger than L, then $L_1 \to L$. The neutron flux in the column falls off exponentially in a direction perpendicular to the face containing the neutron source. Hence the nomenclature 'exponential pile'.

8.8　Elementary Theory of the Chain-Reacting Pile

Absorption and Production of Neutrons in a Pile　We consider the 'pile' which is a mass of uranium spread in some suitable arrangement throughout a block of graphite. Whenever fission occurs in uranium, on an average 2.5 neutrons are produced. The neutrons produced from fission have energies of the order of 1 MeV and are conventionally called 'fast neutrons'. After a neutron is emitted, its energy decreases as a result of elastic collisions with the nuclei of carbon and to some extent also by inelastic collisions with the nuclei of uranium. A large fraction of the fast neutrons will be slowed down to thermal energies in about 100 collisions with carbon nuclei. On attaining thermal energies, the neutrons keep on diffusing until they are finally absorbed. In several cases, however, the neutrons are absorbed before the slow-down process is completed.

The neutrons may be absorbed either in carbon or uranium. But, the absorption cross section in carbon is quite small, being about 5 mb. For graphite of density 1.6 this corresponds to a mean free path of about 25 m. Since σ_a follows the l/v law, the absorption cross section which is already quite small at thermal energies is practically negligible at higher energies. It is, therefore, sufficiently good approximation to neglect the absorption by carbon during the slowing-down process. The absorption cross-section of thermal neutrons in ^{235}U is quite large, being about 650 b. The absorption of neutron in uranium may lead either to fission or to radiative capture—a process in which gamma ray is produced. The relative importance of resonance absorption and fission is dependent on neutron energy. For this purpose, we may broadly consider three energy intervals:

(i) Neutrons of energy >1 MeV above the fission threshold of ^{238}U. These are the so-called fast neutrons, for which the most important absorption process is fission which normally occurs in the abundant isotope ^{238}U, resonance absorption being small but not negligible.

(ii) Neutrons of energy below the fission threshold of ^{238}U but above the thermal energy (0.025 eV). These are called 'epithermal neutrons'. In this energy interval, the most important absorption process is resonance capture. The variation of cross section with energy is quite irregular, and displays the occurrence of a large number of maxima, called resonance maxima, which are explained by Breit Wigner theory, Fig. 8.10. The phenomenon of resonance absorption becomes important in all practical cases at energies of about 200 eV and becomes increasingly more important as the energy decreases.

(iii) Neutrons of thermal energies. For thermal neutrons both the resonance and fission absorption processes are important. In this energy interval, both the absorption processes approximately follow the l/v law; consequently, their relative importance becomes independent of energy. From the preceding discussion we conclude that only a fraction of the original fast neutrons produced will survive, and ultimately lead to fission. Further, in systems of finite size, leakage of neutrons is expected outside the pile.

8.8.1 Life History of Neutrons and Four-Factor Formula

The following sequence of events may take place in one life cycle of a neutron:

1. Suppose that at a given instance, there are available n thermal neutrons which are captured in the fuel. Let η be the average number of fast fission neutrons

emitted per fission, then due to the absorption of n thermal neutrons $n\eta$ fast neutrons will be produced. If v is the average number of neutrons emitted when a thermal neutron is absorbed in the fuel, then $v = (\eta\sigma_f)/(\sigma_f+\sigma_r)$ where σ_r refers to radiative capture. The values of v for various fuels are 2.43 for ^{235}U, 2.47 for natural uranium, 2.5 for ^{233}U, 2.89 for ^{239}Pu.

2. There is a small probability that some of the neutrons may be absorbed in uranium before their energy is appreciably decreased. If this is the case then the absorption in ^{238}U often leads to fission. The available number of neutrons is therefore increased by a factor ε, which is called fast fission factor and is defined by the ratio of the total number of fast neutrons produced by fissions due to neutrons of all energies to the number resulting from thermal neutron fissions. The probability of producing such fast fission neutrons is only a few percent, for uranium fuel, for example, $\varepsilon = 1.03$, with either graphite or D$_2$O as moderater. Indeed, if the system contains little uranium and a large amount of carbon, the elastic collisions with carbon tend to reduce the energy very rapidly to a value well below the fission threshold for ^{238}U. On the other hand, if the system is very rich in uranium then inelastic collisions rapidly reduce the energy of the originally fast neutrons to the point much before the neutrons get a chance to cause fission in ^{238}U. The number of neutrons then becomes $n\eta\varepsilon$.

3. As a result of collisions, mainly elastic, with the moderator, the fast neutrons will be ultimately thermalized. With graphite as moderator, about 14.6 collisions are required to reduce their energy by a factor of 10, and about 110 collisions to reduce 1 MeV neutrons to 0.025 eV While the slow-down process is in progress, the neutrons may be absorbed by resonance capture in uranium. Let p be the probability that neutron escapes resonance capture and is able to reach thermal energy. This is called resonance escape probability. The number of neutrons reaching thermal energies is then $n\eta\varepsilon p$.

4. If the neutrons have not been absorbed then on reaching thermal energies, they will ultimately be absorbed either in uranium or carbon. If uranium and carbon are mixed uniformly, the probability for these two events would be in the ratio of the absorption cross-sections of uranium and carbon for thermal neutrons multiplied by the atomic concentrations of the two elements. Of the thermal neutrons, therefore, a fraction f called thermal utilization factor will be absorbed in fuel material.

Thus

$$f = \frac{\Sigma_a(\text{fuel})}{\Sigma_a(\text{fuel}) + \Sigma_a(\text{moderater}) + \Sigma_a(\text{other material})}$$

The number of thermal neutrons absorbed in fuel becomes $n\eta\varepsilon pf$.

To sum up, n neutrons are multiplied to $n\eta\varepsilon pf$ after the completion of one cycle. Let k_∞ be the multiplication factor or reproduction factor which is defined as the ratio of the number of neutrons in one generation to the number in the previous generation

$$k_\infty = n\eta\varepsilon pf/n = \eta\varepsilon pf \tag{8.110}$$

Table 8.4 Values of p, f, and the product pf for various ratios of atomic concentrations of graphite and natural uranium

N_m/N_U	f	p	fp
100	0.960	0.523	0.5020
200	0.923	0.647	0.5971
300	0.889	0.706	0.6276
400	0.858	0.751	0.6443
500	0.828	0.774	0.6408
600	0.800	0.795	0.6360

This is called the *Four-factor formula*.

If k_∞ is at least equal to 1, then the reactor would work continuously (it is assumed that the system has infinite dimension so that there is no leakage of neutrons) since the life cycle of neutron will be repeated again and again. This repetitive phenomenon is called *chain reaction*. For an assembly of finite dimensions the effective reproduction constant k_{eff} will be less than k_∞ by the non-leakage factor $L(L < 1)$

$$k_{eff} = k_\infty L \tag{8.111}$$

The reactor is said to be critical if $k_{eff} = 1$.

The value of k_∞ will depend on the relative production and loss of neutrons by various processes. In practice, apart from leakage through a system of finite dimensions, neutrons will be lost as a result of non-fission capture by ^{235}U and ^{238}U including both thermal and resonance capture and as a result of parasitic capture of neutrons by the moderator, coolant, structural materials, fission products and any other material, called poison, present in the reactor.

In the special case in which the fuel contains only ^{235}U, both ε and p will be very close to unity. Thus, $k_\infty = \eta f$.

Of the four factors in (8.110), ε and η are more or less fixed by the character of the fuel over which we have no control; and the success of the chain reactor depends upon the values of p and f which vary with geometry, the composition of the reactor elements and the ratio of fuel to moderator. In order to ensure the propagation of chain reaction, p and f should be as large as possible although each of them is always less than unity. Unfortunately such changes in the relative proportion of fuel and moderator which cause f to increase and cause p to decrease and vice-versa. Thus, in order to make f large, it is desirable to have the system very rich in uranium so that the probability of absorption of thermal neutrons by carbon is reduced. On the other hand, the smaller proportion of carbon implies that the slowing-down process will be very slow, and consequently, the probability of resonance absorption is increased. The reverse will be true if the concentration of moderator is large.

Table 8.4 shows the values of p and f, as well as the product pf for various ratios of atomic concentrations of graphite and natural uranium which contains 99.3 % ^{238}U and 0.7 % ^{235}U.

Actually, a homogeneous natural uranium-graphite-moderated reactor cannot become critical since k_∞ is always less than unity. But, the opposing tendency of p

and f is quite obvious from Table 8.4. In any case, in order to make the product pf maximum, we must make a compromise between two conflicting requirements by finding an optimum value for the ratio of uranium to carbon.

So long we deal with homogeneous mixture of uranium and carbon, the values of p and f depend only on the relative concentrations of the two components. However, if we do not restrict ourselves to homogeneous systems, then it is possible to have a more favorable situation by suitable geometrical distribution of the two elements. This is possible is due to the following circumstances. The resonance absorption cross-section which is responsible for the loss of neutrons has sharp peaks of the Breit-Wigner type discussed earlier. Therefore, if uranium instead of being smeared throughout the assembly, is used in sizable lumps, then a thin surface layer would shield the interior of the lump from the action of neutrons with energy close to resonance maxima. Therefore, the resonance absorption of a uranium atom inside the lump will be much less than it would be for an isolated atom. Of course, self protection in a lump reduces not only resonance absorption but also thermal absorption in uranium. Calculations, however, show and experiments have confirmed, that at least up to a certain size of the lumps, the gain obtained by reducing the resonance loss of neutrons, weighs out by a considerable amount the loss due to a lesser absorption of thermal neutrons in the interior of the lump. The typical structure of the pile is a lattice of uranium lumps embedded in a matrix of graphite. The lattice may be, for example, a cubic lattice of lumps or a lattice of rods of uranium. Such an assembly is obviously; heterogeneous. In practice, one deals with an assembly of finite size so that some neutrons will leak from the assembly.

We now give typical figures for the production and losses of neutrons in one complete cycle, as in Fig. 8.18.

Thus numbers of neutrons in the beginning and end of the cycle are almost identical so that the reactor becomes critical.

8.8.2 Fast-Fission Factor (ε)

The value of this quantity can be easily calculated for a very small lump of uranium. In this case, it is obviously given by

$$\varepsilon = \sigma_F n d \tag{8.112}$$

where σ_F is the average value of the fission cross section for fast fission neutrons; n is the concentration of uranium atoms in the lump, and d is the average value of the distance that the neutron produced in the lump must travel before reaching the surface of the lump. Calculations for lump of large size are rendered difficult owing to the multiple collision processes comprising elastic, and inelastic collisions. In particular, the last process for a lump of large size effectively slows down the neutrons below the fission threshold of ^{238}U and brings them down to an energy level in which they are readily absorbed by the resonance process.

Fig. 8.18 Typical figures for the production and losses of neutrons in one complete cycle

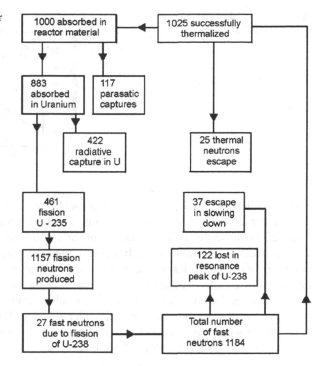

8.8.3 Resonance Absorption

For isolated atoms of uranium contained in the medium of graphite in which fast neutrons are produced and slowed down to thermal energies, the probability per unit time of resonance absorption of neutrons with energy larger than thermal energy is given by the following expression:

$$\frac{q\lambda}{0.158} \int \sigma(E)\frac{dE}{E} \tag{8.113}$$

where q is the number of fast neutrons entering the system per unit time and unit volume, λ is the M.F.P. and $\sigma(E)$ is the resonance absorption cross section at energy E. The integral must be taken between a lower limit just above the thermal energy and an upper limit equal to the average energy of fission neutrons. We should expect the largest contribution to the integral from the Breit-Wigner peaks.

The above formula would be very much in error in case of a lattice of lumps. As we have already pointed out, this is due to the fact that inside a lump there is an important self-screening effect that reduces very considerably the density of neutrons having energy close to a resonance maximum. The best approach to the practical solution to the problem is, therefore, a direct measurement of the number of neutrons absorbed by resonance in lumps of uranium of various sizes.

We consider below a typical empirical relation which is used for the evaluation of resonance escape probability. We note that the effective resonance inte-

Fig. 8.19 Resonance integral
as a function of Σ_s/N_0
for ^{238}U

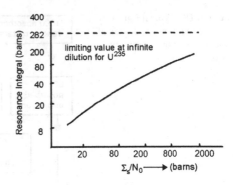

gral (8.63) depends on Σ_s/N_0, which represents the scattering cross section of the mixture per atom of absorber. This quantity is not identical with Σ_s(absorber) because Σ_s is the scattering cross section of the fuel-moderator mixture as a whole, $\Sigma_s = \Sigma_s(\text{fuel}) + \Sigma_s(\text{moderator})$. In general, the scattering contribution due to uranium can be neglected. Experiments show that the value of the effective resonance integral is essentially independent of the mass of the moderator atom, so that the dependence can be taken to be the same for all common moderators. For a given fuel, its value depends only on the fuel/moderator ratio. For ^{238}U, the empirical relation between the effective resonance integral and Σ_s/N_0 that was found to be very satisfactory for the ratio $\Sigma_s/N_0 \leq 1000$ b is:

$$\int_E^{E_0} (\sigma_a)_{eff}\frac{dE}{E} = 3.85(\Sigma_s/N_0)^{0.415} \tag{8.114}$$

The integral is shown as a function of Σ_s/N_0 in Fig. 8.19 for ^{238}U.

For ^{238}U, the integral approaches a limiting value of 282 b at infinite dilation. For the pure ^{238}U metal, the value of the integral is 9.25 b.

The numerical value of the resonance escape probability for these two limits can be obtained directly from (8.60), from which it follows that for infinite dilution, the ratio $N_0/\Sigma_s \to 0$, the exponent approaches zero, so that $p \to 1$. For the pure ^{238}U metal p, of course, approaches zero. The tendency for the variation of p with dilution is in conformity with the previous discussion and is contradictory to the variation of f with dilution.

Example 8.5 Calculate k_∞ for a homogeneous natural uranium-graphite moderated assembly which contains 200 moles of graphite per mole of uranium. Assume natural uranium to contain one part of ^{235}U to 139 parts of ^{238}U, and use these constants

Natural uranium	Graphite
$\sigma_a(U) = 7.68$b	$\sigma_a(M) = 0.0032$ b
$\sigma_s(U) = 8.3$ b	$\sigma_s(M) = 4.8$ b
	$\xi = 0.158$

Solution   If N_u is the number of uranium atoms per cm^3 and N_0 is the number of ^{238}U atoms only per cm^3, then $N_0 = 139 \times N_u/140$. Also, $N_m/N_u = 200$. Therefore

$$N_m/N_0 = 200 \times 140/139 = 201.4$$

Thermal utilization factor (f)

$$f = \frac{\Sigma_a(u)}{\Sigma_a(u) + \Sigma_a(m)} = \frac{N_u\sigma_a(u)}{N_u\sigma_a(u) + N_m\sigma_a(m)}$$

$$= \frac{1}{1 + \frac{N_m\sigma_a(m)}{N_u\sigma_a(u)}} = \frac{1}{1 + \frac{201.4 \times 0.0032}{7.68}} = 0.9226$$

Resonance escape probability p

$$\frac{\Sigma_s}{N_0} = \frac{\Sigma_s(u) + \Sigma_s(m)}{N_0} = \sigma_s(U) + \frac{N_m}{N_0}\sigma_a(M) = 8.3 + 201.4 \times 4.8 = 975 \text{ b}$$

Assuming the validity of the empirical expression (8.114), for this high value of Σ_s/N_0, we get for the effective resonance integral, $3.85(975)^{0.415} = 67$ b.

We are justified in ignoring the contribution of the uranium in the scattering since instead of $\xi\Sigma_s$, the use of the quantity $\xi_u\Sigma_s(u) + \xi_m\Sigma_s(m)$ hardly changes the result

$$\therefore \quad p = \exp-\left(\frac{67}{975 \times 0.158}\right) = \exp(-0.435) = 0.8$$

Taking $= 1$, and $\eta = 1.34$

$$k_\infty = 1 \times 1.34 \times 0.647 \times 0.923 = 0.776$$

where k_∞ means the reproduction factor for infinite dimension. This assembly is, therefore, not capable of sustaining a chain reaction. Similar computations for other ratios of moderator/uranium lead to the result shown in Fig. 8.20 for which the opposing tendency of f and p with varying fuel concentrations is also quite apparent. It is to be noted that since k_∞ is always less than unity, even with the most favorable product pf, it is not possible for a homogeneous natural uranium-graphite-moderated reactor to become critical. Same thing is true of a uranium-water-moderated reactor. This conclusion was reached before the first reactor designed by Fermi was put into commission, and led to the concept of a heterogeneous reactor. It must, however, be stressed that for a homogeneous mixture of natural uranium with D_2O as moderator, a chain reaction can be set up.

In natural uranium ^{235}U constitutes only 0.7 %. If, by isotope separation we use a higher proportion of, ^{235}U then the sample is said to be *enriched*.

It turns out that in a homogeneous assembly, chain reaction is possible, with graphite or light water as moderator provided enriched uranium is used as fuel. Figure 8.21 shows k_∞ for homogeneous mixtures with some common moderators for different molar ratios and various degrees of enrichment.

Fig. 8.20 Computations for the ratios of moderator/uranium

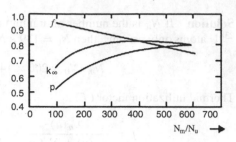

Fig. 8.21 k_∞ for homogeneous mixtures

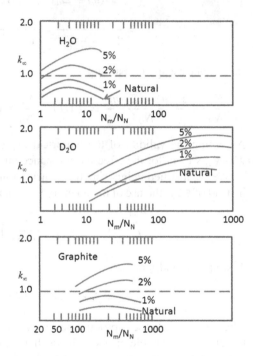

8.9 Neutron Leakage and Critical Size

It was pointed out that for an assembly of finite dimensions, the effective reproduction factor k_{eff} will be less than k_∞ by a factor L, $(L < 1)$ which is determined by leakage of the system

$$k_{eff} = k_\infty L \tag{8.115}$$

It is convenient to separate the total leakage effect into two components, a fast neutron non-leakage factor l_f, and a thermal neutron non-leakage factor l_{th}. This separation is justified by the diffusion theory which treats the diffusion of fast neutrons and that of thermal neutrons separately. W_C, therefore, set

$$L = l_{th} l_f \tag{8.116}$$

Thus

$$k_{eff} = k_\infty l_f l_{th} \qquad (8.117)$$

For a given reactor k_{eff} is defined as the ratio of the number of neutrons (at corresponding stages of the neutron cycle) in successive generations. The self-multiplication of neutrons is the underlying principle of a nuclear chain reaction. The magnitude of k_{eff} determines the speed with which the number of neutrons builds up and the rate at which nuclear fissions occur in the reactor.

If $k_{eff} > 1$, the assembly continues to produce more neutrons than it consumes, the system is said to be super-critical.

If $k_{eff} < 1$, fewer neutrons are produced than consumed. Such an assembly is said to be sub-critical.

If $k_{eff} = 1$, the rate of neutron production is exactly balanced by the rate of their consumption and in this case the assembly is called critical.

If we start with an assembly for which $k_{eff} > 1$, we can decrease k_{eff} by progressively decreasing the size, thereby increasing the neutron loss through leakage from the assembly. If this reduction be continued till $k_{eff} = 1$, then the reactor is said to have critical size, the assembly being critical, in which case the neutron losses due to all causes including leakage is exactly balanced by the rate of production in the assembly. Note that the neutron production is proportional to the volume ($\propto r^3$ for a sphere) while leakage depends on the surface area ($\propto r^2$). The ratio leakage/production $\propto 1/r$, so that by decreasing the size of the reactor it is possible to reach critical dimensions.

8.10 The Critical Dimension of a Reactor

8.10.1 One Group Theory

In the previous sections we have developed equations concerned with the production, absorption, moderation and diffusion of neutrons. We shall now apply these equations to the problem of critical size of a reactor for a given mixture of fuel and moderator. To determine the critical size of a reactor for a given mixture of fuel and moderator, we consider the case of a homogeneous unreflected reactor operating at a steady state so that the neutron flux is constant with time at all points in the assembly.

The balance equation is then

$$S + (\lambda_{tr}/3)\nabla^2\phi = \phi \Sigma_a \qquad (8.118)$$

For the source term S, the simplest assumption is that all the production, diffusion and capture of neutrons occurs at a single energy—an assumption which drastically simplifies matters but is far removed from reality. We need to consider the properties

Fig. 8.22 Infinite slab reactor

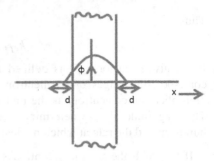

of neutrons of the same energy. If k is the multiplication factor for the system, each neutron absorbed results in the production of k_∞ neutrons as a result of fission. We may therefore, set $S = k_\infty \phi \Sigma_a$. The balance equation can, therefore, be rewritten as:

$$k_\infty \phi \Sigma_a + \frac{\lambda_{tr}}{3} \nabla^2 \phi - \phi \Sigma_a = 0 \quad \text{or} \tag{8.119}$$

$$\nabla^2 \phi + B^2 \phi = 0 \tag{8.120}$$

where

$$B^2 = 3(k_\infty - 1)\frac{\Sigma_a}{\lambda_{tr}} = \frac{k_\infty - 1}{L^2} \tag{8.121}$$

Equation (8.120) is the familiar wave equation. The boundary condition is that the flux mast fall to zero at the boundary. Actually, the flux would fall to zero at a little greater distance corresponding to the extrapolation distance (equal to $0.71\lambda_{tr}$) than the physical boundary. The other requirement imposed on the flux is that it should be finite everywhere including at large distances.

We consider below some examples which illustrate the calculation of the critical dimensions of the assembly, for well defined geometry, and use the known appropriate Laplacian.

8.10.1.1 Infinite Slab Reactor

The appropriate balance equation for the geometry of an infinite slab reactor is (Fig. 8.22)

$$\frac{d^2\phi}{dx^2} + B^2 \phi = 0 \tag{8.122}$$

The solution is obviously

$$\phi = A_1 \cos Bx + A_2 \sin Bx \tag{8.123}$$

Taking the origin at the centre of the slab, from symmetry we would expect $\phi_x = \phi_{-x}$. We must, therefore, drop off the sine function

$$\therefore \quad \phi = A_1 \cos Bx \tag{8.124}$$

Also, $\phi = 0$, when $x = \pm x_0/2$ where $x_0/2$ is one half the actual width of the slab plus the extrapolation distance $d = 0.71\lambda_{tr}$. We set

$$A_1 \cos \frac{Bx_0}{2} = 0 \quad \text{or} \tag{8.125}$$

$$B = \frac{n\pi}{x_0}, \quad +n = 1, \ 3, \ 5, \quad \text{or} \tag{8.125a}$$

$$\phi = A_n \cos \frac{n\pi x}{x_0} \tag{8.126}$$

is the general solution. The first term is called the fundamental eigen value and all the higher terms are the harmonics. For a sub-critical reactor, with an auxiliary source of neutrons, the flux distribution is obtained from (8.126), where all the harmonics can be present. For a critical reactor, all the harmonics drop off and only the fundamental eigen value is needed. Under these conditions, $B = \pi/x_0$, and

$$\phi = A \cos \frac{\pi x}{x_0} \tag{8.127}$$

Note that A is undetermined. Its absolute value is determined only by the power level of the reactor. This is an important conclusion from diffusion theory according to which at constant temperatures and pressure within the system, once a reactor has gone critical its power can be raised to any desired level without requiring any additional fuel. The power is limited only by the efficiency with which heat can be extracted from the reactor. This also means that the reactor must be always under control.

The critical volume of an infinite slab reaction is infinite.

8.10.1.2 Spherical Reactor

The balance equation is

$$\frac{d^2\phi}{dr^2} + \frac{2}{r}\frac{d\phi}{dr} + B^2\phi = 0 \tag{8.128}$$

Change of variable $v = \phi r$ in the above equation gives

$$\frac{d^2v}{dr^2} + B^2v = 0 \tag{8.129}$$

which has the solution

$$v = A_1 \sin Br + A_2 \cos Br \tag{8.130}$$

$$\therefore \quad \phi = \frac{A_1}{r} \sin Br + \frac{A_2}{r} \cos Br \tag{8.131}$$

When $r \rightarrow 0$, ϕ must be finite. Therefore, we must put $A_2 = 0$. We are finally left with

$$\phi = A_1 \frac{\sin Br}{r} \tag{8.132}$$

Imposing the boundary condition that at the critical radius $r = r_0$, $\phi = 0$ (8.132) gives

$$Br_0 = \pi, \ 2\pi, \ 3\pi$$

Taking only the fundamental eigen value for the critical assembly

$$\phi = \frac{A_1}{r} \frac{\sin \pi r}{r_0} \tag{8.133}$$

The radius r_0 is given by

$$r_0 = \pi/B = \pi \sqrt{\frac{\lambda_{tr}}{3(k-1)\Sigma_a}} \tag{8.134}$$

Volume is given by

$$V = \frac{4\pi^4}{3B^3} \cong 130/B^3$$

Actual critical radius of the sphere $= r_0 - 0.71\lambda_{tr}$, taking into account the extrapolation distance.

8.10.1.3 Rectangular Parallel-Piped Reactor

The equation is

$$\frac{\partial^2 \phi}{\partial x^2} + \frac{\partial^2 \phi}{\partial y^2} + \frac{\partial^2 \phi}{\partial z^2} + B^2 \phi = 0 \tag{8.135}$$

The above equation can be solved by the method of separation of variables and is very much the same as for (8.97) we finally obtain

$$\phi = a \cos \frac{x}{x_0} \cos \frac{y}{y_0} \cos \frac{z}{z_0} \quad \text{and} \tag{8.136}$$

$$B^2 = (\pi/x_0)^2 + (\pi/y_0)^2 + (\pi/z_0)^2 \tag{8.137}$$

If the sides are equal, i.e. $x_0 = y_0 = z_0$ then $x_0 = \sqrt{3}\pi/B$. The critical volume is approximately given by $x_0^3 = 161/B^3$.

8.10.1.4 Cylindrical Reactor

The appropriate equation is

$$\frac{\partial^2 \phi}{\partial r^2} + \frac{1}{r}\frac{\partial \phi}{\partial r} + \frac{\partial^2 \phi}{\partial z^2} + B^2 \phi = 0 \tag{8.138}$$

Substituting, $\phi = R(r)Z(z)$, and then dividing throughout by RZ, and re-arranging

$$\frac{1}{R}\frac{\partial^2 R}{\partial r^2} + \frac{1}{Rr}\frac{\partial R}{\partial r} + B^2 = -\frac{1}{Z}\frac{\partial^2 Z}{\partial z^2} \tag{8.139}$$

Obviously, each side of the above equation must be equal to a constant, say A^2

$$\therefore \quad Z = a_1 \cos Az + a_2 \sin Az \tag{8.140}$$

Because of symmetry, $a_2 = 0$. Thus

$$Z = a_1 \cos Az \tag{8.141}$$

Also, the flux must vanish at $z = \frac{1}{2}z_0$, where, z_0 is the total height of the cylinder. We get, $A = \pi/z_0$. We can, therefore, write

$$Z = a_1 \cos \frac{\pi z}{z_0} \tag{8.142}$$

Also

$$\frac{1}{R}\frac{\partial^2 R}{\partial r^2} + \frac{1}{Rr}\frac{\partial R}{\partial r} + \left(B^2 - A^2\right) = 0$$

Multiplying the above equation throughout by Rr^2

$$r^2 \frac{d^2 R}{dr^2} + \frac{r\,dR}{dr} + r^2\left(B^2 - A^2\right)R = 0$$

Put $b^2 = B^2 - A^2$, and $br = x$, $\frac{d}{dr} = b\frac{d}{dx}$, the above equation is transformed into

$$x^2 \frac{d^2 R}{dx^2} + \frac{x\,dR}{dx} + x^2 R = 0 \tag{8.143}$$

This is the Besel equation of zero order

$$\therefore \quad R = a_3 J_0(br) + a_4 Y_0(br) \tag{8.144}$$

The second solution in (8.144) must be rejected, i.e. we must put $a_4 = 0$, since at $r = 0$ the flux must be finite, and positive. Let the flux vanish at $r = r_0$ then

$$J_0(br_0) = 0 \tag{8.145}$$

The first root of the Bessel function gives

$$br_0 = 2.4$$

Therefore, the complete solution is

$$\phi = a \cos \frac{\pi z}{z_0} J(2.4 r/r_0) \tag{8.146}$$

Now

$$B^2 = b^2 + A^2 = (2.4/r_0)^2 + (\pi/z_0)^2$$

Volume of the cylinder is

$$V = \pi r_0^2 z_0 = \frac{5.76 \pi z_0^3}{B^2 z_0^2 - \pi^2}$$

The minimum volume is conditioned by, $\frac{dV}{dz_0} = 0$, yielding, $B^2 = 3\pi^2/z_0^2$. Therefore

$$V_{min} = 148/B^3$$

8.11 Reactor with a Reflector

In practice no rector will be bare. We consider the reactor to be surrounded by a material called reflector which has the property of scattering back neutrons into the reactor which would otherwise escape. The reflector should have the same properties as that of a moderator viz large scattering cross-section, low absorption cross-section for neutrons, and small atomic weight. Suitable material for reflector are graphite, water and heavy water. The use of reflector results in the flattening of neutron flux within the core and fuel saving, and improved power utilization.

The reflector reduces the neutron leakage by scattering back the escaping neutrons into the core. Furthermore, the fast neutrons which have entered the reflector from the core are moderated much more efficiently than in the core itself. This is because the absorbing material free regions in the reflector will reduce neutron losses due to resonance absorption, so that a good fraction of the fast neutrons in the reflector can reach thermal energies than in the fuel containing regions of a moderator. The improved neutron economy reduces the amount of fuel, or for the given core size, the fuel concentration compared to a bare reactor.

The flux flattening across the core for a reactor with a reflector compared to a bare reactor with neutron flux being greater at the core-reflector interface results in a higher average neutron flux for the same maximum neutron flux, Fig. 8.23. Now the power production rate is proportional to the average neutron flux, this then means that the reactor can be operated at a higher power level for the same maximum neutron flux. Also, because of flux flattening effect, the power production rate will

Fig. 8.23 Variation of
neutron flux with the reflector

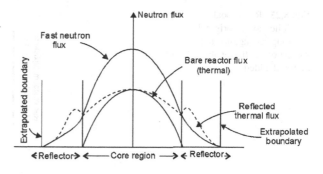

Fig. 8.24 Variation of the
critical core thickness T_C
with the reflector
thickness T_R

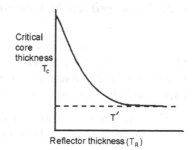

also be more uniform over the core volume, which is highly desirable particularly
for large power reactors.

Figure 8.24 shows the variation of the critical core thickness T_C with the reflector
thickness T_R. The critical thickness decreases with increasing reflector thickness.
After the reflector has reached a certain thickness T', very little reduction in critical
core thickness can be gained by a further increase of reflector thickness.

8.12 Multigroup Theory

On-group calculations are only approximate as the physical properties of core and
reflector for fast and slow neutrons are quite different. A greater accuracy is achieved
by dividing the neutrons into two or more groups and considering the behavior of
each group in the core and reflector separately. In the two-group theory thermal
neutrons comprise one group and the epithermal and fission neutrons the second
group. Average values for the physical properties for the two groups give fairly good
results. Greater accuracy is obtained by increasing the number of energy intervals
into which neutrons are divided. Balance equations can then be written for the group
in the core and reflector to obtain a set of n equations with n arbitrary constants. The
problem is complicated but can be solved with computers. For simple geometry, the
two group theory will predict critical size and mass with an accuracy of 80 % or
greater.

Fig. 8.25 Reciprocal multiplication of original counting rate against the weight of the fissionable material added

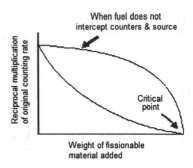

8.12.1 *Experimental Measurement of Critical Size*

Simple diffusion theory gives critical size or mass only to an accuracy of 80 % or 90 %, and because the criticality factor of a reactor must be controlled to a fraction of a percent before a reactor is installed, its critical size must be determined experimentally. There are two methods most widely used.

1. Critical assembly method (for small reactors)
2. Exponential pile method (for large reactors)

In both the methods an auxiliary source of neutrons is used.

8.12.1.1 Critical Assembly Method

For small reactors, the critical size can be determined by increasing the fuel to moderator ratio until criticality is attained. A small neutron source (10^6 neutrons/s) is placed near the centre of the assembly and the neutron intensity at various points is measured by the neutron counters. In the absence of fissionable material in the assembly, the counting rate as measured by the counters is determined by the intensity of the primary neutron source. As fissionable material is added, net neutron intensity increases. As the amount is increased, the multiplication of primary neutrons is increased, until the assembly is exactly critical, the chain reaction takes place as a result of fuel only, and the presence of the auxiliary source makes the assembly supercritical. At this point, the multiplication of the source neutrons becomes infinite. Figure 8.25 shows the reciprocal multiplication of original counting rate against the weight of the fissionable material added. Nature of the curve depends on geometry, position of counters etc. It is necessary to use a neutron source; otherwise, dependence on spontaneous fission neutrons will lead to such statistical fluctuations that whole situation may be underestimated, and too much addition of fuel may endanger the reactor itself. When criticality is reached the neutron source is withdrawn and the chain reaction is allowed to proceed at a very low power. The merit of this method is that the critical mass can be found out without actually going to criticality by simply extrapolating the reciprocal multiplication versus weight of the fissionable material curve (Fig. 8.25).

8.12.1.2 Exponential Pile Experiment

This method is used for large reactors. It involves the construction of any assembly having exactly the same lattice as the reactor under construction but with dimensions of about 1/3 the critical size. Such an assembly cannot go critical, but by the use of an artificial neutron source, a steady state neutron flux distribution throughout the assembly can be realized. The so-called exponential pile is a rectangular column built from the moderator material with one of its faces containing the neutron source. The neutron flux in such a column falls off exponentially in a direction perpendicular to the face containing neutron source, i.e. $\phi \propto \exp(-x/L_1)$ where L_1 is a constant which is related to the diffusion length through (8.109). Thus L can be found out; and since diffusion length is related to the buckling constant and critical size is related to the buckling constant, critical size of the given rectangular pile can be known from

$$B_g^2 = (\pi/x_0)^2 + (\pi/y_0)^2 + (\pi/z_0)^2 \tag{8.147}$$

For cubical geometry $x_0 = y_0 = z_0$ and critical volume in given by $V = \frac{161}{B_g^3}$, where B_g is known as geometrical buckling.

8.13 Fast Neutron Diffusion and the Fermi Age Equation

We shall now consider the diffusion of prethermal neutrons during the slow-down stage. Neutrons at these energies cannot be considered monoenergic as was done for the thermal neutron group since in the course of slowing down the fast and prethermal neutrons undergo considerable energy changes. The neutron density per energy interval, $n(E)$ depends on the difference between the slowing down density $q(E + \Delta E)$ into the energy interval ΔE and the slowing-down density $q(E)$ out of it. For thermalized neutrons, this difference vanishes because a thermal equilibrium between the neutrons and their surroundings is established which simply means that the rate of neutron flow into energy interval ΔE is equal to the rate of neutron flow out of it.

We assume that the slowing-down is a continuous process, which means that a large number of collisions are involved in the course of thermalization of fast neutrons. We further assume that the scattering M.F.P. λ_s is nearly constant. The assumption of continuous energy loss is valid for most moderators except for the very lightest, such as hydrogen and deuterium.

Consider the slowing-down of neutrons in a region of moderator in which neutron absorption is absent ($\sigma_a = 0$) and the sources are also absent, i.e. ($Q = 0$).

Let there be n neutrons in a unit volume with energies between E and $E + \Delta E$, then the only physical processes which can alter n are assumed to be:

 (i) Diffusion of neutrons into or out of the unit volume.
(ii) Slowing down of neutrons into the energy interval ΔE and out of it.

A steady state will be reached, i.e. the number of neutrons in the given unit volume and energy interval will remain constant if the neutrons diffusing out of the volume are compensated by an equal number of neutrons slowing down into and remaining in the energy interval ΔE.

The rate of neutron diffusion is:

$$-D\nabla^2 n = -\frac{\lambda_{tr}}{3} v\nabla^2 n \tag{8.148}$$

where D is called the diffusion coefficient.

The number of neutrons slowing down into the energy interval ΔE and remaining is given by the excess of neutrons flowing into ΔE over the number of neutrons leaving it

$$q(E + \Delta E) - q(E) = \frac{\partial q}{\partial E} \Delta E \tag{8.149}$$

$$\downarrow \qquad\qquad\qquad \downarrow$$

$$\text{Influx} \qquad\qquad \text{Outflow}$$

Hence, (8.148) and (8.149)

$$\frac{\partial q}{\partial E} \Delta E = -\frac{\lambda_{tr}}{3} v\nabla^2 n \tag{8.150}$$

Now, $q(E) = \phi(E)E\xi \Sigma_s$. It follows from successive differentiation of the above equation

$$\nabla^2 q = Ev\xi \Sigma_s \nabla^2 n \tag{8.151}$$

Substituting the value of $\nabla^2 n$ from (8.151) in (8.150) and re-arranging

$$\nabla^2 q = \frac{\partial q}{\frac{-\lambda_s \lambda_{tr}}{3\xi E} \partial E} \tag{8.152}$$

Introduction of a new variable τ, such that

$$d\tau = \frac{-\lambda_{tr} \lambda_s dE}{3\xi E} \tag{8.153}$$

leads to

$$\nabla^2 q - \frac{\partial q}{\partial \tau} = 0 \tag{8.154}$$

This is called Fermi age equation and the variable τ the Fermi age or Neutron age. The subsidiary condition imposed on τ is

$$\tau = \int_E^{E_0} d\tau, \qquad \tau(E_0) = 0 \tag{8.155}$$

It is seen that the dimension of τ is (length)^2. The age has nothing to do with time. It is thus called because τ appears in the same way as time appears in the standard heat-diffusion equation. The age equation contains a complete description of the neutron density distribution in both energy and space co-ordinates for neutrons undergoing moderation. Note that the Fermi age equation is a time-independent or steady-state equation as it does not contain time explicitly.

If λ_s and λ_{tr} are assumed constant over the slowing-down energy range integration of (8.153) yields

$$\int_E^{E_0} d\tau = \tau(E_0) - \tau(E) = \frac{-\lambda_{tr}\lambda_s}{3\xi} \int_E^{E_0} d(\ln E)$$

$$= \frac{-\lambda_{tr}\lambda_s}{3\xi} \ln(E_0/E) \qquad (8.156)$$

But, $\frac{1}{\xi}\ln(E_0/E) = C$, is the average number of collisions a neutron undergoes with the moderator nuclei, the energy being reduced from E_0 to E. We can then write

$$\tau(E) = \frac{\lambda_{tr}\lambda_s C}{3} \qquad (8.157)$$

In this expression, $C\lambda_s$ represents the total ziz-zag path of a neutron between the beginning and end of slow-down. If we set

$$\Lambda_s = c\lambda_s \qquad (8.158)$$

We note that Λ_s is quite analogous to λ_s in (8.91). We can then by analogy define

$$\tau_0 = \frac{\lambda_{tr}\Lambda_s}{3} = L_f^2$$

where L_f is called the fast diffusion length. It is a measure of the distance a fission neutron travels from the point of creation till it is thermalized.

We solve Fermi equation (8.154) for the particular case of a point source emitting fast monoenergetic neutrons of strength Q inside a moderator. Using the Laplacian in spherical co-ordinates

$$\frac{d^2 q}{dr^2} + \frac{2dq}{rdr} - \frac{\partial q}{\partial \tau} = 0 \qquad (8.159)$$

The solution is found to be

$$q(r,\tau) = \frac{Q}{(4\pi\tau)^{3/2}} \exp\left(-r^2/\tau\right) \qquad (8.160)$$

The neutron slowing-down distribution for a given τ is thus Gaussian. It has its maximum value at the origin ($r = 0$), and for different choices of the parameter τ, the maximum, is higher the smaller the value of τ (Fig. 8.26). It is seen from (8.156) that the lower the value of E, the greater the corresponding τ. The neutron age is, therefore a direct measure of the degree of moderation of the neutrons.

Fig. 8.26 Neutron
slowing-down distribution for
a given τ

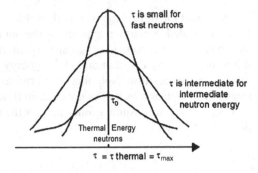

The root mean square distance for a given neutron age can be obtained from

$$\langle r^2 \rangle = \int_0^\infty \frac{r^2 q(r) 4\pi r^2 dr}{\int_0^\infty q(r) 4\pi r^2 dr} = 6\tau \tag{8.161}$$

In particular, if the terminal energy is equal to thermal energy so that we may write the corresponding age as $\tau = \tau_0$; then

$$\langle r^2 \rangle = 6\tau_0 = 6L_f^2 \tag{8.162}$$

Thus, the neutron age τ is equal to $1/6$ of the mean square distance from the point of creation to the point where their energy is reduced to a value E corresponding to that τ. The role of neutron age is therefore, quite analogous to that of L^2 in thermal diffusion. For the respective processes of fast diffusion and thermal diffusion, τ_0 and L^2 are each $1/6$ of the mean square distance traveled by a neutron from the point of its origin to the point of its termination. The sum of τ_0 and L^2 is called the migration area M^2.

8.13.1 Correction for Neutron Capture

In the derivation of the age equation, the neutron absorption by the moderator was ignored. Should the moderator have a relatively weak capture cross-section for neutrons above thermal energies, the differential equation (8.154) would contain an additional term which is linear in q, but the form of the solution of the age equation is not affected. If q is the solution of the age equation with zero absorption, and q' is the solution of the modified equation with absorption, it can be shown that

$$q' = pq \tag{8.163}$$

where, p is the resonance escape probability for the medium in which neutrons slow down.

8.13.2 Application of Diffusion Equation to a Thermal Reactor

Let $Q(x, y, z)$ be the number of fast neutrons produced per unit time and unit volume at each position in a lattice. These neutrons diffuse through the mass and are slowed down. During this process, some of the neutrons are absorbed at resonance. Let $q(x, y, z)$ be the number of neutrons per unit time and unit volume which become thermal at the position x, y, z; q is called the 'density of nascent thermal neutrons'.

The balance equation which expresses the local balancing of all processes where by the number of thermal neutrons at each place tends to increase or decrease may be written as

$$\frac{\lambda_{tr}}{3}\nabla^2\phi - \phi\Sigma_a + q = 0 \tag{8.164}$$

where the source of thermal neutrons is given by the slowing-down density which is a function of the space coordinates r and the neutron age τ. The first term represents the increase in number of neutrons due to diffusion; the second one, the loss of neutrons due to absorption, and the third, the effect of the nascent thermal neutrons. We must, therefore, first find the value of q for thermal energies. This is done by solving (8.154) as indicated below. $\lambda_a = 1/\Sigma_a$ is the absorption mean free path of thermal neutrons.

8.13.2.1 Thermal Neutron Source as Obtained from the Fermi Age Equation

The present treatment applies to homogeneous assembly, but would be equally applicable to a heterogeneous assembly if its unit cell, i.e. the representative unit from which the whole lattice can be imagined to have been built up is very much smaller than the critical dimension of the reactor. By treating such a heterogeneous assembly as a homogeneous system, we neglect the local depressions in the flux density that occur at the location of the fuel lumps and we consider only the large scale variation of the flux across the linear extension of the reactor. We use the method of separation of variables. Let

$$q(r, \tau) = R(r)T(\tau) \tag{8.165}$$

Differentiating twice with respect to space co-ordinates only

$$\nabla^2 q = T(\tau)\nabla^2 R \tag{8.166}$$

and differentiating with respect to τ only

$$\frac{\partial q}{\partial \tau} = R(r)\frac{\partial T}{\partial \tau} \tag{8.167}$$

Substituting (8.166) and (8.167) in the Age equation (8.154), we get

$$T(\tau)\nabla^2 R = R(r)\frac{\partial T}{\partial \tau} \quad \text{or}$$

$$\frac{\nabla^2 R}{R} = \frac{1}{T}\frac{\partial T}{\partial \tau} \tag{8.168}$$

Since each side of (8.168) is independent of the variables of the other side, each side must be equal to the same constant, say $-B^2$, so that

$$\frac{1}{T}\frac{\partial T}{\partial \tau} = -B^2 \quad \text{and} \tag{8.169}$$

$$\frac{\nabla^2 R}{R} = -B^2 \quad \text{or} \tag{8.170}$$

$$\nabla^2 R + B^2 R = 0 \tag{8.171}$$

The solution of (8.169), is

$$T = T_0 \exp(-B^2\tau) \tag{8.172}$$

where T_0 is the initial value of T when $\tau = 0$. Since q decreases with increasing age because of neutron losses, $T < T_0$, so that B^2 must be real and positive number.

The slowing-down density at the beginning of the slowing-down process, q_0 is given by (8.165)

$$q_0 = R(r)T(0) = R(r)T_0 \tag{8.173}$$

We shall now express q_0 in terms of the physical properties of the assembly. The number of neutrons per cm^3 per s that become available for slowing down is given by the rate of production of fission neutrons which is equal to $\Sigma f \eta$ per thermal neutron that is absorbed (at the present, we omit the resonance absorption).

Since the rate of thermal neutron absorption per cm^3 is $\phi \Sigma_a$, the rate of production of fission neutrons per cm^3 is $(\varepsilon f \eta)(\phi \Sigma_a)$. This is also the rate per cm^3 at which fast neutrons become available for slowing down, which is the same as the initial slowing-down density q_0. Hence

$$q_0 = \phi(r)\Sigma_a \varepsilon f \eta \tag{8.174}$$

but

$$q = T(\tau)R(r) = T_0 \exp(-B^2\tau)R(r) = q_0 \exp(-B^2\tau) \tag{8.175}$$

where we have used (8.172) and (8.173)

$$\therefore \quad q = \exp(-B^2\tau)\phi\, \Sigma_a \varepsilon f \eta \tag{8.176}$$

where we have used (8.174). But, we must allow for the absorption during the prethermal stage. The source term, with the necessary correction for resonance escape then becomes

$$q' = qp = \exp(-B^2\tau)\phi\, \Sigma_a \varepsilon f \eta p = K_\infty \Sigma_a \phi \exp(-B^2\tau) \tag{8.177}$$

Substituting this in (8.164)

$$\frac{\lambda_{tr}}{3}\nabla^2\phi - \phi\Sigma_a + k_\infty \Sigma_a \phi \exp(-B^2\tau) = 0 \tag{8.178}$$

Multiplying this equation throughout by $3/\lambda_{tr}$ and using the value of L from (8.91) we may write:

$$\nabla^2\phi + \frac{\phi}{L^2}\left[k_\infty \exp(-B^2\tau) - 1\right] = 0 \tag{8.179}$$

This is the Fermi age equation with correction for neutron capture.

8.13.3 Critical Equation and Reactor Buckling

We shall now show that the numerical value of the constant B^2 is determined by the neutron flux distribution ϕ inside the assembly. For this purpose we evaluate $\nabla^2 q'$ and $\frac{\partial q'}{\partial \tau}$ for the slowing down density q' as given by (8.177)

$$\nabla^2 q' = k_\infty \Sigma_a \exp(-B^2\tau)\nabla^2\phi$$

$$\frac{\partial q'}{\partial \tau} = k_\infty \Sigma_a \phi(-B^2)\exp(-B^2\tau)$$

$$\therefore \quad \nabla^2 q' - \frac{\partial q'}{\partial \tau} = k_\infty \Sigma_a \exp(-B^2\tau)(\nabla^2\phi + B^2\phi) = 0 \quad \text{or}$$

$$\nabla^2\phi + B^2\phi = 0 \tag{8.180}$$

Since the thermal flux distribution $\phi(r)$ across the reactor depends on the size, shape, and the general geometry of the assembly, the value of B^2 is similarly determined by the geometry of the reactor. In fact it is given by (8.180)

$$B^2 = \frac{-\nabla^2\phi}{\phi} \tag{8.181}$$

Because of its intimate connection with the geometry of the nuclear assembly, B^2 as determined by (8.180) or (8.181) is called the geometrical buckling, and is usually denoted more specifically by B_g^2.

The quantity $-\nabla^2\phi/\phi$ is essentially the second derivative of ϕ divided by the function ϕ itself, which describes the curvature or bending of ϕ or buckling of neutron flux. Combining (8.180) and (8.179), we get

$$\frac{k_\infty \exp(-B^2\tau)}{1+L^2B^2} = 1 \tag{8.182}$$

This is the critical equation and the left side of the above expression is equal to k_{eff}, or the criticality

$$k_{eff} = \frac{k_\infty \exp(-B^2\tau)}{1+L^2B^2} \tag{8.183}$$

Equation (8.182) is a transcendental equation for B^2; it determines B^2 in terms of the physical properties of the reactor materials which are involved through k_∞, τ and L^2. The numerical value of B^2 as determined from this equation is, therefore, called the material buckling of the reactor and is designated more specifically by B_m^2.

When the reactor is critical, the geometrical buckling as determined by (8.180) is equal to the material buckling B_M^2 as obtained from (8.182). This assumption was in fact made when we combined (8.179) and (8.180). In general, the choice of B^2 that satisfies the mathematical requirements of (8.180) is not unique, but the smallest numerical value of B^2 that satisfies Eq. (8.180) is the one that has physical significance for our problem. It is also clear from (8.183) that the dimensions of B^2 are that of reciprocal of area, cm², since $B^2\tau$ and L^2B^2 must be pure numbers. We shall see that increasing the geometrical dimensions of a critical reactor causes the numerical value for the geometrical buckling B_g^2 to decrease. But, increasing the size of a reactor beyond its critical size results in a k_{eff} greater than unity. On the other hand, the material buckling B_m^2 as given by (8.182) depends only on the material properties of the assembly and does not change with the reactor size. Hence we conclude that for a super-critical reactor $k_{eff} > 1$, and B_m^2 must be greater than B_g^2.

Similarly, a reduction in size which makes the reactor sub-critical ($k_{eff} < 1$) causes B_g^2 to increase without causing a similar change in B_m^2; so that we have in that case B_g^2 greater than B_m^2.

8.13.4 The Non-leakage Factors

Previously, we had introduced the non-leakage factors l_f and l_{th} for the fast and thermal neutrons respectively. We shall now relate them to the material properties of the assembly. We rewrite Eq. (8.177) as

Thermal Diffusion Rate
$$\downarrow$$

$$-\frac{\lambda_{tr}\nabla^2\phi}{3} + \Sigma_a\phi = k_\infty \Sigma_a\phi \exp(-B^2\tau) \tag{8.184}$$

Thermal absorption rate   Production rate of thermal neutrons

Since $\Sigma_a\phi$ is the rate of thermal neutron absorption per cm^3, it follows that $k_\infty \Sigma_a\phi$ is the rate of fission neutrons produced per cm^3. For a reactor of infinite size all these neutrons will survive through the slowing-down stage and reach thermal energies. Since our reactor has finite dimensions, some of the neutrons on reaching the boundary will leak outside. The non-leakage probability will be given by the ratio of the actual production rate of thermal neutrons, $k_\infty \Sigma_a\phi \exp(-B^2\tau)$ over the maximum possible rate for a reactor of infinite size, $k_\infty \Sigma_a\emptyset$,

$$\therefore \quad \text{Non-leakage factor} = \frac{k_\infty \Sigma_a\phi \exp(-B^2\tau)}{k_\infty \Sigma_a\emptyset} = \exp(-B^2\tau) \tag{8.185}$$

This then is the fraction of fast neutrons that does not leak out of the assembly during slowing down and reaches thermal energies. Therefore

$$l_f = \exp(-B^2\tau) \tag{8.186}$$

The left side of (8.184) represents the rate at which thermal neutrons disappear from the reactor. The thermal non-leakage factor is given by the ratio of absorption rate and absorption + thermal diffusion rate, i.e.

$$l_{th} = \frac{\Sigma_a\phi}{\Sigma_a\phi - \frac{\lambda_{tr}}{3}\nabla^2\phi} \tag{8.187}$$

We shall now use (8.180) and replace $\nabla^2\phi$ by $-B^2\phi$, and also use $\Sigma_a = 1/\lambda_a$. We can write:

$$l_{th} = \frac{1}{1 + \frac{\lambda_{tr}\lambda_a B^2}{3}} = \frac{1}{1 + B^2 L^2} \tag{8.188}$$

It is instructive to note that by multiplying (8.186) and (8.188), we get

$$l_f l_{th} = \frac{\exp(-B^2\tau)}{1 + B^2 L^2}$$

but

$$k_{eff} = k_\infty l_f l_{th} = \frac{k_\infty \exp(-B^2\tau)}{1 + B^2 L^2} \tag{8.189}$$

a formula which is identical with (8.183).

8.13.5 Criticality of Large Thermal Reactors

It was pointed out that B is related reciprocally to the dimensions of the reactor, so that for large reactors, B^2 becomes small enough to permit an expansion of the exponential term, and if τ is not large, to omit terms containing higher orders of B^2 with negligible error. Thus

$$k_{eff} = \frac{k_\infty \exp(-B^2\tau)}{1 + L^2 B^2} = \frac{k_\infty}{(1 + L^2 B^2)\exp(B^2\tau)} = \frac{k_\infty}{(1 + L^2 B^2)(1 + B^2\tau)}$$

$$= \frac{k_\infty}{1 + B^2(L^2 + \tau)} \tag{8.190}$$

The quantity $L^2 + \tau$, is demoted by M^2 and is known as the Migration area, and M as the migration length. Therefore

$$M^2 = L^2 + \tau \tag{8.191}$$

Hence, for large thermal reactor

$$k_{eff} = \frac{k_\infty}{1 + B^2 M^2} = 1 \tag{8.192}$$

8.13.6 The Diffusion Length for a Fuel-Moderator Mixture

When we deal with a mixture of fuel and moderator, the fuel hardly affects the scattering properties of the material, but has a marked effect on the absorbing properties. Since the ratio of the moderator to fuel in a reactor is very large, the slowing-down and diffusion properties of the mixture are those of the moderator, and the value of λ_{tr} to be used in the formula $L^2 = \frac{\lambda_{tr}\lambda_a}{3}$ is that for pure moderator. On the other hand since the neutron absorbing properties are definitely affected by the presence of the fuel, the λ_a to be used must be that for the mixture as a whole. Thus

$$\lambda_a = \frac{1}{\Sigma a} = \frac{1}{\Sigma_{ao} + \Sigma_{am}}$$

$$L^2 = \frac{\lambda_{tr}\lambda_a}{3} = \frac{\lambda_{tr}}{3\Sigma_a} = \frac{\lambda_{tr}}{3(\Sigma_{ao} + \Sigma_{am})}$$

But, the thermal utilization factor

$$f = \frac{\Sigma_{ao}}{\Sigma_{ao} + \Sigma_{am}} \quad \text{or} \tag{8.193}$$

$$1 - f = \frac{\Sigma_{am}}{\Sigma_{ao} + \Sigma_{am}} \quad \text{or}$$

$$\frac{1}{\Sigma_{ao} + \Sigma_{am}} = \frac{(1-f)}{\Sigma_{am}} = (1-f)\lambda_{am}$$

$$\therefore \quad L^2 = \frac{\lambda_{tr}\lambda_{am}(1-f)}{3} = L_m^2(1-f) \tag{8.194}$$

where L_m is the diffusion length for pure moderator. This shows that the diffusion length for a fuel-moderator mixture is smaller than that for the pure moderator by a factor $\sqrt{1-f}$. It can be shown that in practice (8.193) is also valid for heterogeneous assemblies. For the same fuel/moderator ratio, f is smaller for a heterogeneous assembly than for a homogeneous assembly; the diffusion length for the former will be greater than for the latter type of fuel-moderator arrangement.

8.13.7 k_∞ for a Heterogeneous Reactor

In the case of a heterogeneous reactor, not only is the average thermal neutron flux in the regions occupied by fuel and moderator different, but also the volumes occupied by these components are different. This fact must be allowed for, while calculating the absorption rate of thermal neutrons. Figure 8.27 shows a unit cell and its equivalent radius r_1. The fuel rod of radius r_0 is also indicated. Let V_u and V_m represent the volumes of fuel and moderator respectively. Then, the absorption rate in fuel is given by $\Sigma_{au}\overline{\phi}_u V_u$ and for the moderator $\Sigma_{am}\overline{\phi}_m V_m$. The fraction of thermal neutrons absorbed by uranium fuel as compared to the total number of thermal neutron absorptions in the assembly is then

$$f = \frac{\Sigma_{au}\overline{\phi}_u V_u}{\Sigma_{au}\overline{\phi}_u V_u + \Sigma_{am}\overline{\phi}_m V_m} = \frac{1}{1 + \frac{\Sigma_{am}\overline{\phi}_m V_m}{\Sigma_{au}\overline{\phi}_u V_u}} \tag{8.195}$$

Thermal neutron absorption such as in 'poisons', structural materials, coolants, etc., may be taken into account by including similar terms in the denominators of (8.195). Equation (8.195) reduces to the simple formula (8.193) appropriate for homogeneous assembly on putting $\overline{\phi}_m = \overline{\phi}_u$, and $V_m = V_u$. Since the thermal disadvantage factor defined by the ratio $\overline{\phi}_m/\overline{\phi}_u > 1$ and also $V_m/V_u > 1$, it is obvious that $f_{het} < f_{hom}$, that is the thermal utilization factor for a heterogeneous assembly is smaller than that for a homogeneous assembly, both using the same amount of fuel and moderator.

The value of the disadvantage factor varies with the size of the fuel elements and the lattice spacing (also called pitch). For a given fuel-to-moderator ratio, the thermal advantage factor increases with the diameter of the cylindrical fuel elements, and consequently the thermal utilization factor must decrease. Figure 8.28 shows f as function of the fuel element radius r_0 for three values of radius r_1 of the equivalent cylindrical unit cell.

Fig. 8.27 Unit cell and
equivalent cell radius

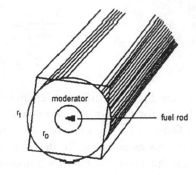

Fig. 8.28 Fuel element
radius r_0 graphs showing the
behavior of $\varnothing$ with fuel
element radius for unit cells
of different radii

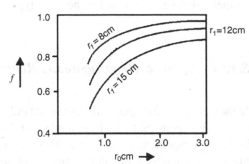

8.13.7.1 Resonance Escape Probability

Previously, we have remarked that an increase in k_∞ in a heterogeneous assembly of
fuel and moderator is primarily caused due to decrease in the resonance absorption
of neutrons by ^{238}U. Theoretical calculations of p are rendered complicated because
all the resonances in ^{238}U are not known with sufficient accuracy, and therefore, the
integral in the expression for p must be replaced as shown earlier by the effective
resonance integral. When evaluating the effective resonance integral, we can proceed on the assumption that it can be considered to consist of two partial contributions (a) a surface contribution which is proportional to surface area of the uranium
fuel element, and (b) a volume contribution which is proportional to the volume of
the uranium lump. The division into two parts means this, that the surface absorption implies the absorption of those neutrons which have been slowed down in the
moderator to an energy corresponding to a strong U^{238} absorption line, whereas all
other absorptions are treated as a volume absorption effect which takes place inside
the uranium lump exclusively.

When both the volume part and the surface contribution are taken into account,
the effective resonance integral is given by

$$\int_E^{E_0} (\sigma_a)_{eff} \frac{dE}{E} = (9.25 + 24.7 \frac{S}{M}) \text{ b} \qquad (8.196)$$

$$\underbrace{}_{\text{Volume part}} \qquad \underbrace{\phantom{24.7 \frac{S}{M}}}_{\text{Surface part}}$$

where S is the surface area expressed in cm^2, and M the mass of the fuel element in grams. Formula (8.196) is limited to uranium rods with diameter greater than 0.6 cm.

Because of the surface screening action of the uranium the neutron flux inside the fuel elements (ϕ_0) is smaller than in the moderator (ϕ_m). Furthermore, the difference in resonance neutron flux for the scattering region (moderator) and the absorption region (uranium) must be taken into account in the expression for p.

We must also take into account the fact that the volume occupied by the fuel elements V_0 and that occupied by the moderator V_m are not equal in the heterogeneous lattice arrangement unlike for the homogenous assembly.

When these weighting factors are included the resonance escape probability for the heterogeneous assembly becomes

$$p = \exp\left[-\frac{N_0 V_0 \phi_0}{\xi \Sigma_s V_m \phi_m} \int_E^{E_0} (\sigma_a)_{eff} \frac{dE}{E}\right] \tag{8.197}$$

It is interesting to note that if $V_0 = V_m$, and $\phi_0 = \phi_m$, expression (8.197) reduces to (8.63) appropriate for the homogeneous arrangement.

As a first approximation we can set $\phi_m = \phi_0$, so that (8.197) simplifies to

$$p = \exp\left[-\frac{N_0 V_0}{\xi \Sigma_s V_m} \int_E^{E_0} (\sigma_a)_{eff} \frac{dE}{E}\right] \tag{8.198}$$

Expression (8.198) gives an error of only 1 %. For fuel elements of uranium with the shape of long cylinders the area of the caps will be negligible compared to the mantle area of the cylinder so that

$$\frac{S}{M} = \frac{2\pi rl}{\pi r^2 l \rho} = \frac{2}{r\rho} \quad (\rho = 18.7 \text{ g/cm}^3)$$

which when used in (8.196) and (8.197) gives for the resonance escape probability for this particular case

$$p = \exp\left[-\frac{V_0 \phi_0 N_0 (9.25 + 49.4/r\rho)}{V_m \phi_m \xi \Sigma_s}\right] \tag{8.199}$$

(Note that the term in parenthesis is in units of barns.) Inspection of this expression shows that increasing the size of the fuel elements by increasing r, but keeping V_0/V_m ratio unchanged, decreases $1/r$ term in the exponent of (8.199), so that p must increase. In addition, increasing r causes an increased resonance neutron absorption by the fuel element because of its greater surface area, and this in turn causes a further depression of $\overline{\phi}_0$, so that as a result, $\overline{\phi}_0/\overline{\phi}_m$ decreases also. This will lead to a further decrease of the exponent, so that p must increase.

As we know, the filtering action of the uranium surface layer causes also a decrease of the thermal flux in the interior of the fuel element, so that increasing the radius r_0 of the fuel element must increase the thermal disadvantage factor. Inspection of (8.195) shows that, as a consequence of this, f must become smaller. This

Fig. 8.29 Variation of p with
r_0 for different cylindrical
radii

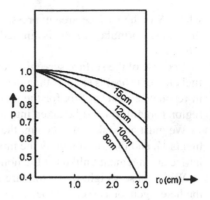

indicates that changing the fuel element size affects p and f in opposite ways, increasing one and decreasing the other. Figure 8.29 shows the variation of p with r_0 for different cylindrical radii.

8.13.7.2 Fast Fission Factor

Figure 8.30 shows the variation of ε with fuel rod radius r_0 for natural uranium. It is seen to increase slowly with r_0. Figure 8.31 shows the variation of k_∞ with the fuel rod radius r_0 (cm). Figure 8.32 shows, the variation of k_∞ for a heterogeneous natural uranium-graphite assembly for various molar ratios of graphite and uranium for different fuel rod radius. It is concluded that by the use of a heterogeneous assembly, chain reaction can be established with natural uranium as fuel and graphite as moderator.

Example 8.6   An unreflected thermal reactor consists of a lattice of uranium rods surrounded by graphite. The appropriate constants are as follows: One neutron captured in one atom of natural uranium produces $\eta = 1.308$ neutrons. The fast fission factor $\varepsilon = 1.0235$. The resonance escape probability $p = 0.893$. The thermal utilization factor $f = 0.8832$. The slowing down length $L_s = 20.05$ cm. The diffusion length $L = 17.03$ cm. If the reactor is to have cubical geometry find out the critical dimension.

Solution

$$k_\infty = 1.308 \times 1.0235 \times 0.893 \times 0.8832 = 1.055$$
$$\tau = L_s^2 = (20.05)^2 = 402 \text{ cm}^2$$

Critical equation is

$$\frac{k_\infty \exp(-B^2\tau)}{1 + L^2 B^2} = 1 \quad \text{or}$$

Fig. 8.30 Variation of ε with fuel rod radius r_0 for natural uranium

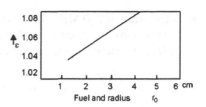

Fig. 8.31 Variation of k_∞ with the fuel rod radius r_0 (cm)

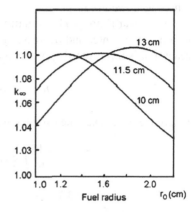

Fig. 8.32 Variation of k_∞ for a heterogeneous natural uranium-graphite assembly for various molar ratios of graphite and uranium for different fuel rod radius

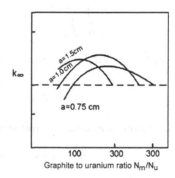

$$k_\infty \simeq 1 + B^2\left(L^2 + \tau\right)$$

$$\therefore \quad B^2 = \frac{k_\infty - 1}{L^2 + \tau} = \frac{0.055}{(17.03)^2 + 402} = 0.0000795$$

but $a^2 = 3\pi^2/B^2 = 3\pi^2/0.0000795$, giving $a = 611$ cm.

8.13.8 Power

It is useful to derive an appropriate relationship between the power produced by a reactor and the intensity of thermal neutrons inside it. Roughly, 50 % of the thermal neutrons absorbed in a reactor, give rise to fission, and the energy released per fission is of the order of 200 MeV. This corresponds to about 1.6×10^{-4} ergs per thermal neutron absorbed. Since the number of thermal neutrons absorbed per unit volume is vn/λ_a, the energy produced is approximately

$$\frac{vn \times 1.6 \times 10^{-4}}{\lambda_a} = vn \times 1.6 \times \frac{10^{-4}}{316} = 5 \times 10^{-7} vn \text{ ergs/cm}^3/\text{s}$$

where we have used $\lambda_a = 316$ cm.

The power is not produced uniformly throughout the reactor because n is a maximum at the center and decreases to zero at the edge of the reactor. For a cubical reactor, n is represented approximately by

$$n = n_0 \cos \frac{\pi x}{a} \cos \frac{\pi y}{a} \cos \frac{\pi z}{a}$$

where we have taken the origin at the center of the cube. The ratio

$$\overline{\phi}/\phi_0 = \frac{1}{V} \int \cos \frac{(\pi x)}{a} \cos \frac{(\pi y)}{a} \cos \frac{(\pi z)}{a} dV$$

but $dV = dxdydz$, and $V = a^3$

$$\overline{\phi}/\phi_0 = \frac{1}{a^3} \int_{-a/2}^{a/2} \cos(\pi x/a) dx \int_{-a/2}^{a/2} \cos(\pi y/a) dy \int_{a/2}^{a/2} \cos(\pi z/a) dz$$

Each of the integrals is equal to $2a/\pi$

$$\overline{\phi}/\phi_0 = \frac{1}{a^3}(2a/\pi)^3 = 8/\pi^3 = 0.256$$

Therefore, we obtain the following formula for the power

$$W = 0.256 \times 5 \times 10^{-7} n_0 v a^3 = 1.28 \times 10^{-7} n_0 v a^3$$

In our example, if we take $a = 611$ cm we obtain:

$$W = 29 n_0 v \text{ ergs/s}$$

When the reactor is operating at a power of 1 Mega Watt, the flux of thermal neutrons at the center is, therefore

$$n_0 v = 10^6 \times 10^7 / 29 = 3.4 \times 10^{11} \text{ neutrons/cm s}$$

8.14   The Chain Reaction Requirements

(a) *Fuel:* ^{235}U, ^{239}Pu, and ^{233}U are the only three fissionable nuclides with slow neutrons. For ^{235}Pu the fraction of delayed fission neutrons is greatest; this fact

is important for reactor control. For ^{239}U breeding gain is high. ^{233}U is poten-
tially the most abundant fuel as the fertile material from which it can be derived
viz thorium which is four times as abundant as uranium. (In the breeder reactor a fertile element is converted into a fissionable element.) Although ^{238}U is
fissionable with fast neutrons, it is not suitable since neutrons given out from
fission do not have enough energy to propagate the reaction.

Pure uranium cannot be used for the construction of nuclear chain reactor, because ^{238}U has very high resonance peaks for absorption through radiative capture. Fission neutrons are generated at about 1 MeV energy, and at this energy
natural uranium has $\sigma_s \simeq 4$ b and $\sigma_f = 0.015$ b. Thus in an assembly of pure
uranium most fission neutrons would simply be scattered until they reach energies corresponding to the resonance region of ^{238}U at which point they would
be rapidly absorbed before they have a chance to fission more of ^{235}U atoms. In
principle one can separate out the isotopes of uranium so that ^{235}U is enriched.
This has the effect of increasing σ_f and reducing radiative captures.

(b) *Moderators*: For natural uranium $\sigma_f = 0.015$ b at 1 MeV, and $\sigma_f = 3.9$ b at thermal energy. Hence, moderation of neutrons is desirable. Three properties determine the choice of a moderator since the moderating ratio is given by $\xi \Sigma_s / \Sigma_a$:

    (i) Large ξ (element should be light say below oxygen in the periodical table)
    (ii) Large Σ_s
    (iii) Small Σ_a

Heavy water is the best liquid moderator because of low atomic weight and
small Σ_a of D_2. Ordinary water has a fairly large cross section for thermal
neutrons and can be used in rectors having enriched fuel. Use of H_2O and D_2O
has the advantage in that by recirculation though an external heat exchange they
can serve as coolant as well as moderator. Beryllium oxide and carbon (graphite)
find considerable application where a solid moderator is needed, although all are
inferior to heavy water. When graphite is used as a moderator, it is important that
it is completely free from foreign matter. Helium is not suitable as a moderator
because it is a gas; and the other two light elements lithium and boron have too
high an absorption cross section.

(c) *Reactor Coolants*: Most nuclear chain reactors operate at high enough power
levels so that some form of cooling is required. Theoretically, a critical reactor
could be run at any power level (as pointed out earlier). Power output is limited
by the efficiency for removal of heat, and thermal tolerances of the materials
of the reactor assembly. The requirements for a reactor coolant are rather rigid.
Such a material must have suitable thermal properties; must be non-corrosive
to materials in the reactor, must be stable to radiations to which it is exposed,
and above all must have a very small σ_a for neutrons. The coolant merely serves
to keep the operating temperature down to a reasonable value. For low power
reactors, the coolants that are in use are air, water, heavy water, and mercury.
For power producing reactors where a high operating temperature is desired,
liquid metals such as mercury, sodium, lead, bismuth and potassium may be
used. The power of a chain reactor is dependent on how rapidly heat is removed

from the system, there being practically no upper limit to the power density. For this reactor coolants with high coefficient of heat transfer are desired.

(d) *Structured Materials*: This may comprise, for example pipes to handle coolants; cladding to protect the fissionable material from corrosion etc. The requirements are, the structural material must be stable against neutron and gamma bombardment, and must have small σ_a for neutrons. Suitable materials are, lead, bismuth, beryllium, aluminum, magnesium, zinc, tin, and zirconium. Among these most important and cheap is aluminum, useful in the form of jackets protecting uranium rods and cooling pipes (in low temperature thermal reactors). For reactors using enriched fuel, stainless steel may be used inspite of its high σ_a.

(e) *Control Rods*: If a reactor were to operate at any appreciable power level the multiplication constant must be greater than one. This excess reactivity is necessary to overcome temperature effects as the neutron flux is raised to the operating level, for overcoming the poisoning effects of fission products that gradually build up in the reactor fuel and to make available additional fuel to compensate for depletion or burn-up of fuel as the reactor operates. While handling such a super-critical reactor, it is important to have it under control. This is easily done by inserting in the reactor a material such as boron or cadmium which has a large σ_a for thermal neutrons. Adjustable control rods of boron or cadmium steel are inserted at the proper distance into the reactor to maintain k at the desired power level. As the fission products accumulate, they act as poison to the chain reaction and it is necessary to pull out the control rods gradually to keep constant reactor power. If it is desired to shut down the reactor, the control rods are merely inserted their maximum distance into the reactor; k drops below unity, and the chain reaction dies out. Normally, three types of control rods are present in a reactor viz shim. Fine control and safety rods. During normal operation of the reactor, most of the excess reactivity is absorbed in shim rods. As operation continues, small changes in temperature, pressure and other variables take place from time to time and require small changes in reactivity to keep the reactor at a steady power. This is accomplished by automatic or manual operation of fine control rods which absorb less reactivity than the shim rods. In case of emergency requiring shut down of reactor, the safety rods enter the reactor and reduce k well below 1. The disadvantage with control with rods is the loss of neutrons.

Other methods of control are:

1. Addition or removal of moderator
2. Motion of reflector
3. Negative temperature coefficient which is useful in the water moderated reactors

(f) *Reactor Shielding*: The radiations from a reactor that must be shielded against include prompt neutrons, delayed neutrons, prompt γ-rays, fission product γ-rays, capture γ-rays from fuel, moderator, coolant and structural elements in the reactor, Bremmstrahlung , annihilation γ-rays from β^+, inelastic scattering of γ-rays, capture γ-rays in shield and photo-neutrons.

Concrete (density 2.3 and composition 50 % oxygen; 19 % calcium; 18 % silicon and small amount of magnesium, carbon, iron etc.) provides an effective shield for neutrons rather than γ-rays. Lead is an effective shield against γ-rays. Radiation damages of the shield can be avoided by surrounding it with thermal shield such as iron, water or boron carbide which can absorb large quantities of heat without damage.

(f) *Reflector*: Surrounding a reactor with a reflector which is a medium of high σ_s and low σ_a has certain advantages over bare reactors:

1. Improved neutron economy: The reflector reduces neutron leakage from the core by reflecting or scattering many of the escaping neutrons back into the core region of the reactor and also acts as a moderator for the fast neutrons that have entered it from the core. In a way the moderation of the fast neutrons in the reflector will be more efficient than in the core itself, since the absence of neutron absorbing material in the reflector will reduce neutron loss due to resonance absorption, so that a larger fraction of the fast neutrons in the reflector can reach thermal energies than is possible in the fuel containing region of a moderator.

2. Possibility of fuel saving: Fuel can be saved to a certain extent if reflector is incorporated in the design of a nuclear reactor. It is possible to reduce the critical dimension of a reactor when it is bare.

3. Improved reactor power utilization: Improvement in the power utilization is a consequence of the flux flattening across the reactor core that occurs when reflector is used. This is desirable as it permits a higher power level operation without at the same time over-heating the central portion of the core, The neutron flux will be markedly greater at the core-reflector interface than what it would be in the absence of the reflector. Since the power production rate is proportional to the average neutron flux, the reactor can be operated at a higher total power output for the same maximum neutron flux. By virtue of the flux flattening effect the power production rate will also be more uniform over the core volume, which is highly desirable from the operation point of view especially with large power reactors.

8.15 The Reactor Period

8.15.1 Thermal Lifetime and Generation Time

In thermal reactors the average slowing-down time is much smaller (10^{-5} to 10^{-6} s) compared to thermal diffusion time (10^{-1} to 10^{-2} s) and can be ignored in the following considerations.

The thermal lifetime can be calculated from

$$t = \frac{\lambda_a}{v} = \frac{1}{v \Sigma_a} \tag{8.200}$$

where v is the average thermal neutron speed of 2200 m/s at 293.6 K.

Note that in an assembly which contains fuel the thermal lifetime will be shorter than that which contains pure moderator because of larger absorption cross-section. Again, the average lifetime in a finite assembly will be shorter than that in an infinite assembly because of leakage.

It t_0 is the thermal lifetime in the presence of leakage then

$$t_0 = t l_{th} = \frac{t}{(1 + L^2 B^2)} \tag{8.201}$$

where we have used (8.200).

Combining (8.201) and (8.188), the thermal lifetime of neutrons is

$$t_0 = \frac{1}{v \Sigma_a (1 + L^2 B^2)} \tag{8.202}$$

For large reactor for which $L^2 B^2 \ll 1$, $t \simeq t_0$. We, conclude that the average time that elapses between two successive generations of thermal neutrons known as generation time is basically equal to the thermal diffusion time.

The Generation time of a neutron includes the time required to fission a ^{235}U nucleus, the slowing down time of the fast neutron produced and the diffusion time of the thermal neutron before it is captured in fuel or impurity. If we consider only the prompt neutrons, they are given off in time of the order of 10^{-14} s after thermal fission, so that the fission process is usually considered to be instantaneous. Slowing-down times are about 10^{-4} to 10^{-5} s and the thermal diffusion times for a natural uranium reactor are about 10^{-3} s, so that the generation time is 10^{-3} s. Suppose we have a reactor operating in a steady state with a criticality factor of unity. If the reactivity c is now suddenly increased, e.g. by pulling out a control rod, the criticality factor will increase by a factor Δc and the neutron flux will start increasing. After one generation time, there will be $\phi_0(1 + \Delta c)$ neutrons. After n generation times, there will be $\phi = \phi_0(1 + \Delta c)^n$ neutrons present, where ϕ_0 is the flux at time zero

$$\therefore \quad \ln \frac{\phi}{\phi_0} = n \ln(1 + \Delta c) \simeq n \Delta c \tag{8.203}$$

$$\therefore \quad \phi = \phi_0 e^{n \Delta c} \tag{8.204}$$

If t is the time after the change in criticality factor is made and if Λ is the generation time of a neutron, then $= t/\Lambda$

$$\therefore \quad \phi = \phi_0 e^{t \Delta C / \Lambda} \tag{8.205}$$

Thus, when the criticality factor is changed suddenly from unity, the flux increases or decreases exponentially with time. Clearly, since Λ is of the order of 10^{-3} s in a very short time t, ϕ will increase enormously even for very small values of Δc, which would make the reactor highly dangerous. If we use an average lifetime of $\sim 10^{-3}$ s for a typical reactor, an excess reactivity of only 0.005, for example, would lead to a 20000 fold increase of neutron flux in about 2 s.

Table 8.5 Details of delayed neutrons from ^{235}U taken from [1]

Delayed neutron groups	Yield (%) n_i	Mean life time τ_i/s	$n_i \tau_i$/s
1	0.0267	0.33	0.0088
2	0.737	0.88	0.0648
3	0.2526	3.31	0.8361
4	0.1255	8.97	1.1257
5	0.1401	32.78	4.5925
6	0.0211	80.39	1.6962
	$\sum_{i=1}^{6} n_i = 0.64$		$\Sigma n_i \tau_i = 8.324$

8.16 Effect of Delayed Neutrons

Fortunately, since some of the neutrons are delayed for as long as several seconds, which means that the average life time of a neutron is much greater than the thermal diffusion time of 10^{-3} s the reactor period is increased. The average time of delay of neutrons (inspite of small percentage of slow neutrons) comes out to be ~ 0.1 s. This can be seen as follows. The prompt neutrons which make up over 99 % of all fission neutrons are emitted within 10^{-3} s after the initiation of fission process. ^{235}U yields the largest fraction of delayed fission neutrons (0.645 %), ^{233}U yields less than half as many (0.266 %), and ^{239}P about 1/3 as many (0.209 %). There are very slight differences in these yields for fast and thermal neutron fissions. For ^{235}U, we use $t_0 = 10^{-3}$ s as the average life for the generation of fission neutrons, their population n_0 being 99.36 %, and using $\sum_{i=1}^{i=6} n_i t_i = 8.324$ corresponding to the six delayed neutron groups, we find the average life t for both the prompt and delayed neutrons as

$$t = \frac{\sum_{i=0}^{i=6} n_i t_i}{\sum_{i=0}^{i=6} n_i} = \frac{8.324 + 99.36 \times 10^{-3}}{100} = 0.084 \text{ s}$$

or about 0.1 s.

The details of delayed neutrons from ^{235}U taken from [1] are shown in Table 8.5.

The result shows that the presence of delayed fission neutrons has lengthened the average life time of a neutron generation by a factor of nearly 100. Instead of a mean life time for prompt neutrons of about 10^{-3} s; the life time of neutrons, generation time has been increased to 10^{-1} s by the delay in the emission of only a minute fraction of the fission neutrons. Since it is the generation time that determines the period of the reactor, this means that the delayed neutrons have effectively lengthened the reactor period by a factor of 100, thus making the reactor control much more manageable and elastic. Suppose the power level of a reactor (and hence the flux) should suddenly double, the production of prompt neutrons would also double. But the production of delayed neutrons would increase only slowly because they are dependent upon the concentration of their percursors as determined by the old power level; so that the effect of delayed neutrons is to slow down the power rise of the reactor. In the case of ^{235}U about 0.73 % of the neutrons are delayed, and by never allowing k to go above 1.0073, a dangerous rise in power can be avoided.

8.17 Classification of Reactors

Nuclear reactors are classified in a variety of ways according to

1. The arrangement of fuel and moderator
2. The neutron energy at which fission mainly occurs
3. Type of fuel used
4. Type of moderator used
5. The purpose of the reactor
6. The methods of heat removal and the coolants employed

There are innumerable types of reactors based on different combinations of the above features. We will now compare some of the types of reactors in detail.

8.17.1 Homogeneous Reactor

The fuel and moderator are intimately and uniformly mixed either as a solid mixture in the form of fine slurry or as a liquid solution of a uranium salt in the moderator, for example uranyl sulfate mixed in H_2O or D_2O. Homogeneous thermal reactors employing natural uranium can reach criticality only with heavy water as moderator. However, k_∞ greater than 1 can be obtained with H_2O or graphite provided one uses enriched uranium (>25 %).

The active solution of uranyl sulfate or nitrate in light or heavy water circulates directly through the heat exchanger and back to the active core contained in a stainless steel sphere. The sphere is large enough to contain the critical mass. The entire system is pressurized in order to permit the liquid to reach an operating temperature well above the normal boiling point of the solvent.

The main advantage of homogeneous reactor is the possibility of continuous processing of fuel to remove fission products, operation with liquids which are easily transported by pumping, simple mechanical design and the non-requirement of expensive metallic elements. Thus, the reactor need not be shut down periodically to replenish the burnt-up fuel or to remove the ^{239}Pu produced. It is fairly stable during operation. High neutron flux and a power level of several megawatts are obtainable with the use of high enriched solution for relatively small reactor size.

When the active core is surrounded by a blanket of fertile material, this type of reactor can also serve the purpose of breeding with a high breeding ratio. Figure 8.33 is the schematic diagram of a homogeneous reactor.

The disadvantage is that a water moderated reactor requires the use of enriched fuel because of high absorption cross-section with hydrogen. As uranium is present

Fig. 8.33 Schematic diagram of a homogeneous reactor

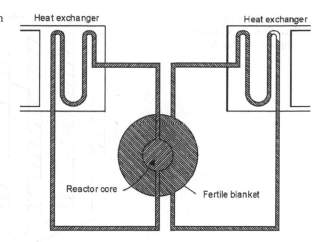

in uranyl nitrate or uranyl sulfate, corrosion is a serious problem. Further, the disruptive action of the fission fragments causes the decomposition of the moderator.

The corrosion effect of the fuel solution on the reactor components can be avoided by protecting them by cladding them with gold or uranium.

In the water-moderator-coolant reactor the need for pressurization can be avoided by employing organic moderator-coolants like polyphenols or their derivatives. Such substances have high enough boiling points to facilitate their use at fairly high temperature without resorting to pressurization. Further, the use of organic moderator-coolants results in low induced radioactivity in the pure materials, which is a significant feature.

8.17.2 Heterogeneous Reactors

In a heterogeneous reactor fissionable material is concentrated in plates, rods or spheres which are arranged in arrays in the form of a matrix throughout the moderator. The spacing between fuel lumps is sometimes called pitch. While a homogeneous reactor cannot become critical with natural uranium as fuel and graphite as moderator, only with D_2O as moderator can the criticality be reached. For this reason, majority of the reactors that have been constructed are of heterogeneous type, notwithstanding the fact that it is much easier to construct a homogeneous type. It was pointed out that a higher resonance escape probability can be achieved compared to a homogeneous arrangement because the resonance captures are reduced considerably with a lattice arrangement. This is due to two reasons.

First most of the slowing down can take place in the moderator which is a region completely free of resonance absorbing material. This circumstance gives the neutrons to reduce their energy below the resonance peaks so that when these neutrons re-enter the uranium lumps they will not be able to interact with ^{238}U nuclei appreciably.

Fig. 8.34 Variation of thermal neutron flux across a heterogeneous assembly

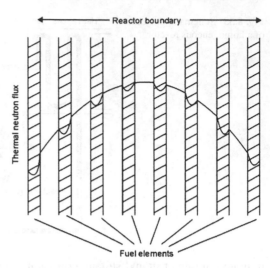

Second, neutrons of resonance energies 6.7, 21, 37 eV and higher can enter the fuel only from outside and are rapidly absorbed in the surface layer of uranium before they get a chance to penetrate deeper into the fuel. A thin surface layer of ^{238}U acts as an impenetrable screen which protects the interior of the uranium lump against neutron resonance absorption. Thus, the neutron resonance absorption becomes mainly a surface effect as the ^{238}U atoms in the interior of lumps have little opportunity to encounter neutrons of resonance energy. The resonance absorption can be kept to minimum by employing large lumps so that surface to volume ratio is reduced.

It must be pointed out that the screening effect on the surface of uranium lumps reduces not only the resonance neutron flux but also the thermal neutron flux. However, theory shows and experiments confirm that the reduction in resonance absorption is by far more significant than that for thermal neutron flux so that a net gain in neutron economy results from the lumping arrangement.

The heterogeneous assembly suffers from the disadvantage that the thermal utilization is lowered because of the decrease of thermal neutron flux within the uranium rods.

Figure 8.34 shows the variation of thermal neutron flux across a heterogeneous assembly. The flux distribution reveals rhythmic depressions across the lattice at the location of fuel elements. They are caused by much larger neutron absorption in the fuel elements as compared to that in the moderator, the neutron flux being about 2/3 of that in the graphite moderator. If we disregarded the minor depressions then the overall flux distribution is essentially the same as for the homogeneous assembly.

The fast fission factor ε makes a contribution of about 1.03 to the production of fast neutrons in uranium before they enter the moderator.

To summarize, the resonance escape probability increases, the thermal utilization factor decreases and the fast fission factor ε is constant at around 1.03. On the whole k_{∞} increases compared to the homogeneous assembly.

Fig. 8.35 Arrangement for the uranium rods embedded in the graphite moderator

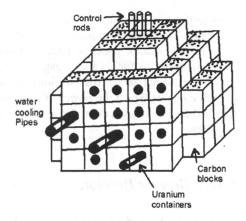

Fig. 8.36 Schematic diagram of a typical heterogeneous reactor

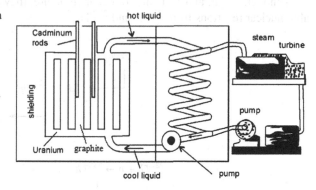

Figure 8.35 shows the arrangement for the uranium rods embedded in the graphite moderator. Figure 8.36 is the schematic diagram of a typical heterogeneous reactor. Among Indian reactors, the Canada-India reactor which is a uranium-graphite-moderated reactor giving a power of 40 MW and located at Trombay (near Mumbai) is of this type.

8.17.3 Fast Reactors

High-energy neutrons (>1 MeV) are used to sustain the chain reaction. Here, the moderator is unnecessary. As the moderator is eliminated the critical core size may not be more than one foot across. This small size presents difficult technical problems for heat removal. Natural uranium cannot be used for the fast reactor because criticality cannot be achieved. Enriched uranium and plutonium-239 are suitable fuels for this type of reactor. The reactor can be operated successfully with an enriched fuel with not less than 25 % of fissionable material content (i.e. ^{233}U, ^{235}U, ^{239}Pu). A large amount of fissionable material is needed to reach criticality. The reflector is usually natural uranium and the coolant is mercury. The main application of fast reactors is for breeding, in particular to produce plutonium.

In the fast reactor parasitic neutron absorption is greatly reduced. Further, the σ_f/σ_c ratio for the fuel is also greater for the fast reactor. Fast reactors suffer from the disadvantage that all cross-sections including the fission cross-section are small at high emerges. Since the power of a reactor is proportional to the product of the flux and fission cross-section for any reasonable power level the fast flux must be very large.

8.17.4 Breeder Reactors

The main function of a breeder reactor is to produce new fissionable material.

Although ^{239}Pu, and ^{233}U do not occur in nature, they can be produced by suitable nuclear reactions from ^{238}U, and ^{232}Th

$$^{238}U + n \rightarrow {}^{239}U + \gamma$$
$$\qquad \longrightarrow \beta^- + {}^{239}Np$$
$$\qquad\qquad \longrightarrow \beta^- + {}^{239}pu$$

$$^{232}Th + n \rightarrow {}^{233}Th + \gamma$$
$$\qquad \longrightarrow {}^{233}Pa + \beta^-$$
$$\qquad\qquad \longrightarrow {}^{233}U + \beta^-$$

^{238}U and ^{232}Th which themselves are not fissionable can be converted into fissionable isotopes and are known as fertile materials. Note that although the two fissionable isotopes ^{233}U and ^{239}Pu are derived from fairly abundant ^{232}Th and ^{238}U, each conversion process requires a source of neutrons which at present must come from ^{235}U. We thus must consume fissionable material (^{232}U) in order to produce new fissionable material ^{235}Pu or ^{233}U in larger quantity. This process is known as breeding.

The number of new fissionable nuclei produced for each ^{235}U nucleus destroyed is termed as the *conversion factor*.

8.17.4.1 Conversion Factor C

Let a ^{235}U nucleus on the average produce η fast neutrons per neutron absorbed. One of these neutrons is required to continue the chain reaction. The maximum possible conversion factor is then

$$C_{max} = \eta - 1$$

When allowance is made for the losses of neutrons due to leakage L and through absorption in poison nuclei

$$C = \eta - 1 - L \tag{8.206}$$

The breeding gain G in defined as the excess of fissionable nuclei produced over the number of fissionable nuclei consumed per nucleus of ^{235}U fuel consumed

$$G = C - 1 = \eta - 2 - L$$

Let there be N nuclei of fissionable material. Then the number of new nuclei that are formed are

$$NC + NC^2 + NC^3 + \cdots$$

If $C \geq 1$; the series is divergent.

If $C < 1$; the limiting conversion of fertile nuclei

$$= \frac{NC}{1 - C} \tag{8.207}$$

The amount of fissionable material available after x stages is nC^x, and if $C < 1$, this decreases with x. If $C = 1$, then this is equal to N at all stages. Thus the condition for breeding is $C > 1$, that is G must be positive or

$$\eta - 2 - L > 0 \quad \text{or} \quad \eta > 2 + L$$

If $L = 0$, the absolute minimum requirement for breeding is $\eta > 2$.

8.17.5 Thermal Breeders

In the case of ^{239}Pu, at thermal energies $\eta = 1.94$ rules out breeding. In the case of ^{235}U, $\eta = 2.12$ and breeding is possible whereby ^{235}U fuel is charged to a reactor containing ^{238}U or ^{232}Th, yielding ^{239}Pu and ^{233}U respectively by an amount greater than the fuel charged. However, the margin of only 0.12 neutron available is so small as to jeopardize the breeding process. Apart from the loss of neutrons by leakage and parasitic absorption, there is another factor which hampers thermal breeding. The transformation of the fertile material to fissionable material is not instantaneous. Consequently, intermediate products which are formed absorb neutrons and act as sinks via formation of non-fissionable products such as ^{233}Th, with $\sigma_a = 1400$ b and $T_{1/2} = 23.5$ min and ^{233}Pa, with $\sigma_a = 37$ b and $T_{1/2} = 27.4$ days. Both of them limit the possibility of breeding by thorium cycle.

8.17.6 Fast Breeders

In the case of fast neutron fission the relative decrease in the fission cross-section is less than the decrease in the capture cross-section, and the value of η is considerably larger than 2.12. As breeding depends upon the value of $\eta - 2$ a marked increase in the breeding gain can be achieved by working in the fast region. Among Indian reactors the one located at Kalpakkam (Tamil Nadu) is of breeder type.

8.17.7 Doubling Time

Let ϕ be the flux of neutrons, N the number of atoms of fuel at time t, σ_a the total microscope absorption cross-section of the fissionable nuclide, then the number of neutrons captured per (s)(cm^2) in fuel is $N\sigma_a\phi$. For each neutron captured by fissionable material, η fast neutrons are produced of which one neutron is needed to continue the chain reaction and one neutron is needed for breeding. The increase of fuel in time T or the breeding gain is

$$\text{Breeding gain} = N\sigma_a\phi t(\eta - 2) \tag{8.208}$$

Equation (8.208) gives the maximum value of the breeding gain since the loss of neutrons by leakage and parasatic capture by moderator, coolant etc., have been ignored.

The time required to double the total amount of fuel is known as the doubling time t_d; setting the breeding gain equal to N in (8.208) gives

$$N = N\sigma_a\phi t_d(\eta - 2) \quad \text{or} \tag{8.209}$$

$$t_d = \frac{1}{\sigma_a\phi(\eta - 2)} \tag{8.210}$$

Thus, the greater is the flux, η and σ_a, the smaller will be the doubling time. Notice that if $\eta < 2$, breeding is not possible and the concentration of the fuel simply continuously decreases with time.

8.18 Other Types of Reactors

We shall now briefly mention other types of reactors.

8.18.1 Power Reactors

The main purpose of a power reactor is to convert the fission energy into useful power. The main types of power reactors are (1) the pressurized water reactor,

(2) the boiling water reactor, (3) the gas-cooled natural uranium, graphite moderated reactor and (4) the homogeneous reactor.

8.18.2 The Pressurized Water Reactor

This is a heterogeneous reactor that uses mildly enriched uranium (1.4–2 % ^{235}U) as fuel and light water as moderator and coolant. The core is kept in a pressure vessel under a pressure of 1000 to 2000 psi. Pressurized water with raised boiling pint (300 to 400 °C) is circulated by means of a pump through the core and external heat exchanger for satisfactory heat transfer.

8.18.3 The Boiling Water Reactor

The boiling water reactor also uses light water as moderator-coolant. Here the steam is generated in the reactor core itself and is passed directly to the turbines avoiding the heat exchanger. It is a comparatively safe reactor and largely self, regulatory. Sudden power surge causes formation of steam bubbles or steam voids in the liquid moderator, leading to the reduction of thermalization of neutrons while increasing the neutron leakage rate. Consequently, the fission rate and hence the power production will be lowered.

The fuel elements (pure uranium metal or uranium oxide UO_2) must be "canned", i.e. they must be enclosed by cladding materials and sealed to protect them against corrosion on coming into contact with the coolants and also prevent the fission products from escaping. Aluminum and to a larger extent zirconium, are widely used as cladding material.

8.18.4 The Gas-Cooled Natural-Uranium-Graphite Reactor

This type of reactor requires graphite moderator of high degree of purity devoid of parasitic absorbers and the fuel employed is uranium metal. The coolant consists of either nitrogen or carbon dioxide. The fuel elements are inserted at regular intervals in the graphite moderator. The gas coolant passes through the fuel channels and carries away heat generated in the fuel elements. Apart from the generation of power, it finds application for the production of ^{239}Pu.

8.18.5 The Homogeneous Reactor

The working of the homogeneous reactor was described in Sect. 8.17.1. High neutron flux and power level of several megawatts are obtained by the use of highly enriched fuel solution of uranyl nitrate or uranyl sulphate. This class of reactors is characterized by a small critical size and high power density. It can also be used as a breeder reactor.

8.18.6 Research Reactors

Here the main concern is to provide relatively high neutron flux densities for experimental work. Unlike the power reactors, the power produced in the form of heat is undesirable and the elaborate cooling arrangements can be dispensed with.

The average thermal neutron flux in approximately given by

$$\phi_{th} = 2.6 \times 10^{10} \times \frac{P \text{ (W)}}{m \text{ (g)}} \tag{8.211}$$

where m is the critical mass of the reactor fuel ^{235}U. Now m is proportional to the volume of the reactor core. Hence in order to obtain a large flux ϕ an enriched fuel with a good moderator like D_2O may be employed and a very compact size of the core with linear dimensions of the order of 1 ft is possible.

Experimental facilities are provided by openings that lead into the reactor core or into the lattice where the entire reactor spectrum is available, the fast neutron flux and the thermal flux being in equal proportion. When openings lead into the fuel free moderator region, the neutron flux will be predominantly thermal with admixture of fast neutrons since fission neutron flux decreases exponentially with distance from the fuel.

One can have access to well-thermalized neutrons by the use of a *thermal* column which is an extension of the moderator against a portion of one side of the reactor from which the reactor shielding has been removed.

The four main types of research reactors are (1) the water boiler, (2) the swimming pool, (3) the tank-type reactor, (4) the graphite moderated reactor.

8.18.7 The Water Boiler

This is usually a homogeneous mixture of a highly enriched uranium salt dissolved in ordinary water contained in a small stainless steel vessel surrounded by a reflector and shield. A neutron flux of $\sim 10^{12}$ can be achieved.

This type of reactor is not to be confused with the boiling water reactor described in Sect. 8.18.3. The term "Water boiler" is used for the reason that a sudden surge in power would cause the formation of steam bubbles in the solution leading to the shutdown of the reactor. Under normal operating conditions boiling does not occur as the temperature of the fuel solution is kept below 80 °C by circulating coolant through the coils inside the core vessel.

8.18.8 The Swimming Pool Reactor

This reactor consists of a concrete tank containing 100 to 200 m^3 of highly purified water. The fuel consists of highly enriched rectangular uranium plates clad with aluminum suspended by a steel frame work spanning the width of the swimming pool and immersed 5 to 7 feet below the water surface.

The water in the pool serves the purpose of moderator, coolant as well as shield. At a power level up to 100 kW the neutron flux available will be of the order of 10^{12}.

The first reactor constructed in INDIA called APSARA is located at Trombay and is of the swimming pool type.

8.18.9 The Tank-Type Reactor

This type of reactor is similar to the swimming pool type except that its size is reduced to that of a tank. The heat transfer which is much more efficient enables the operating power to boost up to provide neutron flux as high as $\sim 10^{14}$. A number of tank-type research reactors use D_2O instead of H_2O as moderator-coolant, with 1 to 2 kg of 90 % enriched uranium at power level of several megawatts.

8.18.10 The Graphite-Moderated Natural Uranium Reactor

All of them are prototype of the first historical reactor built at Chicago under Fermi's direction. All of them use 20 to 50 tons of natural uranium as fuel in the form of rods clad mostly in aluminum embedded in several hundred tons of graphite which serve the purpose of both moderator and reflector. Air or CO_2 at low pressure is used as a coolant. The power level ranges from 100 kW to 30 MW and flux 4×10^{10} to 4×10^{12}.

It D_2O be used instead of H_2O as a moderator then the reactor size is considerably reduced. This is because the amount of D_2O required to moderate fission neutrons is much smaller than that of graphite. A smaller volume implies a higher neutron density and hence greater power density and grater neutron flux. This is the reason for the extensive use of D_2O moderated reactors for research purposes and also for production of plutonium.

8.19 Variation of Reactivity

8.19.1 Fuel Depletion and Fuel Production

The reactivity of a reactor is affected over long spans of time. In case of depletion of fuel or accumulation of poisonous material the reactivity goes down and when surplus fissile material is produced the reactivity increases. Normally, in order to run the reactor continuously it is necessary to build up an excess reactivity into the assembly to compensate for the fuel depletion. For reactors employing highly enriched uranium, the decrease in reactivity is proportional to the fraction of ^{235}U burnt up, and for those which employ natural uranium, the depeletion of ^{235}U is partly compensated through the production of ^{239}Pu from ^{238}U by neutron absorption.

In a Breeder reactor, the reactivity increases because the amount of ^{239}Pu produced exceeds the amount of ^{235}U burnt up. For this reason proper control must be provided to check the build up of reactivity.

8.19.2 Effect of Fission Products Accumulation

When the reactor runs there will be a gradual build-up of fission fragments both in the reactor core and the fuel elements which affects the neutron multiplication and hence the reactivity of the reactor. The most poisonous fission fragments are ^{135}Xe with $\sigma_{th} \sim 3 \times 10^6$ b and ^{149}Sm with $\sigma_{th} \sim 5 \times 10^4$ b. Such nonproductive absorption of neutrons adversely affects the thermal utilization f and hence the reactivity. ^{135}Xe is an intermediate product of the fission product chain which terminates with ^{135}Ba as the stable end-product, and ^{149}Sm is the stable end-product of the fission chain.

When the reactor is shut down, the production of poisonous Xe nuclei does not halt immediately but continues to build up and reaches a maximum in about 11 hours, resulting in a reactivity loss of about 40 %. The build-up of poisonous Xe nuclei after shut down occurs because of the continuing decay of the parent ^{135}I (6.7 hour half-life) into ^{135}Xe (9.2 hour half-life) without the compensating neutron absorption reactions. These reactions are not taking place after the shut down because of rapid decay of neutron flux. It is then the case of chain radioactive decay with $\lambda_{parent} > \lambda_{daughter}$ in which transient equilibrium is established (at 11 hours after the shut down). This then means that extra fuel reserves must be accessible to over-ride the Xe poisoning if the reactor is to run without any break. Otherwise, one will have to wait for several hours before the Xe nuclei substantially die out. This aspect is very important in the design of reactors which drive a submarine or a ship.

8.19.3 Temperature Effects

When a reactor is operated at any appreciable power level the fission energy heats up the fuel, moderator and other material present. An increase in temperature will decrease the density of the material present, and will decrease σ because of high average neutron velocity $(1/v)$ law.

The Parameters of Chain Reaction The *Fast fission factor* ε is expected to be insensitive to small temperature changes since it is determined by the behavior of fast neutrons.

It is reasonable to expect v, the number of neutrons released per fission to be temperature independent. If σ_f and σ_a follow the $1/v$ law, then η the number of neutrons released per capture will also be temperature independent since $\eta = \frac{v\sigma_f}{\sigma_f + \sigma_c}$.

For a homogeneous reactor, the thermal utilization factor

$$f = \frac{\Sigma_{a(\text{uranuim})}}{\Sigma_{a(\text{uranuim})} + \Sigma_{a(\text{modertor})} + \Sigma_{a(\text{other})}}$$

Because of $1/v$ law dependence, all absorption cross-sections will be proportionately decreased and f remains unaffected.

For heterogeneous assembly, however, f will increase with the rise in temperature. This can be explained as follows

$$f = \frac{1}{1 + \frac{\Sigma_{am}\phi_m V_m}{\Sigma_{au}\phi V_u}} \tag{8.212}$$

An increase in temperature will decrease the thermal disadvantage factor $\overline{\phi}_m / \overline{\phi}_u$ because the decreased uranium absorption cross section will be less effective in reducing the neutron flux entering the fuel rods resulting in a more even distribution of the neutron flux across the lattice cell. This causes a decrease in the thermal disadvantage factor and therefore an increase in f.

The *resonance escape probability* p does change with temperature because of Doppler broadening of the resonance levels in ^{238}U nuclei with the temperature rise which implies larger absorption cross section and smaller value of p.

The value of multiplication factor K_∞ is unchanged for all practical purpose for small temperature changes.

The *Thermal diffusion length* L is also affected by temperature changes. From its definition, $L^2 = \lambda_{tr}/3\Sigma_a$. Obviously, L^2 is inversely proportional to the square of density as well as the absorption cross-section σ_a. Ignoring the variation of σ_s with temperature which is small the temperature rise causes decrease in density as well as σ_a, leading to increase in L.

From the definition of the Fermi age equation $d\tau = -(\lambda_{tr}\lambda_s/3\xi)(dE/E)$, $\tau \propto 1/\Sigma_s^2$ which means that the variation of τ is determined by the temperature dependence of the density. The change in the thermal diffusion length is more significant than the change is τ with temperature.

An increase in temperature will cause an increase of the reactor core which would decrease the buckling B^2.

8.19.4 Temperature Coefficient and Reactor Stability

The temperature coefficient may be defined as

$$\frac{1}{k_e}\frac{dk_e}{dT} = \infty \tag{8.213}$$

which is the fractional change in effective multiplicative factor per degree.

As far as the nuclear cross-sections are concerned a rise in temperature leads to a decrease in the reactivity so that the nuclear temperature coefficient is negative. Likewise the temperature coefficient turns out to be negative. However, the volume temperature coefficient turns out to be positive. But the last one turns out to be much smaller than the first two. Consequently the net reactivity decreases with the rise in reactor temperature.

If the reactivity decreases with the rise in temperature then the reactor is said to be stable. On the other hand, if the reactivity increases with temperature then it is unstable. Safe design must ensure a reactor to be stable at all times since it is self regulating. Generally, H_2O or D_2O-moderated reactors have larger negative temperature coefficients than graphite-moderated reactors, and are therefore inherently more stable than the latter.

8.20 Nuclear Fusion

8.20.1 Fusion Reactions

The curve for binding energy/nucleon versus mass number (Chap. 4) shows that energy will be released either when a heavy nucleus like ^{235}U breaks up into two medium heavy fragments as in fission or two very light nuclei below $A = 56$ are fused as in the case of fusion reactions of the type

$$D + D \rightarrow T + p + 4.03 \text{ MeV}$$

$$\rightarrow {}_2^3\text{He} + n + 3.27 \text{ MeV}$$

Since deuterium is present in ordinary water to the extent of one part in 6500, there is an inexhaustible supply of nuclear energy in the oceans of the world. The main problem is to run the fusion reactions in a controlled way. It is necessary that the deuterons acquire sufficient energy to overcome the Coulomb barrier in order to induce the reactions. Just firing a beam of deuterons on to a target of deuterium is not practical. First, it is far too expensive to produce a beam of deuterons. Secondly, the cross-section for the fusion reaction is so small that most of the deuterons would lose all their energy in collisions with the atoms of the target before undergoing fusion reaction with other deuterons. Alternatively, the deuterium may be heated to such temperatures that the electrons are all knocked off the atoms and the gas is reduced to the state of plasma consisting of positive nuclei intermingling with negative electrons. Also, the nuclei acquire such large energies that when they collide a portion of them will fuse together.

In order to produce the plasma for deuterium a temperature of only 10^5 K is required. But to achieve fusion reactions much higher temperature, of the order of 10^8 K is needed, so that the deuteron's thermal energy is so large that they may be permitted to overcome the Coulomb barrier. Such reactions are known as thermonu-clear reactions. The advantages of fusion over fission reactions are (a) light nuclei are easily available, (b) end products are usually light and stable, rather than heavy and radioactive.

The disadvantage is that charged nuclei must penetrate the coulomb's barrier in order to come into contact with each other. This is accomplished either by acceler-ating the particles and bombarding the target nuclei or by raising them to very high temperatures. For fission the question of barrier penetration does not arise as neutron is neutral and slow neutrons can be used. The main difficulty for fusion reactions is the containment of plasma for sufficiently long time.

8.20.1.1 Coulomb Barrier

The Coulomb barrier V_c at the point of contact of the particles (here two deuterons) is given by

$$V_c = \frac{e^2}{4\pi\varepsilon_0}\frac{Z_1 Z_2}{(R_1 + R_2)} \tag{8.214}$$

where $Z_1 e$ and $Z_2 e$ are the charges, and R_1 and R_2 the nuclear radii. For numerical calculations

$$V_c = 1.44\frac{Z_1 Z_2}{R_1 + R_2}\ \text{MeV-fm} \tag{8.215}$$

where R_1 and R_2 are in Fermis.

Using $R_1 + R_2 = 5$ fm for deuterons, $V_c = 280$ keV particles with relative energy greater than 280 keV each will be able to come into contact where nuclear force will take over. Equating this energy to the mean thermal energy we can find the corresponding temperature

$$\frac{3}{2}KT = 2.8 \times 10^5\ \text{eV}$$

$$T = \frac{2}{3} \times \frac{2.8 \times 10^5}{8.15 \times 10^{-5}} = 2.3 \times 10^9\ \text{K}$$

This is a very high temperature. Fortunately, there are two circumstances which lead us to expect that nuclear reactions will take place at much lower temperature.

8.20.1.2 Maxwell's Distribution of Colliding Particles

Particles at temperature T have their energy distributed according to Maxwell's law

$$n(E)dE = \text{const}\frac{\sqrt{E}}{T^{3/2}}e^{-E/kT}dE \tag{8.216}$$

The left hand side is the number of particles with energy E and $E + dE$. A fraction of the particles will have energy in excess of $\frac{3}{2}kT$, the mean energy. Further, the cross-section for fusion increases rapidly as the energy of colliding particle increases. Thus, a few high energy particles from the 'tail' of the distribution are precisely those which produce the most fusion reactions. This is the first circumstance which brings down the temperature at which fusion in deuterium takes place to something well below 10^9 K.

8.20.1.3 Tunneling of Coulomb Barrier

The second circumstance is the ability to tunnel the Coulomb's barrier. For $E \ll V_c$, the fusion reaction

$$\sigma_f(E) \propto \frac{1}{\sqrt{E}} \exp\left[-\text{const} \times \frac{Z_1 Z_2}{\sqrt{E}}\right] \tag{8.217}$$

where E is the relative energy.

The proportionality factor contains the matrix elements and statistical factors which depend on the spin of particles participating in the reaction. The dominant term is the negative exponential which represents the Gamow factor, and accounts for the tunneling effect of the potential barrier.

8.20.1.4 Factors Which Affect the Reaction Rates

We can ignore the presence of electrons in the hot plasma as they do not affect the interaction of nuclei. However, their presence is necessary to preserve electrical neutrality. We have noted that in the case of two deuterons the Coulomb barrier is approximately 280 keV However, at temperature of 10 keV, that is 1.2×10^8 K there are still many deuterons with energies much greater than 10 keV. Also, the tunneling effect well below 280 keV is quite important. In Fig. 8.37, curve A represents the relative number of nuclei as a function of energy, curve B gives the barrier penetration probability. Curve C is the product of these two curves and represents the number of fusion reactions that actually take place. Observe that most of the reactions occur from a small number of particles in the tail of the Maxwellian distribution.

8.20.2 Three Important Fusion Reactions

From Gamow's formula (8.217) it is obvious that the smaller in the product $Z_1 Z_2$ the larger will be the probability for barrier penetration. The best candidates are then the lightest elements, the isotopes of hydrogen. The three important reactions are

$$D + D \rightarrow {}^3\text{He} + n + 3.27 \text{ MeV} \tag{8.218}$$

Fig. 8.37 Factors affecting the reaction rates

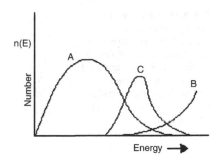

$$D + D \rightarrow T + p + 4.03 \text{ MeV} \qquad (8.219)$$

$$D + T \rightarrow \alpha + n + 1.76 \text{ MeV} \qquad (8.220)$$

Note that the reaction $p + p \rightarrow {}_2^2\text{He}$ does not proceed as ${}_2^2\text{He}$ does not exist.

The temperature requirement can be brought down with the choice of lighter nuclei so that V_c would be smaller. But even then $T \sim 10^8$ K would be required. Such temperatures are available in the interior of stars and fusion via thermonuclear reactions would be possible, effectively 4 protons are converted into an α-particle through a series of reactions and decays.

8.20.3 Reaction Rate

For neutron induced fission reactions $\sigma \propto \frac{1}{v}$ (away from resonance) or $\sigma v = \text{const}$. For fusion this is not so. Also, particle speed distribution is Maxwellian

$$n(v)dv \propto e^{-mv^2/2kT}dv \qquad (8.221)$$

The left hand side is the number of neutrons with speed v and $v + dv$.

The reaction rate R is given by

$$R = \Sigma\phi = \Sigma n v \qquad (8.222)$$

where n is the number of neutrons per unit volume

$$\langle \sigma v \rangle \propto \int_0^\infty e^{-2G} e^{-mv^2/2kT} v\, dv$$

$$\propto \int_0^\infty e^{-2G} e^{-E/kT} dE \qquad (8.223)$$

where $G \simeq \frac{e^2 Z_1 Z_2}{4\varepsilon_0 \hbar v}$ is Gamow's factor and v is the relative velocity. Figure 8.38 shows the variation of σv averaged over a Maxwell-Boltzmann energy distribution for D–T fusion reaction (solid line) and D–D reactions (broken line) extremely

Fig. 8.38 Variation of σv averaged over a Maxwell-Boltzmann energy distribution for D–T fusion reaction (*solid line*) and D–D reactions (*broken line*) extremely high temperatures ($T \sim 10^{10}$ K) correspond to MeV energies

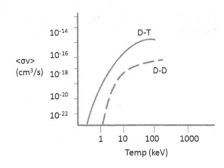

high temperatures ($T \sim 10^{10}$ K) correspond to MeV energies. At such high temperatures D–T reactions are less favored than other types of fusion reactions. But for the practical range of temperatures ($T \sim 10^7$ to 10^8 K) D–T reactions are favored more than the D–D reactions. The simple calculations are actually valid for D–D reactions. For two different types of nuclei as in the D–T reactions σ and R must be calculated taking into account Maxwellian distribution for both the species. Nevertheless the general conclusions remain unchanged.

The reaction rate for fusion reactions

$$R_{DT} = n_1 n_2 \langle \sigma v \rangle \qquad (8.224)$$

where n_1 and n_2 are the densities of the fusing nuclei.

If only one kind of fusing nuclei are present as in D–D reactions, then

$$R_{DD} = \frac{1}{2} n^2 \langle \sigma v \rangle \qquad (8.225)$$

where the factor $\frac{1}{2}$ takes care of double counting of reactions.

8.20.4 Power Density

Adding Eqs. (8.218), (8.219) and (8.220)

$$5D \rightarrow {}^3\text{He} + {}^4\text{He} + p + 2n + 24.9 \,\text{MeV} \qquad (8.226)$$

Thus, on an average, each D–D reaction produces one-half of 24.9 MeV and uses up 2.5 deuterons ultimately. The rate at which this energy is produced is that given in Table 8.6 and Fig. 8.35.

The energy generated per cubic meter per s is then given by

$$\frac{1}{2} n_D^2 \overline{\sigma v_{DD}} \times 12.45 \,\text{MeV}\,\text{m}^{-3}\,\text{s}^{-1} \qquad (8.227)$$

Table 8.6 The values of $\overline{\sigma v}\,\mathrm{m^{-3}\,s^{-1}}$ at various temperatures

Temperature (keV)	$D–D$	$D–T$
1.0	2×10^{-28}	7×10^{-27}
5.0	1.5×10^{-25}	1.4×10^{-23}
10.0	8.6×10^{-24}	1.1×10^{-22}
100.0	3.0×10^{-23}	8.1×10^{-22}

where v_{DD} is the relative velocity of two deuterons. Similarly, from (8.225) and (8.220) we find the corresponding expression for the $D–T$ reaction

$$n_{D}n_{T}\overline{\sigma v} \times 17.6\,\mathrm{MeV\,m^{-3}\,s^{-1}} \tag{8.228}$$

Power density (power per cubic meter) for $D–D$ reaction

$$= \frac{1}{2}n_{D}^{2}\overline{\sigma v}_{DD} \times 12.45 \times 1.6 \times 10^{-13}$$

$$= 1.0 \times 10^{-12}n_{D}^{2}\overline{\sigma v}_{DD}\,\mathrm{W\,m^{-3}} \tag{8.229}$$

Power density for $D–T$ reactions

$$= n_{D}n_{T}\overline{\sigma v}_{DT} \times 17.6 \times 1.6 \times 10^{-13}$$

$$= 2.8 \times 10^{-12}n_{D}n_{T}\overline{\sigma v}_{DT}\,\mathrm{W\,m^{-3}} \tag{8.230}$$

As an example of the $D–D$ reaction, let us choose a temperature of 100 keV, which turns out to be the temperature at which the $D–D$ reactions could provide useful amount of energy (see Fig. 8.36). From Table 8.6 we find $\overline{\sigma v}_{DD} = 3.0 \times 10^{-23}\,\mathrm{m^{3}\,s^{-1}}$. Therefore power density at 100 keV $= 3 \times 10^{-35}n_{D}^{2}\,\mathrm{W\,m^{-3}}$. Suppose that the density of deuterium is that for NTP so that 1 mole or 4×10^{-3} kg occupies $22.4 \times 10^{-3}\,\mathrm{m^{3}}$. Since deuterium is diatomic a mole contains $2 \times 6 \times 10^{23}$ atoms. Thus, the number of nuclei

$$n_{D} = 12 \times 10^{23}/22.4 \times 10^{-3}\,\mathrm{m^{-3}} = 5.4 \times 10^{25}\,\mathrm{m^{-3}}$$

Substituting this in (8.230), power produced

$$= 3 \times 10^{-35} \times \left(5.4 \times 10^{25}\right)^{2} = 9 \times 10^{16}\,\mathrm{W\,m^{-3}}$$

Compared to a fission nuclear reactor of power 1000 MW the thermonuclear material at 100 keV, under the stipulated conditions, would produce 100 million times greater power. This sort of power is produced in an atom bomb explosively, certainly not in laboratory. The 100 keV corresponds to a temperature of 1.16×10^{9} K. Now the pressure of a perfect gas is proportional to the absolute temperature and we have taken the density to be that expected at N.T.P. Therefore the pressure at 100 keV is

$$\frac{4 \times 1.16 \times 10^9}{273} \quad \text{or} \quad \text{17 million atm}$$

The factor 4 accounts for the fact that each molecule of deuterium has split up into two nuclei and two electrons, and each of the four particles exerts its own partial pressure.

If the thermonuclear power is to be produced in a usable form, either the temperature must be lowered drastically or the density decreased by a very large factor. The former cannot be done conveniently mainly because the reaction would be too slow. The only option is to reduce the pressure. The sort of power which can be handled is of the order of $10^8 \, \text{W m}^{-3}$ similar to that generated in a fission reactor.

In a fission reactor at such rates energy can be transported from the region of production and used to generate electricity. If we insist on temperature of 100 keV, the power must be reduced by a factor of about 10^9, and since power depends on n_D^2, the density is lowered approximately by a factor of 3×10^4.

The corresponding pressure will be

$$17 \times 10^6/3 \times 10^4 \quad \text{or} \quad \text{600 atm}$$

At a temperature of 10 keV, the pressure would be lowered to 60 atm which is manageable. Densities of the order of 10^{-4} to 10^{-5} times normal atmospheric density are recommended.

The above discussion for $D–D$ reaction is equally valid for the $D–T$ reaction which is much faster than the $D–D$ reaction at temperature below 100 keV (Fig. 8.35). This then means that the same power can be obtained from $D–T$ reactions at a lower temperature than for a $D–D$ reaction, a fact which is of paramount importance for the production of thermonuclear power.

8.20.5 Thermonuclear Reactions in the Laboratory

Unlike the stars in whose core thermonuclear reactions are believed to occur, in the laboratory we can no longer depend on the forces of gravitation to hold the plasma together. We are faced with a very difficult problem of containment. The difficulty is to prevent the plasma from dispersing rapidly to the walls of the containing vessel. Time is therefore essence of success. The reaction which proceeds with greatest rate is likely to be most successful under the given conditions of temperature and pressure. The best candidate is the $D–T$ reaction which at temperatures below 100 keV proceeds 100 times as fast as the $D–D$ reaction. Tritium is not found in nature as it is radioactive ($T_{1/2} = 12.26$ Y) except for small quantities produced by cosmic rays. In any case it is very expensive. The solution to this problem is to use the neutrons produced in the $D–T$ reaction to get more tritons than are used up in the reaction, so that in principle we can get both energy and replacement of the expensive part of the fuel.

8.20.6 The D–T Reaction

$$D + T \rightarrow He^4 + n + 17.6\,\text{MeV}$$

The rate of reaction is

$$R_{DT} = n_D n_T \overline{\sigma v} \text{ interactions m}^{-3}\,\text{s}^{-1}$$

From (8.231) total power released in D–T reaction is

$$\text{Power} = 2.8 \times 10^{-12} n_D n_T \overline{\sigma v} \text{ W m}^{-3}$$

Using the value of $\overline{\sigma v}_{DT} = 1.1 \times 10^{-22}$ from Table 8.6 we find

$$\text{power} = 3.1 \times 10^{-34} n_D n_T \text{ W m}^{-3} \tag{8.231}$$

Putting $n_D = n_T$ and anticipating the power $= 10^8$ W m^{-3} as in the fission reactor, we find

$$n_T = n_D = 5.7 \times 10^{20} \text{ particles m}^{-3}$$

Of the total energy released, 17.6 MeV per fusion, 14.1 MeV appears as kinetic energy of the neutron and 3.5 MeV as kinetic energy of the ^{4}He nucleus. The collision cross-section of neutrons with matter at this energy is small. As they escape they pose a serious health hazard. Hence the reactor must be surrounded by a thick 'biological shield'. Only the energy of helium nuclei remains within the plasma and is shared by collisions with all other charged particles present (^{4}He, D, T and electrons).

8.20.7 Energy Losses

At the high temperatures the electrons in the plasma moving close to the ions suffer Coulomb scattering and emit bremsstrahlung radiation due to acceleration. Taking into account the Maxwell's distribution into account, the power per unit volume radiated in bremsstrahlung is calculated as

$$P_{br} = 0.5 \times 10^{-36} Z^2_{nne} \sqrt{kT} \text{ W/m}^3 \tag{8.232}$$

where Z is the ion charge, n the density of positive ions, n_e the density of electrons and kT is in keV.

Figure 8.39 shows the power density curves for D–D and D–T reactions. It is seen that fusion output exceeds bremsstrahlung for temperatures exceeding 4 keV for D–T reactions and 40 keV for D–D reactions. We conclude that the D–T reactions are to be preferred to D–D reactions. Other radiation losses including synchrotron radiation (see [2], Chap. 2) from charged particles orbiting around magnetic field lines, can also be neglected.

Fig. 8.39 Power density
curves for D–D and D–T
reactions

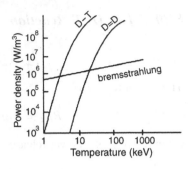

8.20.8 Lawson Criterion

Fusion reactor will have net energy gain if energy released in fusion reactions exceeds the radiation losses and the original energy invested in heating the plasma to the operating temperature. Consider the D–T reactions. Energy released per unit volume from fusion reactions in the plasma is

$$E_f = \frac{1}{4}n^2\overline{\sigma v}Q\tau \tag{8.233}$$

where the densities of D and T are each equal to $\frac{1}{2}n$ (so that total $n = \eta_e$), $Q = 17.6$ MeV is the energy released per reaction, $\tau =$ length of time the plasma is confined during which reactions can occur.

Thermal energy/unit volume required to raise the ions and electrons to temperature T will be

$$E_{th} = 2 \times \frac{3}{2}nkT = 3nkT$$

where

$$n(\text{ions}) = n(\text{electrons}) = n$$

Condition for the reactor to operate is

$$E_f > E_{th} \quad \text{or}$$

$$\frac{1}{4}n^2\overline{\sigma v}Q\tau > 3nkT$$

that is

$$n\tau > \frac{12kT}{\overline{\sigma v}Q} \quad \text{(lawson criterion)} \tag{8.234}$$

For D–T reactor, at $T = 10$ keV, $\overline{\sigma v} \sim 10^{-22}$ m^3 s^{-1} so that

$$n\tau > \frac{12 \times 10}{10^{-22} \times 17.3} \quad \text{or} \quad n\tau > 7 \times 10^{22}$$

Lawson criterion will be different for different operating temperature T.

For $D-D$ reaction, bremsstrahlung losses do not permit the reactor to work at 10 keV (Fig. 8.36). For

$$T = 100 \text{ keV}, \qquad \overline{\sigma v} \simeq 0.5 \times 10^{-22}$$

$$n\tau > \frac{12 \times 100}{0.5 \times 10^{-22} \times 4} \quad \text{or} \quad n\tau > 6 \times 10^{24}$$

We therefore, need 100 fold increase in density of ions or confinement times or a combinations of both to gain energy in $D-D$ reactions.

8.20.9 Ignition Temperature

Observe that as the temperature rises from 2 to 8 keV, the power density in the $D-T$ reactions increases by a factor of order 1000, Fig. 8.35. However bremsstrahlung which is proportional to the square root of temperature goes up by $\sqrt{4}$ or 2 over the same temperature range. Thus, for the temperatures up to 30 keV, the rate of energy generation in the $D-T$ plasma is much more rapid than the loss due to bremsstrahlung. By the time the temperature has reached 4 keV the energy generated and retained in the plasma just balances the loss by bremsstrahlung . The temperature at which the energy gain balances the loss is called the ignition temperature. Below this temperature we cannot expect the fusion reactions to be self-sustained, as the net energy loss would rapidly cool off the plasma. Above the ignition temperature, self sustaining reactions are possible. Now, a temperature of 4 keV is the same as 46 million degrees kelvin which emphasises the difficulties in extracting energy from fusion reactions.

Should the plasma be contaminated with high Z elements the energy losses are increased due to bremsstrahlung which is proportional to the square of nuclear charge. It is, therefore, imperative that the plasma is free from impurities of high Z materials. On the other hand, for the purpose of fusion reactions the presence of high Z material is useless.

8.20.10 Controlled Fusion Reactions

In order to obtain useful energy we must have controlled fusion reactions over reasonably long times. The requirements are:

1. The thermonuclear fuel must be heated to temperatures of the order of 10^8 K corresponding to the mean kinetic energy of 10 keV. At such high temperatures the fuel would be in the form of plasma. (For hydrogen only 13.6 eV energy is needed for ionization.)

2. The density of particles must be high enough for a reasonably long time so that the reaction rate may be high enough, and must be low enough so that an explosive situation does not arise.
3. The $D-T$ reactions are decidedly favourable compared to the $D-D$ reactions as not only are they faster ($\sim$100 times) at around 10 keV but also the energy gain is in excess of bremsstrahlung losses.
4. Plasma must be confined and must be stable.

8.20.11 Confinement

The confinement of plasma is the major problem. Hot fuel would exchange energy with the wall of the container and melt it. Two types of confinement are used.

1. *Magnetic Confinement* Here the plasma is contained by a carefully designed magnetic field.
2. *Inertial Confinement* A solid pellet is suddenly heated and compressed by hitting it from a number of directions with intense beams of photons or particles.

8.20.12 Containment of a Plasma

A hot plasma is a gas which would expand and quickly fill up the container. The mean free path of the nuclei of the plasma is so large that the nuclei are capable of traversing thousands of kilometers. However before they do so, they would hit the wall of the container several times to give up practically all their energy to the molecules of the walls. It is therefore necessary to keep the plasma away from the walls of the container and maintain its temperature for a short time in which the nuclei undergo fusion reactions. In this connection, the product $n\tau$ occurring in Lawson's criterion is important (8.234). For densities which can be handled conveniently ($n \sim 10^{22}$ m^{-3}) the containment time needed is of the order of 1 s.

It can be shown that electric fields cannot be used to cage the plasma with any arrangement of electrically charged conductors whatever. The other option is to employ strong magnetic field to contain the plasma.

8.20.12.1 Magnetic Confinement

Simplest magnetic confinement in two directions is provided by a uniform magnetic field. Charged particles spiral about the field direction. The magnetic field is established by the large current carrying coils, Fig. 8.40.

There will be loss of particles along the field direction in the simple method described above. There are two solutions to circumvent this difficulty.

Fig. 8.40 Magnetic
confinement in two directions
provided by a uniform
magnetic field

Fig. 8.41 A magnetic field in
a torus

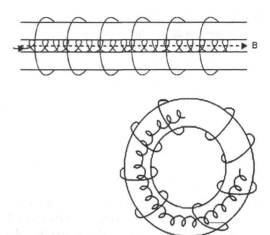

1. A magnetic field is provided in a torus in which particles spiral following the magnetic field lines. As the particles spiral there will be a gradual drift toward the outer wall (Fig. 8.41).
2. Particles are confined in a magnetic mirror and follow magnetic lines. They are reflected from high field region into low field region (Fig. 8.42).

The fusion plasma cannot be maintained at thermonuclear temperature if it is permitted to come in contact with the reactor wall as the material of the wall exposed to the plasma would cool it off. This difficulty is avoided by making the particles spiral around the magnetic lines of force and 'bottled' up. The principle of magnetic confinement is illustrated with reference to Fig. 8.42(a). The charged particles are trapped between magnetic 'mirrors'. The plasma is constrained to be bunched in the space filled by the magnetic lines of force of an electromagnet. Consider the plasma injected in the central region of a hollow solenoid, (Fig. 8.42b). The holes shown by circles represent the windings. The field at the ends, PP' and RR' is much stronger

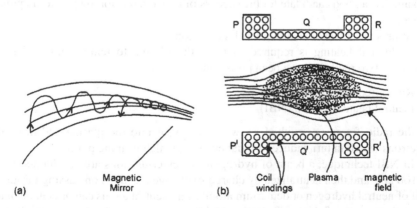

Fig. 8.42 (a) The principle of magnetic confinement. (b) Plasma injected in the central region of a hollow solenoid

Fig. 8.43 Tokamak, a device
which works on the principle
of toroidal field

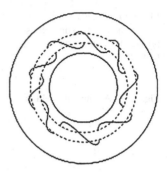

than in the central region QQ'. The plasma is therefore constrained to move inward towards the center. By varying the currents in the field coils PP', QQ' and RR' the shape of the field can be modified and the plasma made to move sideways. By an overall increase in field, the plasma volume is compressed thereby producing higher ionic velocities and hence greater collision frequency leading to higher temperatures conducive for fusion.

8.20.12.2 Tokamak

Tokamak is a device which works on the principle of toroidal field, Fig. 8.43. However the toroidal field is weaker at larger radii so that a particle spiraling drifts to region of lower field towards the wall. This effect is reduced by introducing a polaroid field—a field which has a component along the surface to toroid. It is obtained by passing a current along the axis of toroid through the plasma itself. The current serves double purpose.

1. Heating the plasma
2. Confining the particles

Tokamak is a good candidate for the success of a fusion reactor. It operates in pulsed mode.

At present Tokamak is limited to 1 s duration.

Additional heating is required to raise the plasma to temperatures of 10 to 100 keV. Two methods have been proposed

1. Radio frequency (rf) heating
2. Neutral beam injection (NBI)

In the radio frequency method rf waves are sent into the plasma and drive the electrons which in turn induce toroidal currents which heat the plasma.

In NBI technique, a beam of hydrogen or deuterium ions are accelerated to 10 to 100 keV and then neutralized by charge-exchange reactions on passing through a cell of neutral hydrogen or deuterium atoms. The neutral atoms can pass undeviated through magnetic fields of Todmak and into the plasma where they rapidly lose energy through Coulomb scattering from ions and electrons.

Auxiliary heating systems of tens of Mega Watts are required to achieve ignition of plasma, after which 3.5 MeV α's from D–T fusion will provide necessary heat to sustain the reactions. The α's confined to plasma by magnetic field eventually lose energy to plasma through collisions.

Magnetic confinement in three dimensions, the "magnetic well" the so-called minimum B configuration is also employed.

8.20.12.3 Inertial Confinement

In a different approach a tiny pellet containing deuterium and tritium is suddenly struck with an intense laser pulse. This heats the pellet and compresses it to high density. Fusion is expected to occur before the pellet expands and blow apart. Ten to hundred pellets per second are expected to be used. We can estimate the power produced applying Lawson-s criterion for a D–T mixture for confinement times 10^{-9} to 10^{-10} s. The particle density works out as atleast 10^{29} to $10^{30}/m^3$ which is two order of magnitude greater than ordinary liquid or of solid densities for hydrogen. Energies of 10 keV/particle to be supplied to a spherical pellet of 0.4 mm radius will yield energy

$$E = \frac{4}{3}\pi \left(0.4 \times 10^{-3}\right)^3 \times 10^{29} \times 10^4 \times 1.6 \times 10^{-19} = 4.3 \times 10^4 \text{ J}$$

and the power will be

$$P = \frac{4.3 \times 10^4}{10^{-9}} = 4.3 \times 10^{13} \text{ W}$$

which is too terrific to be of any practical use.

8.20.13 Plasma Diagnostics

8.20.13.1 Spectroscopic Measurements

Temperature measurement of hot plasma is not possible by conventional methods. Furthermore, at such temperatures at which thermonuclear reactions are expected to proceed the deuterium and tritium gas will be fully ionized and therefore the characteristic line spectra of atoms will not be emitted. To make matters worse, measurements must be taken in a small fraction of a second. However, the impurities of high Z elements which might be present in the plasma will not be fully ionized and will emit their own characteristic spectrum and its intensity will be a measure of the amount present. The half width at half maxima is taken as a measure of broadening. One reason for broadening is Stark effect since the atoms are placed in electric field. In a plasma the electrons and ions produce strong and varying electric fields in the neighborhood of neutral atoms and the energy of the line emitted is changed. As the field varies from atom to atom, depending on the configuration of the ions and electrons in the neighborhood, a mixture of wavelength is emitted leading to the broadening of the line. Atomic theory can be applied to the measured broadening of the spectral line from which the distribution of E can be ascertained. It is then possible to calculate the density of ions in the plasma.

Another cause of broadening of spectral lines is the optical Doppler effect. This type of broadening is quite different in shape and can be easily separated from that due to stark effect. As it depends on the velocities of atoms, its measurement establishes the speed distribution of atoms. If the plasma is in a steady state then the observed distribution of all the particles can be fitted with a Maxwellian distribution characterized by a unique temperature for the plasma. Doppler broadening applies to both atoms and ions but not necessarily to electrons.

8.20.13.2 Bremsstrahlung

Usually there is no time for both ions and electrons to reach a common temperature. It is therefore, necessary to measure the temperature of the two constituents separately. Doppler effect fixes the temperature of atoms and ions and measurement of bremsstrahlung spectrum establishes the temperature of electrons. Since this radiation is caused by changes in the energy of fast moving electrons in the field of relatively stationary heavy nuclei and so it depends on the electron temperature rather than that of nuclei. As the temperature of electrons rises, the bremsstrahlung spectrum moves toward shorter wavelength. Assuming a Maxwellion distribution for electrons in the plasma the expected bremsstrahlung spectrum can be worked out at the given electron temperature. The electron temperature can be found out by making the best fit with the measured spectrum.

The total intensity of bremsstrahlung per unit volume of the plasma depends on both its temperatures and density. Measurement of total intensity of light emitted together the knowledge of temperature yields the density of the plasma.

8.20.13.3 Detection of Neutrons Emitted by a Plasma

It is important to establish that the neutrons that are observed from a fusion reactor are indeed of thermonuclear origin. For there may be neutrons which are produced from spurious sources for example deuterons and tritons which get accelerated by the electric fields and collide with other nuclei to produce neutrons. We can, however, distinguish between the genuine and spurious sources since in the former the neutrons are expected to be emitted isotropically with the same energy distribution while in the latter the neutrons emitted in the direction of field would be more energetic than in the opposite direction. Neutrons may be detected by employing BF_3 counters and their energy distribution can be infered from the energy distribution of recoil protons on their collisions with hydrogen nuclei.

In conclusion we may state that the power from fusion reactors is still on experimental stage. In the current experiment "Jet" the output is slightly less than input and lasts for a few seconds. In June 2005, "Inter" Reactor was designed to produce more power than input for plasma, lasting for a few minutes.

8.21 Questions

8.1 Describe the basic design of a power generating thermal nuclear reactor, giving an outline of its principal components.

8.2 Obtain the formula for the average log energy decrement.

8.3 Obtain the Fermi-age equation.

8.4 Set up the diffusion equation which governs the distribution of neutrons throughout the reactor.

8.5 Mention the advantages and disadvantages of homogeneous and heterogeneous reactors.

8.6 Write an essay on the classification of reactors.

8.7 Indicate the use of the following material in nuclear reactors:

(i) Graphite	(ii) Beryllium	(iii) Cadmium
(iv) Stainless steel	(v) Liquid sodium/Potassium	(vi) Lead
(vii) Concrete	(viii) ^{235}U	(ix) ^{238}U
(x) ^{233}Th	(xi) Zirconium	(xii) Aluminum

as

(a) fuel in thermal reactor	(b) fuel in the fast reactor
(c) fertile material in breeder reactor	(d) control rods
(e) coolant	(f) reflector
(g) shield against neutrons	(h) shield against gamma rays
(i) cladding material for uranium rods	(j) moderator

8.8 Define thermal diffusion time and generation time. Obtain an expression for the thermal lifetime in the presence of thermal neutron leakage.

8.9 Distinguish between prompt neutrons and delayed neutrons. What is the effect of delayed neutrons on the working of a reactor?

8.10 Describe the working of a breeder reactor. Define the doubling time in connection with breeders.

8.11 Obtain the four-factor formula in the design of a nuclear reactor.

8.12 Which of the nuclei are fissionable with thermal neutrons: ^{227}Th, ^{233}U, ^{235}U, ^{238}U, ^{239}Pu, ^{242}Pu?

8.13 Explain why a self-sustaining chain reaction cannot be obtained with natural uranium as fuel (except with heavy water as moderator) in a homogeneous assembly.

8.14 Explain in detail the effect of heterogeneous arrangement on p, f and ε.

8.15 Outline the considerations which lead to the Lawson criterion for the minimum value of the product of ion density and confinement time required for net power generation by a fusion reactor.

8.16 Give a brief description of the main components of an inertial confinement fusion reactor.

8.17 The temperature at which deuterons would overcome electrostatic repulsion can be calculated. Due to some physical considerations, the temperature can be lowered for the fusion to occur. What are these considerations?

8.22 Problems

8.1 Calculate k_∞ for a homogeneous natural uranium heavy water-moderated reactor, with $N_m/N_0 = 45$. Use the constants:

For D$_2$O; $\sigma_a = 0.00092$ b, $\sigma_s = 10.6$ b, $\xi = 0.57$
For natural uranium, $\sigma_a = 7.68$ b, $\sigma_s = 8.3$ b, $\eta = 1.34$, $\varepsilon = 1$

[Ans. $k_\infty = 1.11$]

8.2 Calculate the thermal utilization factor for a heterogeneous lattice made up of cylindrical uranium rods of diameter 3 cm and pitch 18 cm in graphite.
     Take the flux ratio ϕ_m/ϕ_u as 1.6

Densities: Uranium $= 18.7 \times 10^3$ kg m^{-3}
Graphite $= 1.62 \times 10^3$ kg m^{-3}
Absorption cross-sections $\sigma_{au} = 7.68$ b
$\sigma_{am} = 4.5 \times 10^{-3}$ b

[Ans. 0.933]

8.3 Explain what is meant by the four factor formula. Discuss how the factors vary with fuel rod diameter and lattice pitch in a natural uranium-graphite reactor and how the optimum pitch is obtained. Given $\sigma_{f235} = 580$ b, $\sigma_{a235} = 680$ b, $\sigma_{a238} = 2.8$ b and $v = 2.5$. Calculate η at 0.71 and 2 % enrichment.
[Ans. 1.35; 1.77]

8.4 Using one group theory, calculate the critical radius of a bare spherical reactor.

Diffusion length $= 23.6$ cm
Slowing down length $= 9.9$ cm
$K_\infty = 1.54$

[Ans. 101.3 cm]

8.5 Compare the moderating ratios for light and heavy water for 2 MeV neutrons. Use the following cross-sections.

Medium	σ_a	σ_s
H_2O	0.66 b	44 b
D_2O	0.46 mb	11 b

8.6 Assuming the energy released per fission of ^{235}U is 200 MeV, calculate the amount of ^{235}U consumed per day in the Canada India reactor 'CIRUS' operating at 40 megawatts of power.
[Ans. 4.23 g]

8.7 Assuming that the energy released per fission of $^{235}_{92}U$ is 200 MeV, calculate the number of fission processes that should occur per second in a nuclear reactor to operate at a power level of 40 MW. What is the corresponding rate of consumption of $^{235}_{92}U$?
[Ans. 12.5×10^{17}/s; 4.88×10^{-4} g]

8.8 Assume that in each fission of ^{235}U, 200 MeV is released. Assuming that 5 % of the energy is wasted in neutrinos, calculate the amount ^{235}U burned which would be necessary to supply at 30 % efficiency, the whole annual electricity consumption in Britain 50×10^9 kWh.
[Ans. 1.92×10^4 tons]

8.9 Assuming that the fission process releases on the average 200 MeV and 2.5 neutrons, what mass of plutonium-239 is produced annually in a 200 MW reactor if no neutrons are lost?
[Ans. 39 kg]

8.10 Assuming that the $n-p$ scattering is isotropic in the CM-system and that E_1 and E_2 are the neutron energies before and after the collision, show that $\ln(E_1/E_2) = 1$.

8.11 Calculate the number of collisions required to reduce fast fission neutrons with an average initial energy of 2 MeV to the thermal energy (0.025 eV) in a graphite moderated assembly.
[Ans. 115]

8.12 Calculate the number of collisions required for neutrons of 2 MeV to lose 99 % of initial energy in graphite.
[Ans. 29]

8.13 What would be the energy of 2 MeV neutrons that have made 50 collision with carbon nuclei?
[Ans. 742 eV]

8.14 Calculate the moderating ratio for heavy water. Given the epithermal cross-sections

$$(\sigma_S)_D = 6.0 \text{ b}, \qquad (\sigma_s)_0 = 4.2 \text{ b}$$

$$(\sigma_a)_D = 0.00046 \text{ b}, \qquad (\sigma_a)_0 = 0.0002 \text{ b}$$

[Ans. 7932]

8.15 Calculate the slowing-down time in beryllium for neutrons starting with an initial energy of 2 MeV and terminating at thermal energies (0.025 eV). $\Sigma_s = 0.57 \text{ cm}^{-1}$.
[Ans. 3.7×10^{-5} s]

8.16 Calculate p for a lattice consisting of natural uranium fuel rods of circular cross-section, diameter 2.4 cm, and spaced 24 cm apart in a graphite moderator. You may assume $\frac{\phi_m}{\phi_0} = 1$, $\rho_m = 1.62 \text{ g cm}^{-3}$, $\rho_u = 18.7 \text{ g cm}^{-3}$, $\sigma_s = 4.8 \text{ b}$, $\xi = 0.158$.
[Ans. 0.933]

8.17 Calculate the slowing-down length for fission neutrons of 2 MeV average energy to thermal energy 0.025 eV in graphite. Assume $\Sigma_s = 0.335 \text{ cm}^{-1}$.
[Ans. 19.2 cm]

8.18 Calculate K_∞ for a homogeneous, natural uranium-graphite moderated assembly which contains 400 moles of graphite per mole of uranium. Assume natural uranium to contain one part of ^{235}U to 139 parts of ^{238}U, and use the following constants:

Natural uranium	Graphite
$\sigma_a(U) = 7.68 \text{ b}$	$\sigma_a(M) = 0.0032 \text{ b}$
$\sigma_s(U) = 8.3 \text{ b}$	$\sigma_s(M) = 4.8 \text{ b}$
$\varepsilon = 1.0$; $\eta = 1.34$	$\xi = 0.158$

[Ans. 0.859]

8.19 Calculate k_∞ for an enriched uranium-graphite-moderated reactor, using 300 moles of graphite to 1 mole of uranium and a $^{238}U/^{235}U$ ratio of 60. Use the constants:

$$\sigma_{a(235)} = 698 \text{ b}, \qquad \sigma_{a(238)} = 2.75 \text{ b}$$

$$\sigma_{f(235)} = 590 \text{ b}, \qquad \nu = 2.46; \ \varepsilon = 1$$

$$\sigma_a(M) = 0.0032 \text{ b}, \qquad \sigma_s(M) = 4.8 \text{ b}$$

[Ans. 1.1]

8.20 Calculate L for thermal neutrons in graphite using the constants: $\sigma_a = 3.2$ mb; $\sigma_s = 4.8$ b; $\rho = 1.62$ g/cm^3.
[Ans. 59.2 cm]

8.21 Calculate the diffusion length for a homogeneous mixture of 1 atom of ^{235}U per 5000 atoms of ^{12}C. $L_m = 52$ cm; $\sigma_a(m) = 0.0032$ b; $\sigma_{a(235)} = 698$ b.
[Ans. 7.89 cm]

8.22 Calculate the neutron age and L_f for fission neutrons of 2 MeV average energy to thermal energy 0.025 eV in beryllium. For Be, $A = 9$ and $\Sigma_s = 0.57$ cm^{-1}.
[Ans. $\tau = 97$ cm^2, $L_f = 9.8$ cm]

8.23 10000 fission neutrons in a critical thermal reactor employ ^{235}U and graphite in an atom ratio of $1 : 10^5$. Use the constants: $\sigma_a(c) = 0.003$ b; $\sigma_{a(235)} = 698$ b; $L_m = 54$ cm; $\tau_0 = 364$ cm^2; For uranium, $\eta = 2.08$. Assume $= 1$, $\varepsilon = 1$. Find the number of neutrons lost by fast diffusion and that by thermal diffusion.
[Ans. 1120 fast and 1970 thermal diffusion]

8.24 Calculate the material buckling of a large critical homogeneous reactor employing ^{235}U and Be in an atomic ratio of $1 : 20000$. For Be, $L_m = 21$ cm, $\tau_0 = 98$ cm^2; $\sigma_a = 0.01$ b. For uranium $\sigma_a(U) = 698$ b, $p = 1$, $\epsilon = 1$, $\eta = 2.08$.
[Ans. 0.00315 cm^2]

8.25 Calculate the critical radius for a spherical reactor using the value $B^2 = 0.002$ cm^{-2}.
[Ans. 70.25 cm]

8.26 Calculate the thermal diffusion time for graphite. Use the constants: $\sigma_a(C) = 0.003$ b; $\rho_c = 1.62$ g cm^{-3}; Average thermal neutron speed $= 2200$ ms^{-1}.
[Ans. 1.87×10^{-2} s]

8.27 Calculate the generation time for neutrons in a critical reactor employing ^{235}U and graphite. Use the following data: $\Sigma_a = 0.0008$ cm^{-1}; $B^2 = 0.000325$; $L^2 = 878$ cm^2; $\langle v \rangle = 2200$ m s^{-1}.
[Ans. 0.442 s]

8.28 Calculate the increase in neutron flux in 1 s in a critical reactor which employs ^{235}U fuel with graphite as moderator, the generation time for neutrons being 4.4×10^{-3} s. The multiplication factor is suddenly increased by 0.5 %.
[Ans. 3.11]

8.29 Show that $f_{(nat)} \eta_{(nat)} = f_{(235)} \eta_{(235)}$.

8.30 Deutrons almost at rest undergo fusion reaction $D + D \rightarrow {}^{3}_{2}\text{He} + n + 3.2\,\text{MeV}$. Find the kinetic energy of neutron.
[Ans. 2.4 MeV]

8.31 It is estimated that fusion reactions would ensue if deutrons come within a distance of separation of 100 fm. Find the energy to overcome electrostatic repulsion.
[Ans. 14.4 keV]

8.32 Calculate the ^{3}He energy in the reaction $D + P \rightarrow {}^{3}\text{He} + \gamma + 5.5\,\text{MeV}$ when initially deuterons are at rest.
[Ans. 5.4 keV]

8.33 In the reaction given in Problem 8.32 the deutron and proton need to approach each other at a distance of 650 fm. Calculate the temperature at which the fusion reaction can proceed.
[Ans. 1.7×10^{7} K]

8.34 It is believed that the fusion reaction $D + P \rightarrow {}^{3}\text{He} + \gamma + 5.5\,\text{MeV}$, occurs in the interior of the sun where the temperature is approximately 1.7×10^{7} K. Assuming that the deuteron and the proton are at rest, how close they must approach for the reaction to occur?
[Ans. 650 fm]

References

1. G.R. Keepin, T.F. Wimett, R.K. Zeigler, J. Nucl. Energy **6**(1), 1 (1957)
2. A.A. Kamal, *Particle Physics* (Springer, Berlin, 2014)

References

1. C. Albajar et al. (UA1 Collab.), Search for B0 anti-B0 oscillations at the CERN proton–anti-proton collider. 2. Phys. Lett. B **186**, 247 (1987)
2. Alitti et al. (UA2 Collab), An improved determination of the ratio of W and Z masses at the CERN $\bar{p}p$ collider. Phys. Lett. B **276**, 354 (1992)
3. R. Ansari et al. (UA2 Collab.), Measurement of the standard model parameters from a study of W and Z bosons. Phys. Lett. B **186**, 440 (1987)
4. G. Armison et al. (UA1 Collab.), Experimental observation of isolated large transverse energy electrons with associated missing energy at $\sqrt{s} = 540$-GeV. Phys. Lett. B **122**, 103 (1983)
5. J.N. Bahcal, Solar neutrino experiments. Rev. Mod. Phys. **50**, 881 (1978)
6. J.N. Bahcal, *Neutrino Astrophysics* (Cambridge University Press, Cambridge, 1989)
7. C. Baltay et al., Antibaryon production in antiproton proton reactions at 3.7 BeV/c. Phys. Rev. B **140**, 1027 (1965)
8. J.D. Barrow, The baryon asymmetry of the universe. Surv. High Energy Phys. **1**, 182 (1980)
9. W. Bartel et al. (JADE Collab.), Tau lepton production and decay at PETRA energies. Phys. Lett. B **161**, 188 (1985)
10. H.J. Behread (CELLO-Collab.), A measurement of the muon pair production in e^+e^- annihilation at 38.3 GeV $\leq \sqrt{s} \leq$ 46.8 GeV. Phys. Lett. B **191**, 209 (1987)
11. H.A. Bethe, *Elementary Nuclear Theory* (Wiley, New York, 1947)
12. A.A. Bethe, J. Ashkin, Passage of radiations through matter, in *Experimental Nuclear Physics*, vol. 1, ed. by E. Segre (Wiley, New York, 1953), p. 166
13. H.A. Bethe, P. Morrison, *Elementary Nuclear Theory* (Wiley, New York, 1952)
14. J.B. Birks, *Scintillation Counters* (McGraw-Hill, New York, 1953)
15. J. Blatt, V.F. Weisskopf, *Theoretical Nuclear Physics* (Wiley, New York, 1952)
16. E. Bleur, G.J. Goldsmith, *Nucleonics* (Holt, Rinehart and Winston, New York, 1963)
17. M.H. Blewett, Characteristic of typical accelerators. Annu. Rev. Nucl. Sci. **17**, 427 (1967)
18. R.J. Blin-Stoyle, *Nuclear and Particle Physics* (Chapman and Hall, London, 1991)
19. A. Bohr, B.R. Mottelson, *Nuclear Structure* (Benjamin, New York, 1969)
20. H. Bradner, Bubble chambers. Annu. Rev. Nucl. Sci. **10**, 109 (1960)
21. D.M. Brink, *Nuclear Forces* (Pergamon, Oxford, 1965)
22. H.N. Brown et al., Observation of production of a e^-e^+ pair. Phys. Rev. Lett. **8**, 255 (1962)
23. W.E. Burcham, *Elements of Nuclear Physics* (Longman, Harlow, 1979)
24. W.E. Burcham, M. Jobes, *Nuclear and Particle Physics* (Longman, Harlow, 1995)
25. J. Button et al., Evidence for the reaction, $p^-p \to \Sigma^0 \Lambda$. Phys. Rev. Lett. **4**, 530 (1960)
26. N. Cabibbo, Unitary symmetry and leptonic decays. Phys. Rev. Lett. **10**, 531 (1963)
27. C.G. Callan Jr., D.J. Gross, High-energy electroproduction and the constitution of the electric current. Phys. Rev. Lett. **22**, 156 (1969)

A. Kamal, *Nuclear Physics*, Graduate Texts in Physics,
DOI 10.1007/978-3-642-38655-8, © Springer-Verlag Berlin Heidelberg 2014

28. G. Charpak, F. Sauli, High resolution electronic particle detectors. Annu. Rev. Nucl. Part. Sci. **34**, 285 (1984)
29. F.E. Close, *An Introduction to Quarks and Partons* (Academic Press, New York, 1979)
30. F.E. Close, The quark parton model. Rep. Prog. Phys. **42**, 1285 (1979)
31. B.L. Cohen, *Concepts of Nuclear Physics* (McGraw-Hill, New York, 1971)
32. E.D. Commins, *Weak Interactions* (McGraw-Hill, New York, 1973)
33. E.D. Courant, Accelerators for high intensities and high energies. Annu. Rev. Nucl. Sci. **18**, 435 (1968)
34. C.L. Cowan Jr., F. Reines et al., Detection of the free neutrino: a confirmation. Science **124**, 103 (1956)
35. R.H. Dalitz, On the analysis of τ-meson data and the nature of the τ-meson. Philos. Mag. **44**, 1068 (1953)
36. R.H. Dalitz, Strange particle resonant states. Annu. Rev. Nucl. Sci. **13**, 339 (1963)
37. M. Danysz, J. Pniewsky, Delayed disintegration of a heavy nuclear fragment: I. Philos. Mag. **44**, 348 (1953)
38. A. Das, T. Ferbel, *Introduction to Nuclear and Particle Physics* (Wiley, New York, 1993)
39. G.G. Eichholz, J.W. Poston, *Principles of Nuclear radiation detection* (Ann Arbor Science, Ann Arbor, 1979)
40. J. Ellis et al., Physics of intermediate vector bosons. Annu. Rev. Nucl. Part. Sci. **32**, 443 (1982)
41. L.R.B. Elton, *Introductory Nuclear Theory* (Pitmans, London, 1959)
42. H.A. Enge, *Introduction to Nuclear Physics* (Addison-Wesley, Reading, 1966)
43. R.D. Evans, *The Atomic Nucleus* (McGraw-Hill, New York, 1955)
44. C.W. Fabjan, H.G. Fischer, Particle detectors. Rep. Prog. Phys. **43**, 1003 (1980)
45. C.W. Fabjan, T. Ludlam, Calorimetry in high energy physics. Annu. Rev. Nucl. Part. Sci. **32**, 335 (1982)
46. G. Feinberg, L. Lederman, Physics of muons and muon neutrinos. Annu. Rev. Nucl. Sci. **13**, 431 (1963)
47. E. Fermi, *Nuclear Physics*, 5th edn. (University of Chicago Press, Chicago, 1953)
48. R.P. Feynman, *Photon-Hadron Interactions* (Benjamin, New York, 1972)
49. Firestone et al., Observation of the anti-omega. Phys. Rev. Lett. **26**, 410 (1971)
50. W.R. Fraser, *Elementary Particles* (Prentice Hall, Englewood Cliffs, 1966)
51. J.I. Friedman, H.W. Kendall, Deep inelastic electron scattering. Annu. Rev. Nucl. Sci. **22**, 203 (1972)
52. S. Gasiorowicz, *Elementary Particle Physics* (Wiley, New York, 1966)
53. M. Gell-Mann, Y. Ne'eman, *The Eight Fold Way* (Benjamin, New York, 1964)
54. M. Gell-Mann, A. Pais, Behavior of neutral particles under charge conjugation. Phys. Rev. **97**, 1387 (1955)
55. D. Glaser, The bubble chamber, in *Encyclopaedia of Physics*, ed. by S. Fluegge (Springer, Berlin, 1955), p. 45
56. S.L. Glashow, Towards a unified theory, threats in tapestry. Rev. Mod. Phys. **52**, 539 (1980)
57. M. Goldhaber, L. Grodzins, A. Sunyar, Helicity of neutrinos. Phys. Rev. **109**, 1015 (1958)
58. M. Goppert-Mayer, J.H.D. Jensen, *Elementary Theory of Nuclear Shell Structure* (Wiley, New York, 1955)
59. K. Gottfried, V.F. Weisskopf, *Concepts of Particle Physics*, vol. 1 (Clarendon, Oxford, 1984)
60. K. Gottfried, V.F. Weisskopf, *Concepts of Particle Physics*, vol. 2 (Clarendon, Oxford, 1986)
61. A.E.S. Green, *Nuclear Physics* (McGraw-Hill, New York, 1955)
62. O.W. Greenberg, Quarks. Annu. Rev. Nucl. Sci. **28**, 327 (1978)
63. D. Griffiths, *Introduction to Elementary Particles* (Wiley, New York, 1987)
64. C. Grupen, *Particle Detectors* (Cambridge University Press, Cambridge, 1996)
65. D. Halliday, *Introduction to Nuclear Physics* (Wiley, New York, 1955)
66. T. Hamada, L.D. Johnson, A potential model representation of two nucleon data below 315 MeV. Nucl. Phys. **34**, 382 (1962)

67. W.D. Hamilton, Parity violation in electromagnetic and strong interaction processes. Prog. Nucl. Phys. **10**, 1 (1969)
68. H. Harairi, Quarks and leptons. Phys. Rep. **42**, 235 (1978)
69. E.M. Henley, Parity and time reversal invariance in nuclear physics. Annu. Rev. Nucl. Sci. **19**, 367 (1969)
70. P.E. Hodgson, E. Gadioli, E. Gadioli Erba, *Introductory Nuclear Physics* (Clarendon, Oxford, 1997)
71. R. Hofstadter, Nuclear and nucleon scattering of high energy electrons. Annu. Rev. Nucl. Sci. **7**, 231 (1957)
72. I.S. Hughes, *Elementary Particles*, 2nd edn. (Cambridge University Press, Cambridge, 1987)
73. G. Hutchinson, Cerenkov detectors. Prog. Nucl. Phys. **8**, 195 (1960)
74. J.D. Jackson, *The Physics of Elementary Particles* (Princeton University, Princeton, 1958)
75. L.W. Jones, A review of quark search experiments. Rev. Mod. Phys. **49**, 717 (1977)
76. G. Kallen, *Elementary Particle Physics* (Addison-Wesley, Reading, 1964)
77. N. Kemmer, J.C. Polkinghorne, D. Pursey, Invariance in elementary particle physics. Rep. Prog. Phys. **22**, 368 (1959)
78. K. Klein-Knecht, *Detectors for Particle Radiation* (Cambridge University Press, Cambridge, 1999)
79. M. Kobayashi, T. Maskawa, CP violation in the renormalizable theory of weak interaction. Prog. Theor. Phys. **49**, 652 (1973)
80. J.J. Kokkedee, *The Quark Model* (Benjamin, New York, 1969)
81. E.J. Konipinski, Experimental clarification of the laws of β-radioactivity. Annu. Rev. Nucl. Sci. **9**, 99 (1959)
82. K. Krane, *Introductory Nuclear Physics* (Wiley, New York, 1987)
83. R.E. Lapp, H.L. Andrews, *Nuclear Radiation Physics* (Prentice Hall, New York, 1972)
84. C.M.G. Lattes, H. Muirhead, G.P.S. Occhialini, C.F. Powell, Processes involving charge mesons. Nature **159**, 694 (1947)
85. J.D. Lawson, M. Tigner, The physics of particle accelerators. Annu. Rev. Nucl. Part. Sci. **34**, 99 (1984)
86. L. Lederman, Neutrino physics, in *Pure and Applied Physics*, vol. 25, ed. by Burhop (Academic Press, New York, 1967)
87. L.M. Lederman, Lepton production in hadron collisions. Phys. Rep. **26**, 149 (1976)
88. L. Lederman, The upsilon particle. Sci. Am. **239**, 60 (1978)
89. T.D. Lee, C.S. Wu, Weak interactions. Ann. Rev. Nucl. Sci. **15**, 381 (1965)
90. M.S. Livingston, J.P. Blewett, *Particle Accelerators* (McGraw-Hill, New York, 1962)
91. M.S. Livingstone, *High Energy Accelerators* (Interscience, New York, 1954)
92. P. Marmier, E. Sheldon, *Physics of Nuclei and Particles*, vols. I and II (Academic Press, San Diego, 1969), (1970)
93. B.R. Martin, G. Shaw, *Particle Physics* (Wiley, New York, 1992)
94. I.E. McCarthy, *Nuclear Reactions* (Pergamon, Elmsford, 1970)
95. E.M. McMillan, Particle accelerators, in *Experimental Nuclear Physics*, ed. by Segre (Wiley, New York, 1959), p. 3
96. W. Meyarhof, *Elements of Nuclear Physics* (McGraw-Hill, New York, 1967)
97. M. Moe, P. Vogel, Double beta-Decay. Annu. Rev. Nucl. Part. Sci. **44**, 247 (1994)
98. L. Montanet et al., Review of particle properties. Phys. Rev. D **50**, 1173 (1994)
99. A. Muirhead, *The Physics of Elementary Particles* (Pergamon, London, 1965)
100. R.L. Murray, *Nuclear Reactor Physics* (Wiley, New York, 1960)
101. P.E. Nemirovskii, *Contemporary Models of the Atomic Nucleus* (Pergamon, Elmsford, 1963). Student eds.
102. L.B. Okun, *Weak Interactions of Elementary Particles* (Pergamon, London, 1965)
103. J. Orear, G. Harris, S. Taylor, Spin and parity analysis of bevatron τ mesons. Phys. Rev. **102**, 1676 (1956)
104. P.J. Ouseph, *Introduction to Nuclear Radiation Detectors* (Plenum, New York, 1933)
105. C. Pelligrini, Colliding beam accelerators. Annu. Rev. Nucl. Sci. **22**, 1 (1972)

106. D.H. Perkins, Inelastic lepton nucleon scattering. Rep. Prog. Phys. **40**, 409 (1977)
107. D.H. Perkins, *Introduction to High Energy Physics* (Addison-Wesley, Reading, 1987)
108. M.L. Perl, The tau lepton. Annu. Rev. Nucl. Part. Sci. **30**, 299 (1980)
109. B. Povh, K. Rith, C. Scholz, F. Zetsehe, *Particles and Nuclei* (Springer, Berlin, 1995)
110. C.F. Powell, P.H. Fowler, D.H. Perkins, *The Study of Elementary Particles by the Photographic Method* (Pergamon, Elmsford, 1959)
111. M.A. Preston, *Physics of the Nucleus*, 2 edn. (Addison-Wesley, Wokingham, 1963)
112. W.J. Price, *Nuclear Radiation Detectors* (McGraw-Hill, New York, 1964)
113. D. Prowse, M. Baldo-Ceolin, Anti-lambda hyperon. Phys. Rev. Lett. **1**, 179 (1958)
114. N.F. Ramsey, Electric dipole moments of particles. Annu. Rev. Nucl. Part. Sci. **32**, 211 (1982)
115. F. Reins, Neutrino interactions. Annu. Rev. Nucl. Sci. **10**, 1 (1960)
116. G.D. Rochester, C.C. Butler, Evidence for the existence of new unstable elementary particles. Nature **160**, 855 (1947)
117. M.N. Rosenbluth, High energy elastic scattering of electrons on protons. Phys. Rev. **79**, 615 (1950)
118. B. Rossi, *High Energy Particles* (Prentice Hall, Englewood Cliffs, 1952)
119. E.G. Rowe, E.J. Squires, Present status of C-, P-, and T-invariance. Rep. Prog. Phys. **32**, 273 (1969)
120. R.R. Roy, B.P. Nigam, *Nuclear Physics* (Wiley Eastern, New Delhi, 1993)
121. C. Rubbia, Experimental observation of the intermediate vector bosons W^+, W^- and Z^0. Rev. Mod. Phys. **57**, 699 (1985)
122. R.G. Sachs, *Nuclear Theory* (Addison-Wesley, Reading, 1953)
123. J.J. Sakurai, *Invariance Principles and Elementary Particles* (Princeton University Press, Princeton, 1964)
124. A. Salam, Gauge unification of fundamental forces. Rev. Mod. Phys. **52**, 525 (1980)
125. G.R. Satchler, *Introduction to Nuclear Reactions*, 2nd edn. (MacMillan, London, 1990)
126. L.I. Schiff, *Quantum Mechanics* (McGraw-Hill, New York, 1949)
127. E. Segre, *Experimental Nuclear Physics, 3 Vols.* (Wiley, New York, 1963)
128. E. Segre, *Nuclei and Particles* (Benjamin, Elmsford, 1977)
129. J.R. Stanford, The Fermi national accelerator laboratory. Annu. Rev. Nucl. Part. Sci. **26**, 151 (1976)
130. R. Stephenson, *Introduction to Nuclear Engineering* (McGraw-Hill, New York, 1954)
131. G. 't Hooft, Gauge theories of the forces between elementary particles. Sci. Am. **243**, 90 (1980)
132. J.M. Taylor, *Semiconductor Particle Detectors* (Butterworths, Stoneham, 1989)
133. R.D. Tripp, Spin and parity determination of elementary particles. Annu. Rev. Nucl. Sci. **15**, 325 (1965)
134. S. Van der Meer, Stochastic cooling and accumulation of antiprotons. Rev. Mod. Phys. **57**, 699 (1985)
135. S. Weinberg, Unified theories of elementary particle interactions. Sci. Am. **231**, 50 (1974)
136. S. Weinberg, Conceptual foundations of the unified theory of weak and electromagnetic interactions. Rev. Mod. Phys. **52**, 515 (1980)
137. G.B. West, Electron scattering from atoms, nuclei and nucleons. Phys. Rep. C **18**, 264 (1975)
138. G.C. Wick, Invariance principles of nuclear physics. Annu. Rev. Nucl. Sci. **8**, 1 (1958)
139. W.S. Williams, *An Introduction to Elementary Particles* (Academic Press, New York, 1971)
140. S.S.M. Wong, *Introductory Nuclear Physics* (Prentice Hall, New York, 1990)
141. C.S. Wu et al., Experimental test of parity conservation in beta decay. Phys. Rev. **105**, 1413 (1957)
142. H. Yukawa, On the interaction of elementary particles. Proc. Phys. Math. Soc. Jpn. **17**, 48 (1935)

Index

A. Kamal, *Nuclear Physics*, Graduate Texts in Physics,
DOI 10.1007/978-3-642-38655-8, © Springer-Verlag Berlin Heidelberg 2014

Printed in the United States
By Bookmasters